Laboratory Methods in Food Microbiology

3rd edition

This book is dedicated to my wife, Rita

Laboratory Methods in Food Microbiology

3rd edition

by

W. F. Harrigan

San Diego London Boston
New York Sydney Tokyo Toronto

This book is printed on acid-free paper.

Academic Press
525 B Street, Suite 1900, San Diego, California 92101-4495, USA
http://www.apnet.com

Academic Press Limited
24–28 Oval Road, London NW1 7DX, UK
http://www.hbuk.co.uk/ap/

ISBN 0–12–326043–4

A catalogue record for this book is available from the British Library

Typeset by LaserScript, Mitcham, Surrey
Printed in Great Britain by WBC Book Manufacturers, Bridgend, Mid-Glamorgan

98 99 00 01 02 03 WB 9 8 7 6 5 4 3 2 1

Contents

PART II: Techniques for the Microbiological Examination of Foods

PART III: Microbiological Examination of Specific Foods

PART IV: Schemes for the Identification of Microorganisms

Preface

This third edition continues to be written for a dual audience – first, food microbiologists working in the food industry and in microbiological quality assurance laboratories, and, second, final-year students taking degree and diploma courses and postgraduate students, in food science, food technology and allied subjects. Since the publication of the second edition, there has been great attention paid to quality management systems in general, and to Hazard Analysis Critical Control Point systems in particular. The subject of microbiological quality in this context has been discussed in another text (*Making Safe Food* by Harrigan and Park). In a company practising HACCP, many of the modern analytical methods could provide an opportunity for reasonably rapid monitoring. Nevertheless, there are still no microbiological methods that will permit on-line monitoring with feed-back control of the critical control points (CCPs) in a food processing line. However, microbiological monitoring of the effectiveness of the CCPs is essential for verification and validation, and many of the methods described in this edition are suitable for this purpose.

There has been increasing attention paid to the need to achieve international agreement on methodology and criteria for food quality. In food microbiology this started in earnest with the setting up of the International Committee (now Commission) on Microbiological Specifications for Foods and the start in the 1960s of publications by this organization. The uniting role of the ICMSF has made easier the task of the International Organization for Standardization in developing international standard methods (ISO standards). Nevertheless, many countries also use national standard methods that may or may not be in accord with ISO methods. On an international scale, in most quality assurance laboratories in the food industry and in government control agency laboratories, analyses still predominantly rely on the use of conventional methods for determining microbiological quality. With this edition, I am trying to encourage the use of ISO methods where appropriate. However, for many microbiological analyses an ISO method has not been agreed, and in these cases I have attempted to present the currently preferred method. Sometimes a national standard method is totally different from the ISO method, and occasionally an ISO method still requires modification and improvement. Where I have considered it necessary I have discussed the differences between ISO methods and standard methods of other origin (for example the methods of the Association of Official Analytical Chemists of the US).

In many quality assurance laboratories, rapid and instrumental methods are used – for example, electrometric methods, immunomagnetic capture, DNA probes and polymerase chain reaction (PCR) techniques. These methods are outlined in this edition. However, from my work in a number of countries (for example in Southeast Asia and South America), in both industrial consultancy and in presenting short courses for food microbiologists from industry and government regulatory agencies, it is obvious that in many countries such methodologies, especially those that involve either very expensive capital equipment (e.g., electrometric methods) or expensive diagnostic reagents (e.g., gene probe techniques) cannot be routinely applied, so that conventional methods will continue to be used for some time. International quality criteria will need to reflect this in the methods that are described within specifications. This edition, therefore, continues to describe the conventional methods of microbiological analysis, but easily adopted rapid diagnostic methods have also been included.

The book is intended to be capable of use as a bench-top reference for the vast majority of microbiological analyses, with adequate detail presented in Parts I to IV, and detailed recipes for reagents and media in Appendix 1.

Part I on 'Basic Techniques' includes sections outlining the management and safe operation of the food microbiology laboratory; the fundamentals of microscopy and staining; culture techniques; viable and total counts, including DEFT and its use with membrane filtration, ATP measurement, and impedance/conductance measurements; biochemical and physiological tests; serological methods including ELISA and magnetic immunocapture.

Part II provides the general (horizontal) techniques used to examine foods. The sections on detection and enumeration of pathogenic and toxigenic organisms have been extended to include, for example, the foodborne pathogenic vibrios, *Campylobacter jejuni*, *Listeria monocytogenes*, *Escherichia coli* O157, *Yersinia enterocolitica*.

Part III discusses the microbial ecology of different types of food, and the specific methodologies required by those different food commodity types. The examination of starting materials and end-products to establish conformance to specification is addressed.

Part IV provides schemes for the identification of bacteria, yeasts and moulds isolated from foods using non-selective media. When selective media are used to detect particular organisms, such as *Escherichia coli* or *Salmonella*, a shorter series of tests will often permit identification of the target organism; such identification procedures are described in the appropriate sections of Part II.

The Appendix of recipes for reagents and culture media will permit the food microbiologist to prepare most media and reagents from their basic ingredients.

I am indebted to my wife Rita, not only for her tremendous encouragement and assistance, but also for the patience shown as our house became knee-deep in paper, journals and reference books.

PART I

Basic Methods

1
Introduction

As long ago as 1955 Wilson observed that 'it is far more important to lay down a strict code for the preparation or processing of food and see that it is carried out properly than to rely on bacteriological sampling of the finished product.'

Microbiological examinations in the laboratory must be seen in the wider context of the entire quality management system. This, in relation to the microbiological quality of foods, is the subject of *Making Safe Food* by Harrigan and Park (1991). Consequently it is suggested that the reader and user of this present book needs to set the scene and perform those microbiological examinations which can throw light on the efficacy of the quality management and hygienic management of the food production–storage–distribution system. Only in end-product examinations being performed to assess batches of food solely against microbiological specifications will the laboratory work be dissociated from a study of the plant operations.

1.1 THE ROLE OF THE MICROBIOLOGIST IN THE FOOD INDUSTRY

There are a number of aspects to the job of a food microbiologist. Some will be undertaken as teamwork with, for example, marketing personnel, chemists, engineers and so on:

1. Training factory personnel in the need for hygienic practices and proper procedures for cleaning and disinfection.
2. Production of relevant 'in-house' codes of practice and hygiene manuals, involving as necessary the assessment of detergent–disinfectants and sanitizers.
3. Involvement in new product development and in the development of new and modified production processes.
4. Surveying the microbiological condition of (i) raw materials and (ii) water.
5. Process control; also quality control on the finished products (i.e. control of distribution, if possible), or quality assurance on the finished products.
6. Provision of detailed storage instructions for the wholesaler and retailer, and of detailed storage and/or cooking instructions for the consumer.
7. Investigation of customer complaints of a microbiological nature.

Additionally, the food microbiologist may find himself/herself involved in:

(a) maintenance and preparation of microbial cultures used in the production of fermented foods;
(b) microbiological assay of vitamins, amino acids and protein quality, or of antimicrobial constituents, additives and contaminants;
(c) examination and control of factory effluent quality.

It can be seen from this list that a number of these duties require microbiological analysis of foods or components. At present microbiological analyses are destructive, so that 100% inspection of a batch is impossible. Analysis of casually taken samples will provide information about the microbiological condition of the samples themselves, but it will be difficult to draw conclusions from such results about the condition of the unexamined items of the same batches. Sampling procedures should be designed statistically; only then can the results of microbiological analysis provide a basis for statistically valid conclusions about the microbiological quality of the batches of foods from which the samples were drawn.

1.1.1 Microbiological quality criteria

These may be applied for the purposes of:

1. An assessment of spoilage potential and keeping quality/shelf-life of a food.
2. An assessment of the public health hazard of a food in terms of the presence of pathogens and also the presence of toxins or of toxigenic organisms.
3. Microbiological counts which can be related to the hygienic standards of production, and/or which may be considered in an aesthetic sense to be undesirable in the food.
4. Microbiological quality that relates to the food conforming to legal standards or specifications.

A consideration of any specific quality criterion in the first two categories above will show that the value taken as representing the boundary between an acceptable and an unacceptable microbial content depends not only on the past history of the foodstuff but also on assumptions concerning the likely handling and treatment of the rest of the batch in the future. Thus, for example, the tolerance for a population concentration of a given pathogen will be related to:

1. The minimum effective dose required to establish an infection. This depends on the susceptibility of the individual consumer: certain sectors of the community may be more readily infected than others (e.g. babies, old people, people recently under antibiotic therapy, and immunodeficient or immunocompromised people).
2. The amount of foodstuff normally consumed.
3. Whether the pathogen dies, survives or multiplies in the foodstuff.

4. The normal and possible storage conditions to which the foodstuff may be subjected before consumption.

1.1.2 Quality control and quality assurance

For true quality control, the primary objective is to control the production process, and to attempt to ensure that the process is producing material of satisfactory lot quality. Obviously, in this type of feedback system for production control, the more rapidly an inspection technique can be completed, the more rapidly any remedial action may be taken. Up to the present microbiologists have found themselves at a considerable disadvantage here, with most microbiological techniques requiring at least 18 h to complete. Even today, most rapid microbiological analyses that are commercially applicable will take at least an hour to complete and the actual procedure must be performed off-line. This is in contrast to many inspection techniques for examining physical and chemical parameters (e.g. weight, colour, temperature, pressure, flow rate, viscosity, water content) which can be applied on-line and provide an answer in seconds or fractions of a second. Thus, for production control we would require microbiological analyses that can be completed in seconds or less.

1.2 MICROBIOLOGICAL SPECIFICATIONS FOR FOODS

Another type of quality control could be considered to be the examination of product before release to the consumer and diversion of below-standard product, i.e. distribution control. Two problems arise here.

The first problem, as already observed, is that, because microbiological analyses are destructive and 100% inspection is therefore impossible, complete control (i.e. a guarantee of intercepting all faulty goods) is not attainable. Consequently these end-product analyses must be performed as part of a scheme for quality assurance, in which an acceptance sampling plan is applied to allow a decision to be made on whether to accept or reject a batch on the basis of an examination of a sample drawn from that batch. An acceptance sampling plan in fact allows the inspector to quantify the risk of accepting batches of less than a given quality. Although we may call this application of analyses quality assurance, the customer cannot, in the absence of 100% inspection, be protected from receiving batches that are of unsatisfactory quality and yet have been accepted. Even with this type of examination, the microbiologist comes up against the second problem. This is that many conventional analyses (e.g. an examination for the presence of *Salmonella*) take so long that perishable and semi-perishable product will already have been released and consumed before the analysis is complete. Rapid methods that will provide a result in a few hours will be sufficiently rapid to be applied in such situations.

1.3 MICROBIOLOGICAL ANALYSES IN THE MONITORING OF QUALITY MANAGEMENT AND OF HACCP SYSTEMS

A more reliable way to achieve the distribution of good quality product is to set up a quality management system that can be monitored effectively. A quality management system aimed specifically at hazard control in the food industry is the HACCP (Hazard Analysis Critical Control Point) system. The HACCP system is discussed by Harrigan and Park (1991) and Mortimore and Wallace (1994). It should be noted that a full HACCP system will address *all* types of hazard – microbiological, chemical and physical (e.g. metal fragments and other foreign bodies contaminating the food). Laboratory-based microbiological analyses of samples (of food components, of food products, of swabs of equipment, etc.) can be important in two stages of implementing a HACCP system. The first stage in which such analyses are likely to prove valuable is in the identification of *hazards* of microbial origin. The second stage is in the evaluation of the effectiveness of monitoring of the *critical control points* (CCPs). The CCP should offer positive *control* of the identified hazard, but nevertheless the ideal monitoring activity of the CCP would enable rapid feedback adjustment into the production process. So, for example, the appropriate monitoring of a High Temperature Short Time heat treatment applied to a liquid product such as milk would correct deviations from the appropriate processing parameters, *and* divert the flow of inadequately heated product by means of a flow diversion valve. Obviously electrical or electronic temperature sensing permits such feedback activity to occur automatically. Laboratory analyses (such as the phosphatase test on milk, or examination for the presence of heat-sensitive bacteria such as coliforms) would in such a situation be used to detect whether the monitoring system just described is operating properly. In other words, the laboratory analyses have the function of monitoring the monitoring. A number of the analytical procedures described in this book are of use in this role. In this application of laboratory analyses, control charts may be useful (see Harrigan and Park, 1991).

1.4 LABORATORY REPORTS

All work carried out in the laboratory should be recorded fully in a laboratory report book. If you do not record work in the report book at the time it is performed, very full notes should be taken in a laboratory records book. It cannot be emphasized too strongly that the notes made in a laboratory records book should be as full as possible. Often, during the course of an experiment, observations are made that are not required in a report bearing on a particular aspect of the work. It may be that, at a much later date, these apparently superfluous observations will prove extremely useful. This is particularly the case when a research project is being carried out, as a rather narrow line must frequently be followed, with the necessity at the time to ignore many observations and results that suggest follow-up experiments.

It is always a good policy not to destroy laboratory records books. In the quality assurance (QA) laboratory, this will enable reports to be substantiated to a certain extent should the need arise. In this connection, it is perhaps advisable to use bound books, not loose-leaf folders, and to date all entries.

Reports of experiments and tests performed should be written up in a standard form. For example:

1. Title of the experiment or test performed and the date.
2. (a) The object of the experiment or test.
   (b) Summary of results and conclusions.
3. An outline of the methods used and, if stock cultures were employed, the names and brief descriptions of the organisms. When the methods depart from those detailed in the manual of methods customarily used, they must be described in full.
4. A description of the results obtained. This is often best given as a written report with tables or graphs used to clarify the results, but in some cases tables with any necessary footnotes may be more suitably employed to give the results obtained.
5. A statement of the conclusions that can be drawn from the results.

Alternatively, in QA laboratories performing a number of standard analyses on large numbers of samples, pre-printed pro forma report books can be used.

Laboratory reports on quality assurance or quality control work which are being issued to factory managers should be drafted in a form that will provide easy access to the substance of the report and to any recommendations. Thus it is useful to write a summary of the findings, and recommendations for action, towards the beginning of the report rather than at the end. The laboratory methods used will ordinarily not be described at all, but full information should be given of the sources of the samples and the reasons for choosing them. Results are best given as grades (e.g. A, B, C, D), with the actual counts being retained in the laboratory records books. The preferred system for recording counts is discussed in Sections 5 and 6 and in relevant sections of Part II.

In research reports references should be given in a manner similar to that used in this book. If the original paper has not been seen, then the reference of the source of the citation should be given.

There should be every effort to ensure traceability in all aspects of the laboratory work. There is a need both to validate the choice of the source for dehydrated media, media components and the like, and also to validate batches of medium. However, such validation using positive and negative controls may not always detect the effects of changing sources. Consequently, there should be procedures for recording the sources of media ingredients, dehydrated media and reagents, and for permitting correlation of any changes in counts, detection rates, etc. with any changes in sources of media, ingredients or reagents.

2

Management and Operation of the Microbiological Laboratory

To provide a reliable service of microbiological analyses, many establishments seek accreditation of the laboratory by the relevant national accreditation agency. Since the criteria adopted in assessments for accreditation will vary from country to country, and as these will usually be affected by the national safety legislation, readers are recommended to contact the appropriate accreditation body for advice on criteria and assessment procedures; this topic is discussed further in Section 2.5 below.

2.1 SAFETY PRECAUTIONS IN THE MICROBIOLOGICAL LABORATORY

The laboratory worker is exposed to many hazards. In the laboratory there are chemicals that are toxic, flammable, explosive, corrosive or carcinogenic, and on occasions there are dangers from the use of high voltages, ultraviolet and other radiation. In most countries there is substantial legislation to protect both workers and visitors from these hazards. In addition to these hazards, microbiologists are also exposed to hazards from the microorganisms with which they are working.

It should always be assumed that the microorganisms with which you are working are capable of causing disease – the assumption will often be true. Great care should therefore be taken in handling cultures, slides and all material that may contain, or have been in contact with, living microorganisms. The main routes of entry of infection to the body are by inhalation, by ingestion, through cuts and abrasions, and by infecting the eyes (the eye can serve as the portal of entry for infections that do not produce local pathology). A few types of microorganisms can even enter through the unbroken skin.

2.1.1 Precautions against microbiological hazards

2.1.1.1 Personal protection

1. Laboratory coats must be worn. In the UK a particular style of laboratory coat, known as the 'Howie' coat, is usually worn for microbiological work. This has side or back opening, a high neck, and elasticated or close fitting cuffs. Alternatively, if this style of coat is not available, a surgical gown may be worn. Ideally the laboratory coat should be removed before leaving the laboratory area. Other protective clothing should be used when entering a processing area for the purpose of taking samples, in order to minimize the risk of cross-contamination from the laboratory to the food production area. Some larger companies employ sampling staff in the production areas who are separate from the laboratory staff in order to reduce the risk of cross-contamination still further.
2. Used laboratory coats should be autoclaved before being sent for laundering.
3. When highly infectious organisms are likely to be encountered, disposable surgical gloves should be worn; these are then decontaminated by autoclaving after use.
4. Do not eat, drink or smoke in the laboratory.
5. Labels should be of the self-adhesive type to avoid the temptation of moistening gummed labels with the tongue. Do not place the ends of pens or pencils in the mouth whilst working.
6. Existing cuts and abrasions should be adequately covered with a water-impermeable dressing, i.e. a dressing that does not permit the access of liquid to the wound. The blue-coloured dressings available for use in food processing areas are also suitable for laboratory use. (The perforated wash-proof dressings usually sold for ordinary first-aid purposes, often termed 'waterproof', are not suitable for this purpose.) Alternatively, disposable surgical gloves can be worn, which should be decontaminated before disposal.

2.1.1.2 Accidents and spillages

Accidents such as spilled cultures, cuts and abrasions should be reported or written into a book kept for the purpose. If a cut or abrasion is received in the laboratory, make sure that suitable first-aid treatment is obtained. A spilled culture should be flooded with a suitable disinfectant solution (e.g. an iodophor such as 0.5% Wescodyne or a phenolic such as 2.5% Hycolin), and this left for 15–30 min before clearing up. Meanwhile the area should be properly identified by a notice to prevent other workers from coming into contact with the spillage. Broken glass should then be collected with forceps and placed in an autoclavable sharps container.

2.1.1.3 Risks from aerosols and airborne spores

1. It must be emphasized that the absence of breakage does not imply that no danger exists, as the dropping of a culture in a plastic Petri dish, for example, can result in

the release of a microbial aerosol into the atmosphere. The microbial aerosol is dangerous because it can reach a highly susceptible target – the lung – undetected, and in the lung can produce maximum effect in low dosage. When a culture is dropped on the floor you should therefore not bend down to clear it up immediately, but treat the area with a bactericidal solution and leave time for a reduction in the concentration of any aerosol that may have been generated.

2. Rapid and forceful ejection of the contents of a pipette can produce an aerosol. Aerosols can also be generated whilst using wire loops, and even during the removal of a screw cap or a rubber bung from a culture (Darlow, 1969, 1972). In general, slow and unhurried movements are to be preferred in microbiological work.
3. Sporulating cultures of fungi offer hazards of respiratory infection or allergic reaction even in the absence of aerosol generation and should be handled slowly and without sudden movement in a draught-free atmosphere. (Pathogenic fungi should be handled in suitable inoculation chambers; see below.)
4. Homogenizers and blenders must not be used in conjunction with bacterial *cultures* without adequate precautions against the spread of airborne contamination (see below). Their use to obtain dilutions of food samples will rarely present a hazard because of the low microbial concentrations encountered. The preferred method for the preparation of dilutions from food samples is using the 'Stomacher' (see Section 17.4.1), and this has the added advantage of minimizing aerosol production.

2.1.1.4 Laboratory practices

5. The use of a wire loop requires considerable skill if risk of contamination of the air and the working area is to be avoided. Inoculating needles and loops must be sterilized before and *after* use, by heating in a Bunsen flame until red hot along the entire length of the wire. Spattering of material from the wire should be avoided by very gradual introduction into the Bunsen flame if hooded burners or safety loop sterilizers are not being used.
6. Test-tube cultures should always be kept in test-tube racks. Never lay the test-tubes on the bench top.
7. A plate count on a food sample prepared using an 'innocuous' medium such as nutrient agar or milk agar cannot be regarded as harmless merely because the food was suitable for consumption. Many pathogens such as *Staphylococcus* and *Salmonella* can grow on such media. (Desoxycholate agar being selective for *Salmonella* does not make milk agar or plate count agar selective against *Salmonella*.) The microorganisms originally present in the food in small numbers are now in the plate culture in billions, and can cause infection by inhalation.
8. Because of the frequent culturing of pathogens that occurs in microbiological laboratories, there must be an absolute ban on pipetting by mouth. The cottonwool plug in a pipette is there to prevent the contamination of the liquid being pipetted, and not to prevent infection of the user. A safety pipetting device or a rubber teat

or bulb should always be used. When inserting a pipette into a safety pipette filler, first check that there is no broken glass in the filler. Always hold the pipette close to the end being inserted and use the minimum of force with a gentle twisting action.

9. Used pipettes must be placed in pipette jars containing disinfectant solution. Microscope slides and coverslips must also be discarded into jars of disinfectant solution, coverslips first being separated from the slides. All of these items must be totally immersed in the disinfectant solution.
10. Homogenizers and blenders must not be used with bacterial cultures except in a safety cabinet, because of the hazard from the creation of aerosols. The design of the Stomacher ensures that aerosol creation into the laboratory is minimized.
11. Positive-pressure inoculation chambers are used for sterility testing. They *must not* be used for working with microbial cultures. (The more infectious pathogens must be handled in either completely closed inoculation chambers or in *negative* pressure inoculation hoods whose air outlets incorporate suitable filters to remove organisms and to render the exhaust air safe.)

2.1.1.5 Work with pathogenic microorganisms

In many countries there has been government guidance and legislation on the classification of pathogens, and the design and operation of laboratories in which these pathogens are cultured. For example, in the UK the Advisory Committee on Dangerous Pathogens (ACDP, 1990) has categorized microorganisms and provided Codes of Practice for the containment of the various categories. It should be remembered that most such guidance and legislation has been initiated by concern about work in pathology laboratories and microbiology laboratories in hospitals and similar situations, and the possible infection of workers in these establishments. For microbiologists working in the food industry there is the additional danger of cross-contamination from laboratory to food production area. Thus, ideally there should be a high level of physical isolation of the laboratory from the food handling areas. This topic, and laboratory design and management, is dealt with in greater detail in *Making Safe Food* (Harrigan and Park, 1991).

The food microbiology laboratory should be designed, equipped and run on the assumption that it will deal with ACDP Category 2 organisms (e.g. *Escherichia coli, Campylobacter, Staphylococcus aureus, Listeria* and most *Salmonella*). As isolations will be from samples containing an unknown mixed microflora, it is possible that an ACDP Category 3 pathogen is cultured accidentally. For example, if the laboratory routinely examines samples for the presence of salmonellae, there is always the possibility of the isolation and culture of *Salmonella typhi* or *Salmonella paratyphi*. Consequently, there should be written procedures in place for the handling and disposal of such cultures.

Quality assurance tests on food samples involving colony counts, most probable number (MPN) counts or any other cultural procedures must never be performed in laboratories in which starter cultures are prepared for use in food fermentations. This ban should apply even when the colony counts are intended to quantify the viable population of the starter organisms, as there is no assurance that a single potentially pathogenic cell will not be selected and cultured.

Many microfungi hitherto considered suitable for culture in the open laboratory are now suspected of being allergenic or potentially pathogenic, and because conidiospores are so light and so easily airborne, special precautions should be taken. Indeed, in many microbiological quality assurance laboratories in which food samples are examined for the presence of moulds using mycological media, the examination of plates is performed in a safety cabinet, and this practice is to be encouraged.

2.1.1.6 Procedures to be followed on leaving the laboratory

1. On completion of work in the laboratory, the work area should be cleared. Contaminated items and cultures no longer required should not be left on the bench but placed on trolleys designated for items requiring decontamination by autoclaving, etc. The work area should finally be swabbed down with an appropriate disinfectant solution (e.g. an iodophor such as 0.5% Wescodyne).
2. The hands should be washed thoroughly before leaving the laboratory for whatever reason.

2.1.1.7 First aid

There should be at least one staff member on duty at any given time who has passed a recognized course in first aid and who is trained to provide both respiratory resuscitation by mouth-to-mouth and Silvester methods and external heart compression in the event of cardiac arrest. Every laboratory worker should know the identity and location of 'first aiders'. It is good practice for such people to be identifiable by an appropriate badge worn on their laboratory coats.

The location of first-aid cabinets, eye irrigation bottles and fire extinguishers should be known by all laboratory workers. A list of the specific hazards presented by the chemicals and the microorganisms in the laboratory should be determined and made available, so that in the event of any accident full information can be provided to the doctor or hospital.

2.1.2 Precautions against chemical hazards

In many countries there is specific legislation to protect workers against exposure to hazardous substances. For example, in the UK *The Control of Substances Hazardous to Health Regulations 1988* introduced a legal framework to control people's exposure to hazardous substances. One of the requirements is for an assessment to be made of the health risks and of the measures to be taken. In the case of the legislation just mentioned, microbiological as well as chemical hazards are covered.

2.1.2.1 Hazard and risk

A *risk* assessment combines estimates of *hazard* (the potential of a substance to damage health) and *exposure* to the hazardous substance. Thus this assessment considers not only the nature of the substances being worked with, but also the amounts involved, and the operations and procedures being used.

An extremely hazardous substance (e.g. one legally defined as a carcinogen) will require a stringent containment regime, whatever the procedures, and national legislation is likely to require a written assessment to be made, together with the preparation of written 'handling protocols', which will include descriptions of appropriate disposal and emergency procedures. These documents need to be approved by the appropriate senior manager (e.g. safety officer, technical manager). In these cases, copies of the written protocols should be held by safety officers, first aiders and, if relevant to on-site operations by emergency services such as the fire brigade, copies should be placed with the security office for the site.

In each laboratory there should be an inventory of the chemicals used and stored in the laboratory, listing for each chemical the nature of the hazards and the relevant handling and disposal procedures. There are standard pictograms in use which identify the type of hazard presented by a chemical (e.g. toxic, irritant, corrosive, flammable); unfortunately at present these pictograms have not been universally adopted. If you are working in a country that does not have such a standard labelling system, then you may wish to use a labelling system of your own within the laboratory.

The risk assessment should apply to all stages in the use of the chemicals. Thus the labelling system should be followed through on to bottles of working solutions of reagents where appropriate. For example, working solutions of Voges–Proskauer reagents or Kovacs's indole reagent are hazardous in use and should bear warning labels. However, although a bottle of powdered dehydrated medium may present a hazard because of the presence of selective agents, the made-up medium would not normally be particularly hazardous, and would therefore not require warning labels. If the washing up of used equipment is performed by separate staff, then they should be made aware of the hazards presented by anything that is discarded. There have been many examples of washing up staff being injured or harmed by exposure to unidentified substances left in unmarked bottles.

2.1.2.2 A structured approach to risk assessment

Consider all the aspects shown below.

1. Establish the need to use the substance. Is the particular work or procedure necessary? If the answer to this is yes, then is there a possible non-hazardous or less hazardous substitute for the hazardous substance? For example, for the preparation of polyacrylamide gels, liquid reagents are now widely available, thus avoiding the use of the more hazardous powdered version. Another good example is the availability of Kovacs's oxidase test reagent in touch-stick (Oxoid) or dry-slide

(Difco) format, thus avoiding the risks of weighing out *NNN'N'*-tetramethyl-*p*-phenylenediamine dihydrochloride to produce a solution of the reagent.
2. What is the physical form of the substance (powder, crystals, liquid, vapour)?
3. Choose the least hazardous form available. For example, crystals are easier to contain than a fine powder.
4. Acquire a Safety Data Sheet for the material. These should be available from the supplier of the material. In addition, safety databases are now widely available in book form or on CD-ROM. Look out for any particular properties of the substance. For example, a few toxic chemicals normally used in solid form may nevertheless give off significant amounts of toxic vapour at room temperature.
5. List the physiological and toxicological effects. Pay particular attention to long-term or irreversible effects such as suspected carcinogenicity, teratogenicity, mutagenicity and chronic tissue damage (liver damage, etc.). (See also Luxon, 1992.)
6. Note the use or purpose for the substance.
7. Note the normal quantities of the substance to be stored or used.
8. Determine the possibility of hazardous products arising from reactions occurring between separate compounds.
9. Can a Code of Good Laboratory Practice cope with the risk, or is a specific written protocol necessary?

2.1.2.3 Weighing out toxic and hazardous substances

It is difficult to use an analytical balance within a fume cupboard because of the air disturbance and vibration. There are now commercially available weighing stations that maintain a slight negative pressure within the container holding the balance, with exhausted air being passed through a high efficiency particulate filter. However, if such a contained weighing facility is not available, hazardous chemicals can be weighed out following the procedure described below. Examples of toxic or irritant compounds that should be weighed out in this way are acrylamide powder, thallium compounds, selenium salts, *NNN'N'*-tetramethyl-*p*-phenylenediamine dihydrochloride.

Procedure

1. Weigh an empty clean screw-capped container.
2. Take the preweighed container and the bottle of the chemical to a fume cupboard.
3. Using appropriate precautions to prevent contact with the skin, transfer an amount of the chemical to the preweighed container that approximates to the amount required. Replace the screw cap of the container.
4. Take the closed receptacle containing the chemical back to the balance and weigh.
5. Determine the weight of the chemical transferred, and then calculate the amount of solvent needed to give a solution of the desired concentration.
6. Return with the closed container, solvent, etc. to the fume cupboard and, using appropriate precautions, make up the required solution of the chemical.

2.1.3 Good laboratory practice

In addition to the topics covered above, there are certain other aspects that should be covered in a Code or Manual of Procedures to be followed by both the laboratory workers and visitors to the laboratory.

1. Certain laboratory activities should be performed only by staff competent in those procedures; this may require specific training from the manufacturers of the equipment. Examples are the use of centrifuges, autoclaves, high-performance liquid chromatography (HPLC) and gas chromatography (GC) equipment, and electrophoresis equipment.
2. Appropriate personal protection should be used; this will normally be a laboratory coat and eye protection, but for some work may also involve the use of gloves, full-face protection, etc. Wearers of contact lenses should always wear eye protection, as if a chemical is splashed in the eye it can seep under the contact lens and cause damage before the contact lens is removed. Even quite dilute solutions of alkalis can cause permanent corneal damage.
3. Take care when diluting or dissolving corrosive reagents, and when pouring liquids. When diluting concentrated acids and alkalis, always add the reagent to the water with continuous stirring, as some admixtures are substantially exothermic (e.g. sulphuric acid and water).
4. Never allow electrical leads to trail across work areas, and keep regulators, junction boxes, etc. above the surface of the bench. Never use equipment that is damaged, or that has a frayed lead or with a damaged plug.
5. Glassware assemblies should be supported adequately with clamps and stands. When filling a burette always lower the top of the burette to below eye level and use a funnel.
6. Gas cylinders must never be left free-standing, but should be either clamped to the bench, or supported stably in a cylinder holder.
7. Lubricate glass tubing (e.g. with aqueous detergent) before attempting to pass it through rubber bungs. Use a cloth to hold the bung and gently push with a twisting action.
8. Never carry a bottle by the neck or cap. If carrying more than one bottle, use a bottle carrier. Containers of strong acids and alkalis should be carried in shatterproof (e.g. plastic) secondary containers. Bottles containing ammonia solutions are usually under pressure. Take care when opening. Bottles containing hydrogen peroxide solutions should not be stoppered tightly.

2.2 CLEANING OF GLASSWARE AND APPARATUS

2.2.1 Treatment of new glassware

Borosilicate glass (e.g. Pyrex) or factory-washed soda-glass apparatus needs no special treatment before being used, other than normal washing up. New unwashed soda glass should be soaked in N hydrochloric acid overnight to neutralize partially the alkali contained in the glass.

2.2.2 Treatment of used glassware and other apparatus

All glassware containing microbial cultures or otherwise contaminated by microorganisms should be sterilized by autoclaving for 40 min at 121°C, which will also liquefy any solid media and allow easy removal.

The sterilized glassware should be emptied and rinsed in tap water. It should then be either washed in a laboratory glassware washing machine using an appropriate washing machine detergent (e.g. Decomatic, made by Decon Laboratories), or washed manually.

To wash manually, soak in a suitable detergent solution and finally clean with a brush. Glassware that has contained Vaseline, paraffin wax or liquid paraffin should be washed separately to avoid spreading greasy films. It is advisable to use a detergent specifically formulated for use in microbiological laboratories, as the detergent must (a) completely remove the most tenacious residues (e.g. proteins and fatty materials); (b) be capable of being removed easily and completely from the glass by rinsing; and (c) neither cause deterioration of any of the materials from which the apparatus is constructed, nor be harmful to the skin.

Examples of such detergents are Pyroneg (Diversey) and Decon 90 (Decon Laboratories). If such a detergent is used, it is possible to wash metal, rubber and plastic components (e.g. screw-caps and tube closures) with the glassware. During washing, the rubber liners must be removed from metal screw-caps so that all surfaces receive adequate washing; for this reason autoclavable polypropylene screw-caps which do not possess separate liners (e.g. those made by Sterilin) may be preferred.

After the detergent wash, rinse the apparatus adequately. If the glassware is to be used for microbiological assay or nutritional studies it should be washed five times in hot tap water and then three times in distilled water. Drain and dry in a heated drying cabinet.

2.2.3 Used pipettes

These should be discarded at the bench into jars containing a disinfectant solution. If the pipettes have been used for liquid cultures, or for pipetting water, quarter-strength Ringer's solution or other non-soiling liquids, a self-indicating hypochlorite solution can be used in the jars. In the case of pipettes used for blood, serum, milk and dilutions of many foods, hypochlorite is unsuitable. Hypochlorite must not be used if there is a danger of contact with formaldehyde. It is then necessary to run tests on detergent–sanitizers which are readily available to determine: (a) their ability to remove residues, and (b) their bactericidal and bacteriostatic effects on the types of bacteria most often used or encountered. Suitability of preparations is best determined using the capacity test method (see Section 2.4.4). A suitable disinfectant is Hycolin.

If pipettes are used for the transfer of dangerous pathogens it is best to use disposable pipettes placed into jars of Hycolin, Virkon or other suitable disinfectant, the pipettes being removed to autoclavable disposal bags and autoclaved before final disposal.

2.2.4 Disposable apparatus

Plastic Petri dishes, and pipettes and bottles that are disposable, should be sterilized by autoclaving for 40 min at 121°C before discarding. They may be placed in autoclavable plastic bags (e.g. as supplied by Sterilin), the sealed bags with sterilized contents being more readily and hygienically discarded. The bags are often placed in trays to catch molten medium should the bags leak. Such trays should be relatively shallow (not more than about 20 cm deep) and contain some free water, to generate free steam at the base of the tray during heating. This avoids trapping air within the tray, which would lead to under-processing.

Microscope slides and coverslips, having been discarded into jars of disinfectant on the work bench, can be disposed of without further sterilization, but, if known pathogens are involved, the jars and contents should be autoclaved (in which case a disinfectant other than hypochlorite should be used).

2.3 STERILIZATION OF GLASSWARE, CULTURE MEDIA, ETC.

The sterilization of culture media, containers and instruments is essential in bacteriological work for the isolation and maintenance of pure cultures. Bacteriological tests for sterility may be used to confirm the efficacy of the procedures used.

2.3.1 Sterilization by heat in the absence of moisture

Although heat is much more effective in the presence of moisture, both dry heat methods and moist heat methods have their uses.

2.3.1.1 Red heat in the Bunsen flame

This is used for sterilizing inoculating wires, loops and metal instruments that are not damaged by heat. If infective pathogens are being cultured, hooded burners or properly designed loop sterilizers are essential to contain spattered material.

2.3.1.2 Flaming after dipping in ethanol

This method is frequently used for scalpels, spatulae, etc., with the instruments not being heated to red heat. It does not necessarily achieve sterilization.

2.3.1.3 Hot air oven

The hot air oven is heated by electricity or gas and is thermostatically controlled. Sterilization in the hot air oven is usual for dry glassware such as test-tubes, glass Petri dishes, flasks and pipettes. Glass Petri dishes and pipettes are packed most easily in copper, alloy or stainless steel containers made for the purpose. Glassware should be dry before placing in the oven. The hot air oven is also used for sterilizing dry materials in sealed containers (e.g. chalk) and for mineral oils used in the preservation of stock bacterial cultures. Loading should take place when the oven is cold and spaces should be left between and around the items for circulation of air through the load. The holding time should be a minimum of 2 h beginning when the oven thermometer indicates 160°C. The oven should be allowed to cool before the door is opened as otherwise the glassware may crack.

The efficacy of sterilization can be checked by using a sterilization monitor such as Browne's sterilizer control tubes, type III (green spot) (Browne Health Care, Leicester, UK), which should be packed to simulate the most thermally protected material being treated.

2.3.2 Sterilization by heat in the presence of moisture

2.3.2.1 Boiling water bath

Boiling for 5–10 min is sufficient to kill non-sporing organisms, but many bacterial spores will survive. This method is useful where sterility is not essential (as is the case with many selective media) or where better methods are not available or are unsuitable. If used for items of equipment that cannot be exposed to higher temperatures, distilled water should be used, particularly in districts where the water supply is hard, otherwise the instruments will become covered with a film of calcium salts.

2.3.2.2 Koch's steam sterilizer

Sterilization in a Koch's steam sterilizer ('steamer') in steam at atmospheric pressure and approximately 100°C is used for media or constituents that are damaged by exposure to temperatures above 100°C (e.g. sugars) and milk. The steamer may be used in two ways: (i) a single exposure at 100°C for 90 min; (ii) intermittent heating (Tyndallization) by heating at 100°C for 30 min on each of three successive days interspersed with incubation under conditions in which the medium will be subsequently used. In (ii) the first exposure kills vegetative forms, and between the heat treatments the spores germinate and are thus killed in the subsequent heating.

2.3.2.3 Autoclave

Autoclaving is the most efficient method of sterilizing culture media, and should be used for all media capable of withstanding the high temperature without decomposition. It is also used for glassware and instruments, and for sterilizing cultures and contaminated material before washing. The actual temperature inside the autoclave depends on the steam pressure (Table 2.1).

TABLE 2.1 Autoclave temperatures achieved with pure saturated steam (no entrapped air)

Gauge pressure		
kPa	lb in^{-2}	Temperature (°C)
0	0	100
34.5	5	109
68.9	10	115
103.4	15	121
137.9	20	126

Sterilization in the autoclave is usually achieved by autoclaving at 121°C for 15–30 min in pure saturated steam at 103.4 kPa (15 lb in^{-2}) above atmospheric pressure.

The total exposure time required at the desired temperature will depend on a number of factors including:

1. *The microbial load of the material being sterilized, and the nature of the contaminants*. Ingredients of culture media, and dehydrated media, are usually manufactured in a way that ensures very little contamination. But very heat resistant spores of thermophilic *Bacillus* spp., for example, *may* be encountered in dehydrated media or ingredients at levels so high that the media cannot be decontaminated without considerable thermal denaturation. This failure to obtain sterility would be noticed only when such media were used for thermophilic studies. Most manufacturers produce special media of guaranteed low spore counts for such purposes.
2. *The size of the containers and the thickness of the wall.* About twice as long is required for heat to penetrate a 500-ml medical flat bottle full of medium, as for it to penetrate a 100-ml medical flat bottle full of the same medium. It requires a heating time about half as long again for the heat to penetrate 500 ml of water, diluent or nutrient broth contained in a thick-walled transfusion bottle as for the heat to penetrate the same amount of liquid in a medical flat bottle.
3. *The nature of the contents*. The presence of agar increases the heat penetration time to nearly double that of water.

It is not possible to generalize on the process times necessary, partly because some heat penetration occurs during venting and whilst the steam pressure reaches the required level. The more efficient the autoclave in these respects, the more consideration which must be given to the factors just mentioned. Controlled experiments using spore suspensions, spore strips (obtainable from Oxoid and other manufacturers of dehydrated media, etc.), wandering thermocouples (which can be used in a few autoclaves), or ampoules of chemical indicators (e.g. Browne's tubes, manufactured by Browne Health Care, Leicester, UK) may all help to establish the correct processing treatment.

It is important to ensure that all air is removed from the autoclave before sterilization begins, as a mixture of air and steam results in a lower temperature for any given pressure.

Also, the presence of air tends to prevent steam penetration and causes uneven heating in the different parts of the autoclave. If one-third of the air remains in the autoclave as a result of inefficient venting, the temperature inside the autoclave will only reach 115°C at a pressure of 103 kPa. In the absence of a thermometer in the steam drain, a check on the presence of air in the steam can be made by arranging for the vented steam to pass through a container of cold water. It should be noted that the design of some manually operated autoclaves (especially those designed for connecting to a steam main) *may* be such that any air not vented by downward displacement at the beginning of the operation will remain in the inner chamber and *not* escape through the pressure valve.

Some autoclaves provide the opportunity to generate a thermographic record of the temperature and time of exposure at one or more points in the load using in-chamber thermocouples.

Autoclave indicator tape should be applied to batches of material being subjected to autoclaving, to indicate that the batch has been autoclaved. Note, however, that such tape does *not* indicate the efficacy of the process.

See also Rubbo and Gardner (1965) and Report (1959).

2.3.2.4 The inspissator

This is used for the preparation of media such as Loeffler's serum medium. The medium is distributed in containers placed in a sloping position in special racks. The temperature is slowly raised to 85°C and maintained for 2 h, causing the medium to be solidified completely.

When coagulation of the material is not required and lower temperatures (about 56°C) are used, it is necessary to repeat the process on several successive days.

2.3.3 Sterilization by filtration

This method is used for sterilizing fluids and solutions that would be adversely affected by heat. Such components may then be added aseptically to other materials which have been heat sterilized. Sterilization by filtration is also used when a source of sterile air is required, for example for fermenters, although it may be difficult to preclude the entry of bacteriophage. The filter more or less universally used for this purpose is the membrane filter.

Membrane filters are usually made of highly porous cellulose acetate, and are available in a wide range of different porosities. In contrast to a depth filter, a membrane filter retains particles above a certain size *on its surface* because the size of the largest pore is smaller than the smallest of the retained particles. Particle retention is effected by pore size and not by electrostatic attraction or adsorption. Thus the pore size rating given by the manufacturer is an absolute one and, although the pores will vary somewhat in size, the manufacturer can quote a range of pore size outside of which no pores will occur. For example, a typical filter intended to remove bacteria may have a quoted pore size range of 0.22 ± 0.02 μm. Such filters can be used in positive-pressure filtration equipment with relatively high pressure differentials, permitting the achievement of short filtration times without affecting the sterilizing efficiency of the filter.

For the bacteriological sterilization of media, etc., the final sterilizing filter in any procedure should have a pore size of 0.22 μm, although grades with much smaller pore sizes are available from most manufacturers of membrane filters. Note that the 0.45-μm filter used for counting and isolation work in the bacteriological and mycological analysis of samples is not suitable for the provision of a filtrate of guaranteed bacteriological sterility.

Filtration can be achieved by the application of negative pressure (i.e. suction) or positive pressure. In the microbiological analysis of samples, negative pressure is nearly always used. When media, sugar solutions, sera, etc., are to be sterilized by filtration, it is better to employ positive pressure for two reasons. Firstly, the use of negative pressure requires either the receiving vessel to possess a side arm to which the suction is applied (e.g. a filter flask) or use of a bypass filter attachment. In the former case the sterile liquid will need to be transferred aseptically to the final sterile container. A bypass filter attachment will allow filtration directly into the final sterile container, but there must still be adequate protection against blowback of non-sterile air, water or oil from the vacuum line or leakage of non-sterile air into the container through improperly sealed joints – and this last is quite difficult to ensure. The second drawback associated with the use of negative pressure is that the absolute (and unattainable) maximum pressure differential is 1 atmosphere. With a positive-pressure filter unit, rapid filtration can be obtained by the use of much higher pressure differentials.

There is available a wide range of membrane filtration apparatus for sterilizing filtration by positive pressure. In most laboratories engaged in bacteriological and mycological analytical work, sterilizing filtration will be used for carbohydrate solutions, alcoholic solutions, yeast extract solutions, sera, etc., which are all mostly required in fairly small quantities, say up to 100 ml batch size. One of the most useful pieces of equipment for the small microbiological laboratory is the Swinnex filter-holder (Millipore Filter Corporation) for use with 13-, 25- or 47-mm membrane filters; these holders attach to any hypodermic syringe employing the usual Luer type of connection. Such filters are perfectly capable of sterilizing 1, 10 or 100 ml of expensive media supplements or of rarely used media with practically no wastage. The Swinnex holder is also extremely useful for in-line sterilization of air or media being supplied, for example, to fermenters.

Carbohydrate (not polysaccharide) solutions, alcohols, salts solutions, vitamin solutions and water can be filtered fairly readily through the sterilizing grade of filter with or without (but preferably with) a disposable fibreglass prefilter pad preceding it. Media incorporating proteins, and some other liquids, can be filtered more readily if two membrane filters are used after the fibreglass prefilter, the final sterilizing grade of filter being preceded by a coarser grade (e.g. Millipore DA, pore size 0.65 ± 0.03 μm). Liquids such as serum and plasma can be filter-sterilized by employing a number of filters with progressively smaller pore sizes; for example, a fibreglass prefilter followed by a filter of pore size 1.2 μm, a filter of pore size 0.65 μm, a filter of pore size 0.45 μm, and finally the sterilizing grade of filter with a pore size of 0.22 μm. Because the membrane filters are so thin, a stack of filters can quite easily be placed in a single filter-holder; the use of Terylene net separators between each pair of filters is recommended to improve flow rates. The stack of filters with separators is sterilized *in situ* in the filter-holder by autoclaving.

2.3.4 Chemical disinfectants

Chemical disinfectants are used mainly for disinfecting the skin, floors, buildings, apparatus, and for articles that cannot be heated effectively without damage. In the laboratory, pipettes and slide preparations containing living cells should be discarded into jars containing suitable disinfectants (see p. 16). Any cultures spilled in the laboratory should be covered with absorbent cottonwool soaked in disinfectant before removal. An iodophor such as 0.4% Wescodyne is suitable for these purposes.

2.4 EVALUATION OF DISINFECTANTS

2.4.1 Introduction

In the microbiological quality assurance laboratory it will seldom be necessary to perform evaluations on disinfectants used either in the laboratory or in the food processing factory; it is recommended that disinfectants and sanitizers be purchased from reputable manufacturers who can provide certificated evaluation test results for their products. However, just occasionally the quality assurance microbiologist may find it necessary to carry out such an evaluation. Standard procedures have been described by the Association of Official Analytical Chemists (AOAC, 1995), by the International Standards Organization (ISO) and by various national standards organizations (e.g. British Standards Institute).

There are four types of test: (a) the phenol coefficient tests, such as the Rideal–Walker test, in which the disinfectant under investigation is compared with standard dilutions of phenol; (b) suspension tests, which determine the survivor–time curves for a disinfectant; (c) capacity tests, such as the Kelsey–Sykes test, which determine the ability of the disinfectant to be soiled with microorganisms and organic or other material and still remain effective; (d) hard-carrier tests, which determine the ability of a disinfectant or sanitizer to render a test surface free of the test organism in the presence of a standard soiling substance (whether by disinfection or detergency).

2.4.2 Phenol coefficient tests

In this type of test, typified by the Rideal–Walker test, the activity of the test disinfectant is compared with that of phenol as a standard; the result obtained is known as a phenol coefficient. The Rideal–Walker test, however, is valid only in comparing disinfectants similar in chemical composition to phenol, and is therefore less useful in assessing the efficiency of disinfectants such as hypochlorites and quaternary ammonium compounds. For the procedure, refer to AOAC (1995), British Standard BS 541:1985, or Croshaw (1981).

2.4.3 The suspension test (International Dairy Federation, 1962)

In principle, this method (also known as a survivor-curve method) consists of adding a known number of microorganisms in suspension to a solution of disinfectant at the required concentration and then determining the number of survivors after given time intervals. An advantage of the suspension test is that concentrations of disinfectant and exposure times can be chosen which simulate those used in practical conditions. The suspension test has therefore proved extremely useful in studying the activity of disinfectants used in the food and dairy industries. The effect of additional organic matter on disinfectant activity can be studied by adding organic matter (e.g. milk) to the disinfectant at the time of adding the cell suspension. The British Standard BS 6471:1984 describes a suspension test procedure for determining the antimicrobial activity of quaternary ammonium compounds.

2.4.3.1 Preparation of cell suspension

A suitable test organism may be chosen as appropriate, e.g. *Escherichia coli*, *Lactococcus lactis*, and incubated for 24 h at the optimum growth temperature in a suitable medium before harvesting. Cell suspensions are prepared from solid media by washing off the growth with sterile diluent (quarter-strength Ringer's solution). Cells can be separated from liquid media by centrifuging, discarding the supernatant liquid and resuspending the sedimented cells in sterile diluent. The resulting cell suspension is shaken to disintegrate clumps and can then be standardized to the desired strength by diluting with sterile diluent. A concentration of approximately 10^{10} cells per ml is recommended, and can most conveniently be found for routine purposes by first determining the relation between the numbers of microorganisms and the optical density of the suspension. In subsequent work, it is then sufficient merely to adjust the suspension to the required opacity, for example with McFarland's or Brown's opacity tubes (see Section 6.6.2).

2.4.3.2 Test procedure

1. Distribute the disinfectant solution at the required concentration in 99-ml quantities in 250-ml conical flasks. Concentrations tested should include those likely to be encountered under practical conditions.
2. Add 1 ml of cell suspension to the disinfectant solution, taking care that the tip of the pipette is held just above the surface of the disinfectant while delivering the suspension. Note the precise time of adding the suspension and mix well by rotating the flask.

3. After exposure periods of 30 s, 2 min, 5 min, and 10 min, remove 1 ml from the flask and transfer to tubes containing 9 ml of a sterile solution of inactivator. Mix well by rotation.

   The inactivator must be appropriate to the disinfectant under test. Sodium thiosulphate (0.5%) is used for tests with hypochlorite and iodophors. For quaternary ammonium compounds, the following inactivators may be satisfactory but suitability should be checked for the particular disinfectant (see Section 2.4.4): (1) a mixture of 2% egg lecithin in a 3% aqueous solution of Cirrasol ALN-WF (ICI); or (2) 2% Tween 80; or (3) 10% serum. Organic mercurial compounds can be inactivated by the addition of 0.25% sodium thioglycollate solutions (Sykes, 1965).
4. Determine the numbers of survivors for each exposure period by taking out 1 and 0.1 ml from each tube of inactivator solution on to a suitable agar medium and incubating at the optimum growth temperature of the test organism.

2.4.4 The capacity test

This type of test, typified by the Kelsey–Sykes test, provides information on the capacity of a use-dilution of a disinfectant to be soiled with microorganisms and organic material without losing disinfectant activity (Croshaw, 1981).

This test procedure is appropriate, for example, for testing disinfectants or detergent–disinfectant mixtures to be used in used-pipette containers in microbiological laboratories, and determining the required concentrations and required intervals for renewal of the disinfectant in such containers. The capacity test is more suitable for this purpose than the hard-carrier type of test. A modified Kelsey–Sykes test forms the basis of BS 6905:1987.

2.4.4.1 Outline test procedure

The details of the experiment should be determined by the conditions under which the disinfectant is used.

The test organisms are grown in nutrient broth or other suitable medium at optimum temperature for 24 h. (Alternatively suspensions washed from agar slopes may be used.) Suitable test organisms include:

Pseudomonas fluorescens
Escherichia coli
a capsulate strain of *Enterobacter*
Proteus sp.
Staphylococcus epidermidis

Clumps of organisms are broken by shaking with sterile glass beads for 1 min. Prepare an appropriate dilution of the culture or suspension (about 10^8 organisms per ml) using an inorganic diluent. Add 1 ml of sterile nutrient broth or 1 ml of sterile (UHT) milk to 10 ml of the dilution to be used, to provide a standard organic content. Determine the viable count of this dilution. The test temperature employed is commonly 22°C, but other temperatures can be used. Place 6 ml of the required concentration (e.g. the recommended use-dilution) of disinfectant into a sterile jar. At 10-min intervals 1 ml of the bacterial suspension is added, and the mixture mixed by swirling (avoid foam or bubble formation). A sample is withdrawn 8 min after each addition, to determine the presence or absence of viable organisms.

Place five single drops (0.02 ml) on the surface of a poured, dried nutrient agar plate, and one drop into each of two nutrient broths containing an appropriate inactivator. This procedure is continued for 1 h, which allows six additions of organisms with the concentration of disinfectant being cut to approximately half (actually 0.49).

The end point of the test with respect to the activity of the disinfectant is the highest number of additions that gives fewer than five colonies from five drops or fewer than two positive broths. The concentration of disinfectant, number of bacteria added, and concentration of organic matter at this point can be calculated. In practice a disinfectant use-dilution can be regarded as satisfactory if three or more increments can be added before a positive culture is obtained.

Preparation of disinfectant dilutions

It is suggested that three dilutions be examined: the manufacturer's recommended concentration, one-half the recommended concentration, and 1.5 times the recommended concentration.

Standard hard water

It is recommended that a standard hard water is used to prepare the dilutions of the disinfectants.

17.5 ml of 10% (w/v) solution of $CaCl_2 \cdot 6H_2O$ and 5 ml of 10% (w/v) solution of $MgSO_4 \cdot 7H_2O$ are added to 3.3 litres of distilled water. Sterilize by autoclaving.

Inactivators

These should be incorporated in nutrient broth at the concentrations given for the suspension test (see Section 2.4.3.2 above).

To test the suitability of an inactivator

Prepare:
(a) inactivator and disinfectant in test ratios and concentrations
(b) inactivator and water
(c) water alone

Inoculate 10 ml of each with 1 ml of suspension containing 10^3 to 5×10^3 organisms per ml. Immediately plate 1 ml of each with nutrient agar. Repeat after 30 and 60 min of contact. The inactivator is suitable if the counts obtained in the three systems are not significantly different.

2.4.5 Hard-carrier tests

These tests determine the extent of the reduction of numbers of viable microorganisms on a standard surface, which can be due to killing, detergency (washing still-viable organisms from the surface) or a combination of the two effects. The tests are therefore very sensitive to both the surfaces used and the precise methodology of exposure of the inoculated surface to the substance being evaluated. It is extremely important to use a highly standardized and reproducible method. Methods that can be used are AOAC Official Methods 991.47 and 991.48 (AOAC, 1995), or the modified Lisbôa test (Blood *et al.*, 1981).

2.5 QUALITY CONTROL AND QUALITY ASSURANCE IN THE LABORATORY

2.5.1 Introduction

A laboratory may run its own quality management and quality assurance scheme on its own activities, without reference to any third-party audit or external validation. This used to be quite common for quality assurance laboratories in the food industry. However, as the results of laboratory analyses and other activities of quality assurance staff become increasingly used in determining conformance to contractual obligations between supplier and customer, it is more likely that a laboratory will be expected to demonstrate its suitability and diligence by having its procedures and quality systems externally audited or monitored. In addition, such external checks are of value if laboratory results are called on in any litigation. (See also International Dairy Federation, 1993; Wilson and Weir, 1995, Günzler, 1996.)

There are three types of scheme in relation to the quality of activities in a laboratory.

Firstly, the laboratory could demonstrate that it has appropriate quality management *systems* in place, by obtaining certification to the ISO 9000 series of standards for quality management systems. These standards are general, and do not address the specific laboratory activities. (For further discussion of the ISO 9000 series of standards see Harrigan and Park, 1991, or Hoyle, 1994.)

Secondly, a laboratory may seek accreditation in relation to its conformance on measuring and calibration equipment and procedures. This will require the laboratory to carry out regular and approved checking of the reliability of incubators, water-baths, autoclaves, thermometers and microscopes, for example; in the case of microbiological laboratories, the *procedures* used for examining for the presence of microorganisms will also be included, but the *effectiveness* of the laboratory in performing the procedures will not normally be determined.

Thirdly, a laboratory may participate in a proficiency testing scheme. In such a scheme, a central agency will send out samples to the participating laboratories for them to analyse. Such samples may be spiked with known concentrations or types of organisms, in which case there is a theoretical absolute value against which their analytical results will be judged. A laboratory achieving a low count on a sample will be considered to have an inefficient analytical procedure; too high a count may well indicate poor laboratory procedures permitting growth of microorganisms in the samples or dilutions. As an alternative to spiked samples, if a large number of laboratories participates, natural samples may be used, and the performance of a given laboratory then judged against the norm (mean or mode of the results). In the case of natural samples, higher recovery rates than the norm *may* mean that the laboratory in question is better at analytical techniques than the majority, rather than having poorer technique – without further investigation it will not be possible to determine whether that laboratory is better or poorer in its work. Thus the best proficiency testing scheme is likely to be one that combines the use of both spiked and natural samples.

In any laboratory engaged in *official* analysis of food samples there is likely to be a national mandatory requirement that the laboratory conforms to certain quality management procedures. In many countries the same analytical laboratories will be involved in official analysis of clinical specimens associated with outbreaks of food poisoning or food-borne disease. In these cases, there may well be a further requirement for that laboratory in respect of its clinical work. This is true, for example, in the UK, where such laboratories are required to obtain clinical pathology laboratory accreditation (CPA).

Official laboratories in many countries (in the UK and USA, for example) will require both:

1. to be accredited in relation to their measuring and calibration apparatus and their measurement procedures, by the national accreditation body (for example, in the UK this is the National Measurement Accreditation Service (NAMAS) of the United Kingdom Accreditation Service (UKAS)); *and*
2. to participate in a proficiency testing scheme. In the UK this may be the Food Examination Performance Assessment Scheme (FEPAS) operated by the Ministry of Agriculture, Fisheries and Food's Central Science Laboratory (see Scotter, 1996), or the Public Health Laboratory Service Food Microbiology External Quality Assessment Scheme.

In the European Union, the Official Control of Foodstuffs Directive 89/397/EEC requires laboratories that officially examine foods to be accredited by their national accreditation organization and to participate in a proficiency testing scheme, conforming to the EN 45001 standard.

For international harmonization, testing laboratories can consider following the guidelines of ISO/IEC Guide 25 (see also FAO, 1991).

2.5.2 Testing culture media

Each batch of medium that is autoclaved or filter-sterilized should be checked for sterility by preincubating a sample of the batch at 25° and 37°C.

2.5.2.1 Diagnostic and biochemical test media

Each batch of such media should be tested using stock cultures of organisms known to be positive and negative to the test in question. National Type Culture Collection Catalogues often list the relevant biochemical test results for strains; alternatively the Curator of such a Culture Collection will be able to recommend an appropriate culture.

2.5.2.2 Agar media

Selective isolation media in particular should be checked: the selectivity and sensitivity of such media can vary from batch to batch, both as a result of the ingredients and the handling, preparation and storage of the medium. If dehydrated media are being used, most manufacturers can provide a certificate for a batch of dehydrated medium to indicate the quality assurance results they have obtained. However, it should be remembered that the care with which a dehydrated medium is rehydrated, sterilized and stored may have a considerable effect on its efficacy, so even when quality assurance certificates are obtained from the supplier, a quality assurance check should be carried out to determine that the medium is being prepared and used properly.

2.5.2.2.1 Surface drop count method

For this test, decimal dilutions (in 0.1% peptone) of a fresh, stationary phase culture of the relevant organism should be used. Each plate should be marked into quarters, and drops of 0.1 ml should be placed in each quarter. Place four different dilutions on each plate and use multiple plates to obtain replicates. With a sterile glass spreader, spread each drop over its quarter of the plate, starting with the highest dilution, and taking care not to run the inocula into each other. Incubate as appropriate, and determine counts and any response to differential characteristics. The selectivity and sensitivity of a selective medium can be determined by comparing the counts obtained on the selective medium with those obtained on a non-selective medium chosen for optimal growth. In the case of selective media, this should be done for both the 'wanted' or *target* organisms (i.e. those the medium is designed to isolate) and 'unwanted' organisms (i.e. those the medium is designed to inhibit). In general, the more efficacious selective medium will show the greatest difference between the highest dilution at which the target organism will grow and that at which the 'unwanted' organism will grow (at a sufficiently high population concentration an organism may be able to grow on a medium normally inhibitory for that organism). The results can be determined quantitatively.

2.5.2.2.2 *The ecometric method of Mossel* (see Mossel *et al.*, 1983, 1995)

This uses a standardized streaking procedure to obtain progressive dilution of a suspension of the test organism on the plate of medium being tested. Using a standardized procedure provides a semi-quantitative result. A plate of poured medium is marked off into quarters with five lines being marked in each quarter as shown in Fig. 2.1.

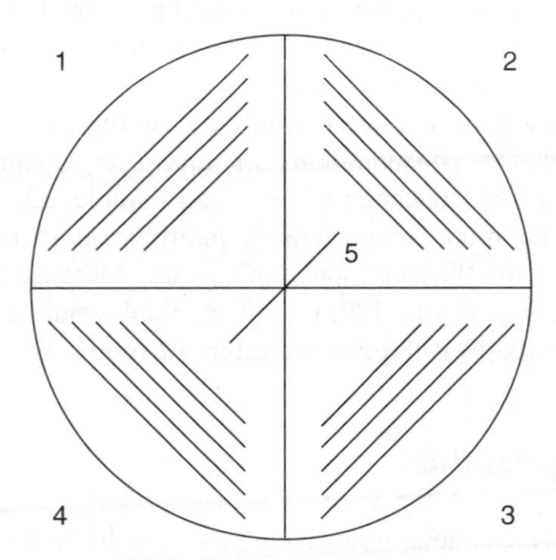

FIG. 2.1 Streaking pattern for Mossel's ecometric method.

A loopful of an overnight broth culture is streaked successively on the five streaks in section 1, then in section 2, and so on, finishing with a single streak in the centre. Mossel's Absolute Growth Index (AGI) is given by the last streak in which 'significant' growth occurs (with the streaks being interpreted as decimal fractions); for example, if growth occurs on the five streaks of section 1 and the first three streaks of section 2, this would be recorded as an AGI of 1.6. Such AGIs can be used to compare different batches of the same medium. They can also be used to determine the inhibitory action of the selective medium against the *target* organism as the ratio of the AGI on the selective medium to the AGI on a non-selective optimal growth medium (this is termed by Mossel the Relative Growth Index, or RGI). As important in the case of selective media is the *target* organism displaying a high AGI compared with the AGIs of 'unwanted' organisms.

2.5.2.2.3 *Use of the spiral plater* (see Section 5.1.3)

This is the procedure of choice for any laboratory possessing a spiral plater. The spiral plater provides a quick and convenient semi-automated method of achieving the same dilution test as the two methods described above, but giving a truly quantitative result, with quantitative comparisons between selective media and non-selective media, permitting determination of the effects of heavy concentrations of 'unwanted' organisms on a medium's ability to inhibit them.

2.5.2.3 Liquid media

Quality assurance checks on non-selective liquid media can be based on growth rate determinations (see Section 12.2).

A simple quality assurance check on a liquid selective enrichment medium is to inoculate a sample of the medium with a small number of the *target* organism and then demonstrate its recovery on the subsequent solid selective medium.

However, more comprehensive quality assurance checks on liquid media used for selective enrichments are quite difficult to undertake, as the efficacy of the medium is judged by the change in the ratio of *target* organisms to 'unwanted' organisms over the period of incubation. This will be affected not only by the medium itself, but also *inter alia* by the initial population concentrations of *target* organisms and 'unwanted' organisms, and the nature and concentration of food sample added to the medium. Readers wishing to perform a more comprehensive quality assessment are advised to use Method E in Appendix 1 of the 'Pharmacopœia of culture media for food microbiology' (Baird *et al.*, 1987, or Corry *et al.*, 1995). This method would require still further modification if it were desired to determine the effect of food sample components.

2.5.3 Checks on apparatus

Operating temperatures of incubators, water-baths, etc. should be regularly checked and recorded. If working to a specified standard that defines the tolerance for incubation temperature for an analysis, it should be noted that the temperature range within which the incubator should be permitted to operate will be reduced by the tolerance range for both the working thermometer (the thermometer installed in the incubator or water-bath) and the reference thermometer against which the working thermometer is checked. Details of temperature-monitoring procedures are given in national standards, international standards and the guidance documents issued by national accreditation bodies.

Similar consideration needs to be given to checking the operation of balances, with annual calibration.

Checks to be carried out on autoclaves and hot-air ovens are discussed in Section 2.3.

3

Basic Microscopic Techniques

In this Section are described principally the methods used for normal white-light optical microscopy of stained specimens of bacteria and of wet mounts of bacteria to detect motility. Microscopic methods used for microfungi are described in Section 14. A good general discussion of light microscopy is given by Murray and Robinow (1994).

A microscopic technique increasingly used in food microbiology is the direct epifluorescent technique (DEFT), a brief outline of which is given in the section on fluorescence microscopy (Section 3.5).

3.1 USE OF THE OPTICAL MICROSCOPE

When examining stained dry preparations, the microscope should be set up as described in Section 3.1.1. To examine for motility by hanging drop, the illumination will need to be set up slightly differently, as described in Section 3.3.

3.1.1 Procedure for the use of a microscope with an oil-immersion objective

1. First ensure that objectives, eye-piece, condenser and mirror are clean, removing any dust with a soft brush and lens tissue.
2. Align the microscope and artificial light source. If an external light source is used, place the lamp about 15 cm from the mirror. Ensure that no frosted glass filters are inserted between lamp and specimen.
3. Adjust the draw tube to the correct length (usually this will be 160 mm when a microscope with a rotating nose-piece is used).
4. If an external light source is used, with the $\times 10$ objective in position and the eye-piece removed adjust the plane side of the mirror to send the light centrally up the microscope draw tube. If an internal light source is used, check that the lamp is centred. Next check that the condenser is properly centred.

5. Replace the eye-piece and place the object slide on the microscope stage, with the specimen uppermost. Focus the ×10 objective on the object using the coarse adjustment.
6. Focus the light source on the object by placing a pencil close against the light source and racking the condenser up and down to get a sharp image of the pencil in the same field as the focused object. (This is termed 'critical illumination', the image of the light source being in the same plane as the focused specimen.)
7. Using ×10 and/or ×40 objectives select a suitable field for subsequent viewing with the high-power (×90 or ×100) oil-immersion objective. (Oil immersion objectives are usually identified by 'H.I.', 'OIL' or 'OEL' being engraved on the barrel.)
8. (a) Rack up the objective (this may be unnecessary if parfocal objectives are fitted) and rotate the oil-immersion objective into position, taking care that the rotating nose-piece locates at its click-stop.
   (b) Place one drop of immersion oil (refractive index of 1.51–1.52) on the microscope slide.
   (c) Using the coarse adjustment, gently and very slowly lower the ×90 (or ×100) oil-immersion objective until the oil layer is flattened without the front of the lens touching the slide.
   (d) Whilst viewing the object through the microscope, rack very slowly upwards with the coarse adjustment until the specimen comes into view. The restriction in the use of the coarse adjustment to upward travel only, while looking down the microscope, lessens the danger of slide breakage and consequent damage to the objective lens. Focus sharply with the fine adjustment.
9. Remove the eye-piece again and adjust the mirror or condenser centring to ensure that the back lens of the objective is symmetrically filled with light. To avoid glare, close the iris diaphragm until the back lens of the objective is about three-quarters full of light.
10. Replace the eye-piece and refocus with the fine adjustment if necessary.

After use, all immersion oil *must* be wiped off the objective lens and elsewhere with lens tissue. Failure to remove the immersion oil may allow the oil to penetrate the lens mount and cause a severe loss of definition.

Procedures for setting up phase-contrast microscopes and dark-field microscopes are described by, for example, Quesnel (1971).

3.1.2 Problems in observation of slides

1. One of the most commonly encountered problems in observing stained dried slides of bacteria derives from insufficient care taken in the preparation and staining of the smears. Suspensions of bacteria should be dilute and mixed evenly to give a *thin*, even, distribution of bacteria on the slide. If the smear is too thick or of uneven density, it will often be difficult to visualize and interpret the results. This is especially true with the decolorization and counterstaining steps in the Gram stain.

2. If the front surface of the objective lens is dirty then there can be a severe loss of definition and of contrast, so that it seems that the bacteria are being viewed through a mist. Commonly this is caused by a failure to clean immersion oil from the lens at the end of a session of microscopy. Immersion oil can gradually harden on the surface of the lens and may even require a lens tissue moistened with xylene to remove it. Care should be taken not to use too much xylene for too long, as this may weaken the lens mount sealants. Prolonged omission to remove immersion oil from the objective can result in the oil penetrating the lens mount and contaminating the back face of the front lens of the objective. The only solution here is to obtain a replacement objective.
3. Dirt on eyepiece lenses or condenser lenses usually appears as particles in the field of view. Eyepiece dirt can be identified easily by rotating the eyepiece whilst viewing the slide. Dirt on the condenser lenses will become more or less sharp as the condenser focusing is adjusted. Dirt and dust should be removed carefully with a lens tissue. Dirt and dust can sometimes collect on the interior lens surfaces of an eyepiece. However, eyepieces are relatively easy to disassemble and reassemble, and a puffer-brush (obtainable from photographic suppliers) can help to remove the dirt from the eyepiece tube.
4. From time to time check that the objectives are fully screwed into the nose-piece.
5. Slides and coverslips should be of the correct thickness (consult the microscope manufacturer's instruction manual).
6. If the field of view is unevenly or inadequately illuminated, check that the condenser is properly mounted. Some substage condensers have slip-in friction mountings and these can easily become misplaced. The condenser and light bulb should also be checked to ensure that they are centred correctly.
7. If the stained preparation can be readily focused with the $\times 10$ or $\times 40$ objective, but fails to come into focus with the $\times 90$ or $\times 100$ objective, check that the slide has not been placed on the microscope stage upside down.

3.2 STAINING METHODS

3.2.1 Preparation of smears for staining

Smears of bacteria from cultures on solid media are made on clean glass slides as follows:

1. The slide may be divided up into sections, one section for each smear, using a wax pencil.
2. With a wire loop, place a small drop of water on each section of the slide.
3. Sterilize the wire by holding it vertically in the Bunsen flame until it is heated to redness along its entire length.
4. Allow the loop to cool.
5. Holding the loop like a pen, pick off a little of the bacterial growth. Try to avoid transferring any agar medium.

6. Transfer to a water drop on the slide and, using the flat of the loop, emulsify the growth, finally smearing it evenly over the area of the slide allocated to it. Aim at obtaining *thin*, evenly spread, smears which are almost too thin to be seen when dry. The individual bacteria will then be well spaced for examination under the microscope. If the suspension does not spread evenly over the slide, but collects in small droplets, the slide is greasy and should be discarded.
7. Sterilize the loop.
8. Prepare the other smears in the same manner.
9. Leave the slide to dry in the air, then heat-fix by passing the slide twice through the Bunsen flame. This coagulates the cell contents and causes the bacteria to adhere firmly to the slide.

Smears of bacteria from cultures in liquid media are made in a similar manner except that dilution with water is not required. The preparations may be less clean because some ingredients of the medium will remain and be stained. If the liquid culture medium or suspending liquid is known to be high in sugar (as in the case of media for lactic acid bacteria, for example), a more severe heat-fixing is necessary to prevent bacteria from washing off the slide as the sugar dissolves in the staining solutions or wash-water.

3.2.2 Simple stains

Place the heat-fixed smear on a staining rack over the sink and flood with Loeffler's methylene blue solution. Stain for 5 min. Rinse the slide gently in water and blot dry with clean blotting paper or filter paper.

3.2.3 Gram's staining method

This is a differential double-staining method which forms the basis of most examinations and the preliminary identification of bacteria. In this method, bacteria are first stained with crystal violet and are then treated with iodine solution. The bacterial smears are next treated with ethanol, acetone or an ethanol–acetone mixture, which entirely removes the violet stain from Gram-negative bacteria, but not from Gram-positive bacteria. Safranine may be used as a counterstain, to stain the Gram-negative bacteria an orange-red.

This method divides bacteria into two classes:

(a) *Gram positive*. These do not decolorize with ethanol, thus appearing *purple*. Examples of Gram-positive bacteria are *Staphylococcus, Bacillus*.
(b) *Gram negative*. Gram-negative bacteria appear *pink*. Examples of Gram-negative bacteria are *Pseudomonas, Escherichia*.

Note that the test should be carried out on young cultures (18–24 h), or cultures of various ages including young cultures should be examined, as some bacteria change in their Gram reaction as well as in their morphology as the cultures age.

There are in existence many modifications of Gram's staining method (see, for example, Silverton and Anderson, 1961; Balows *et al*., 1991). The reagents and times employed depend on the nature of the specimens being examined most commonly in a particular laboratory, but user preference also plays a large part in the choice of the modification. The reagents used in the method described below are chosen for having a greater latitude than many others towards deviations from the recommended staining times, and they are thus particularly suitable for routine use. Nevertheless, every effort should be made to adhere to the times recommended, particularly for the decolorization stage.

Procedure

1. Prepare a heat-fixed smear from an 18–24-h culture in the usual way.
2. Stain with crystal violet solution for 2 min. Take care that the *entire* smear is covered with the liquid at this and at all subsequent stages.
3. Holding the slide at an angle of 45°, wash the crystal violet solution off the smear. When all traces of crystal violet solution have been removed, allow the iodine solution to act for 1 min.
4. Pour off the iodine, blot dry and, holding the slide at an angle of 45°, wash the slide with 95% ethanol (or industrial methylated spirits) until no more violet stain runs from the slide (only 5–15 s in the case of well prepared thin smears).
5. Rinse under the tap and stain with a 1% aqueous solution of safranine O for 15 s.
6. Wash the slide well and blot dry.

3.2.4 Staining of bacterial flagella

The diameter of a bacterial flagellum is below the limit of resolution of the light microscope, and for this reason flagella are normally not visible. By the use of special staining techniques, an appropriate stain can be made to build up around each flagellum, thus increasing its apparent diameter. This enables the flagellum to be visualized under the light microscope, and the arrangement of the flagella on the bacterial cell to be determined. Specifically, the flagella stain is of use in differentiating members of the Pseudomonadaceae, which have polar flagella, from members of the Enterobacteriaceae, which have peritrichous flagella (when motile).

Good results in flagella staining are obtained only with difficulty, as the extremely delicate flagella easily become detached from the bacterium. In addition, most of the staining methods result in the production of a background precipitate which, on occasions, renders observation of the flagella difficult.

The flagella stains most commonly used fall into two categories: firstly, those in which a silver salt is used to deposit silver on the flagella (Rhodes, 1958) and, secondly, those in which basic fuchsin is deposited on the flagella (see Leifson 1951, 1958).

The silver deposition method described below is a modified Fontana method (Rhodes, 1958).

Procedure

Only slides that are completely clean (and therefore grease-free) should be used. This is best achieved by using new grease-free slides. If slides are reused, it may be necessary to degrease by treatment with chromic acid (see p. 387), followed by rinsing in distilled water.

The bacteria should be grown on agar slopes at 3–5°C below the optimum growth temperature, the slopes having first been moistened with one or two drops of sterile distilled water.

1. Flame a microscope slide in a Bunsen burner for about 5 s. Place on a staining rack to cool, and then draw a halfway division with a wax pencil (one end of the slide can be used for holding, using forceps).
2. With a pipette or Pasteur pipette add 2 ml of sterile distilled water to a young, actively growing, slope culture (usually about 18 h old) and gently suspend the growth by careful agitation and rotation of the test-tube. It is advisable not to use a wire loop to suspend the growth if it can be avoided. Transfer to a *clean* test-tube, check for motility with a hanging drop preparation, and dilute the suspension with distilled water until only slightly turbid. Place in an incubator at 20–30°C for 30 min and then remove a large loopful of the suspension to one end of the cool microscope slide. Tilt the slide until the drop runs to the central pencil line. Dry in air at room temperature. Do *not* heat-fix the film.
3. Cover with mordant for 5 min.
4. Rinse gently but thoroughly with distilled water to remove all traces of mordant.
5. Cover with hot Fontana silver solution, and stain for 5 min, renewing the stain once a minute. (The Fontana silver solution should be heated over a boiling water-bath.) Do not allow the staining solution to evaporate from any areas of the smear – keep this covered with liquid at all times.
6. Wash with water, allow to dry in air, and examine.

Alternative procedure using Leifson's stain

If Leifson's flagella stain is used substitute steps 3–6 above by the following:

3. Cover the smear with 1 ml of Leifson's flagella stain and allow to act until a very fine rust-coloured precipitate has formed (about 10 min).
4. Rinse gently but thoroughly with distilled water.
5. Counterstain for 5–10 min with 1% methylene blue.
6. Wash with water, allow to dry in air, and examine. Without a counterstain the cells and their flagella stain pinkish-red. When the counterstain is used, the cells stain blue, and the flagella are red.

Note. Since flagella are often detached rather easily, a culture showing large numbers of peritrichously flagellate cells indicates that the strain is peritrichously flagellate, but an apparent polar arrangement of flagella does not *necessarily* contraindicate a peritrichous organism.

3.2.5 The staining of bacterial spores

Bacteria in the genera *Bacillus* and *Clostridium* produce endospores, which are highly resistant to high temperature, lack of moisture and toxic chemicals. The endospores are also resistant to bacteriological stains and, in a smear stained by Gram's method, they can be seen as colourless areas in the vegetative organisms which stain Gram positive. However, once stained, the spores tend to resist decolorization.

Procedure for Bartholomew and Mittwer's spore staining method (Bartholomew and Mittwer, 1950)

1. Prepare a smear in the usual way, but heat-fix very thoroughly by passing through a Bunsen flame 20 times.
2. Stain for 15 min with a saturated aqueous solution of malachite green.
3. Wash gently with cold water for 10 s.
4. Counterstain with a 0.25% solution of safranine for 15 s.
5. Wash with water and blot dry.
6. Examine under the oil-immersion objective.

Note. The times in steps 2 and 4 may need modifying when staining some species.

3.2.6 The demonstration of bacterial capsules

The capsules of bacteria examined in blood or animal tissues can often be seen as unstained haloes separating the bacteria from the stained background, even when ordinary staining procedures (e.g. Gram's method) are used. To visualize the capsules of bacteria that are being studied in pure culture, special techniques are usually necessary. Leifson's flagella stain (Leifson, 1951, 1958) may be used for bacterial capsules, the capsules staining red and the bacteria blue when methylene blue is used as a counterstain. A simpler method of demonstrating capsules is by the use of a wet Indian ink film.

Production of a detectable capsule is often dependent on the composition of the medium and the age of the culture (Wilkinson, 1958). Most capsules are of carbohydrate and are best detected when grown to stationary phase in the presence of an excess of a utilizable sugar (Duguid and Wilkinson, 1961).

Procedure for the wet Indian ink film

1. Place a loopful of Indian ink on a *very clean* microscope slide.
2. Mix into the Indian ink a little of the bacterial culture or suspension.

3. Place a coverslip on the mixture, avoiding air bubbles, and press firmly with blotting paper until the film of liquid is very thin.
4. Examine with the high-power dry objective or the oil-immersion objective. The capsule will be seen as a clear area around the bacterium.

Note. A film of the ink alone should always be prepared and examined as a control, as Indian ink may occasionally become contaminated with capsulate bacteria.

3.2.7 Negative staining

This is a very simple and effective method for demonstrating the external shape of bacteria in a smear preparation. The bacteria are surrounded by a thin film of black dye and appear as white objects upon a grey background.

Procedure

1. Prepare a very thin smear in the usual way, using a clean, grease-free slide.
2. At one end of the slide place one drop of nigrosin solution (2%).
3. Take another microscope slide, lay one end on the first slide at an angle of 30° touching the drop of nigrosin, and use it to push the nigrosin across the surface of the first slide. The smear will thus be covered with a thin, even, film of dye.
4. Allow the dye to dry and examine the preparation under the oil-immersion objective.

This procedure can be used in conjunction with a simple staining technique to demonstrate the presence of capsules. In this case, the smear should be stained with dilute carbol fuchsin, washed and dried, before treating with nigrosin.

3.2.8 Ziehl–Neelsen method for staining acid-fast bacteria

Members of the genera *Mycobacterium* and *Nocardia* can be differentiated from many other organisms by this staining technique.

The Ziehl–Neelsen method consists firstly of staining the organisms with a hot, concentrated dye. Once stained, the cells resist decolorization with acid; they are thus 'acid-fast'. Decolorization is effected with suitably strong acid and the smear is then counterstained with methylene blue solution. Acid-fast bacteria stain red; other bacteria and the background stain blue. This method (with 20% sulphuric acid) is used clinically for the detection of *M. tuberculosis* in body tissues and fluids (e.g. lungs, liver, sputum, urine). When 1–5% sulphuric acid (or hydrochloric acid) is used instead of 20% sulphuric acid, the technique is useful for identifying saprophytic mycobacteria, nocardiae and certain coryneform organisms.

Procedure

1. Cover the slide with strong Ziehl–Neelsen's carbol fuchsin and heat the underside of the slide with a lighted alcohol-soaked swab. Stop heating when the slide steams. Keep the slide hot and replenish the stain if necessary, taking care not to allow the smear to become dry. Heat for 5 min, not allowing the staining solution to boil.
2. Wash well.
3. Decolorize with acid-alcohol or with 1, 5 or 20% sulphuric acid. The excess stain is removed as a brownish solution, and the smear will become brown. Rinse in water, when the film will appear pink once more. Apply more acid and repeat the rinsing several times until the film appears faintly pink upon washing.
4. Wash well.
5. Counterstain with Loeffler's methylene blue for 5 min.
6. Wash well and carefully remove the stain deposits from the back of the slide with filter paper. Blot dry and examine.

3.3 EXAMINATION OF CULTURES FOR MOTILITY BY 'HANGING DROP' PREPARATIONS

Cultures for examination should be broth cultures 18–24 h old. Alternatively, a small amount of culture from an 18–24-h agar slope can be emulsified gently in a drop of broth or normal saline, taking care that the emulsion is not too dense. By the following procedure, a drop of culture is suspended from a coverslip over the depression in a hollow-ground slide:

1. First place a little immersion oil round the edge of the depression in the slide. Then, with a wire loop, transfer a small loopful of the culture to a clean dry coverslip laid on the bench. Do not spread the drop.
2. Invert the cavity slide over the coverslip so that the drop is in the centre of the cavity and press the slide down gently but firmly so that the oil seals the coverslip in position. Invert the slide quickly and smoothly, and the drop of culture should now be in the form of a hanging drop. The preparation should be examined without delay and as quickly as possible.
3. When examining hanging drop preparations the substage condenser of the microscope should be racked down slightly from its normal position and the iris diaphragm should be almost completely closed. Not only does excessive illumination render the unstained organisms invisible, but also the heating effect may cause them to lose their motility.
4. First, use the low-power objective to focus on the edge of the drop, moving the slide until the edge of the drop appears across the centre of the field. This should be recognized easily as minute droplets of condensed water can usually be seen on the other side of the line which represents the edge of the hanging drop. Then place the high-power ($\times 40$) *dry* objective in position and refocus the edge of the drop. If

necessary, open the iris diaphragm slightly. The bacteria should now be seen easily, particularly towards the edge of the drop where the reduction in the depth of liquid assists in keeping the organisms within the depth of field. The oil-immersion objective should *not* be used because the focusing movements of the objective would be mechanically transmitted to the coverslip causing streaming in the culture liquid which would seriously hinder observation and may even be misinterpreted by an untrained observer as motility of the organisms.

It is necessary to distinguish between Brownian movement (a continuous agitation of very small particles suspended in a fluid, which is caused by unbalanced impacts with molecules of the surrounding fluid) or drift in one direction caused by the slide being slightly tilted, and true motility.

Sometimes the characteristics of motility are used for identification. For example, *Listeria* displays a tumbling motion, *Cytophaga* a gliding motion. If using type of motility for identification it is important to become well acquainted with this, by observing known stock cultures.

3.4 EXAMINATION OF MICROBIAL COLONIES

It is sometimes useful to examine microbial colonies microscopically. This is especially valuable in the case of microfungi (see Section 14). The preferred microscope for this purpose is a binocular stereomicroscope, and an incident light source is usually best. Many stereomicroscopes have an adjustable incident light source built in to the nose-piece, with a choice of objectives (×10, ×20 and even ×40). Usually there is a clear-glass microscope stage, permitting light to be passed through the specimen from beneath, if required.

If it is intended to use Henry illumination (see below) to identify *Listeria* (see Part II), choose a stereomicroscope without a built-in transmitted light source, but which has a removable substage mirror for use with an external source of transmitted light.

3.4.1 Henry illumination

Colonies of *Listeria* show a characteristic appearance when illuminated obliquely from below (Wood, 1969; ISO 11290-1:1996). Grow the colonies on a clear medium such as tryptose agar. Naked-eye observation is possible, but a stereomicroscope is preferred.

Remove the substage mirror from the stereomicroscope and place it on the bench, plane mirror uppermost, about 10 cm from the microscope (Fig. 3.1). Place a source of white light in such a position that a beam of light is directed at an angle of 45° on to the mirror and reflected at 45° on to the glass stage of the microscope. The colonies of *Listeria* appear bright blue-green to blue-grey with a finely granular surface resembling ground glass.

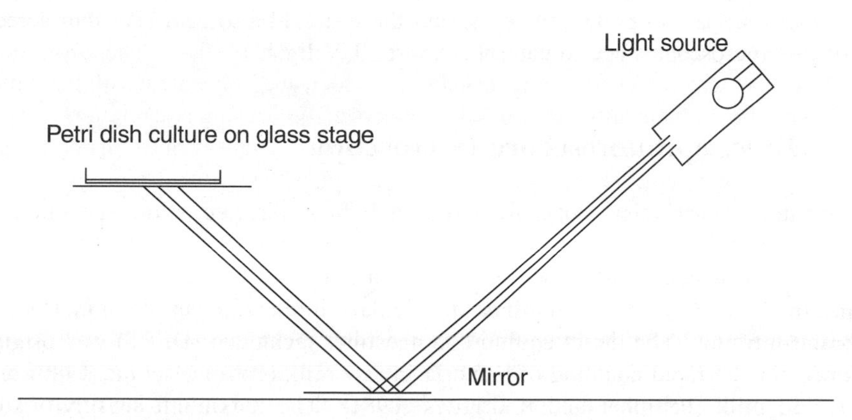

FIG. 3.1 Henry illumination.

3.5 FLUORESCENCE MICROSCOPY

Fluorescence is a phenomenon in which light energy at one wavelength is absorbed by a substance and immediately re-emitted as light of a longer wavelength. A substance showing such a re-emission is known as a fluorophore. Some naturally occurring biological substances (e.g. chlorophyll) display fluorescence; this is called autofluorescence. However, in microbiological microscopy specimens are usually stained with fluorescent dyes (called fluorochromes). The particular wavelength shift of use is when the fluorochrome emits light in the visible region of the spectrum whilst being irradiated with ultraviolet light. Examples of such dyes are acridine orange, fluorescein isothiocyanate and acid fuchsin. The fluorochrome will display typical peak wavelengths for excitation and emission. In the case of acridine orange, it acts as a metachromatic stain, because whereas the monomer fluoresces green (around 525 nm) linked dye molecules fluoresce red (around 650 nm). The dye–dye interactions resulting in red fluorescence can occur when the dye attaches to RNA, whereas when the dye attaches to DNA such dye–dye interactions are absent and the resulting fluorescence is green.

The light source for a fluorescence microscope can be a halogen-quartz lamp, a high-pressure mercury vapour lamp or a xenon arc.

The fluorochromes can be conjugated with antibodies to give immunofluorescent stains which permit the detection of specific organisms. Such a technique could be employed, for example, to study the location of specific pathogenic organisms within the host tissues.

3.5.1 Transmitted UV-light fluorescence microscopy

Because glass absorbs UV light, the condenser lens system should be made of quartz. The objective and eyepiece lenses can be made of glass as usual, as it is the visible light re-emitted by the specimen that will be observed. However, there needs to be a steep-cut UV filter in the light path to protect the eyes, and there should also be a UV filter screen in front of the microscope stage to control scattered UV light.

3.5.2 Direct epifluorescence microscopy

In an epifluorescence microscope, the exciting light is directed at the specimen from above through the objective lens system. The light path is so designed that focusing the objective on the specimen also focuses the exciting light.

This method of microscopy finds particular application in combination with membrane filtration. The direct epifluorescence filter technique (DEFT) was originally developed for the rapid counting of bacteria in raw milk (Pettipher *et al.*, 1980) and in heat-treated milk (Pettipher and Rodrigues, 1981). The maximum sensitivity of the method is around 10^3–10^4 bacteria per g (Pettipher *et al.*, 1989).

Membrane filtration is most readily applied to samples that are liquid and which, by appropriate treatment, can be made to pass through the membrane filter. For example, milk is treated with proteolytic enzyme to break down bovine somatic cells, and with a surfactant to disperse fat globules. However, methods have also been described for other types of food product.

4

Cultivation of Microorganisms

The characteristics of bacteria that are used in bacterial study and identification depend on the behaviour of populations or cultures rather than of individual organisms. Also, a large proportion of determinations of the microbiological quality of foods depends on detection of viable organisms and, at present, this in turn depends on providing those organisms with the cultural opportunity to metabolize and multiply. Consequently, materials and methods for achieving growth and multiplication of the individual organisms are a basic requirement in bacteriology. Nutrient materials provided in a form suitable for growth are known as culture media. Examples of some commonly used media are given in Section 7. The composition of the growth medium can substantially affect the microscopic appearance and metabolic activities of a microorganism. For example, production of endospores may be enhanced or diminished by specific constituents. Luckily, the Gram-staining reactions of most microorganisms are less often affected by the composition of the medium, although the period of incubation may have a profound effect (see Part IV).

Culture media may be distributed in various ways in test-tubes, flasks or screw-capped bottles, depending on the method of inoculation to be used. Test-tubes or flasks are stoppered with closely fitting plugs of non-absorbent cottonwool, metal caps, plastic caps or specially designed rubber bungs. An advantage of screw-capped containers is that evaporation is prevented and the medium therefore does not dry out on storage.

4.1 TYPES OF CULTURE

1. *Liquid batch cultures*. Cultures in liquid media, to which no fresh nutrient is added during growth.
2. *Agar slope (or slant) cultures*. Test-tubes or small bottles containing about 5 ml of solid medium (e.g. nutrient agar), dissolved and allowed to cool in a sloping position. The inoculum is either spread over the surface of the medium or applied in a thin streak using a wire loop.
3. *Stab cultures*. Tubes or bottles containing an agar or gelatin medium are allowed to solidify in the upright position. The medium is inoculated by plunging a long straight wire, charged with inoculum, vertically into the centre of the tube.

4.4.2 Maintenance of lactic-acid bacteria

Lactic-acid bacteria will not grow well on the surface of solid media incubated aerobically, and are more suitably maintained in a liquid medium (e.g. yeast glucose chalk litmus milk or Robertson's cooked meat medium), being subcultured at intervals of 2–4 months. Media containing milk or added sugars should also contain chalk to buffer against the development of too low a pH. After bacteria have died in the coagulated milk layer in yeast glucose chalk litmus milk, viable organisms may still remain associated with the chalk sediment.

4.4.3 Maintenance of cultures of anaerobic bacteria

Anaerobic bacteria will not grow on the surface of solid media incubated aerobically and are best kept in a medium providing reducing conditions (e.g. Robertson's cooked meat medium), and subcultured at intervals of up to 1 year.

4.4.4 Glass beads stored at −60° to −85°C

This method permits frequent access to stored cultures. It is very useful, for example, in providing cultures to give known positive and negative reactions to biochemical tests. However, the capital and energy costs are quite high compared with freeze-dried cultures, for example. Glass beads of 2–4 mm in diameter are distributed in small screw-capped bottles (up to around 30 beads per bottle) and sterilized by autoclaving. A turbid suspension of a culture is made in suspension medium. This is added aseptically to a sterile bottle of beads. This is then agitated to coat the beads with the suspension, and excess suspension removed with a sterile pipette. The screw-capped bottles are then labelled and placed in compartmented trays in a freezer at −60° to −85°C. Coloured beads can be used to aid identification. It is highly recommended that a storage index is made, in order to minimize the time for which the freezer is open whilst retrieving the required culture. To obtain an active culture, it is merely necessary to remove one bead aseptically, place it into a tube or bottle of growth medium and then incubate. Suitable suspension media are obtained by incorporating 15% glycerol into a normal liquid medium (nutrient broth for normal aerobes, reinforced clostridial medium (RCM) for anaerobes, salt nutrient broth for halophiles, and so on). Complete systems are commercially available (e.g. Microbank from Pro-Lab Diagnostics).

4.4.5 Preservation under oil

Cultures are first grown on agar slopes as in (2) above or in stab cultures (as in (3)), and then completely covered with sterile liquid paraffin or mineral oil. (To sterilize liquid paraffin, dispense in flasks in shallow layers and sterilize in the hot air oven at 160°C for 1–2 h.) Cultures maintained in this way will generally remain viable for several years without subculturing. This method is particularly useful for maintaining cultures of microfungi, although Onions and co-workers (1981) warn that some fungi (e.g. *Fusarium*) may produce atypical cultures as a result.

4.4.6 Freeze-dried cultures

In this process the cultures are freeze-dried or lyophilized and stored in sealed glass ampoules under vacuum. Ampoules may be stored at room temperature or in the refrigerator, and cultures can be expected to remain viable over several years. This method is particularly useful for the storage of culture collections and for despatch of cultures.

4.4.7 Desiccated serum suspensions

Laboratories not possessing freeze-drying apparatus can obtain good results with the procedure described by Alton and Jones (1963). Suspend a loopful of growth from a 24–48 h culture in 2 ml of sterile serum, and place one drop into each of a number of sterile, plugged, 50 × 6-mm tubes. Dry in an evacuated desiccator over phosphoric oxide for 7 days at 5°C, re-evacuating once after 24 h. Place each tube (containing a thoroughly dry suspension) in a larger soda glass test-tube, evacuate and seal in a Bunsen flame. Store refrigerated or in a deep freezer. Such cultures may remain viable for years.

4.5 PLATE CULTURES

4.5.1 Preparation of plates for streaking

Where a large surface area of medium is necessary, as in the separation of organisms from mixtures, the agar medium is allowed to solidify as a thin layer in a Petri dish. Some 10–15 ml of the melted sterile medium are poured into a sterile Petri dish, care being taken to avoid contamination. The outside of the test-tube or bottle should be wiped dry before pouring to avoid water droplets falling into the dish. The mouth of the flask or tube containing the medium should be flamed after removal of the cottonwool plug, and the lid of the Petri dish raised only enough to allow easy access.

When the plates have been poured and the medium allowed to solidify, they must be dried, as the presence of moisture on the surface of the medium will interfere with the production of discrete colonies. The plates are dried in an incubator at 37°C for from 20 min to 1 h. The lid of the Petri dish is first laid in the incubator and the part of the dish containing the medium is inverted and placed with one edge resting on the lid (Fig. 4.1).

This method of drying helps to avoid contamination from dust. Alternatively, the plates may be dried in the incubator for one or more days until the condensed water *mycoides*, *Serratia marcescens*, fluorescent *Pseudomonas* and *Proteus* (when swarming) are quickly recognized.

4.6.1 Morphological characters

(a) Gram reaction.
(b) Shape, size and arrangement of organisms.
(c) Motility.
(d) Presence of endospores, capsules and flagella (detected by appropriate stains).
(e) Reaction of Ziehl–Neelsen and any other special stains.

A drawing of typical organisms and their arrangement should be made. The size may also be recorded.

4.6.2 Cultural characters (Wilson *et al.*, 1984)

4.6.2.1 Surface colonies on solid media

1. *Shape*. Circular, irregular, rhizoid (Fig. 4.4).
2. *Size*. Record diameter in millimetres. Punctiform (pinpoint): less than 1 mm in diameter.
3. *Chromogenesis*. Colour of pigment, soluble or insoluble in medium. Pigmentation is often difficult or impossible to detect on media containing dyes (as indicators of pH or redox potential, or inhibitors).
4. *Opacity*. Transparent, translucent, opaque.
5. *Elevation*. Flat, raised, convex, umbonate (Fig. 4.5).

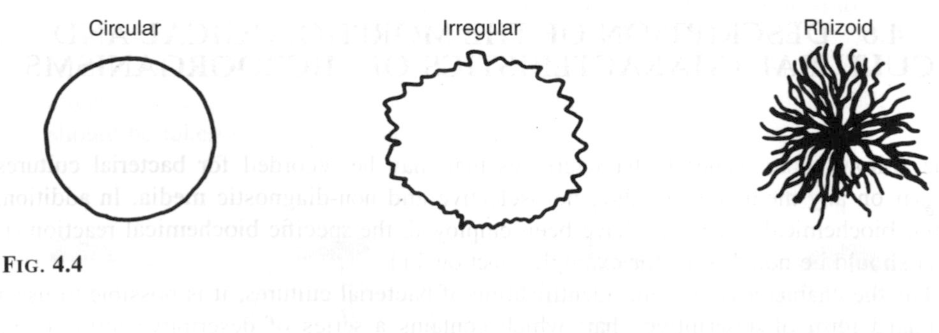

FIG. 4.4

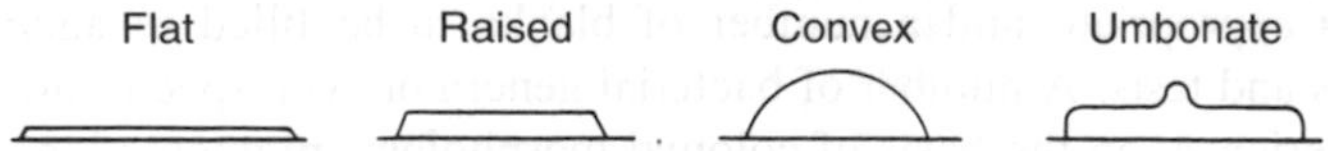

FIG. 4.5

6. *Surface*. Smooth, rough, dull, glistening.
7. *Edge*. Entire, undulate, lobate, dentate, rhizoid (Fig. 4.6).

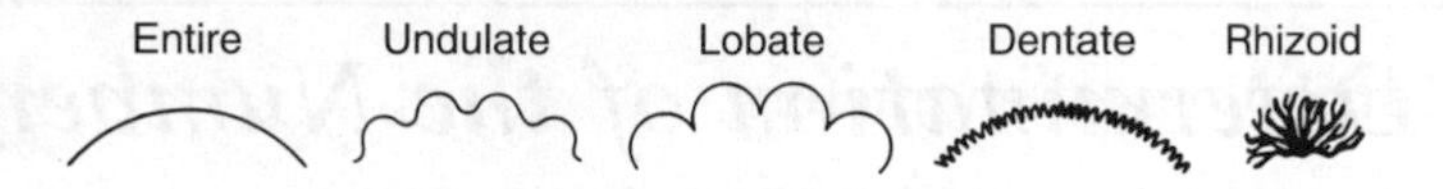

FIG. 4.6

A drawing of a typical colony, plan and elevation should be made.

8. *Consistency* (tested by touching with a sterile wire loop). Butyrous, viscid, granular.
9. *Emulsifiability*. Easy or difficult in water; forms uniformly turbid suspension; forms granular suspension; does not emulsify.

4.6.2.2 Broth culture

(a) *Amount of growth*. None, scanty, moderate, profuse.
(b) *Surface growth*. Present or absent; formation of a ring; pellicle that disintegrates or not on shaking.
(c) *Turbidity*. Uniform, flocculent or absent.
(d) *Deposit*. Amount; granular, flocculent, viscid, disintegrates or not on shaking.

4.6.2.3 Shake culture

Growth. On surface; in depth of tube (record depth to which growth occurs, bearing in mind that insufficient mixing of the inoculum may have stopped the organisms reaching the bottom); position of optimal growth.

5

Determination of the Number, and Detection, of Viable Microorganisms in a Sample

Anthony Sharpe has observed (Sharpe, 1980) that the 'true tragedy of agar is that it stimulated the development of a whole science around a unique type of analysis – the *plate count*', and that the plate count 'from its very nature, provides data so unique that they can be related to no other analytical data than those from other plate counts'. It is true that colony counts can be difficult to interpret. For example, most bacteria tend to occur as microcolonies in clumps or chains, and even in a thoroughly mixed dilution one *colony-forming unit* consists of an indeterminate number of bacteria from one to perhaps 100 or more. The longer the sample is mixed in the diluent, the more these clumps and chains will be broken up. Thus the colony count will tend to increase with further mixing. However, the longer the mixing time, the greater the chance of mechanical damage causing death of some bacteria. Consequently there is no 'true' colony count 'waiting' to be discovered by an experimenter. It is even more difficult to interpret a colony count of microfungi. Fungal conidia are very small, and of little immediate metabolic significance to the food, yet each can give rise to a colony. Fungal hyphae will be broken on mixing with diluent, and the more the hyphae are broken, the smaller the individual pieces and the higher the colony count, until the hyphal pieces become so small as to be rendered non-viable.

Also, the amount of microbial metabolite (e.g. toxin or spoilage compound) in a sample correlates more closely with the cell mass than with the number of colony-forming units.

Often, some bacteria in a sample will be in a metabolically damaged state so that they are unable to recover and produce colonies on media that would be suitable were the bacteria undamaged. This problem is encountered particularly when selective media are used, and in selective isolations resuscitation stages may be essential (see Andrew and Russell, 1984). In these situations when viable *counts* are required, the resuscitation stages must be designed carefully so that they do not interfere with quantitation.

One topic that has provoked much discussion in recent years has been the possible occurrence and status of 'viable but non-culturable organisms' (VNCOs) (see, for example, Xu *et al.*, 1982; Roszak *et al.*, 1984; Turpin *et al.*, 1993). The suggested existence of VNCOs, which are presumably by definition not recoverable by simple resuscitation techniques, is based on the comparison of colony counts with microscopic counts of 'metabolically active' cells as determined by a response after incubation with yeast extract plus either nalidixic acid (Kogure *et al.*, 1979) or piromidic, pipemidic and nalidixic acids (Kogure *et al.*, 1984). The debate becomes metaphysical: if a microorganism is capable of being resuscitated either *in vitro* or *in vivo*, then it is presumably not 'non-culturable', but if it is not capable of being resuscitated, does the existence of a residual metabolic activity mean that the organism is viable? This debate is not the same as the debate on 'clonal viability' that took place in the 1960s, and which had great relevance to the research work then being done to shorten the incubation period needed to determine viable counts.

Nevertheless, in spite of all these questions attached to viable count techniques in general, and to colony counts in particular, most national and international microbiological specifications for foods are still based on various types of viable counts (colony counts, most probable number (MPN) counts) so that for the foreseeable future a working quality assurance microbiologist should have an awareness of:

(a) the way is which such viable counts should be properly performed;
(b) the precautions necessary to minimize the introduction of the various types of experimental, systematic error; and
(c) the limitations in the interpretation of such counts.

5.1 COLONY COUNT METHODS

5.1.1 Pour-plate method

This enables the number of living organisms or clumps of organisms (i.e. colony-forming units) in a sample to be counted, subject to the appropriate medium and incubation conditions being used. Solid materials that are water soluble or give fine suspensions in water (e.g. soil, flour, dried milk, sugar) can be examined by shaking a known weight of the sample in sterile diluent and proceeding as below. Some solid materials, including some foodstuffs (e.g. cheese, meat) need to be macerated in sterile diluent in order to prepare the suspension (see Section 17.4.1). Commonly used diluents include peptone water diluent, peptone saline diluent and quarter-strength Ringer's solution.

A measured amount of the suspension, or of a known dilution of the suspension, is mixed with molten agar medium in a Petri dish. After the medium sets, the plates are incubated and the number of colonies is counted. Counts should be made on plates that contain fewer than 300 colonies. Whenever possible, duplicate or triplicate sets of plates should be incubated at each temperature.

The procedure described below is a standard method for liquid samples. The precise details of procedure may vary for different substances or products (see Parts II and III).

5.1.1.1 Mixing the sample

It is important that the sample be mixed thoroughly in a standardized way before proceeding further. In the case of a liquid sample, if the sample bottle is only partly filled, the sample should be mixed by shaking the bottle 25 times with an excursion of 30 cm. If the sample bottle is full it should be inverted 25 times with a rapid rotary motion to mix the contents thoroughly, about one-quarter of the contents should then be poured away and the sample shaken 25 times as described above.

5.1.1.2 Preparing the dilutions

1. Holding a sterile 1 ml blow-out pipette vertically, introduce the pipette tip not more than 3 cm below the surface of the sample and, using a safety pipetting device, suck up and down 10 times to the 1-ml mark. Withdraw 1 ml of the sample, touching the tip of the pipette against the neck of the bottle to remove the excess liquid adhering to the outside of the pipette. Transfer the pipette to the first tube of the dilution series with the tip touching the side of the tube 2–3 cm above the level of the diluent. The pipette must not contact the diluting fluid. Blow out the contents of the pipette, allow 3 s to elapse, then blow out the remaining drops. Discard this pipette into a used-pipette jar containing a suitable disinfectant, and label the first dilution tube 1/10, or 10^{-1}.
2. Using a fresh sterile pipette, mix the contents of the first dilution tube by sucking up and down to the 1-ml mark 10 times (or by rotating the tube between the hands). The tip of the pipette should be not more than 2–3 cm below the surface of the diluent. Then withdraw 1 ml of this first dilution and transfer to a second tube of sterile diluent, expelling the contents of the pipette as described above. Discard this pipette and label the second dilution tube 1/100 or 10^{-2}.
3. Taking a fresh sterile pipette and a further dilution tube, prepare in the same way a 1 in 1000, or 10^{-3} dilution.
4. Further dilutions of 10^{-4}, 10^{-5}, etc., can be made similarly as required, depending on the probable bacterial content of the sample. It is useful to assume that most bacteria have a volume greater than 1 μm^3 so that a solid bacterial mass would be unlikely to give a count of more than 10^{12} per ml. Spoilt meat could give a colony count of about 10^9 per cm^2 of surface. River water could give a colony count of 10^4–10^6 per ml.

5.1.1.3 Preparing the plates

1. Using a fresh sterile pipette, mix the contents of the final dilution tube, e.g. 1/1000, by sucking up and down ten times. Withdraw 1 ml of the dilution, touching the tip of the pipette against the side of the tube to remove excess adhering to the outside, and transfer the contents to a sterile Petri dish. Allow 3 s to elapse, then touch the tip of the pipette against the dish away from the previous inoculum, and gently blow out the remaining drops.

2. The same pipette can be used to transfer 1 ml from the 1/100 dilution to a sterile Petri dish, but before taking the sample raise and lower the 1/100 dilution in the pipette three times to rinse the sides of the pipette and also to give the dilution a final mixing.
3. In the same way, the same pipette may be used to transfer into Petri dishes 1 ml of the 1/10 dilution and then of the original sample.
4. An alternative method, which is somewhat quicker, is to inoculate the plates at the same time as the dilutions are made. Thus, as soon as the particular dilution has been mixed by sucking up and down 10 times, transfer 1 ml to a sterile Petri dish before carrying over 1 ml to the next tube of sterile diluent.

5.1.1.4 Pouring the plates

1. To each plate, add 10 ml of molten agar medium at 45°C and immediately mix the medium and inoculum by a combination of to-and-fro shaking and circular movements lasting 5–10 s. The exact procedure consists of movements five times to and fro, five times clockwise, five times to and fro at right angles to the first and five times anticlockwise. This procedure ensures complete dispersal of the sample. Take care not to get the agar on the lid of the Petri dish.

   No more than 15–30 min should elapse between dilution of the sample and its admixture with the medium, because suspension in the diluent may be lethal or may encourage growth or separation of clumps.
2. Allow the plates to set, then invert and incubate at the appropriate temperature. It is essential to label the plates adequately. It is worth giving some thought to labelling. Use codes chosen with care to avoid ambiguity. Remember that more writing takes time. Consistent use of different colours reduces the need for extensive detail. To facilitate rapid equilibration of temperature, the plates should not be stacked in piles of more than six.

Note. At all times, take full precautions to prevent contamination. The pipettes should be rapidly passed through the Bunsen flame before use. The mouths of the sample bottle and test-tubes should be flamed after removing the stopper or plug and *again* before replacing the stopper or plug. The lids of the Petri dishes should be raised only enough to allow easy access. Before pouring in the molten agar, dry the outside of the tubes or bottles containing the medium, and flame the necks of the containers after removing the stoppers (and again before replacing the stoppers if the containers held multiples of 10 ml).

5.1.1.5 Counting the plates

Select a dilution that yields fewer than 300 colonies per plate (in the case of standard plates of 8–9 cm, as with colonial concentrations exceeding this the count will usually be depressed to an unknown degree by overcrowding and microbial antagonism. If only a single plate at each dilution has been prepared, the dilution chosen for counting should also provide more than 30 colonies on the plate, as the statistical error involved in counting fewer than 30 colonies becomes overwhelmingly great. When replicate plates have been prepared at each dilution, the arithmetic mean of the colony counts at the chosen dilution is used to calculate the microbial concentration in the original sample. The count, N, of c.f.u. per g or per ml of sample is then determined by dividing by the dilution, d, (e.g. 10^{-2}) used.

$$N = \frac{(n\bar{x})}{n \cdot d}$$

In ISO standard methods (see, for example, ISO 4833:1991) another procedure has been adopted for estimating the colony count:

$$N = \frac{\sum c}{(n_1 + 0.1 n_2) \cdot d}$$

where $\sum c$ is the sum of all colonies counted on all dishes containing at least 15 colonies but not more than 300 colonies; n_1 is the number of dishes retained in the first dilution; n_2 is the number of dishes retained in the second dilution; and d is the dilution (e.g. 10^{-2}) corresponding to the first dilution counted.

In the ISO standards an example is given where, at the first dilution retained (10^{-2}), two dishes contained 168 and 215 colonies respectively, and, at the second dilution retained (10^{-3}), two dishes contained 14 and 25 colonies respectively, so that:

$$N = \frac{168 + 215 + 14 + 25}{[2 + (0.1 \times 2)] \times 10^{-2}} = \frac{422}{0.022} = 19\,182$$

$$= 1.9 \times 10^4 \text{ per ml or per g, rounded to two significant figures}$$

However, it is not really statistically permissible to obtain an average count by using the data from more than one dilution in this way. The colony counts at each dilution are derived from different statistical universes and have different sources of variability and therefore different variances. For example, a 10^{-3} dilution has been subjected to a further dilution step (with its inherent errors) compared with the 10^{-2} dilution. It is possible to use data from more than one dilution only if you have estimates of these different variances, so that you can then calculate a *weighted mean*, rather than a simple arithmetic mean.

The significance of the count can be assessed by determining the 95% confidence limits (that is, the range in which the true count should lie in 95 cases out of 100). This range is given by:

$$\frac{(n\bar{x}) \pm 1.96\sqrt{(n\bar{x})}}{n}$$

where $(n\bar{x})$ is the total number of colonies counted on all the plates at the chosen dilution, and n is the number of plates at that dilution. ($\bar{x}$ is the arithmetic mean of the colony counts at the chosen dilution.)

The fewer the colonies counted, the wider will be the 95% confidence limits. Maximum practical precision is obtained by counting 600–1000 colonies (Meynell and Meynell, 1970).

When the colonies are being counted, a hand lens or bench magnifier should be used with a good source of illumination (preferably oblique or lateral illumination), the plate being placed against a dark background. Ideal conditions are provided by various commercially available plate-counting chambers, but it is quite easy to construct at very low cost plate illuminators that perform just as well, using a matt-black plastic laminate viewing platform with a small tungsten strip light, shaded from view, at each side of the plate.

All colonies should be counted, including 'pinpoint' colonies. Spreading colonies (e.g. *Bacillus* colonies spreading over the agar surface) may cause trouble because of the suppression of the growth of other organisms as well as causing possible masking of other small colonies, so plates containing spreading colonies should preferably not be used for counting, but if there is no choice but to use them the presence of spreading colonies should be noted in the record.

Occasionally compact clusters of very small colonies occur at the agar–glass or agar–plastic interface on the bottom of the dish. It is customary to count each cluster as one colony as it is usually assumed that they derive from individual organisms contained within a clump (e.g. *Staphylococcus* or *Micrococcus*) or chain (e.g. *Streptococcus*) which have spread at an early stage but which in the depths of the medium would have acted as a single viable unit and resulted in only one colony.

5.1.1.6 Recording the results

In daily laboratory records counts should initially be recorded as a colony count at the given dilution, for example '182, 200 and 260 colonies at 10^{-3}'. This information is then used to calculate the microbial concentration in the original sample by multiplying the arithmetic mean by the dilution factor; in our example this would be 210 000 per g or per ml. Note that the final estimate should be given to only two significant figures in the report. In more demanding investigations both this estimate and the 95% confidence interval (c.i.) should be quoted in the report thus: '210 000 per g; 95% c.i. 197 000–231 000 per g'. Unfortunately in some laboratories it is the practice to record the counts directly as the microbial population in the original sample, often to the usual two significant figures; if replicate plates are prepared the arithmetic mean is worked out in the head or on a piece of scrap paper, and multiplication by the dilution factor is similarly performed. It is obvious that this means that data are irretrievably lost at the time of counting. The inaccuracies inherent in counting only one, two or three colonies at a given dilution are completely masked, and a count of one colony at 10^{-4} dilution on one occasion will wrongly appear to be the exact equivalent of a count of 300 colonies on triplicate plates at the 10^{-2} dilution on another occasion. It is therefore advisable to record full data in the laboratory records book.

Other frequent omissions are the details concerning the medium used and the temperature and time of incubation. These details are of great importance because they obviously affect the count obtained, sometimes profoundly. It can be argued that in any one laboratory the techniques are standardized so that a count performed, for example, on sliced roast pork will be a count on plate count agar at 37°C. This may be true, but alterations and modifications in the methodology should be very well documented. If at some future date a retrospective survey of trends is required then the precise points will be known at which the standardized method was modified; such modifications may have a profound effect on the apparent counts. Consider how easy it would be to deal with the following questions: 'When did we replace the plate count agar prepared from basic ingredients in the laboratory by Oxoid dehydrated Plate Count Agar?' 'When was the temperature of incubation altered from 37°C to 35°C?'. Perhaps a change in the long-term mean count on similar samples coincides with a change in the source of a dehydrated medium from manufacturer A to manufacturer B. Fig. 5.1 shows an example of the type of entry recommended for laboratory records books.

Date	*Sample*	*Type of Count*	*Dilutions examined (duplicate)*	*Incubation*
16/2/98	Cooked ham	General viable count (Plate count agar)	10^{-2}; 10^{-3}; 10^{-4}	3 days at 30°C

Diluent: 0.1% peptone
Blended: by Colworth Stomacher

Numbers of colonies found
>300, >300; 150, 162; 12, 17

Mean count per g of ham: 156×10^3
Report as: 1.6×10^5/g (95% confidence limits $\frac{312 \pm 1.96\sqrt{312}}{2}$)

Grade: C

FIG. 5.1

5.1.2 Miles and Misra surface colony count

The method (Miles and Misra, 1938) consists of placing drops (0.02 ml) of serial dilutions on the surface of poured agar plates and counting the colonies that develop on incubation of the plates.

This method is useful when the bacteria are best grown in surface culture (e.g. when the presence of obligate aerobes is suspected) or when an opaque medium is employed. Surface counts are also indicated where the microbial population in a sample is likely to include bacteria that are killed by the brief exposure to 45–50°C that occurs when using melted agar for the pour-plate techniques (Mossel, 1964).

1. Pour agar plates and dry the surface of the medium for 24 h at 37°C with the lids closed, followed by 2 h at 37°C with the lids and bases separated as described in Section 4.5.1. This enables the medium to absorb the water of the inoculum quickly.

2. 0.02-ml drops of tenfold dilutions are dropped on to the surface of the agar from a height of 2.5 cm, using either a push-button multiple pipetter with sterile plastic tips (e.g. Gilson or Eppendorf) or a calibrated dropping pipette (prepared as described below). For six dilutions, six plates are used for each count, one drop from each dilution being placed on each plate. Allow the drops to be absorbed fully before inverting and incubating the plates. Complete absorption should occur within 15 min (longer times are indicative of inadequate drying).
3. After incubation, counts are made of the drops showing colonies without confluence, the maximum number obtainable depending on the size of the colony and ranging from 20 to 100. The total count of the six drops of the appropriate dilutions is divided by six and multiplied by 50 to convert the count to colonies per 1 ml of the *dilution used.*

A less precise method of counting involves the recording of growth or no growth, and the use of probability tables to give a MPN (Harris and Somers, 1968). This is particularly useful for counting groups defined by physiological characteristics detectable on solid media (e.g. counting of starch-hydrolysing bacteria).

Preparation of pipettes

Pull and cut Pasteur pipettes of a sufficiently fine bore to enable the capillary to be inserted through a standard wire gauge of diameter 0.91 mm. Insert the capillary of a pipette in the gauge and, with a glass-cutting knife, score the capillary at the point of insertion. It should then be possible to snap the capillary level with the gauge in such a way as to obtain a square-cut end (a bending and slightly pulling movement will be found to give the best result).

Pipettes prepared in this manner will give drops of about 0.02 ml when the drops are delivered at a rate of one per second, with the pipette held vertically. Nevertheless, it is recommended that pipettes should be calibrated for the liquids being handled as solutions, suspensions and emulsions differ in their effect on the actual volume delivered. For example, Wilson (1922) found that, when used to deliver dilutions of bacterial culture, pipettes gave drops of more or less constant volume irrespective of the dilution used. He suggested that this was probably because the increase in density of the emulsion counterbalanced the increase in viscosity, causing the drop volume to remain more or less constant. On the other hand, when using calibrated pipettes to deliver dilutions of milk, the volume of the drop depended on the dilution (Wilson, 1935). Table 5.1, prepared by the late Dr T. Richards (unpublished work), indicates the correction factor required for the pipettes prepared as described above when used with varying dilutions of milk.

5.1.3 Surface colony counts: the spiral plater

The spiral plating machine (e.g. Spiral Systems, Cincinnati, Ohio, USA or Don Whitley Scientific, Shipley, UK) provides a way of distributing a small known amount of liquid (a liquid sample or a dilution of a sample) across the surface of an agar plate.

TABLE 5.1 Correction factors for Pasteur pipettes when used to dispense dilutions of milk

Liquid being delivered	Volume of one drop	No. of drops per ml
Tap water	0.0209	48
Milk 10^{-3}	0.0193	52
Milk 10^{-2}	0.0181	55
Milk 10^{-1}	0.0164	61
Skimmed milk	0.0153	65

The Petri dish containing solidified agar medium is rotated at constant speed on a turntable while the dispensing pipette (stylus) moves automatically from the centre to the edge of the rotating plate. Consequently, if the liquid is being dispensed at constant rate, there will be a dilution effect because the liquid distributed during a single rotation will be spread over about four times the track length on the outermost spiral compared with the innermost spiral. However, in addition, the dispensed volume is controlled by a cam, which results in the sample volume dispensed per rotation being reduced as the dispensing stylus moves across the plate.

The microbial population in the liquid is determined, after incubation in the usual way, by counting the colonies on a part of the track on the Petri dish where the colonies are easily counted. The plate is placed over a grid template, which enables the experimenter to determine the volume of liquid that was dispensed on to that part of the spiral track used for the colony count. Automated colony counting is possible using equipment designed for the purpose.

On one plate, colony counts can be performed on microbial concentrations over three orders of magnitude (e.g. from 500 to 500 000 c.f.u. per ml).

The two major benefits of the system are, firstly, in reducing the amount of repetitive and inherently boring work in making dilutions and plates, and, secondly, in considerable savings on the costs of labour and of media and reagents (relatively few dilutions need to be produced and few plates need to be prepared). The decision to acquire and use the relatively expensive spiral plating equipment in a quality assurance laboratory is likely to be on economic grounds.

Statistical analysis has shown no significant differences between results produced by this technique compared with the conventional Miles and Misra method (Jarvis *et al.*, 1977). Use of the spiral plater has been approved for official analyses in some countries (see, for example, FDA, 1992; AOAC, 1995).

5.1.4 Use of agar droplets

Considerable savings in plates and media compared with the normal pour-plate method can be obtained by performing colony counts on small droplets of agar. Such a system has been described by Sharpe and Kilsby (Sharpe, 1973; Sharpe and Kilsby, 1971). Although the mechanized automated apparatus is no longer commercially available, the manual technique is still used in some laboratories because of the savings in media, plates and incubator space.

In use, a number of bottles or tubes containing 9 ml of agar medium are held on the bench in a water bath at 45–50°C. One millilitre of the sample, or the first dilution of it, is pipetted with a 1 ml pipette into the first bottle and mixed. A row of five 0.1-ml drops of the inoculated agar is dispensed into a sterile Petri dish, followed by 1 ml into the next bottle of agar to provide the next decimal dilution. A standard Petri dish can hold three rows of five droplets and thus one dish allows the preparation of quintuplicate counts from three successive dilutions.

The principal disadvantages are that the system does not allow good differentiation of colony types when used with diagnostic media (e.g. brilliant green agar) and smaller colonies at the edges of the droplet can be difficult to see.

5.1.5 Use of dry rehydratable films

An example of this is the use of the Petrifilm (from 3M Company). The Petrifilm consists of a sandwich of plastic films containing a disc of dehydrated medium. The film is placed on a flat and horizontal surface, the top layer of plastic film peeled back and an inoculum of 1 ml of a dilution of a sample added to the disc of dehydrated medium, resulting in rehydration of the medium. After replacing the top plastic film, a plastic former is pressed down on to the Petrifilm, distributing the dilution across the medium and producing a uniform depth of medium across the disc. The Petrifilm is left on the bench for a few minutes to allow it to equilibrate.

Petrifilms are available with a number of media: for example for counts of aerobic mesophilic bacteria (an AOAC Official Method; see AOAC, 1995); yeasts and moulds; Enterobacteriaceae; and coliforms and *Escherichia coli* (also AOAC Official Methods; see AOAC, 1995). A principal advantage of the system is the small amount of space taken up by the films, both in storage before use and during incubation. However, one Petrifilm that I have found provides a more impressive advantage is that used to count *E. coli*, because in this case the top plastic film traps gas, so that gas production is detected while performing a colony count (see also Section 20.1).

5.2 MEMBRANE FILTRATION

The membrane filter consists of a thin, very porous, disc composed of cellulose acetate, cellulose nitrate or mixed cellulose esters. Membrane filters are available in a range of pore-size grades ranging from 10 nm to 8 μm or more in pore diameter. Filters recommended for most microbiological isolations and counts are those with a pore size of 0.43–0.47 μm. A few bacteria with a very small cell diameter require a filter with a pore size of 0.2–0.22 μm.

Membranes can be obtained which carry a printed square grid to assist in the counting of bacterial colonies. The other type of grid-printed membrane, the Hydrophobic Grid Membrane Filter, is discussed in Section 5.3. Various manufacturers produce a very wide range of apparatus for use with their membrane filters, including multiple filter manifolds, to facilitate the handling of many samples and dilutions.

The membrane will allow large volumes of water or aqueous solutions to pass through rapidly when under positive or negative pressure, but the very small pore size prevents the passage of any bacteria present. The bacteria, which remain on the surface of the membrane, can be cultivated by placing the membrane on an absorbent pad saturated with a liquid medium. The capillary pores in the membrane draw up the nutrient liquid and supply each bacterium with nutrient. The bacteria will give rise to individual colonies after incubation. The media used are specially designed for the membrane filter method and contain ingredients in concentrations different from those in standard media. Such media can be made up in the laboratory, or can be obtained in dehydrated or ready-prepared form from, for example, Oxoid, Difco Laboratories and Baltimore Biological Laboratories. Alternatively the membrane filter may be placed on the surface of a set agar medium. In the case of selective agar media, their suitability for use with membrane filters should be determined, as the different diffusion rates of the medium constituents may significantly affect selectivity.

The membrane filter colony count has the advantage that, provided the sample will pass through the filter readily, small numbers of organisms can be detected in large amounts of sample. Membrane filter counts thus may achieve the sensitivity of the multiple tube count whilst retaining the accuracy of the colony count method.

Alternatively, a microscopic examination can be carried out if the bacteria are stained and the membrane filter then rendered transparent by treating it with a liquid such as immersion oil or cottonseed oil. Suitable stains are Loeffler's methylene blue and crystal violet.

Membrane filters can be used for the routine examination of water, air, sugar solutions and some beverages. Milk and dairy products can be examined if fat globules are first broken down by the use of an appropriate wetting agent such as iso-octylphenoxy-polyethoxyethanol (Triton X-100).

Membrane filters are also used for a variety of other purposes such as the sterilization of fluids including microbiological media and gases; the sterility testing of sterile liquids for clinical use; and the separation of phage from bacteria.

The general procedure for performing colony counts is as follows.

1. *Sterilization of equipment.* The membrane filters should be placed between sheets of filter paper, wrapped in paper and autoclaved for 15 min at 121°C. At the start of every day, the filter-holder should be loosely assembled, wrapped in paper and autoclaved for 15 min at 121°C. Between samples, the funnel top can be sterilized by swabbing with alcohol and flaming.

   Incubating tins or dishes, each containing an absorbent pad, should be wrapped and sterilized by autoclaving or by dry heat.

2. *Preparation of incubating dishes*. To the sterile absorbent pad in each incubating dish, add aseptically sufficient sterile medium to achieve saturation of the pad without an excess of liquid being visible.
3. *Filtration of a sample and incubation of the membrane*. Attach the sterile filter-holder to a filter flask that is connected to a suction pump. Using sterile forceps place a sterile membrane filter, grid side up, on the platform of the filter unit, after first removing the funnel top (taking care not to contaminate the interior of the funnel). Replace the funnel and lock into place.

   Pour the liquid sample into the funnel, and draw the sample through the membrane filter by applying suction. After all the sample has passed through the filter, rinse the funnel with sterile quarter-strength Ringer's solution. Reduce the suction and, after removing the funnel top aseptically, transfer the membrane filter with sterile forceps to an incubating dish containing an absorbent pad previously saturated with an appropriate liquid medium. The filter should be placed on the absorbent pad with a rolling action in order to avoid trapping air bubbles between the filter and the pad. After replacing the lid of the incubating dish, incubate with the lid uppermost. The period of incubation normally required is somewhat less than that for poured plates.

   After incubation, the numbers of colonies are counted and the viable count calculated per millilitre or gram of sample. Sometimes, there may be insufficient contrast between the colour of the colonies and the colour of the membrane, and in this case a staining procedure (staining either the membrane or the colonies) may assist accurate counting:

   (a) *Malachite green staining of the membrane*. After incubation, the membrane may be flooded with a 0.01% aqueous solution of malachite green for 3–10 s, and then the excess stain poured off into a jar containing disinfectant (Fifield and Hoff, 1957). In this method, the membrane is stained, while the colonies remain unstained.

   (b) *Methylene blue staining of bacterial colonies*. Saturate a fresh pad with a 0.01% aqueous solution of methylene blue and transfer the incubated membrane bearing the colonies to this pad for 5 min. Then transfer the membrane to a pad saturated with water. The methylene blue tends to be removed from the membrane more rapidly than from the colonies, and the colonies can be counted as soon as there is sufficient contrast with the membrane.

5.3 MOST PROBABLE NUMBER (MPN) COUNTS

In these counts, the concentration of viable organisms or propagules is inferred from examining multiple cultures prepared from aliquots of a dilution series, and determining the proportions of such cultures that show growth and do not show growth in a suitable growth medium.

5.3.1 Multiple tube count

The multiple tube count is the commonest form of MPN count. It provides an estimate of the number of living organisms in a sample that are capable of multiplying in a given liquid medium. A liquid medium is chosen that will support the growth of the bacteria under investigation. Sometimes the count is based on the production of a certain reaction in the medium (e.g. coliform counts determined by acid and gas production from lactose).

In the usual type of MPN count, tubes containing the liquid medium are inoculated with, for example, 1-ml quantities of serial dilutions of the material or suspension of material being examined (see Section 5.1.1.2). After incubation, the highest dilution (i.e. the lowest concentration) giving growth (or the appropriate reaction, such as acid and gas production) is noted and this enables an *estimate* of bacterial numbers in the original sample to be made.

In the example illustrated (Fig. 5.2), growth has occurred in tubes inoculated with 1-ml quantities of dilutions of 1 in 10 and 1 in 100, but there has been no growth at the level 1 in 1000. Therefore, there were probably at least 100 but not as many as 1000 bacteria per millilitre of the original sample.

In practice, several tubes are inoculated at each dilution and the set of three decimal dilutions is chosen that includes the dilution at which around half are positive and half are negative for growth or reaction, together with the dilution on each side of this 'target' dilution. The most probable number (MPN) to result in this combination of positive and negative tubes is obtained by reference to probability tables (see Appendix 2).

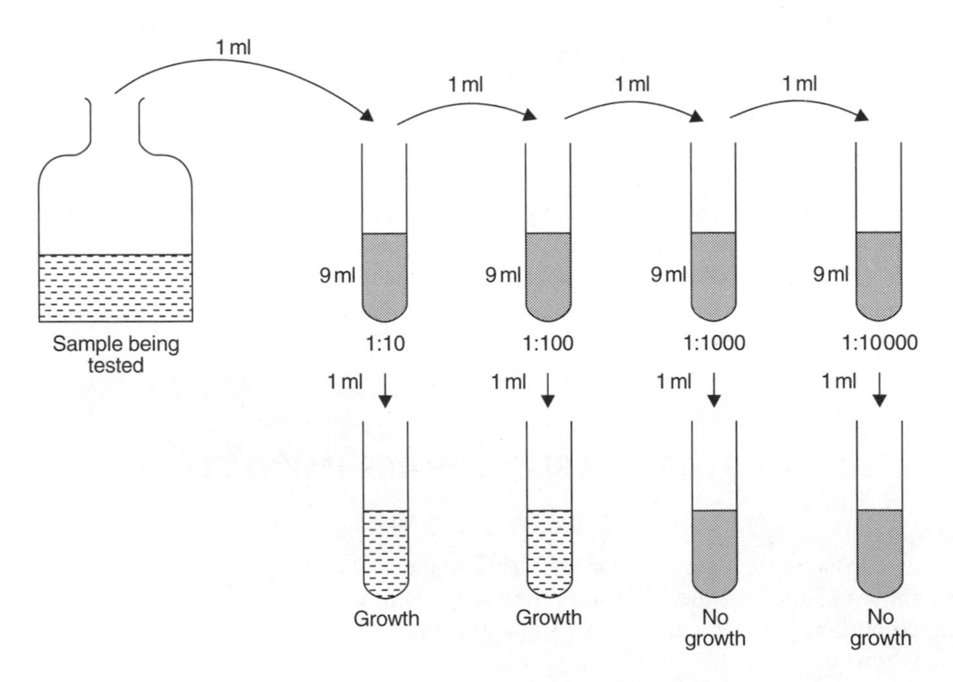

FIG. 5.2

It should be noted that the tables indicate the MPN, that is the microbial concentration most likely to give that result, *given the correctness of certain assumptions about, for example, the randomness of distributions of propagules in the samples and dilutions thereof.* Some MPN tables indicate the 95% c.i. on the MPN count, but once again this 95% c.i. assumes that the experiment is in statistical control. Many MPN tables do not indicate whether such a combination of positive and negative results is *itself* probable. Woodward (1957) listed the likelihoods of the various combinations, in the form of four categories of result, and these have been incorporated into the MPN tables in Appendix 2. The MPN tables appended to ISO standards now indicate a categorization of probability for the combinations (see, for example, ISO 4831:1991).

Commonly, five tubes are inoculated at each of a number of ten-fold dilutions. The MPN table for this arrangement is given in Table A.3 of Appendix 2.

After deriving the MPN from the appropriate probability tables, the 95% c.i. should be indicated. Cochran (1950) provided a simple technique for obtaining a rough estimate of the 95% c.i. as MPN/x to MPN·x, where the value of x depends on both the dilution factor and the number of tubes inoculated at each dilution. For five tubes at each of a number of tenfold dilutions, $x = 3.30$; for three tubes at each of a number of tenfold dilutions, $x = 4.68$; for two tubes at each of a number of tenfold dilutions, $x = 6.61$. In fact, the 95% c.i. will depend in a complex manner on the numbers and positions of the positive tubes. Accurate 95% and 99% c.i.s have been calculated by de Man (1975, 1977), and these are now incorporated into the MPN tables appended to ISO standards.

Figure 5.3 shows an example of the type of entry recommended for laboratory records books.

The multiple tube method of counting is particularly useful (and may be the only method available) when an estimate is required of numbers present in very low concentrations (e.g. less than one per millilitre), when the large amount of material to be examined makes the plate count method impracticable. A comparatively large volume of sample will be required and the presence or absence of microorganisms in the various quantities tested (e.g. 5 × 100 ml, 5 × 10 ml, 5 × 1 ml) may be determined. In such sampling regimes it is convenient to add the sample or its dilution to an equal volume of double-strength medium.

Date	*Sample*	*Type of Count*	*Dilutions examined*	*Incubation*
16/2/98	Cooked ham	Coliform count (MTC: MacConkey broth) positives:	(×5) 10^0; 10^{-1}; 10^{-2}; 10^{-3} 5 5 1 0	2 days at 30°C

Diluent: 0.1% peptone
Blended: by Colworth Stomacher

MPM per g of ham: 35 (95% confidence limits $\frac{35}{3.3}$ to 35 × 3.3)

* Grade: C

FIG. 5.3

The multiple tube count procedure may be used in conjunction with selective and diagnostic media to determine the numbers of particular groups of organisms, for example the numbers of coliform bacteria in food samples.

5.3.2 MPN counts using the hydrophobic grid membrane filter

The hydrophobic grid membrane filter (HGMF) was developed by Sharpe and colleagues in the 1970s as a means of obtaining estimates of counts of c.f.u.s (or propagules) with a minimum of dilutions (see Sharpe and Peterkin, 1988).

The Iso-Grid® HGMF (QA Life Sciences, San Diego, California, USA) is a square membrane filter with a filtration area of 5×5 cm. This is divided into 1600 grid-cells by a black hydrophobic print of grid-lines on the membrane.

During incubation, microbial growth will occur within, and be bounded by, the grid-lines surrounding a grid-cell. After incubation, a count can be made of the number of grid-cells showing growth (or, when a differential medium is used, the appropriate colour reaction). However, this is not a 'colony count', as a positive grid-cell can result from its being inoculated with more than one c.f.u. or propagule (Sharpe prefers the term 'growth unit' to c.f.u. in this context). The HGMF instead provides the equivalent of an MPN using replicate growth compartments at a single dilution. The MPN is calculated as

$$N \log_e \frac{N}{N - x}$$

where N is the total number of grid-cells (i.e. $N = 1600$ in the case of the Iso-Grid®) and x is the number of positive grid-cells.

The range of counts in the sample inoculum capable of being estimated with reasonable precision by the 1600-cell Iso-Grid® is from 1 to about 5000. Thus fewer dilutions need to be prepared and tested than is the case with normal colony counting by either plates or membrane filters. Maximum precision occurs when half of the grid-cells are positive.

The HGMF method has been recognized as an acceptable official method for the microbiological analysis of food in the USA (see FDA, 1992; AOAC, 1995) and Canada.

Sharpe and Peterkin (1988) provide operational instructions for the use of HGMFs to detect and estimate aerobic mesophilic counts and a wide range of specific counts including, for example, *Escherichia coli, Salmonella* and *Listeria monocytogenes*.

Because of the regular arrangement of the grid-cells, this method lends itself to the application of automatic electronic counting devices which are able to count HGMFs carrying a large proportion of positive cells. Since maximum precision is obtained when half of the grid-cells are positive, this would involve much laborious counting of around 800 cells, if counted manually, so that errors from operator fatigue can then occur. Automatic counters are therefore not just labour-saving but will also reduce experimental error.

An alternative to hydrophobic grid membrane filters to achieve a similar single-dilution MPN is to use a microtitre plate to obtain multiple inocula into the wells of the plate. Once again, if the medium is designed so that growth of the target organism results in a colour change, the microtitre plate can be examined automatically after incubation. Such systems are available commercially. One example is the Quanti-Tray used in conjunction with Colilert (IDEXX Laboratories, Westbrook, Maine, USA) to perform MPN counts of total coliforms and *E. coli*.

5.4 DYE REDUCTION METHODS

These methods depend on the ability of microorganisms to alter the oxidation–reduction potential of a medium. They are in consequence a measure of the activity of microorganisms in the test system rather than of the numbers in the sample. Suitable indicator dyes include methylene blue and resazurin. The length of time taken to reduce the dye depends on the mass and activity of bacteria present in the sample: the greater the number present, the shorter the time required for reduction. However, many other factors are important, including the nature of the sample, the medium used and the types of organisms present. The organisms must be capable of metabolism and growth in the medium to which the dye is added, and if the sample itself is incapable of supporting growth the dilution liquid should be a nutrient liquid. For reproducible and interpretable results the test system, including the sample, must be of a sufficiently constant chemical composition to have an invariable effect on the microorganisms present.

The test procedure adopted is dependent on the type of sample being examined, and because it must result in standardized conditions a dye reduction test method can be applied only to foods or food constituents that have a reasonably constant and predictable composition. Tests may be based on determining either the time taken for decolorization or the amount of reduction obtained after a set time. The test methods allow interpretation relative to arbitrary grades only.

In Part III there are examples of standard test methods as applied to raw milk; these methods are useful as a platform rejection test for raw milks in countries where raw milk is delivered to the creamery or factory in cans or churns. Dye reduction tests are much less useful in assessing the quality of raw milk that has been refrigerated in bulk on the farm and then transported in a refrigerated bulk carrier.

In designing a standard test method to be used with a particular food, the following need to be defined:

1. The *objective* of the dye reduction test. For example, although it is a dynamic metabolic test, the methylene blue test was once used in the UK on ice-cream samples purportedly to assess hygienic quality (Report, 1950). The test measures the ability of microorganisms present in the ice-cream to grow in a dilution of the ice-cream when kept at a temperature of 20–37°C. Yet ice-cream will be kept frozen until consumption, and many microorganisms that could be present, especially certain pathogens, would not affect the test result. In these circumstances it is difficult to see how the methylene blue test assesses either the keeping quality of the ice-cream or its potential health hazard.

2. The *constituent species* of the expected microflora, and their relative effects on redox potential.
3. The *incubation temperature* to be chosen having due regard to (1) and (2).
4. The *levels of microbial population* in the sample to be detected as separate grades. Can these be related to a set of reduction times needed for decolorization to occur at the test temperature?

5.5 ELECTROMETRIC METHODS

Impedance/conductance methods (Firstenberg-Eden and Eden, 1984; Stannard *et al.*, 1989) are among the most successful rapid techniques, and are applicable both to quality assurance microbiology and to research into the effect of the food environment on microorganisms. These methods depend on detecting the decrease in the electrical impedance of the growth medium as the microorganisms metabolize and multiply.

The great value of the methods derives from a number of characteristics of the commercially available equipment. The principal systems available include Malthus (Malthus Instruments), RABIT (Don Whitley Scientific), Bactometer (bioMérieux) and BacTrac (Sy-Lab).

Firstly, all the systems are able to handle a large number of samples concurrently, to monitor the electrical changes automatically, and then, using software incorporated into the equipment, to interpret these changes against calibration curves so that the initial microbial loads in the samples can be deduced. A variety of microbiological criteria can be assessed, by using appropriate selective media.

For quality assurance purposes, the equipment is set up to measure the time taken for the impedance to be reduced to a predetermined level (this is known as the detection time); this time will be inversely proportional to the $\log_2$ of the initial microbial load.

The second virtue of these systems is that the higher the initial microbial load, the shorter the detection time; that is, the more unsatisfactory a sample, the quicker this unsatisfactoriness is identified. As the identification of unsatisfactory detection times can be handled by the software channel by channel, applying appropriate analysis to each channel (i.e. to each sample tube), different microbiological criteria (e.g. aerobic mesophilic count, coliform count, etc.) can be judged in respect of any sample type for which calibration curves have been established and entered into the software.

Thirdly, the equipment's computer is programmed to flag up on the visual display unit (VDU) those samples that had initial microbial counts in excess of the maximum permissible under the relevant specification. In some cases these will be alerted in detection times of the order of 6 h. This means that the equipment can be used for positive release of batches of perishable and semi-perishable foodstuffs. For example, the equipment can be positioned in a laboratory possessing glass windows on to an adjacent corridor, with the VDU facing the corridor. The perishable and semi-perishable foods produced during the day will be sampled and put under test in the apparatus. The batches of food will be stored in cold stores ready for loading into refrigerated lorries or refrigerated containers for despatch to supermarkets and shops during the night following their production. Immediately before loading commences, the person in charge of the cold store can walk along the corridor and inspect the VDU to determine whether any batches of any of the products have been identified as unsatisfactory (commonly by the batch number turning red or even flashing red). These batches will then not be despatched to the supermarkets or shops, but will be held in the cold store for the quality assurance staff to make further investigation on them during the next day.

Thus, a fourth virtue of the commercially available systems is that the colour-coded flagging of unsatisfactory samples does not require the presence of laboratory staff.

Some microorganisms such as yeasts do not produce adequate changes in conductance in the culture medium. The problem has been addressed by an approach developed by Owens and colleagues (Owens *et al.*, 1989; Druggan *et al.*, 1993) known as indirect conductimetry, in which the electrodes are not in the culture medium but instead are placed in an absorbent solution that detects the evolution of carbon dioxide from the culture. The indirect conductimetric technique has also been used for the detection of pathogens using selective media which are not susceptible to direct conductance measurements. A number of companies such as Malthus Instruments Ltd, and Don Whitley Scientific Ltd, have developed commercial systems and media for such purposes based on indirect electrometric measurements.

5.6 NUCLEIC ACID PROBES AND THE POLYMERASE CHAIN REACTION

Gene probe assays originally depended on radionuclide labelling, so that they were not suitable for the routine quality assurance laboratory. Since then, the development of colorimetric assays has widened the potential for use in food microbiology laboratories. The sequences targeted in a nucleic acid probe assay can be on the chromosomal DNA, ribosomal RNA (rRNA), messenger RNA (mRNA) or extrachromosomal DNA. Targeting rRNA improves the sensitivity because there are many thousands of copies of the rRNA molecule in each bacterial cell (Grant *et al.*, 1993).

One of the advantages inherent in the technique is that, provided the nucleic acid sequence to be targeted is chosen carefully, it should be possible not only to detect the sought-after organism (e.g. *Salmonella*) against a background of related organisms (e.g. *Citrobacter*, *Proteus*, *Enterobacter* in the case of *Salmonella*), but also to detect strains of the sought-after organism that display atypical biochemical characteristics (e.g H_2S-positive or lactose-positive salmonellae).

The Gene-Trak® system (Gene-Trak Systems, Hopkinton, Massachusetts, USA) uses a sandwich hybridization technique targeted to rRNA. The organisms still need to be increased in numbers from the in-food concentrations by selective enrichment. At the time of writing the Gene-Trak assays for *Salmonella* and *Listeria* were the best validated of commercially available nucleic acid probe assay kits (Mozola, 1997) and had the status of AOAC Official Methods (AOAC, 1995).

If a nucleic acid probe assay targets a sequence on chromosomal DNA, there is a lower sensitivity because there is only one target sequence in the cell. The polymerase chain reaction (PCR) provides a way of replicating a target sequence of DNA, and in principle can greatly increase the sensitivity of a gene probe assay. Atlas and Bej (1994) have discussed the principles of the methodology, and Jinneman and Hill (1997) have outlined the problems involved in using the PCR technique for detecting pathogens in foods. After an interlaboratory trial to evaluate the detection of Norwalk virus in shellfish tissue (Atmar *et al*., 1996), it was concluded that the fairly complex protocol could be performed in government laboratories '*provided personnel received sufficient training*' (my italics). Thus, it will require further development before the technique can be used in routine quality assurance laboratories. The PCR is generally inhibited if applied directly to foodstuffs (Grant *et al*., 1993). The need still to employ enrichment culture leads to prolonging the time taken for a result to be obtained. Sharpe (1994, 1997) believes that in future truly rapid methods are likely to involve techniques for physical (or chemical) separation of the target microorganisms, followed by PCR applied to the purified microbial DNA. Work by scientists at the USFDA has resulted in a proposal for a general protocol using a universal culture medium for PCR detection of 13 species (in nine different genera) of foodborne pathogens (Wang *et al*., 1997).

6

Determination of the Total Number of Microorganisms in a Sample

6.1 THE BREED'S SMEAR METHOD FOR DIRECT MICROSCOPIC COUNTS

This is a method in which the number of bacteria or yeasts in a sample (e.g. broth culture, cell suspension, milk) may be determined by direct microscopic examination. As the staining procedure does not differentiate between living and dead organisms, the bacterial count obtained is known as a 'total count'. Any suitable staining technique may be used (e.g. simple stain or Gram's method). In the case of samples of milk and other foods with a high fat content, it is necessary first to defat (e.g. with xylene). Newman's stain conveniently combines both defatting and staining processes, although not allowing the determination of the Gram staining reaction. The Breed's smear method is also used for counting animal cells – predominantly leucocytes – which are found in much larger numbers in mastitis milk (see Section 28.1.1.1) than in milk from healthy animals and so can be diagnostic of mastitis.

It is essential in this procedure that the microscope is first calibrated so that the area of the microscopic field is known. A known volume (0.01 ml) of the sample or an appropriate dilution is then spread over a known area (1 cm^2), on a glass slide. An alternative procedure often used, which gives the same sample concentration on the slide, is to draw two straight lines across the width of the slide to enclose a length of 2 cm. As the standard microscope slide is 2.5 cm wide, this marks off an area of 5 cm^2, over which 0.05 ml of sample is spread. The sample is allowed to dry, then stained and examined microscopically. The average number of bacteria or cells per field is determined and it is then possible to calculate the number of bacteria or cells per millilitre of original sample.

6.1.1 Determination of the area of the microscopic field

1. Set up the microscope and, using the $\times 10$ eye-piece and low-power objective, adjust the stage micrometer so that the graduated scale (1 mm divided into 100 units of 10 µm each) is in the centre of the field.

2. Place a drop of immersion oil on the stage micrometer and focus with the oil-immersion objective. Determine the diameter of the microscopic field in micrometres (μm), using the micrometer scale, adjusting the tube length of the microscope slightly if necessary. In subsequent observations this same tube length must be maintained.
3. Calculate the area of the microscopic field in millimetres (mm) for the oil-immersion lens and ×10 eye-piece using the formula:

$$\pi r^2 = \text{Area of field}$$

where r is the radius of the field. Knowing the area of the microscopic field, it is then possible to determine the microscope factor (MF) by calculation:

(a) MF = number of fields in 1 cm^2 (100 mm^2) = 100/area of field in mm^2.
(b) Average number n of organisms or cells per field = number counted/number of fields counted.
(c) MF × n = number of organisms or cells present in 1 cm^2.
(d) Since 0.01 ml of sample was spread over 1 cm^2 (or 0.05 ml over 5 cm^2), MF × n × 100 = number of cells or organisms present in 1 ml of sample.
For most microscopes as used above, one organism per field represents approximately 500 000 organisms per millilitre of sample.

6.1.2 Preparation of smear

1. Thoroughly mix the sample to be examined. Prepare dilutions as necessary, to produce smears with no more than 20 organisms per field. (If pure cultures are being examined, a barely detectable turbidity will be given by a suspension containing about 10^6 cells per ml.)
2. Deliver 0.01 ml of sample on to a clean glass slide and spread over an area of 1 cm^2 using either a guide card or a marked slide. The sample may be delivered either (a) by means of a capillary pipette calibrated to deliver 0.01 ml or (b) by using a wire loop. A closed loop of 4 mm in internal diameter will hold a drop of approximately 0.01 ml of sample, if withdrawn with the plane of the loop perpendicular to the surface.
3. Dry the smear immediately by placing on a warm level surface or in an incubator at 55°C. Drying should be complete within 5 min to prevent possible bacterial multiplication.
4. Stain by an appropriate method.
5. Examine, using the oil-immersion objective.

6.1.3 Counting

1. Count each single bacterium as one. Chains and clumps of bacteria are also counted as one. Bacteria further removed from the clump than the longest dimension of the

constituent bacteria should be counted as separate. In this way the 'direct count' is also a 'clump count' and bears a closer relationship to the probable plate count, as each clump or chain would probably give rise to one colony only.

2. The precision of the count depends on the number of bacteria counted and, as in the plate count technique, maximum practical precision would be obtained by counting some 600 items (bacteria in this case, colonies in the case of the plate count). However, unless an automated system is being employed, operator fatigue will occur sooner when performing microscopic counts, so it is recommended that a number of microscope fields are scanned and the numbers of bacteria observed in each field summed, until the total number of bacteria observed is 150 or more. Counting such a number will still provide acceptable precision (Meynell and Meynell, 1970). Note the number of fields scanned. The final field to be examined that achieves the target should be counted in its entirety. Owing to the way in which the drop dries on the slide during the preparation of the smear, the microorganisms tend to become concentrated towards the centre of the smear (Hanks and James, 1940). To make some allowance for this non-random distribution of organisms on the smear, fields should not be sampled at random but should be sampled and counted during edge-to-edge traverses of the smear, successive traverses to be taken at right angles.
3. Determine the average number of organisms per field and hence calculate the number per millilitre or per gram of sample.

6.2 DIRECT MICROSCOPIC COUNTS BY MEMBRANE FILTRATION

See Section 5.2 for the general procedure for filtration of the sample. After filtration, turn off the vacuum pump but do not remove the membrane filter from the filtration apparatus. Pour sufficient staining solution (e.g. Loeffler's methylene blue) into the funnel to cover the membrane filter completely. Allow to stain for the usual time (5 min in the case of Loeffler's methylene blue) and then start the vacuum pump once more to remove the staining solution. Pass distilled water through the membrane filter until either the filter or the filtrate is colourless. Then remove the membrane filter and allow to dry in air. When completely dry, saturate the filter with immersion oil by floating the filter on a small amount of immersion oil placed in a Petri dish. This procedure renders the membrane filter transparent. Remove the filter and mount on one or more microscope slides (in the case of the usual 4.7 or 5 cm filters it will be necessary first to cut the filter in half), after which it is ready for microscopic examination with the oil-immersion objective. If the microscope factor is known, the number of bacteria in the sample can be calculated from the effective area of the membrane filter and the amount of sample passed through the filter.

Note. As staining solutions and washing water are required to pass through the membrane filter during the staining procedure, it is necessary to filter the reagents before use first by passage through a fine filter paper, then by membrane filtration.

6.3 DIRECT EPIFLUORESCENT FILTER TECHNIQUE (DEFT)

A substantial improvement in direct microscopic counting for certain types of sample was achieved by combining membrane filtration, fluorescent staining and microscopy to give the direct epifluorescent filter technique (DEFT) (Pettipher, 1983, 1986; Pettipher *et al.*, 1989). A sample is filtered through a membrane filter, which will retain all the bacteria, moulds and yeasts present. A fluorescent dye is then applied to stain the microbes and excess stain washed through the filter. Microscopic examination is then undertaken with the specimen illuminated with ultraviolet light from above. Specially designed epifluorescent microscopes are available to achieve this optimally. Microorganisms are seen fluorescing against a black background. It is possible to couple the microscope to an image analyser to obtain counts automatically.

Amongst the precautions needing to be taken are: the use of suitable membranes (e.g. Nuclepore polycarbonate membrane filters, available from Corning-Costar, High Wycombe, UK); the prior filtration of all staining and washing reagents through a filter with a smaller pore size, to remove filtrable particles; and the use of non-fluorescent immersion oil. Some manufacturers (e.g. Difco) market complete DEFT kits of membranes and reagents.

DEFT was initially designed for examining milk samples (Pettipher *et al.*, 1980; Pettipher and Rodrigues, 1981), but procedures have been developed for other foods such as meat and poultry (see, for example, Shaw and Farr, 1989).

6.4 FLOW CYTOMETRY

In its simplest form a liquid sample is passed through a flow cell illuminated by a laser, and particles (in this case microorganisms) are detected automatically through a microscope focused on the flow cell. For food and beverage applications, it is obviously necessary to detect and identify only those particles that are microorganisms. This can be achieved by differential staining techniques, one of the most useful being use of fluorescent dyes such as fluorescein isothiocyanate, fluorescein diacetate. Fluorogenic esters can be used to detect viable organisms; these substances are non-fluorescent, but may be taken up by a viable organism, and cleaved by non-specific esterases within the cell to release the fluorescent dye (Diaper and Edwards, 1994). See Brailsford and Gatley (1993) and Pinder *et al.* (1993) for a general discussion of flow cytometry techniques.

Methodology is at present highly specific to the equipment being used, so equipment manufacturers' protocols for analysis need to be followed. ChemUnex (Cambridge, UK), for example, has described methods for using the ChemFlow® apparatus for examining *inter alia* dairy products including yoghurt and dairy desserts, fruit juices and soft drinks (see also Pettipher, 1991), wine and beer. There is potential for using fluorescent-labelled antibodies to detect specific microorganisms, but such applications will need careful validation.

Another potential application is in the sorting of microorganisms which could lead to its use in concentration, but at the time of writing this is in the very early stages of development (see, for example, Nir *et al.*, 1990).

6.5 ATP DETERMINATION BY BIOLUMINESCENCE

Adenosine 5′-triphosphate (ATP) is a component of the living cells of all organisms which is involved in energy transfer. A system capable of detecting its presence is available in the luciferin–luciferase reaction. Detection depends on a bioluminescent reaction in which an oxidizable substrate, luciferin, is affected by an enzyme (luciferase), in the presence of ATP, to generate small amounts of light which are proportional to the amount of ATP present, as long as the substrate and enzyme are present in excess. The light is detected by a very sensitive photomultiplier system. The amount of light detected is recorded and compared with a standard produced using known amounts of ATP. An assumption that ATP constitutes 0.4% of the bacterial dry weight permits an estimation of the biomass present. The method is useful when the total biomass is required to be known, for example in the ecological study of aquatic environments.

One obvious application in the food industry where a determination of total biomass is potentially helpful is in the field of hygiene monitoring, where the cleanliness and microbial load of a sanitized food equipment surface will correlate with detected ATP. A number of manufacturers (including Lumac, Landgraaf, The Netherlands, and Biotrace, Plainsboro, New Jersey, USA) have developed hygiene monitoring kits and equipment (see also Griffith *et al.*, 1994). A swab of the surface is taken and, after swabbing, the swab is placed into a specially designed bioluminometer that will then immediately measure ATP content.

However, one of the drawbacks in the analysis of *food* samples is that ATP is common to all forms of life and therefore many foods will contain ATP derived from the plant or animal material used in that food, as well as ATP derived from microorganisms. The early attempts to overcome this problem (e.g. Sharpe *et al.*, 1970) were based on incubating the food for a short period. Whereas innate food ATP would *decrease* during incubation, microbial ATP (from microorganisms able to metabolize actively in the food) would *increase* during incubation. Therefore, if a calibration curve is produced for a given food with known microbial contents (as determined by conventional colony count procedures, for example), a given ATP value before and after incubation can be used to obtain a value for a presumptive microbial count. There are many problems associated with the approach, one of the principal problems being the need for the innate food ATP concentration to be fairly constant and highly predictable for each food to be assessed. ATP measurement thus seems of limited application for detection and quantitation of viable organisms in foods, and thus the technique is not listed in Section 5.4. Some equipment manufacturers offer bioluminometer equipment for sterility testing on, for example, UHT-sterilized foods or filter-sterilized beverages. However, electrometric techniques seem to offer distinct advantages for such applications (see Section 5.5).

A useful collection of papers on ATP luminescence techniques has been presented by Stanley and co-workers (1989).

6.6 TURBIDIMETRIC METHODS

6.6.1 Introduction

These techniques depend on the microorganisms in a suspension blocking a light beam by scattering or absorption, causing the suspension to appear turbid. The greater the concentration of organisms, the less the light can penetrate the suspension and the more light is scattered.

It is obvious, therefore, that turbidimetric methods can be used only for estimating concentrations of microorganisms that are suspended in liquids (e.g. distilled water, saline, nutrient broth) that have a low innate turbidity. In addition the light-blocking power of an organism will depend on its size, shape and transparency (Powell, 1963; Koch, 1984). Therefore, a turbidity measurement can be correlated with microbial population only when a pure culture is examined and when the growth conditions of the culture have been standardized to ensure reproducibility of size and shape of the organisms in the population.

The main applications of turbidimetric methods are:

(a) The standardization of suspensions of pure cultures of bacteria to be used in laboratory experiments (e.g. for evaluation of disinfectants or of preservation processes). Such standardization ensures that the bacterial concentration is within the desired range, but it will still be necessary to determine the actual viable count of the suspension used (i.e. the count at t_0).
(b) The assay of vitamins and other growth factors, and of antibiotics and other growth inhibitors.
(c) The determination of the effects of environment (e.g. temperature) on growth.

Photoelectric colorimeters and spectrophotometers can be used for measuring the amount of light lost from a reference light beam as the result of passing it through a turbid suspension, but they are relatively insensitive to low bacterial concentrations. Another approach is to measure directly the amount of light scattered using a nephelometer, this instrument being much more sensitive to low turbidities than are most colorimeters and spectrophotometers.

6.6.2 Standard (McFarland's or Brown's) opacity tubes

This is a simple method for rapidly determining the approximate number of bacteria in a suspension by comparing the turbidity (opacity) of the suspension with the graded turbidities of a series of ten standard tubes. Such turbidity standards are used primarily for preparing inocula with a standard and consistent population concentration (e.g. as required when inoculating bioMérieux API identification kits). These can be purchased from a number of manufacturers (e.g. bioMérieux), but can alternatively be prepared using barium sulphate solution and sulphuric acid (see Appendix 1).

Method of use

1. Transfer the bacterial suspension, or a suitable known dilution, to a tube of dimensions similar to the opacity tubes, to give a column 5–7 cm high.
2. Place the tube against a clearly printed page in a good light and compare the opacity tubes one by one with the sample tube.
3. Select the opacity tube that most nearly matches the suspension.
4. The approximate number of organisms per millilitre is obtained by multiplying the number on the opacity tube by 300×10^6 in the case of bacteria, or in the case of yeasts by 10×10^6.

For more accurate determinations, the tubes can be calibrated for the particular organism and growth conditions being studied by reference to some other counting method.

6.6.3 Nephelometry

When using a nephelometer, two reference standards are required: a turbidity standard to set the sensitivity of the galvanometer and a blank (distilled water or uninoculated medium) to set zero. A ground perspex or ground glass turbidity standard can be purchased for use with this instrument.

The limit of sensitivity of nephelometers typically is about 10^5 organisms per millilitre (the actual figure being dependent on species).

Some nephelometers can be used with standard size test-tubes; others require the sample to be transferred into a cuvette. However, in the former type, the test-tube forms part of the optical system in the nephelometer, the hemispherical base of the tube acting as a condenser lens for the light beam. Consequently the test-tubes need to be standard not only in their diameter, wall thickness and glass colour but also in their base configuration. Specially matched nephelometer tubes can be purchased but these are very expensive. The author prefers to order a number of boxes of borosilicate glass test-tubes and, using a standard turbid suspension, to check the effect of the tubes on the turbidity reading. Such test-tubes have been found to be very uniform within a batch, so it is a simple matter to obtain a large number of optically matched tubes. These should then be kept exclusively for nephelometer use.

The sensitivity and range of the instrument is adjusted by using a blank to set zero (this blank may be a tube of distilled water or of uninoculated broth), and a turbidity standard to set maximum deflection.

It will be found that most nephelometers require a warm-up time before stable readings are obtained, and some makes of instrument require to be checked and adjusted frequently against the blank and turbidity standards.

If using test-tubes, it may be found that the test-tube closure affects the turbidity reading, depending on the amount of light reflected by the closure back down the tube. If this is the case, standardize your procedure, preferably by taking all readings with the tube closures removed. In growth-response experiments requiring continuing incubation, the author uses black rubber seals (e.g. Astell seals) as they do not seriously affect the readings.

The accuracy of the readings is affected by the optical cleanliness and standard dimensions of the tubes. Avoid scratching or abrading the test tubes, and ensure that the tubes, at the end of the experiment, do not enter the general washing-up but are kept separately.

It is necessary to use sufficient liquid in the tubes to ensure that the meniscus is above the photoelectric cells.

As response is not completely linear, a reference curve can be made using a series of dilutions of a dense suspension.

7

Composition of Culture Media

7.1 INTRODUCTION

A *culture medium* is any nutrient liquid or solid that can be used in the laboratory for the growth of microorganisms. Such a medium may resemble the natural substrate on which the microorganisms usually grow (e.g. blood serum for animal pathogens, milk for milk microorganisms, soil extract for soil microorganisms) – a habitat-simulating medium. Whatever the medium, it must include all the necessary requirements for growth, which vary according to the organism it is desired to grow but will include:

(a) water;
(b) nitrogen-containing compounds (e.g. peptides, proteins, amino acids, nitrogen-containing inorganic salts);
(c) energy source (e.g. carbohydrate, peptides, amino acids, protein);
(d) accessory growth factors.

The nutritional requirements of bacteria range from the simple inorganic requirements of autotrophs to the many vitamins and growth factors required by some of the more fastidious bacteria (including pathogens and the lactic acid bacteria). Therefore, it is not possible to formulate a medium capable of supporting the growth of all microorganisms. However, the commonly used *empirical* media, such as nutrient broth and nutrient agar, are capable of supporting the growth of many bacteria. Furthermore, a medium such as nutrient agar can be used as a *basal medium* to which is added, for example, blood to 5–10%, serum or milk, to provide the complex growth factors needed by the more fastidious bacteria. Lactic-acid bacteria require B-group vitamins, which can be provided by the addition of yeast extract.

A nutrient medium can be made selective or biochemically diagnostic by the addition of suitable compounds.

The few media described in this section are given to demonstrate the basic principles of media making. Information on the method of preparation of specific media is given in Appendix 1. For a more comprehensive description of the design and formulation of media, see Bridson and Brecker (1970).

Preparation

1. Weigh out ingredients and heat in steamer until dissolved.
2. When cool, adjust pH to 7.6, with N sodium hydroxide (approximately 5–10 ml N sodium hydroxide are usually required).
3. Autoclave at 121°C for 15 min, to precipitate phosphates.
4. Filter. (The use of filtration will ensure a transparent medium.)
5. Adjust pH with N hydrochloric acid to pH 7.2.
6. Distribute in bottles (screw caps should be left slightly loose except where bijou (5 ml) bottles are employed) or in test-tubes as required.
7. Autoclave at 121°C for 20 min, or the relevant time (see Sections 2.3.2.3 and 7.7). Tighten any screw caps on cooling.

7.4.2 Nutrient agar

This medium is based on nutrient broth but solidified with agar, and is of general use as indicated above. The use of filtration will ensure that a transparent medium is produced.

Ingredients

Nutrient broth (pH 7.2)	1000 ml
Agar	15 g

Preparation

1. Weigh out the agar and dissolve in nutrient broth in the autoclave at 121°C for 20 min.
2. Check that the pH is 7.2, and adjust if necessary.
3. Prepare a filter in a Buchner-type filter funnel by first soaking two large (46 × 57 cm) sheets of Whatman No. 1 filter paper in water, mashing to a pulp, bringing the mixture to the boil in a large beaker and then pouring into the filter funnel while applying suction. This will result in a layer of hot paper pulp being evenly distributed over the base of the filter funnel and also will pre-warm the funnel.
4. Immediately place into position the filter flask to be used for the collection of the medium, and filter the nutrient agar while still hot.
5. Distribute in bottles (any screw caps should be left slightly loose except where bijou (5 ml) bottles are used) or test-tubes as required.
6. Autoclave at 121°C for 20 min, or other relevant time. Tighten any screw caps on cooling.

7.4.3 Litmus milk

This is a natural medium, used to indicate the effect of pure cultures of bacteria on milk constituents, and to detect milk spoilage organisms in rinses, etc.

Ingredients

1000 ml of skim milk, reconstituted from skim-milk powder. Alternatively, good quality separated milk may be used if available.
Litmus solution.

Preparation

1. Add sufficient litmus solution (approximately 10 ml of a 4% aqueous solution) to give a pale mauve colour.
2. Distribute in test-tubes or bottles as required.
3. Sterilize in a steamer at 100°C for 30 min on each of three successive days (intermittent sterilization or Tyndallization). *Check sterility by preincubation before use*, preferably at two incubation temperatures and including that to be used in the investigations.

7.5 SEPARATION OF MIXED CULTURES: ENRICHMENT PROCEDURES, ELECTIVE AND SELECTIVE MEDIA

A mixed culture can generally be separated into its constituent organisms (provided the individual strains are present in approximately equal numbers) by using a streak plate method as described on Section 4.5.1. Isolated colonies can then be picked off and subcultured as required. Some of the problems in the use of selective media to detect organisms (especially pathogens) in foods are discussed further in Parts II and III.

7.5.1 Enrichment procedures

When the *target* organism (the organism being sought) is outnumbered or is accompanied by many other species, as in soil, water and most food samples, it may be necessary to use an enrichment technique to increase the numbers of the required organism. The methods adopted should be chosen to take advantage of any known physiological characters of the particular organism, for example optimum temperature or pH, nutritional requirements, tolerance of added inhibitors.

In some cases, pretreatment of the material itself is appropriate, for example heat treatment of suspensions of the sample for the isolation of sporing or other heat-resistant organisms, followed by plating on solid media. Advantage may also be taken of the differing optimum temperatures of microorganisms in a mixed population, so that incubation of material at the optimum temperature for the required organism should increase its numbers in relation to other organisms present. Since, however, enrichment takes place in liquid media, the particular enrichment procedure adopted is unlikely to yield a pure culture, and the final isolation procedure still involves isolation of separate colonies on solid media.

7.5.2 Elective media

In other cases, particularly in the isolation of soil microorganisms, a medium that satisfies the minimum nutritional requirements of the organisms concerned – known as an elective medium (van Niel, 1955) – may be useful, particularly if the required organisms have unusual nutritional characteristics. For example, 'wild' yeasts can be isolated using lysine agar, on which organisms cannot grow unless they can use lysine as a sole source of nitrogen (Morris and Eddy, 1957).

7.5.3 Selective media

Another important method of separating mixed cultures is to make use of a selective medium. This is a basic medium that may support growth of many types but which has been modified to include one or more inhibitory agents, thereby restricting the growth of organisms not required. The choice of a selective medium must therefore be appropriate for the isolation of the particular organism concerned with reference to the nature of the samples being examined. Inhibitory substances used in the preparation of selective media include dyes, antibiotics, bile salts and various inhibitors affecting the metabolism or enzyme systems of particular species.

One form of a selective medium is that in which the pH of the medium has been modified so that it is suitable for the growth of only acid-tolerant or alkali-tolerant species. For example, yeasts, moulds and lactobacilli are acid-tolerant organisms and can grow on media at pH 4–5, whereas less acid-tolerant organisms are unable to grow.

7.5.3.1 Selective media for Gram-negative bacteria

Crystal violet used in a medium at a final concentration of 1:500 000 inhibits the growth of many Gram-positive bacteria (though not streptococci) while permitting that of Gram-negative bacteria (Holding, 1960). Penicillin incorporated at a final concentration of 5–50 units per ml inhibits many Gram-positive organisms, and the medium thus becomes selective for Gram-negative bacteria.

MacConkey's agar is a selective medium used for the detection and isolation of coliforms. In this case the bile salts act as the selective agent, intestinal and coliform organisms being inhibited to a lesser extent than other organisms. Part of this selectivity derives from the surfactant properties of the bile salts. Other media designed to be selective for the same organisms contain, instead of bile salts, a synthetic surfactant (e.g. sodium dodecylsulphate) to achieve a similar selectivity, without the variability that can occur between batches of a naturally derived material such as bile salts.

7.5.3.2 Selective media for Gram-positive bacteria

Potassium tellurite, thallium acetate and sodium azide added to media to give a final concentration of 1:2000 to 1:10 000 have been found to inhibit the growth of Gram-negative bacteria, and these substances are therefore frequently used in selective media for Gram-positive bacteria. Glucose azide broth, for example, is used in the detection of faecal streptococci in water supplies. Similarly, thallium acetate in a glucose agar has been useful for the isolation of lactococci from sour milk, and potassium tellurite for the isolation of *Corynebacterium*.

7.5.4 Differential media

A differential medium is one in which certain species produce characteristic colonies which can easily be recognized. For example, haemolytic and non-haemolytic species can be distinguished by the examination of colonies formed on blood agar, a non-selective medium. In many cases, however, a medium may be both selective and differential. For example, lactose-fermenting coliforms produce red colonies on MacConkey's agar (as a result of acid production affecting the neutral red indicator), whereas non-lactose-fermenting intestinal organisms such as *Salmonella* spp. produce colourless colonies (see comments in Section 8).

7.6 EXAMPLES OF SELECTIVE CULTURE MEDIA

7.6.1 MacConkey's broth (bile salt broth)

This is a combined selective and differential medium used in the enrichment and detection of coliforms (i.e. the lactose-fermenting Enterobacteriaceae) in, for example, milk and water.

Ingredients

Peptone	20 g
Bile salts (Oxoid L55)*	5 g
Sodium chloride	5 g
Lactose	10 g

*Different proprietary preparations of bile salts may not have equivalent activity at the same concentration.

Bromcresol purple (a 1% ethanolic solution)	1 ml [†]
Distilled water	1000 ml

Preparation

1. Add all the ingredients except lactose and heat in a steamer for 1–2 h.
2. Add the lactose and dissolve by steaming for a further 15 min.
3. Cool and then filter.
4. Adjust the reaction to pH 7.4 with N sodium hydroxide.
5. Add 1 ml of 1% ethanolic solution of bromcresol purple.
6. Distribute in 5-ml amounts in 150 × 16-mm test-tubes provided with Durham tubes.
7. Sterilize in the autoclave at 121°C for 15 min.

7.6.2 MacConkey's agar (bile salt neutral red lactose agar)

This is of general use in the detection and isolation of members of the Enterobacteriaceae.

Ingredients

Peptone	20 g
Bile salts (Oxoid L55)*	5 g
Sodium chloride	5 g
Lactose	10 g
Neutral red (a 1% aqueous solution)	7 ml
Agar	20 g
Distilled water	1000 ml

Preparation

1. Dissolve the peptone, bile salts and sodium chloride in the water by steaming.
2. Cool and adjust the pH to 7.4–7.6 with N sodium hydroxide.
3. Add the agar and dissolve in the autoclave at 121°C for 15 min.
4. Prepare a filter in a Buchner-type funnel (see Section 7.4.2 above) and warm with hot water.
5. Filter the medium while hot.
6. Check the pH and adjust to 7.4 if necessary.

*Different proprietary preparations of bile salts may not have equivalent activity at the same concentration.
[†]Recipes used formerly for milk testing specified the addition of 2.5 ml of a 1.6% solution of bromcresol purple in place of the amount indicated above.

7. Add the lactose and 7 ml of neutral red solution.
8. Dissolve in the steamer.
9. Distribute in bottles and test-tubes as required.
10. Sterilize in the autoclave at 115°C for 15 min or other relevant time.

On this medium, after 24 h incubation, colonies of lactose-fermenting organisms are pink or red, whereas those of non-lactose-fermenters are colourless. Typical appearances are as follows: *Escherichia coli* – red, non-mucoid; *Klebsiella aerogenes* – pink, mucoid; *Salmonella* – colourless.

Modifications of MacConkey's agar

The basic MacConkey's agar medium given above may be modified to give improved selectivity for, and definition of, coliform colonies. In the medium prepared in dehydrated form by Difco as Bacto MacConkey Agar and by Oxoid as MacConkey Agar No. 3 the modification consists of:

(a) the addition of crystal violet at 0.001 g per litre (1 p.p.m.); and
(b) by including a more refined and more effective bile salt[‡] at lower concentration.

In *violet red bile agar* (also available in dehydrated form) the MacConkey's agar has been modified by:

(a) lowering the concentrations of peptone and bile salt (a more refined and more effective preparation[‡] is used);
(b) the addition of crystal violet to $0.002\,g\,l^{-1}$ (2 p.p.m.); and
(c) the addition of yeast extract.

In the pour-plate technique, many workers prefer to cover this medium when poured and set with a further 5 ml of melted medium at 50°C, so as to restrict surface colony formation, this being particularly advisable if the medium is being used for the quantitative estimation of coliforms. Submerged coli-aerogenes colonies appear dark red and are usually 1–2 mm in diameter.

7.7 HEAT STERILIZATION OF MEDIA AND DILUENTS

The constituents of media, and especially agar media, should be dissolved in the water with stirring, before heating to sterilize.

[‡]Different proprietary preparations of bile salts may not have equivalent activity at the same concentration.

As mentioned in Section 2.3, the autoclave time required to achieve sterilization depends on the initial microbial concentration, the size of the containers, the volume of material in each container, and the spacing of the containers in the autoclave. In the preparation of media, the concentration of contaminating microorganisms will usually be very low, so the principal variables that influence the autoclave time will be the others just mentioned. The autoclave time is required normally to achieve a 'holding time' of 15 or 20 min at 121°C, this time beginning when the 'heating centre' of the load reaches 121°C. This should be judged by the use of a thermocouple placed in the load, inserted into the centre of the innermost container. A 100-ml volume in a bottle may require about 10 min to reach 121°C if the bottle or flask is located on its own in the autoclave; if a number of such bottles are closely packed in a wire basket, the contents of the bottle may require 20 min to reach 121°C. A large volume of medium or diluent will also require longer; for example, 1 litre in a bottle may require 20 min or more to reach 121°C and if a number of such bottles are packed in a wire basket, the contents of the container at the centre may not reach 121°C for 30 min or more. Obviously the only sure way of autoclaving media and diluents for sterility is by use of thermocouples placed in the load.

When media are distributed in large volumes (1 litre or more) there will be some thermal degradation as a result of the extended heating time required to achieve sterilization. For this reason it is always preferable, after heating to dissolve the ingredients, to dispense the batch of medium into relatively small volumes (e.g. 100 ml in screw-capped 'medical flat' bottles), before sterilizing by autoclaving.

8

Sampling Methods for the Selection and Examination of Microbial Colonies

8.1 NON-SELECTIVE MEDIA

When samples of material containing mixed populations, such as milk, soil, water or foods, are plated on to non-selective media for an estimate of numbers, it is frequently also required that an estimate should be made of the relative numbers of the different kinds of microorganisms that develop on the plates. It is then possible to calculate the percentage distribution of the various organisms present in the original sample. Representative samples of colonies cannot be secured by uncontrolled picking from Petri dishes and various methods have been devised to ensure that a truly random sample of colonies is obtained. In this exercise it should be appreciated that a minor, but still possibly important, component may be overlooked completely. For example, amongst a population of 10^9 pseudomonads there may be 10^7 *Aeromonas*, but on average 100 colonies would have to be examined to reveal one *Aeromonas*. The particular method selected will depend on the numbers of colonies required for examination and on the numbers of colonies present on the isolation plates. Some of the methods in use are as follows.

Method 1

Select every colony on a plate. This is a suitable method provided the numbers present are roughly equivalent to those required for examination.

Method 2

Select every colony occurring either in a single sector or in opposite sectors of a plate. Various methods for marking off sectors are available:

1. The sectors (e.g. quadrants) may be drawn with felt-tipped pen on the surface of the plate.

2. The plate may be superimposed on a Harrison's disc, and the appropriate sectors or colonies marked. Harrison's disc was devised by Harrison (1938) during a study of the numbers and types of bacteria occurring in cheese. It provides a convenient method of obtaining a representative sample when only a few colonies can be studied (Fig. 8.1).
   The Petri dish to be examined is placed concentrically on the disc. First, all colonies occurring in the areas marked 1 are selected for examination. If this does not provide a sufficient number, all colonies occurring in the areas marked 2 are selected. The selection can be continued if necessary into areas 3 and 4, until the required number is obtained. Once a series of numbered sectors has been started, selection must continue until all colonies in this series have been marked off. All colonies lying over lines should be ignored.

Method 3: replica plating

If the plates have been surface inoculated, all the colonies can be transferred by a replicator device. The simplest type of replicator is a sterile velveteen pad as described by Lederberg and Lederberg (1952). A suitable cylinder (e.g. of alloy) of a diameter slightly smaller than a Petri dish is used. Over one end of the cylinder is stretched a piece of sterile velveteen (when autoclaving, reduce the pressure rapidly to ensure drying of the velveteen). This can be pressed lightly on to the surface of the medium bearing the colonies, and then pressed successively on plates of appropriate sterile media and finally on to a control medium to check successful transfer. Usually transfers to nine plates plus a control can be made. In this way every colony can be examined simultaneously for a number of physiological characteristics. If the velveteen is to be reused, it should be discarded after use into a container of water. It is then decontaminated by autoclaving, washed, rinsed, dried and brushed, before sterilizing by autoclaving.

8.2 SELECTIVE AND DIFFERENTIAL MEDIA

Most selective media also have some differential capability to distinguish between the target organisms (those being sought) and other organisms capable of growth on the media. It is recommended that for confirmatory tests a number of presumptive colonies (colonies displaying the differential characteristics typical of the target organism) be chosen. The standard methods of the International Organization of Standardization (ISO methods) using selective media followed by confirmatory tests usually specify that five colonies be chosen from a plate.

It should be noted that this procedure of choosing only 'typical' colonies for subsequent confirmation and identification may lead to atypical strains being overlooked. For example, sucrose-fermenting and/or lactose-fermenting salmonellae are known to occur. Choosing only non-fermenters from media containing lactose and/or sucrose will overlook such strains of *Salmonella*, and their true significance and frequency of occurrence will be impossible to determine.

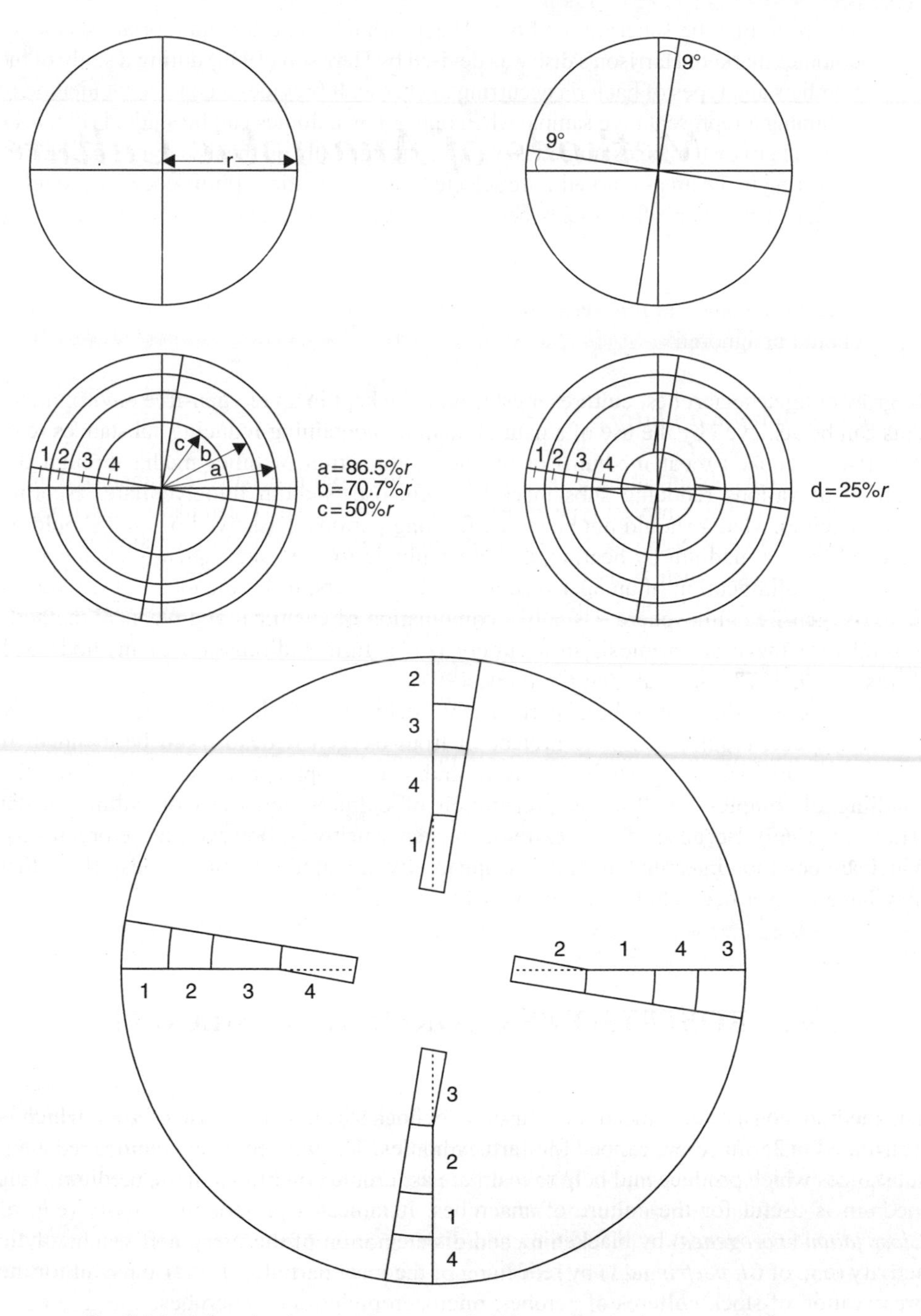

FIG. 8.1 Drawing a Harrison's disc.

9

Methods of Anaerobic Culture

To grow obligate anaerobes, cultures must usually be kept in an oxygen-free environment. This can be achieved by the use of a natural medium containing reducing substances (e.g. Robertson's cooked meat medium) or by the modification of other media through the addition of various reducing substances (e.g. glucose, sodium thioglycollate, ascorbic acid, cysteine). Media should not be stored for long periods. Dissolved oxygen should be removed from a medium by heating, and atmospheric oxygen subsequently excluded by sealing, by solidification of an agar medium in deep layers, or by incubating the cultures in an oxygen-free atmosphere. Usually a combination of chemical and physical methods is used to achieve anaerobiosis in a culture (for a further discussion of methods, see Willis, 1969, 1977; Breznak and Costilow, 1994).

Some very strict anaerobes, particularly members of the asporogenous family Bacteroidaceae, require complete protection from oxygen if death is to be avoided. If such organisms are being studied it is desirable to adopt special techniques for the handling of samples and for the preparation of cultures and even of culture media (Hungate, 1969). Because of their extreme oxygen sensitivity, however, these organisms, which are common intestinal inhabitants, apparently die rapidly in most foodstuffs so that they have not yet been studied extensively by food microbiologists.

9.1 ROBERTSON'S COOKED MEAT MEDIUM

The medium consists of minced meat (usually bullock's heart) in a nutrient broth, which is distributed in 25-ml screw-capped McCartney bottles. The minced meat contains reducing substances which produce and help to maintain anaerobic conditions in the medium. This medium is useful for the culture of anaerobes. It indicates proteolytic activity (e.g. of *Clostridium sporogenes*) by blackening and disintegration of the meat, and saccharolytic activity (e.g. of *Cl. perfringens*) by reddening of the meat particles. It is also useful for the preservation of stock cultures of aerobes, microaerophiles and anaerobes.

Although this medium is available in dehydrated form, it has been the author's experience that such preparations are not as reliable for the cultivation of anaerobes as cooked meat medium prepared in the laboratory.

Preparation (Lepper and Martin, 1929)

Mince 500 g of fresh, fat-free bullock's heart and simmer for 20 min in 500 ml of boiling 0.05 N sodium hydroxide. After cooking, adjust the pH to 7.4. Strain off the liquid and dry the meat by spreading on filter paper. Distribute the meat in 25-ml McCartney bottles to a depth of 2 cm. Add 10 ml of peptone water or nutrient broth. Sterilize by autoclaving for 20 min at 121°C with the screw caps slightly loose. After autoclaving, tighten the caps. Immediately before use boil for 10 min to remove dissolved oxygen and cool rapidly.

9.2 SHAKE CULTURES

These are test-tubes or bottles containing a solid medium. The melted medium is inoculated and then allowed to solidify. The simplest medium to use is nutrient agar with the addition of 1% glucose.

1. Liquefy the medium and maintain at 100°C for 10 min to drive off any dissolved oxygen. Cool to 45°C and use immediately.
2. Add the inoculum and mix the contents of the tube well by rotation between the hands, avoiding the introduction of air bubbles.
3. Solidify the agar by placing the tube in cold water.
4. Incubate at the required temperature.

Since oxygen will diffuse only very slowly through the solid medium, the conditions in the lower parts of the tube will be suitable for the growth of anaerobes. This method is also suitable for the growth of micro-aerophiles and facultative anaerobes.

9.3 SEMI-SOLID MEDIA

Semi-solid media contain 0.02–0.3% agar, which is sufficient to prevent convection currents and therefore helps to prevent the diffusion of oxygen into the medium.

One of the most suitable media is prepared from a nutrient medium (e.g. tryptone soya broth or nutrient broth) with the addition of 0.5% glucose (a reducing compound as well as being an energy source), 0.1% sodium thioglycollate and 0.02–0.2% agar to prevent convection currents, dispensed in McCartney bottles or in deep layers in test-tubes. As an indicator of oxidation–reduction potential 0.0002% methylene blue (0.2 ml of a 1% aqueous solution of methylene blue per litre of medium) can be added.

The additional substances should be added to the nutrient broth, mixed well and steamed until dissolved. The medium is then distributed in 12-ml amounts in 150 × 16-mm test-tubes, which are plugged and sterilized by autoclaving at 121°C for 15 min. The methylene blue acts as an indicator of redox potential, being blue in the oxidized and colourless in the reduced condition. Should the medium show signs of turning greenish-blue after storage, it should be reheated before use to remove the oxygen, 10 min in a boiling water bath usually proving sufficient.

9.3.1 Reinforced clostridial medium (RCM)

This is a semi-solid medium developed by Hirsch and Grinsted (1954) which, although it does not allow the growth of all obligate anaerobes, is useful for the enumeration of anaerobes in, for example, food samples. RCM contains glucose and cysteine as reducing compounds and, when distributed into 25-ml screw-capped McCartney bottles, provides anaerobic conditions without the need for a paraffin or other seal. It is particularly recommended that RCM be used also as the *diluent* in the determination of viable counts of anaerobes.

9.4 VASELINE, PARAFFIN WAX AND AGAR SEALS

By the use of sterile sealing compounds, liquid media can be maintained anaerobically. The tubes containing the liquid medium should be heated in boiling water for 10 min to drive off the dissolved oxygen and sterile melted Vaseline or paraffin wax added to each tube to seal the medium from the air. The tubes are inoculated by means of a capillary pipette after melting the Vaseline. (Alternatively, the tubes can be inoculated immediately after heating and cooling, and *then* the seal added.)

This method can be used to demonstrate the presence of *Clostridium perfringens* in water or milk by the 'stormy-clot' reaction, in which the medium is sterile milk rendered anaerobic and the surface sealed as above.

Vaspar is a useful sealing compound prepared by melting together approximately equal amounts of petroleum jelly (Vaseline) and paraffin wax. After adding the molten Vaspar and allowing it to solidify, the outside of the tube in the area of the Vaspar layer should be warmed *gently* in a Bunsen flame to remelt the Vaspar to eliminate any air channels next to the glass surface. Sterile liquid paraffin can also be used, but it will not allow the detection of gas production, and there is some diffusion of oxygen across a liquid paraffin seal.

9.5 THE ANAEROBIC JAR

Petri dishes or tubes containing any medium to be incubated in an oxygen-free atmosphere for the isolation and growth of anaerobes can be enclosed in an anaerobic jar.

Such a jar also enables viable counts of anaerobes to be performed more readily than by other methods. Counts of the less exacting anaerobes (e.g. *Cl. perfringens*), can often be carried out without the necessity for using anaerobic jars by employing deeply poured plates or shake tubes using suitable media.

Anaerobic jars to be used with external gas supplies have a lid carrying an inlet valve and an outlet valve, and preferably also have a built-in manometer gauge. The air in the anaerobic jar may be replaced by oxygen-free nitrogen. Alternatively the anaerobic jar makes use of the catalytic non-explosive reaction between hydrogen and oxygen to remove free oxygen from the atmosphere inside the jar, leaving the cultures in an oxygen-free atmosphere composed largely of hydrogen. There are two types of jar available: one uses a cold palladium catalyst to bring about the reaction; the other (known as McIntosh and Fildes' jar) employs a hot platinum catalyst. The catalyst in the McIntosh and Fildes' jar is heated by the passage of an electric current, the heating coil being surrounded by wire gauze to prevent the occurrence of an explosive reaction between the hydrogen and oxygen. To minimize further the risk of explosion, it is advisable to remove most of the air from the jar with a vacuum pump before passing in hydrogen. The area in which the jars are being filled with hydrogen should have suitable safety screening. Such safety screens can be obtained from many laboratory furnishers and suppliers (e.g. Fisher Scientific, A. Gallenkamp). Windows and partitions of ordinary glass in existing laboratories can be rendered safe by the application of self-adhesive polyester film.

While the cold catalytic jar carries a lower risk of explosion, it has the slight disadvantage of needing occasional renewal of the catalyst because the catalyst gradually becomes 'poisoned'. Nevertheless, the advantages of the cold catalytic jar – its safety, and the fact that the reaction will continue between hydrogen and oxygen as long as both gases are present, rather than be terminated when an electric current is removed – outweigh this one slight disadvantage. The following procedures therefore describe the use of a cold catalytic jar.

9.5.1 Procedure using external hydrogen supply

Place the cultures inside the jar. If Petri dishes are included, separate the two halves of each dish slightly with a narrow strip of filter paper to prevent sealing by condensed water.

In addition to the cultures, also include a redox indicator. The indicator can consist of a tube of the thioglycollate medium described in Section 9.3. Alternatively, an alkaline glucose solution containing methylene blue can be prepared by mixing in a plugged tube equal quantities of 0.006 N sodium hydroxide, a 0.015% aqueous solution of methylene blue, and a 6% solution of glucose. This requires boiling until it becomes colourless and then it is immediately placed inside the anaerobic jar.

Next, clamp the lid in position. If the jar and lid are not fitted with a rubber or silicone-rubber gasket, the rim of the jar should be first smeared with high-vacuum silicone grease.

Connect one of the two taps in the lid of the jar to a filter pump fitted with a vacuum gauge. The hydrogen supply must be connected to the jar via a variable reducing valve with pressure gauge and fine adjustment, which allow the hydrogen to be delivered at a pressure of $14\,\mathrm{kN\,m^{-2}}$ or less. If a wash-bottle is included in the line between the reducing valve and the anaerobic jar, passage of the hydrogen may be observed. An alternative to using a hydrogen cylinder is Kipp's apparatus.

First apply suction with the filter pump until the jar is evacuated to a pressure of approximately 5 cmHg and then close the tap. Next, adjust the flow of hydrogen through the wash-bottle to approximately one bubble per second and connect to the second tap of the anaerobic jar; open the tap and allow the hydrogen to pass slowly into the jar for 5 min after the pressure inside the jar has reached atmospheric pressure. Close the tap, remove the hydrogen supply and vacuum line, and incubate at the desired temperature. The indicator should decolorize after a few hours, and remain colourless until the jar is opened at the end of the incubation period. Failure of the indicator to decolorize may be due to a leak through side arm, lid gasket or taps, or to poisoning of the catalyst.

Usually, it will be found after incubation that there is a slight negative pressure inside the jar, which should be released by opening one of the taps.

9.5.2 The gas-generating envelope

This system, manufactured, for example, by Baltimore Biological Laboratories (BBL) as GasPak® and by Oxoid as the Oxoid Gas Generating Kit, provides a hydrogen supply within the anaerobic jar by the use of a disposable foil envelope. The envelope in one section contains sodium borohydride, which generates hydrogen when exposed to water, and in a second section a mixture of citric acid and sodium bicarbonate which generates carbon dioxide when water is added. This system was described by Brewer and Allgeier (1966).

The hydrogen is reacted with the oxygen present in the air within the jar by using a cold catalyst. However, the jar is not evacuated first, and because of the high oxygen content of the jar a palladium catalyst can become very hot, so it is important to use a double layer of wire gauze to hold the catalyst (preventing flash-over, on the principle of the Davey lamp). Therefore, gas-generating envelopes should not be used in a traditional anaerobic jar unless the catalyst container is suitably modified. The GasPak polycarbonate jar offers a distinct advantage over other types of jar, of automatic venting. However, a methylene blue indicator *must* be used to monitor catalyst activity as the lack of gas inlet and outlet prevents the use of a manometer, and also incidentally means that the jar cannot be used with external gas sources. The Oxoid Anaerobic Jar follows a different approach, with valves and manometer fitted so that jars can be evacuated and/or used with external gas sources if so desired.

When anaerobic counts are to be performed infrequently, the GasPak system is convenient, as although the system is more expensive, it takes up very little space, does not require special precautions against fire hazards, and cylinder hire charges are avoided. For occasional anaerobic incubation of very few plates, a number of manufacturers produce disposable gas-impermeable bags with gas-generating envelopes (e.g. 'Generbag anaer' from bioMérieux, Anaerocult® from Merck, and GasPak Pouch™ from BBL).

It is unfortunate that the manufacturers do not offer hydrogen-only envelopes. Although many anaerobes are stimulated by carbon dioxide concentrations of up to 5 or even 10%, this is not an invariable rule. For example Futter (1967) found that the presence of 5% carbon dioxide suppressed the recovery rate for *Clostridium perfringens* spores to 53% compared with the corresponding carbon dioxide-free anaerobic atmosphere.

9.5.3 Replacement by nitrogen

After placing the cultures in an anaerobic jar (which possesses gas outlets) and clamping down the lid, apply suction with a filter pump until the jar is evacuated to a pressure of 4 cmHg and then close the tap.

Fill the jar with oxygen-free nitrogen to atmospheric pressure. Repeat evacuation and refilling twice more.

The level of residual oxygen (i.e. derived from the original air in the jar) after three cycles of evacuation and replacement by nitrogen will be approximately 0.003%. Futter (1967) reported that commercially available 'oxygen-free nitrogen' may contain 0.3% v/v oxygen, but this can be removed by passing the gas over heated copper turnings.

9.5.4 Selection of the method to be used

Which of these three methods is to be employed will depend partly on the probable frequency of use, the size of laboratory, the ability to take adequate safety precautions if hydrogen cylinders are to be used, and the proximity to the laboratory of a supplier of hydrogen or nitrogen cylinders.

In addition there may be microbiological requirements to be met. For example, the nitrogen replacement method is unsuitable for very oxygen-sensitive anaerobes, but can be used with the more aerotolerant species. At present comparatively little is known about the growth responses of most anaerobes to the different gaseous atmospheres that result from these three methods. Futter (1967) and Futter and Richardson (1970, 1971) found that the effect of different hydrogen/nitrogen mixtures on the recoverability of *Cl. perfringens* spores depended on the previous treatment of the spores. Recoverability of undamaged spores was independent of the hydrogen/nitrogen mixture used. Spores that had survived heat treatment showed poorest recovery (50–60%) in a 1:1 hydrogen/nitrogen mixture, and good relative recovery in 100% hydrogen or nitrogen. In the case of spores surviving either ethylene oxide treatment or γ-irradiation, progressively lower recoveries were obtained as the hydrogen content of the mixture was increased from 0 to 100%. These results suggest that, if *Cl. perfringens* is being studied in heat-processed foods, gas cylinders, or non-catalyst oxygen-absorbing envelopes, or nitrogen replacement should be used in preference to gas-generating envelopes used with catalysts, as the last will in normal use result in a nitrogen/hydrogen/carbon dioxide mixture of uncertain composition. However, Futter and Richardson's work does not seem to have been explored further.

10

Cultivation in Microaerobic and Carbon Dioxide-enriched Atmospheres

10.1 CULTIVATION IN MICROAEROBIC ATMOSPHERES

Campylobacter is a microaerophile, and is grown in atmospheres containing 5–10% oxygen and about 10% carbon dioxide (see Section 21.8). For isolation of some other bacteria, a similar atmosphere may be required (see Section 21.9).

10.1.1 Use of an external gas supply

The required atmosphere can be obtained from an external cylinder containing the required gas mixture, using an anaerobic jar with inlet and outlet valves and manometer, but no catalyst. Partially evacuate the jar to a pressure of about 260 torr (or mmHg) (about 35 kN m^{-2}). Restore to atmospheric pressure by passing in a gas mixture of 95% nitrogen and 5% carbon dioxide (Skirrow, 1982).

10.1.2 Use of gas-generating or oxygen-absorbing envelopes

There are two types of system available. One system is based on a hydrogen/carbon dioxide-generating envelope, to be used in a jar *of the correct size*, with a cold catalyst. The amount of hydrogen generated is designed to result in an *incomplete* removal of the oxygen present, so that the final atmosphere can support the growth of microaerophiles. Examples of this are the Oxoid Campylobacter Gas Generating Kit (BR56 for use in the Oxoid 3.4-litre jar, or BR60 for use in a 2.5-litre jar), and the BBL Campypak®. The BBL Campypak Plus™ incorporates an integral catalyst so that the problem of 'poisoning' of a permanent in-jar catalyst does not arise.

The other system is based on an envelope containing a substance that will combine with oxygen (e.g. ascorbic acid or finely divided iron) and a carbon dioxide generator. The amount of oxygen-combining compound must also be carefully designed for the volume of container, so that an appropriate concentration of oxygen results. Examples of this system are CampyGen (Oxoid), Campy Pouch® (BBL), Anaerocult C (Merck) and 'Generbag microaer' (bioMérieux). These *must* be used with a container with the volume for which they are designed.

10.2 CULTIVATION IN A CARBON DIOXIDE-ENRICHED ATMOSPHERE

A few bacteria are capnophilic, that is they will grow *only* or grow *better* in an atmosphere that contains a high concentration of carbon dioxide. Some anaerobes and microaerophiles also prefer or require an enhanced concentration of carbon dioxide, and the methods described in Sections 9 and 10.1 are usually designed to provide for this.

It is relatively easy to provide a carbon dioxide-enriched aerobic atmosphere in the laboratory. Place the Petri dish or test-tube cultures in a 'half-size biscuit tin' (approximately 22 × 20 × 12 cm) or container of similar volume, together with a beaker containing about 15 ml of 2 N hydrochloric acid. Drop into the acid a marble chip of 1–1.2 g and put the lid in place. Seal the lid with a PVC tape. The marble will react with the hydrochloric acid to give a concentration of about 5% carbon dioxide inside the tin.

Alternatively carbon dioxide-generating envelopes are available (e.g. Oxoid, BBL, bioMérieux).

11

Biochemical Tests for Identification of Microorganisms

The tests described in this section represent a fairly limited selection of general usefulness in the characterization and identification of bacteria. The biochemical mechanisms involved in many of these tests have been discussed by MacFaddin (1980). Some organisms require the use of a medium specially adapted to particular nutritional or osmotic requirements. In other cases, a selective medium may be employed which not only allows the isolation of the organism under investigation but also incorporates one or more biochemical tests. Thus, where necessary, the appropriate special media are described in the various sections of Parts II and III of this manual.

It is important to check that each batch of a medium is satisfactory by using a strain of bacterium known to give a positive reaction. The ninth edition of *Bergey's Manual of Systematic Bacteriology* (particularly Volumes 1 and 2: Krieg *et al.*, 1984; Sneath *et al.*, 1986) lists the reactions of many type strains, and the sources of suitable cultures. In addition, where appropriate, incubate an uninoculated tube or plate with each test in order to detect false-positive reactions due to impurities in, or a deterioration of, medium or reagents.

11.1 REACTIONS INVOLVING PROTEIN, AMINO ACIDS AND OTHER NITROGEN COMPOUNDS, INCLUDING TESTS FOR PROTEOLYTIC ACTIVITY

11.1.1 Hydrolysis of gelatin

(a) Nutrient gelatin

Medium: Nutrient broth with the addition of 10–15% gelatin, final pH 7.2. Sterilize by autoclaving for 20 min at 115°C.

Stab inoculate a tube of nutrient gelatin and incubate at 20–25°C for up to 30 days. If growth is poor at 25°C, tubes may be incubated at the optimum temperature.

Recording result: Liquefaction of the test medium when an uninoculated tube has remained solid indicates that hydrolysis has occurred. Record the shape and extent of the liquefied portion of the gelatin. When tubes have been incubated above 25°C, the culture must be immersed in iced water for 5 min before being examined for hydrolysis.

(b) Frazier's gelatin agar (modified) (Smith *et al.*, 1952)

Medium: Nutrient agar plus 0.4% gelatin, final pH 7.2. Sterilize by autoclaving for 20 min at 115°C.

Inoculate a poured, dried plate of the medium by streaking *once* across the surface or placing a spot of inoculum and incubate at the optimum growth temperature for 2–14 days.

Test reagent: 1% Hydrochloric acid. Hydrochloric acid does not produce as quick and as clear a precipitation as the mercuric chloride solution originally described, but it is recommended here in order to avoid the toxic and environmental hazard of using a mercuric salt.

Recording result: Flood plates with 8–10 ml of test reagent. Unhydrolysed gelatin slowly forms a white opaque precipitate with the reagent. Hydrolysed gelatin appears therefore as a clear zone. Record the width of the clear zone in millimetres from the edge of a colony to the limit of clearing.

11.1.2 Hydrolysis of casein

Medium: Milk agar, which consists of agar with the addition of skim-milk to 10%, sterilized by autoclaving for 20 min at 115°C. A more opaque milk agar medium may be made by mixing 10 ml of hot sterile 2.5% water agar with 5 ml of hot sterile skim-milk (giving a 30% milk agar) immediately before pouring the plates. This is best used in a double-layer plate, with a thin layer of the 30% milk agar overlaid on to 10 ml of water agar previously poured and set.

Inoculate a poured dried plate of the medium by streaking *once* across the surface and incubate at the optimum growth temperature for 2–14 days.

Test reagent: 1% Hydrochloric acid or 1% tannic acid solution.

Recording result: Clear zones which are visible after incubation of the plates are presumptive evidence of casein hydrolysis. However, false positives may occur; to confirm that clearing is a result of casein hydrolysis, flood the plates with 1% hydrochloric acid which is a protein precipitant. Record width of clear zone in millimetres. (See Smith *et al.*, 1952; Chalmers, 1962.)

11.1.3 Hydrolysis of coagulated serum

Medium: Loeffer's serum, consisting of three parts serum plus one part glucose nutrient broth. As comparatively low temperatures are used in the preparation of this medium, the glucose nutrient broth should be sterilized by autoclaving before mixing with sterile serum in a sterile flask, and then the medium distributed aseptically into sterile small (5 ml) screw-capped bottles. The medium is then coagulated in the sloping position by heating slowly to 85°C (see also Section 2.3.2.4). The medium may be sterilized by heating in the inspissator at 85°C for 20 min on each of three successive days.

Inoculate the serum slope by streaking across the surface in the usual way and incubate at the optimum growth temperature for 2–14 days.

Recording result: Visual examination will indicate whether hydrolysis (liquefaction) of the coagulated serum has occurred.

11.1.4 Production of indole from tryptophan

Medium: Tryptone water (tryptone 1–2%, sodium chloride 0.5%, final pH 7.2) dispensed in 5-ml amounts in test-tubes and sterilized by autoclaving for 15 min at 121°C. Tryptone is the peptone to be recommended for this test, as it is rich in tryptophan, one of the amino acids destroyed by the usual methods of preparing peptones. Inoculate as for a broth culture, preferably from young agar slope cultures, and incubate at the optimum growth temperature for 2–7 days.

Test reagent: Several reagents and methods are available for the detection of indole. Kovacs' reagent is to be preferred, being rather more sensitive due to the higher solubility of the dye complex in the pentanol layer.

Recording result: Add 0.5 ml Kovacs' indole reagent, shake tube gently and then allow to stand. Tryptophan can be metabolized to indole, skatole (methylindole) or indoleacetate. In the presence of indole a deep red colour develops which separates out in the alcohol layer (see Report, 1958). A red or pink lower (aqueous) layer may be indicative of the presence of skatole. Only a red or pink *upper* layer should be recorded as an indole-positive result.

11.1.5 Production of ammonia from peptone or arginine

(a) Peptone water

Inoculate a tube of peptone water (peptone 1%, sodium chloride 0.5%) and incubate with a sterile control tube at the optimum growth temperature for 2–7 days.

Test reagent: Nessler's reagent.

Recording result: Add a loopful of culture to a loopful of Nessler's reagent on a slide or glazed porcelain tile, or add 1 ml of culture to 1 ml of Nessler's reagent in a clean tube. The development of an orange to brown colour indicates the presence of ammonia. The sterile control tube should be tested at the same time and should turn pale yellow or show no colour reaction.

(b) Arginine broth (Abd-el-Malek and Gibson, 1948)

This medium is used mainly for the differentiation of streptococci. Inoculate a tube of arginine broth and incubate with a sterile control tube at the optimum growth temperature for 2–7 days.

Test reagent and recording result: As for method (a) (peptone water).

(c) Thornley's semi-solid arginine medium (Thornley, 1960)

This medium contains phenol red as a pH indicator; it should be dispensed in 5-ml (bijou) screw-capped bottles to a depth of about 2 cm. It is used mainly for the differentiation of Gram-negative rods. Inoculate a bottle of medium and then seal the surface with sterile liquid paraffin or Vaspar. Incubate with a sterile control at the optimum growth temperature for 2–7 days.

Recording result: Hydrolysis of arginine, with the formation of ammonia, results in alkalinity and is indicated by a change in colour of the medium to red from salmon pink. *Pseudomonas* and *Aeromonas* produce ammonia from arginine in the sealed medium, whereas members of Enterobacteriaceae do not. The former can grow *anaerobically*, by being able to use arginine for the generation of ATP without needing oxygen.

11.1.6 Decarboxylation of lysine, ornithine or arginine

Medium: The usual medium for the detection of these amino acid decarboxylases is Møller's medium. This contains a small concentration of glucose to initiate growth, a larger concentration of the selected amino acid, and a pH indicator, with anaerobic conditions provided by a sterile paraffin layer. After an initial lowering of pH by acid produced from the glucose, metabolism of the amino acid leads to a pH reversion.

Lightly inoculate the organism, preferably with a straight wire, into a series of four tubes of Møller's decarboxylase medium. The series consists of a control (no added amino acid), medium plus lysine, medium plus ornithine, and medium plus arginine. Ensure that the wire penetrates beneath the layer of liquid paraffin. Incubate at 25°C and examine daily for up to 7 days.

Reading the results: A positive result is indicated by the colour of the medium changing to violet after an initial change to yellow. Controls and negative reactions are yellow in colour. In the case of a positive reaction in the arginine-containing medium, if ammonia is found in the medium (tested for by the use of Nessler's reagent), and the organism does not possess a urease, the reaction is due to arginine dihydrolase.

11.1.7 Production of hydrogen sulphide

This may result from the decomposition of organic sulphur compounds (e.g. cysteine and cystine) or from the reduction of inorganic sulphur compounds (e.g. sulphite or thiosulphate).

(a) Cystine or cysteine broth

A basal medium of peptone water or nutrient broth is used, with the addition of 0.01% cystine or cysteine. Indicator papers are used which consist of filter paper soaked in saturated lead acetate solution, dried, cut in strips and sterilized.

Inoculate, and then insert a strip of the indicator paper between the plug and the glass with the lower end above the medium. Incubate at the optimum growth temperature for 2–7 days, together with an uninoculated control.

Recording result: Production and liberation of hydrogen sulphide causes blackening of the lead acetate paper. If no blackening has occurred by the end of the incubation period, add 0.5 ml of 2 N hydrochloric acid and replace the plug and lead acetate paper immediately. If any sulphide has been produced but has remained in solution, the addition of the acid will cause the liberation of hydrogen sulphide (Skerman, 1959). Treat the uninoculated control tube similarly to check for possible false positives.

(b) Ferrous chloride gelatin

The medium is prepared by adding freshly prepared 10% ferrous chloride solution to boiling nutrient gelatin to give a final concentration of ferrous chloride of 0.05%, followed by dispensing into sterile narrow tubes, quick cooling and sealing with sterile air-tight (e.g. rubber) stoppers. Inoculate by stabbing. Incubate at 20–25°C for 7 days. Production of hydrogen sulphide is shown by blackening of the medium.

Note. This medium will also indicate liquefaction of gelatin. This is the recommended method for differentiation within the Enterobacteriaceae (see Report, 1958).

(c) Kligler's iron agar

This is a complex medium containing 0.03% ferric citrate and sodium thiosulphate, which may be used for the differentiation of members of the Enterobacteriaceae. It is available in dehydrated form. After sterilization, the medium is slanted with a deep butt

(3-cm butt, 4-cm slant). Stab inoculate the butt of the medium, and streak the slant. Incubate at optimum temperature for up to 7 days. Production of hydrogen sulphide is shown by blackening of the medium. In addition to the detection of hydrogen sulphide production, the medium also contains lactose and glucose for the differentiation of organisms on the basis of sugar fermentation.

Triple sugar iron (TSI) agar has been used instead of Kligler's iron agar to detect hydrogen sulphide production; however, with some hydrogen sulphide-producing strains of *Citrobacter* and *Proteus*, hydrogen sulphide production is masked in TSI agar by sucrose fermentation, so Kligler's iron agar is a more reliable test medium.

11.1.8 Production of ammonia from urea

Christensen's urea agar (Christensen, 1946)

The basal medium is distributed in bottles or test-tubes, heat-sterilized and cooled to 50°C. Sufficient 20% urea solution, previously sterilized by filtration, is then added to give a final concentration of 2%. The medium is slanted, allowed to set, and is then ready for use.

Inoculate as for a slope culture. A control of basal medium containing no added urea should also be inoculated at the same time to check that ammonia is produced from urea and not from peptone. Incubate at the optimum growth temperature for 1–7 days.

Recording result: Urease production and subsequent hydrolysis of urea results in the production of ammonia, which increases the pH as shown by a change in colour of the medium from yellow to pink or red.

11.1.9 Reduction of nitrate

Nitrate peptone water is the medium employed in this test, consisting of peptone water with the addition of 0.02–0.2% potassium nitrate (analytical reagent grade). The medium is distributed in tubes, each with an inverted Durham tube, and sterilized by autoclaving for 15 min at 121°C.

Inoculate as for broth culture and incubate together with a sterile control tube at the optimum growth temperature for 2–7 days.

Test reagents: Griess–Ilosvay's reagents (modified) or commercially available nitrite test strips.

Recording result: Add 1 ml of each of the two reagents to the culture and to the control tube. The presence of nitrite is indicated by the development of a red colour within a few minutes. The control tube should show little or no coloration. Alternatively nitrite test strips may be used.

A negative result should be confirmed by the addition of a small quantity of zinc dust to the tube. This reduces to nitrite any nitrate still present. Thus the development of a red colour indicates that some nitrate remains. If the addition of zinc does not result in the development of colour, no nitrate remains, the nitrate having been reduced by the culture beyond the nitrite stage. The presence of gas in the Durham tube indicates the formation of gaseous nitrogen.

The details of nitrate reduction by bacteria are complex, and interested readers are referred to the review by Payne (1973).

11.1.10 Action on litmus milk

Inoculate as for a broth culture and incubate at the optimum growth temperature for up to 14 days.

Recording result: Examine tubes daily and record any changes in the medium. A number of different reactions and combinations of reactions may occur involving (1) lactose, (2) casein, (3) other milk constituents.

1(a) Acid production shown by a change in the colour of the litmus to pink.
1(b) If sufficient acid is produced the milk will clot. This is known as an acid clot.
1(c) Reduction of the litmus and loss of colour may occur. This may precede or follow other changes.
1(d) Gas may also be produced and show as gas bubbles in the medium, although normally this is visible only if clotting has occurred.

2(a) Coagulation of the milk may occur as a result of proteolytic enzyme activity affecting the casein, the litmus colour remaining blue. This is known as a sweet clot.
2(b) Hydrolysis of the casein as a result of proteolytic enzyme activity causes clearing and loss of opacity in the milk medium, usually referred to as peptonization. Proteolysis may also result in an alkaline reaction due to ammonia production.

3. Utilization of citrate in the milk medium results in the production of an alkaline reaction, shown by the change to a deep blue colour in the litmus medium.

11.1.11 Deamination of phenylalanine

Inoculate a phenylalanine agar slope and incubate at optimum growth temperature (37° or 30°C) for 24 h. Also inoculate a known positive culture (e.g. a *Proteus*) and a known negative culture (e.g. *Escherichia coli*).

Test reagent: Acidic ferric chloride solution.

Recording result: Add four or five drops of acidic ferric chloride solution to the phenylalanine agar slope culture, and gently rotate the tube. A positive reaction is shown by a green colour developing on the surface of the slope and in the syneresis liquid within 1–5 min. (The colour will fade quickly.)

11.2 REACTIONS INVOLVING CARBOHYDRATE AND OTHER CARBON COMPOUNDS

11.2.1 Hydrolysis of starch

Starch hydrolysis may be tested with solid or liquid media, although starch agar is perhaps more convenient. This medium consists of a nutrient agar with the addition of 0.2% soluble starch. The best results are obtained by preparing layer plates which are prepared by pouring 10 ml of nutrient agar into each plate, allowing it to set, and then overlaying this with 5 ml of starch agar.

Inoculate a poured dried plate of the medium by streaking *once* or spot-inoculating across the surface and incubate at the optimum growth temperature for 2–14 days.

Test Reagent: Gram's iodine solution as used for Gram's stain.

Recording result: Flood plates with 5–10 ml of iodine solution. Unhydrolysed starch forms a blue colour with the iodine. Areas of hydrolysis therefore appear as clear zones and are the result of amylase activity. Record the width of any clear zone in millimetres from the edge of the colony to the limit of clearing. Reddish-brown zones around the colony indicate partial hydrolysis of starch (to dextrins).

11.2.2 Production of acids from sugars, glycosides and polyhydric alcohols

To peptone water or other basal medium add 0.5–1.0% of substrate. An indicator is also incorporated in the medium to detect acid production. The indicator may be Andrade's indicator 1%, phenol red 0.01%, or bromcresol purple 0.0025%. Durham tubes are included to detect gas production. Sterilize by steaming for 30 min on three successive days. Substrates that may be excessively decomposed by heat sterilization should be sterilized as 10% solutions by filtration and added aseptically to tubes of previously heat-sterilized basal medium to give the correct final concentration of substrate (see pp. 401–402.)

Incubate at the optimum growth temperature for up to 7 days.

Recording result: Acid production is shown by a change in the colour of the indicator: Andrade's indicator to pink, phenol red to yellow, and bromcresol purple to yellow. Gas, if produced, accumulates in the Durham tube. On continued incubation some organisms can cause a pH reversion as a result of producing ammonia from the peptone. Total utilization of glucose by bacteria that produce very little acid can be detected by the use of glucose-indicating test strips (Park, 1967). The test strip changes colour if glucose is present.

11.2.3 Differentiation of oxidation and fermentation of carbohydrates

For this test Hugh and Leifson's medium (Hugh and Leifson, 1953) is employed. It is sometimes modified for particular groups of organisms, such as lactic acid bacteria (Whittenbury, 1963), and staphylococci and micrococci (see p. 425).

The basal medium, which contains a pH indicator, is dispensed in 5–10-ml amounts in 150 × 16-mm test-tubes and sterilized by autoclaving. The carbohydrate is prepared separately from the basal medium as a 10% solution and sterilized by autoclaving or filtration. The sterile carbohydrate solution is added aseptically to the sterile melted basal medium to give a final concentration of 1%.

For each carbohydrate, stab-inoculate two tubes of medium. Cover the surface of the medium in one tube with sterile liquid paraffin, Vaspar or agar. Incubate at the optimum growth temperature for up to 14 days. One tube only can be used provided it is examined daily.

Recording result: Acid production is shown by a change in the colour of the medium from blue to yellow (or from blue-green to red in the case of the double-indicator version). Fermentative organisms produce acid in both tubes. Oxidative organisms produce acid in the open tube only and usually, or at least initially, only at the surface of the open tube.

11.2.4 Production of carbon dioxide from glucose

The medium used is Gibson's semi-solid tomato-juice medium (Gibson and Abd-el-Malek, 1945). It consists of four parts skim-milk plus one part nutrient agar with the addition of 0.25% yeast extract, 5% glucose and 10% tomato juice, final pH 6.5. The medium is distributed in tubes to give a depth of 5–6 cm. Studies by Stamer and co-workers (1964) indicate that the requirement of lactic acid bacteria for tomato juice may be met by manganese. This may be conveniently added as manganese sulphate (1–10 ml of 0.4% $MnSO_4 \cdot 4H_2O$ solution) to give a final concentration of 1–10 p.p.m. Mn^{++} in place of tomato juice.

Dissolve the medium by heating at 100°C until molten, then cool to 45°C. Inoculate with approximately 0.5 ml of young broth culture, mix by rotation of the tube, then cool in tap water. When set, pour into the tube molten nutrient agar at about 50°C to give a layer 2–3 cm deep above the surface of the medium. Incubate at the optimum temperature for up to 14 days.

Recording result: The semi-solid medium and agar seal trap any carbon dioxide gas produced in the medium. This is shown by disruption of the agar seal and by the presence of gas bubbles in the medium. It may be necessary to place the tube in hot water to release the gas.

11.2.5 Methyl red test

The medium used is glucose phosphate broth.

After inoculation, incubate at the optimum growth temperature for 2–7 days.

Test reagent: Methyl red solution (0.1 g of methyl red in 300 ml of 95% ethanol, made up to 500 ml with distilled water).

Recording result: Add about five drops of the indicator to 5 ml of culture. A red colour, denoting a pH of 4.5 or less, is described as positive. A yellow coloration is recorded as negative.

11.2.6 Voges–Proskauer test

This is a test for the production of acetylmethylcarbinol from glucose. To the inoculated medium after incubation, alkali is added, in the presence of which any acetylmethylcarbinol present becomes oxidized to diacetyl. The diacetyl will combine with arginine, creatine or creatinine, to give a rose coloration. The medium used is glucose phosphate broth, as in the methyl red test or other suitable medium.

After inoculation, incubate at the optimum growth temperature for 2–7 days.

Test reagents – Barritt's modification (Barritt, 1936): A 5% ethanolic solution of α-naphthol and a 16% solution of potassium hydroxide.

Recording result: To a 5 ml culture add 0.5 ml of 5% α-naphthol solution and 0.5 ml of 16% potassium hydroxide. Shake the tube gently. Development of a red coloration at the surface, usually within 30 min, constitutes a positive reaction.

11.2.7 Utilization of citrate as the sole source of carbon (see Report, 1958)

Medium: Simmon's citrate agar; this contains bromthymol blue as a pH indicator, and agar. It is used as slopes, with a 1-inch butt.

The slope culture is inoculated by streaking over the surface with a loopful of peptone water culture or preferably with a wire needle of a saline suspension prepared from a young agar slant culture. Incubate at the optimum temperature for up to 7 days.

Recording result: Utilization of citrate and growth on the citrate agar results in an alkaline reaction, so that the bromthymol blue indicator in the medium changes from green to bright blue. When no growth occurs and citrate is not utilized, the colour of the medium remains unchanged.

11.2.8 Production of polysaccharide from sucrose (Evans *et al.*, 1956; Garvie, 1960)

The medium used, sucrose agar, consists of nutrient agar with the addition of 5–10% sucrose, and is sterilized by autoclaving.

Inoculate a poured dried plate by streaking to obtain separate colonies. Also set up a control on a medium containing only 0.1% sucrose. Incubate at 20–25°C or at the optimum growth temperature for 1–14 days.

Recording result: Synthesis of dextran or laevan from sucrose is indicated by the development of growth of a mucoid character.

11.2.9 Action on litmus milk

Inoculate as for a broth culture and incubate at the optimum growth temperature for up to 14 days.

Recording result: Examine tubes daily and record any changes in the medium. A number of different reactions and combinations of reactions may occur involving (1) lactose, (2) casein, and (3) other milk constituents.

1(a) Acid production shown by a change in the colour of the litmus to pink.
1(b) If sufficient acid is produced the milk will clot. This is known as an acid clot.
1(c) Reduction of the litmus and loss of colour may occur. This may precede or follow other changes.
1(d) Gas may also be produced and show as gas bubbles in the medium, although normally this is visible only if clotting has occurred.

2(a) Coagulation of the milk may occur as a result of proteolytic enzyme activity affecting the casein, the litmus colour remaining blue. This is known as a sweet clot.
2(b) Hydrolysis of the casein as a result of proteolytic enzyme activity causes clearing and loss of opacity in the milk medium, usually referred to as peptonization. Proteolysis may also result in an alkaline reaction due to ammonia production.

3. Utilization of citrate in the milk medium results in the production of an alkaline reaction, shown by the change to a deep blue colour in the litmus medium.

11.3 REACTIONS INVOLVING LIPIDS, PHOSPHOLIPIDS AND RELATED SUBSTANCES

11.3.1 Hydrolysis of tributyrin

Medium: Tributyrin agar, which consists of yeast extract agar at pH 7.5, with the addition of tributyrin followed by emulsification, which is best carried out in an electrical mixer or blender. Sterilize by steaming for 30 min on three successive days. The medium may be used in the form of layer plates, 5 ml of molten tributyrin agar being overlaid on to previously poured and set plates of yeast extract agar. The medium may also be available commercially as a ready-prepared medium.

Inoculate a poured dried plate of the medium by streaking *once* across the surface; incubate at the optimum growth temperature for 2–14 days.

Recording result: Hydrolysis of tributyrin results in clearing of the medium and formation of a clear zone. Record the width of the zone in millimetres from the edge of the colony to the limit of clearing. This reaction is usually regarded as being specific for lipase, but Sierra (1964) reported a bacterial proteolytic enzyme that is capable of hydrolysing tributyrin, although not more complex fats.

11.3.2 Hydrolysis (lipolysis) of butter fat, olive oil and margarine

(a) Butter-fat agar, olive-oil agar (Berry, 1933)

The medium consists of yeast extract agar, pH 7.8, plus butter fat or olive oil to 5%. Emulsification of this medium can be achieved by shaking vigorously.

Inoculate a poured dried plate of the medium by streaking *once* across the surface and incubate at the optimum growth temperature for 2–14 days.

Test reagent: Saturated copper sulphate solution.

Recording result: Flood the plates with 8–10 ml of reagent and allow to stand for 10–15 min. Pour off the reagent and wash the plates gently in running water for 1 h to remove excess copper sulphate. Where lipolysis has occurred, a bluish-green coloured zone appears, due to the formation of the insoluble copper salts of the fatty acids set free on lipolysis.

(b) Victoria blue butter-fat agar (Jones and Richards, 1952) and ***Victoria blue margarine agar*** (Paton and Gibson, 1953).

Use of margarine gives a more stable medium than does butter. The choice of margarine used may affect the results observed, as some margarines are produced from a single source of lipid.

The media contain Victoria blue as an indicator of the presence of free fatty acids. Some strains of bacteria are inhibited by Victoria blue, so these media are not suitable for primary isolation of lipolytic organisms from mixed populations.

Recording result: Where lipolytic activity has occurred, the free fatty acids combine with the Victoria blue to form deep blue salts. Deep blue zones surrounding or beneath microbial growth are thus an indication of lipolytic activity. The background colour of the medium should be pinkish-mauve.

11.3.3 Hydrolysis of Tween compounds (Sierra, 1957)

Tweens are polyoxyethylene sorbitan monoesters of fatty acids. In addition to the Tween, the medium must include a soluble calcium salt (usually calcium chloride); released fatty acid is then detectable as the precipitated calcium salt. Tween 20 (a lauric acid ester), Tween 40 (a palmitic acid ester), Tween 60 (a stearic acid ester) and Tween 80 (an oleic acid ester) are amongst the Tweens that can be used, but Tween 80 is usually chosen.

Inoculate a poured dried plate of the Tween agar medium by streaking *once* across the centre or by spot-inoculating and incubate at the optimum growth temperature for 1–7 days.

Recording result: Opaque zones surrounding microbial growth consist of calcium salts of the free fatty acids and are usually taken as being indicative of a positive lipolytic activity. It should be noted, however, that the Tween forms micelles which are pervaded by water. Current nomenclature defines a lipase as an enzyme that acts on water-insoluble esters or fats at the ester–water interface. Tween/water mixtures probably provide suitable conditions for the activation of both lipases and other esterases.

11.3.4 Hydrolysis of lecithin

(a) Egg-yolk agar

This consists of a nutrient agar with the addition of sodium chloride to 0.9% and egg-yolk emulsion to 10%.

Inoculate a poured dried plate of the medium by streaking *once* across the surface and incubate at the optimum growth temperature for 1–4 days.

Recording result: Lecithinase activity (i.e. hydrolysis of the lecithin of the egg-yolk medium) results in the formation of opaque zones around the region of microbial growth. Record the width of the opaque zone in millimetres.

(b) Egg-yolk broth

A nutrient broth with the addition of 0.9% sodium chloride and 10% (v/v) egg-yolk emulsion.

Inoculate and incubate at the optimum growth temperature for 1–4 days.

Recording result: Lecithinase activity results in opacity in the egg-yolk broth medium, usually with a thick curd.

11.4 TESTS FOR THE PRESENCE OF ACTIVE ENZYMES

11.4.1 Catalase test

Most organisms growing on aerobically incubated plates possess the enzyme catalase, although in differing amounts depending on species and strain (see, for example, Taylor and Achanzar, 1972). The lactic acid bacteria (including *Streptococcus*, *Lactococcus*, *Leuconostoc*, *Lactobacillus*) do not normally produce a detectable catalase, but Whittenbury (1964) reported a number of strains capable of giving a positive reaction in the catalase test, particularly when grown on media containing heated blood. On the other hand, obligate anaerobes (e.g. *Clostridium*) are usually catalase negative.

Test reagent: Hydrogen peroxide (10 vol. concentration, i.e. approximately 3%). (This should be freshly prepared each day, and stored in the refrigerator between tests.)

Method:

(a) Pour 1 ml of hydrogen peroxide over the surface of an agar culture. Alternatively, a loopful of growth may be emulsified with a loopful of hydrogen peroxide on a slide.

(b) Place 1 ml of hydrogen peroxide in a small clean test-tube and add 1 ml of culture withdrawn aseptically from a broth culture.

Recording result: Effervescence, caused by the liberation of free oxygen as gas bubbles, indicates the presence of catalase in the culture under test.

11.4.2 Oxidase test (Kovacs, 1956; Steel, 1961)

This test is particularly useful for differentiating pseudomonads from certain other Gram-negative rods.

Test reagent: 1% aqueous solution of tetramethyl-*p*-phenylenediamine hydrochloride. This may be kept in a *dark* bottle in the refrigerator but auto-oxidation will cause the solution gradually to become purple, when it should be discarded. Auto-oxidation can be retarded by addition of ascorbic acid to 0.1% (Steel, 1962). The reagent is irritant and toxic, so a good alternative is to use commercially available test sticks (e.g. from Oxoid), test discs (e.g. from bioMérieux), or test strips (e.g. from Difco); in this case the manufacturer's instructions for use should be followed.

Method:

(a) Pour the reagent over the surface of the agar growth in a Petri dish.

Recording result. Oxidase positive colonies develop a pink colour which becomes successively dark red, purple and black in 10–30 min.

Method:

(b) Add a few drops of reagent to a piece of filter paper in a Petri dish. With a platinum loop or glass rod (other materials may give false positive results) smear some bacterial growth on to the impregnated filter paper.

Recording result: A purple coloration is produced within 5–10 s by oxidase-positive cultures. A delayed positive is indicated by a purple coloration within 10–60 s, any later reaction being recorded as negative.

11.4.3 Coagulase test

This test is used to differentiate pathogenic (e.g. *Staphylococcus aureus*) from non-pathogenic staphylococci. An easy alternative to the traditional methods described below is to use a commercially available kit. For example, Staphytect from Oxoid is a latex slide agglutination test which detects clumping factor ('bound coagulase') and/or Protein A.

(a) Slide method

This test is carried out on 18–24-h nutrient agar cultures.

Test reagent: Dried rabbit plasma, reconstituted and diluted 1 in 5 has been found satisfactory.

Method: Mark a slide into two sections with a grease pencil. Place a loopful of normal saline (0.85% sodium chloride in aqueous solution) on each section and emulsify a small amount of an 18–24-h agar culture in each drop until a homogeneous suspension is obtained. Add a drop of the reconstituted dilute rabbit plasma to one of the suspensions and stir for 5 s.

Recording result: A coagulase-positive result is indicated by clumping which will not re-emulsify. The second suspension serves as a control.

(b) Tube method

In this modification, 18–24-h nutrient broth cultures are used.

Test reagent: Reconstituted dilute rabbit plasma is used as above.

Method: Place 0.5 ml of reconstituted diluted plasma into each of two small test-tubes. To one tube add 0.5 ml of an 18–24-h broth culture. Incubate both tubes at 37°C and examine after 1 h and at intervals for up to 24 h.

Recording result: Clotting indicates that the strain under test is coagulase positive. Coagulation normally takes place within 1–4 h. The second tube serves as a control and should show no coagulation. The tube test is rather more reliable than the slide method.

11.5 MISCELLANEOUS TESTS

11.5.1 Phosphatase test

Coagulase-positive staphylococci usually produce the enzyme phosphatase (Barber and Kuper, 1951). The production of phosphatase can be detected by cultivation on a nutrient agar medium containing 0.01% phenolphthalein phosphate. Polymyxin can be incorporated to make the medium more selective for the growth of staphylococci (Gilbert *et al.,* 1969), so that the medium may be used to enumerate potentially pathogenic staphylococci in foodstuffs, etc., by using the Miles and Misra surface count technique or spread plates.

Medium: Liquefy the nutrient agar, cool to 50°C and add aseptically sterile 1% phenolphthalein phosphate solution to a final concentration of 0.01%, mix and pour plates.

Test reagent: Concentrated ammonia solution.

Method: Inoculate poured plates and incubate overnight at 37°C. Expose each plate to ammonia vapour by adding a few drops of ammonia to a filter paper inserted in the lid of the dish (use a fume cupboard).

Recording result: Pink or red colonies indicate the presence of free phenolphthalein set free by phosphatase activity, and are therefore phosphatase positive.

11.5.2 Haemolysis of blood agar

Medium: Nutrient agar containing 0.85% sodium chloride and 5% (v/v) defibrinated or oxalated blood. Horse blood is suitable for streptococci, but for other organisms (e.g. staphylococci) the blood of other animals (e.g. sheep, rabbit, ox) may give better results. Liquefy the nutrient agar medium, cool to 50°C and add the sterile blood (0.5 ml to 10 ml of agar) aseptically. This is mixed by rotation between the hands or inversion of the tube, and poured into a Petri dish. Alternatively, 5 ml of blood agar may be poured on top of a thin layer (5–10 ml) of nutrient agar previously poured into the Petri dish and allowed to solidify.

Method: In the case of staphylococci, streak a dried plate of the medium so as to produce separate colonies. If streptococci are being examined, either prepare pour plates or incubate streak plates anaerobically. Incubate at the optimum growth temperature for up to 2 days.

Recording results: Clear zones around the colonies indicate haemolytic activity. *β-Haemolysis* is the term given to this complete clearing of the blood agar when caused by streptococci. Zones of β-haemolysis possess sharply defined edges. α-Haemolysis is the term usually given to a greenish coloration produced around the colonies of some streptococci. These greenish zones have hazy outlines. It should be noted that the terms *α-haemolysin* and *β-haemolysin* have a significance which depends on whether they are used with reference to streptococci or staphylococci. In the case of staphylococci, α-haemolysin produces clear zones and β-haemolysin produces dark hazy zones. Nevertheless the haemolysis produced by staphylococcal α-haemolysin may sometimes be called β-haemolysis. Thus, to avoid confusion, it is recommended that the type of haemolysis is *described* rather than given a designation which may be misinterpreted.

11.5.3 Potentiated haemolysis (the CAMP test)

As originally described by Christie, Atkins and Munch-Petersen (1944) this detects the production of an extracellular, diffusible polypeptide (called the 'CAMP' factor after the authors of the paper) produced by certain species and strains of *Streptococcus* Lancefield group B, which acts synergistically with staphylococcal β-haemolysin to produce enhanced lysis at the junction of streaks of culture of the two organisms grown on sheep-blood agar. Even non-haemolytic group B streptococci will be CAMP positive. Subsequently, the method has also been found to differentiate between certain species of *Listeria*.

Medium: Sheep blood (5% v/v) agar. When testing *Listeria*, it may be found necessary to use saline-washed sheep erythrocytes resuspended in sterile saline, as sheep blood often carries antibodies against *Listeria*.

Method:

(a) *Testing* Streptococcus. Across a plate of sheep-blood agar place a single streak of a stock culture of a β-haemolysin-producing *Staphylococcus aureus* (NCTC 1803 or ATCC 25923 is recommended). Streak the culture under investigation at right angles to the streak of *Staph. aureus*, finishing the streak 1–2 mm from the other streak (see Figure 11.1A). Incubate at 37°C for 18–24 h.

(b) *Testing* Listeria. Across a plate of washed sheep-blood agar place a single streak of a stock culture of *Staph. aureus* (NCTC 1803 or ATCC 25923), and about 4 cm from it a parallel single streak of *Rhodococcus equi* (NCTC 1621 or ATCC 6939). Streak the culture under investigation at right angles to the streaks of *Staph. aureus* and *R. equi*, finishing the streak 1–2 mm from the other streaks (see Fig. 11.1B). Incubate at 37°C for 18–24 h.

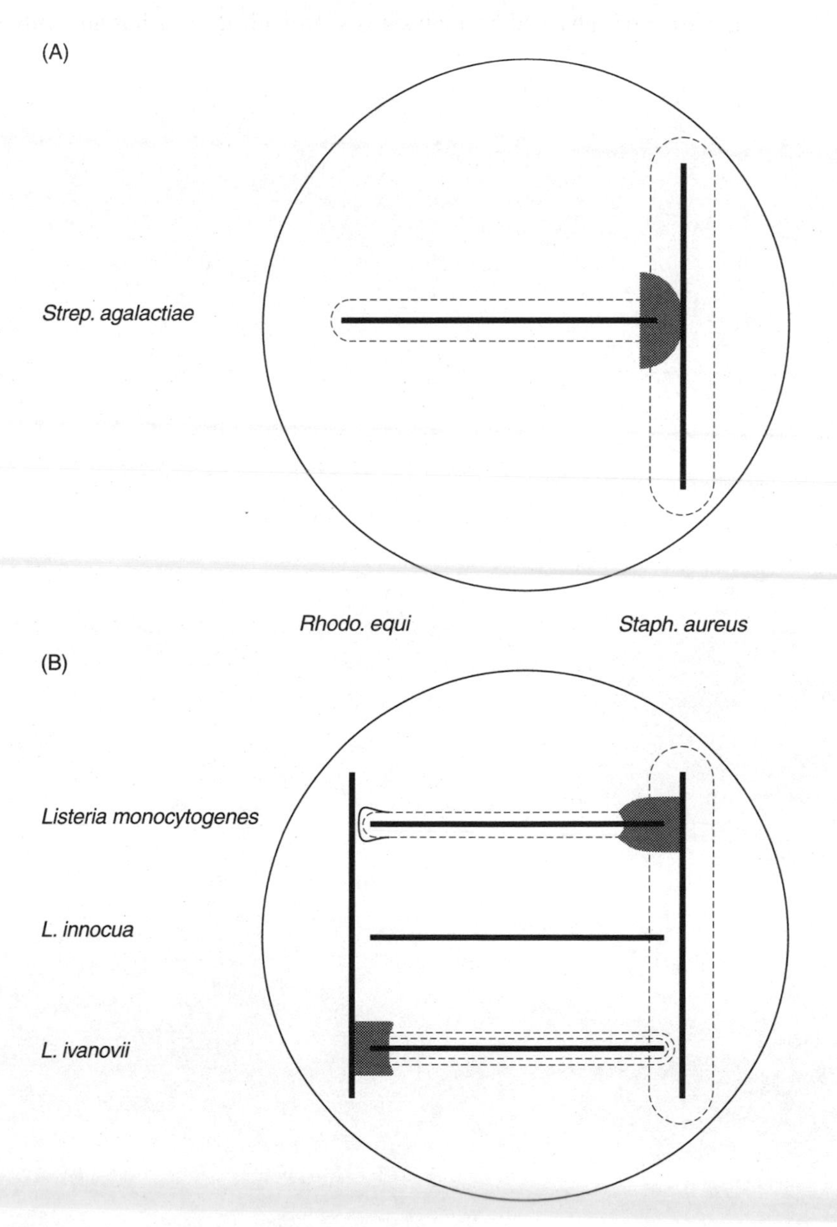

FIG. 11.1 Reactions in the CAMP test.

Reading results:

(a) If the culture produces a CAMP factor, an arrowhead-shaped area of enhanced haemolysis is observed at the junction of the two streaks (see Figure 11.1A).

(b) If the culture produces a CAMP factor, an area of enhanced haemolysis is seen. In the case of *Listeria monocytogenes* this reaction is seen primarily with *Staph. aureus*, although a small response may be seen with *R. equi*. In the case of *L. ivanovii*, a zone of enhanced haemolysis is seen with *R. equi* but not with *Staph. aureus*. *Listeria innocua* shows no CAMP reaction. (See Figure 11.1B.)

12 Physiological Tests

12.1 INTRODUCTION

The determination of the response of microorganisms to temperature, available water (a_w) and salt/sugar concentration, and pH are important in the context of growth, survival and death of microorganisms in foods. To obtain reliable quantitative information on the effect of the environment (e.g. concentration of sodium chloride) on growth, growth rate determinations are necessary. A pragmatic approach, however, is to perform shelf-life determinations on the foods themselves. There has been much discussion of the validity of so-called accelerated shelf-life tests; these usually involve storage of the food at a higher temperature than it will normally experience, so that development of microorganisms is detected more quickly. This approach has for long been adopted for canned foods. Readers interested in this topic are referred to the works by Man and Jones (1994) and the Institute of Food Science and Technology (IFST, 1993).

12.2 GROWTH RATE DETERMINATIONS ON PURE CULTURES

It may occasionally be useful in food spoilage investigations to obtain more precise information about the rate of growth of particular organisms than can be achieved by routine keeping quality tests or incubation of whole food samples. The following procedure, which is similar to that described by Barnes and Impey (1968), can be modified to simulate particular food environments.

Procedure

Whenever possible, use recent isolates from the foodstuff concerned. The inoculum may be either a culture in a liquid medium (when this medium is also the test medium) or a washed cell suspension. Dispense and sterilize 100 ml amounts of culture medium in conical flasks and equilibrate at the desired incubation temperature before inoculation. Add sufficient test culture to give a concentration of cells in the medium of approximately 10^4 per ml, mix the flask by rotation, and withdraw a 1-ml sample for the preparation of decimal dilutions and determination of the viable count at the start of the experimental period (t_0).

Further 1-ml quantities should be withdrawn and viable counts determined at frequent intervals, so that the number of results obtained before the start of the stationary phase is adequate for the construction of meaningful growth curves and the calculation of mean generation time. In any particular system, preliminary outline experiments are likely to be required to establish the most satisfactory detailed procedure. The results obtained may be presented graphically by plotting $\log_{10}$ counts as ordinate against time (minutes) as abscissa.

The mean generation time may be calculated using the formula:

$$G = \frac{t \log 2}{\log b - \log a}$$

where G is the mean generation time, a and b are the numbers of bacteria at two points in the logarithmic growth phase separated by the time interval t.

This technique can be particularly helpful when, for example, comparing the effects of temperature or the presence of food preservatives on the growth of the component organisms in a microflora. In the case of some foodstuffs (e.g. milk, sugar syrups) the sterilized foods can be used as the growth medium. Other foodstuffs (e.g. cooked meats) will require to be prepared as sterile homogenates, although in the case of raw foods filter-sterilized aqueous extracts may be suitable. However, it will be appreciated that such processing may affect significantly the growth characteristics of the organism under investigation, compared with the growth obtained in the original food environment.

It is also possible to study microbial interactions by the determination of counts and generation times when the isolates are grown in the test system as mixtures of the strains under investigation.

12.3 TEMPERATURE

Microorganisms have evolved to cope with an extremely wide range of conditions found in the natural environment. Thus, microorganisms have been described that can grow at the temperatures of up to 90°C or more found in hot springs (Brock, 1971; Brock and Brock, 1971), and yet others have been isolated from various environments that have the ability to grow at temperatures just above or even below 0°C (Farrell and Rose, 1967). It is with these latter microorganisms that we are concerned in this text. The International Commission on Microbiological Specifications for Foods (ICMSF, 1980a) defined psychrophiles, mesophiles, thermophiles and psychrotrophs as shown in Table 12.1.

Over the years there has been considerable confusion in the terminology used by microbiologists, especially in relation to growth at 'low' temperatures, with the terms

TABLE 12.1 Cardinal temperatures according to the International Commission on Microbiological Specifications for Foods (1980a)

Group	Growth temperature (°C) Minimum	Optimum	Maximum
Psychrophiles	–5 to +5	12–15	15–20
Mesophiles	– 5–15	30–45	35–47
Thermophiles	40–45	55–75	60–90
Psychrotrophs	–5 to +5	25–30	30–35

cryophile, psychrophile, obligate psychrophile, facultative psychrophile and psychrotroph being used with little agreement as to their definitions. This is not a trivial matter, of interest only to etymologists, as careful use of correctly defined terms can help to distinguish between organisms that have different growth characteristics and which therefore have different significances in foods preserved at low temperatures. An organism with an optimum growth temperature of around 10°C could display a maximum growth temperature below 20°C and show a loss of viability at 20°C and above, and in consequence it would be of little significance in spoilage of refrigerated foods derived from raw materials that have been exposed to ambient temperatures, although, of course, it might have contributed to microbiological deterioration of the raw materials before they had been exposed to ambient temperatures. An organism that is a psychrotrophic mesophile could, on the other hand, be of great significance even though its growth rate is much diminished at the refrigeration temperature, as survival and multiplication could be occurring over the entire temperature range to which the food materials are being exposed, whether refrigeration or ambient.

Organisms capable of growing at low temperatures can be divided into two distinct classes. The first consists of those with a low optimum growth temperature and which are in consequence termed psychrophiles (from the Greek ψυχροσ, *psychros*, meaning cold, and φιλοσ, *philos*, meaning loving).

The second class of organisms are those that have optimum growth temperatures in the normal range for mesophiles (from the Greek, μεσοσ, *mesos*, meaning middle, and φιλοσ). In this case it is their minimum temperature for growth that makes them of significance at low temperatures. These organisms can therefore be correctly termed psychrotrophs (from the Greek, ψυχροσ, and τροφη, *trophy*, meaning nourishment), implying that they are able to metabolize and multiply at the lower temperatures although showing greater growth rates at the higher temperatures. An organism such as *Pseudomonas fluorescens* can therefore be referred to as a psychrotrophic mesophile.

As long ago as 1960 Eddy (Eddy, 1960) pointed out that the use of the terms ‘obligate psychrophile’ and ‘facultative psychrophile’ (see, for example, Ingraham and Stokes, 1959) was not logical. Morita (1975) made a plea for a uniform terminology, with the definition of psychrophiles being based on the optimum growth temperature and that of psychrotrophy on minimum growth temperature.

It is a pity that the ICMSF (1980a) definitions show inconsistency. The definitions may or may not be suitable for categorizing the majority of organisms so far isolated, but they certainly do not cover all possibilities that we can isolate or imagine isolating. In particular, what should we call an organism with an optimum growth temperature falling between 15 and 25°C, or an organism capable of growing at –5 to +5°C but which has an optimum growth temperature above 30°C? Indeed, in the same chapter of that work (ICMSF, 1980a) reference is made to *Clostridium botulinum* type E which has an optimum temperature for growth around 35°C, and yet can grow down to 3.5°C. It is obvious that we cannot use the ICMSF (1980a) definitions unless some modifications are made. It is illogical to specify minimum and maximum growth temperatures for the '-philes' as part of the definition, just as it is illogical to specify the maximum growth temperature for a psychrotroph (or a minimum growth temperature for a 'thermotroph'). Consequently the food microbiologist should perhaps use definitions similar to those shown in Table 12.2, which have the advantage of inclusivity and of semantic logic. It is perhaps of value to distinguish between psychrotrophs capable of growing just around 0°C, perhaps down to a few degrees below 0°C (for example, *Pseudomonas fluorescens*) and those very few psychrotrophs that show considerable cold tolerance, being able to grow down to –10° or even –12°C (e.g. *Cladosporium herbarum*). These latter could be called cryotrophs (from the Greek *κριοσ*, *cryos*, meaning frost, and τροφη).

It should be emphasized that it is necessary to standardize on the procedure to be adopted for the determination of optimum growth temperature. This should be taken as the temperature at which the growth rate is maximum (i.e. the generation time is minimum) when the microorganisms are in the exponential phase of growth. This cannot be determined, for example, by performing turbidity measurements on tubes of broth cultures after these may have reached stationary phase. In such a growth environment, the growth rate of an aerobe may be limited by some factor other than temperature, or maximum yield may be affected independent of growth rate. For example, Hess (1934) reported that although *Pseudomonas fluorescens* grew more rapidly at 20°C than at 5°C, maximum yield occurred at 5°C. This effect was probably due to the greater solubility of oxygen at lower temperatures plus the lower oxygen demand at the lower temperature preventing the cultures from becoming oxygen-limited, as Sinclair and Stokes (1963) found that vigorous aeration of cultures incubated at higher temperatures in the growth range enabled yields to be obtained that were equal to those obtained at the lower temperatures. It would therefore seem to be best to determine the optimum temperature for growth by determining growth rate in the exponential phase; if the final yield is not

TABLE 12.2 Definitions of microorganisms characterized by cardinal temperatures

	Growth temperature (°C)		
Group	Minimum	Optimum	Maximum
Psychrophiles		<20	
Mesophiles		20–50	
Thermophiles		>50	
Psychrotrophs	<5		
Cryotrophs	<–5		

greatest at this temperature it should be regarded as a manifestation of another limiting factor such as oxygen supply.

Studies of growth temperature relationships can best be performed using a temperature gradient incubator. These are available commercially. Alternatively one can be constructed relatively easily from a solid block of aluminium alloy, drilled to provide banks of wells into which test-tubes can be inserted. One end of the block will be heated, and the other end cooled (e.g. by a refrigerant liquid circulated through a refrigeration unit). The block should be thermally insulated, for example with slabs of expanded polystyrene. Temperatures across the block can be determined by means of thermocouples inserted into test-tubes containing sterile medium. For further constructional details see Baker and Orr (1979).

Most of the microorganisms isolated as representative of the low-temperature spoilage flora of chilled foods have proved to be psychrotrophic mesophiles. However, Farrell and Rose (1967) suggested that in the case of food materials derived from low-temperature environments (e.g. marine fish) this may be a false impression due to the use of conventional methodology causing the death of any psychrophiles if they have maximum growth temperatures only just above their optima. Stanley and Rose (1967) found that when samples from Antarctic lakes were examined with care being taken never to expose the samples, dilutions or cultures to temperatures above 10°C, more than three-quarters of the isolates had optimum growth temperatures below 20°C. It should be noted that more than half of the isolates had optimum growth temperatures between 15° and 20°C, and were therefore undefinable using the criteria listed in Table 12.1.

Generally, the maximum temperature for growth of a microorganism (i.e. the temperature above which multiplication ceases) is more well-defined and easier to determine than the minimum temperature for growth. The supraoptimal temperature at which multiplication ceases is often only a few degrees above the optimum, the growth rate falling sharply with a sharp cut-off point. However, at suboptimal temperatures, the growth rate falls much more gradually.

12.4 EFFECT OF HEAT ON MICROORGANISMS: THE DETERMINATION OF DECIMAL REDUCTION TIMES (D VALUES) AND *z* VALUES

12.4.1 Introduction

Heating a bacterial culture for increasing periods of time results in a progressive reduction in the viable population. The more organisms present initially, the longer the time required at a given temperature to kill the whole population.

Plotting the logarithm of the numbers of survivors against time gives a curve that tends to a straight line. The reciprocal of the slope of this line is the D value, which is defined as the time taken at a given temperature (T) to effect a reduction of 90% in the microbial population. The higher the temperature, the smaller will be the D value. If the logarithms of the D values obtained at various temperatures are plotted against temperature, and the best straight line is drawn through the points, the reciprocal of the slope of this line is the z value. The z value is defined as the number of degrees by which the temperature has to be raised or lowered to bring about a 90% reduction or tenfold increase in the D value.

Theoretically (if these straight-line relationships were to be maintained over the whole lethal temperature range) the effect on a given bacterium of a typical heat treatment, including heating, holding and cooling times, can be determined provided the temperature–time curve of the heat treatment, the z value, and at least one D value for the organism are known. A discussion of these mathematical calculations and of the limitations involved are outside the scope of this book, but readers are referred to, for example, Stumbo (1973), Moats (1971), Moats *et al.* (1971), Cerf (1977), Stephens *et al.* (1994) and Anderson *et al.* (1996).

Two examples of simple experimental procedures that may be used are described here. These can be used for combinations of microorganisms and temperatures where the D values are of the order of minutes. The choice of media, suspending liquids, temperatures and times of heat treatment, and incubation temperatures and times depend on the organism and the heat treatment system under investigation, and may require a simple trial experiment to determine the details of the experimental procedure.

These methods are inadequate for the study of D values of a few seconds or a fraction of a second – this is particularly the case when D values of spores are being determined at UHT temperatures, but will also apply to the determination of D values of Gram-negative bacteria at temperatures of around 70°C and above. Perkin *et al.* (1980) described some methods suitable for work at UHT temperatures. Brown (1994) has listed the range of methodologies appropriate for UHT studies. Mallidis and Scholefield (1985) have described a technique for heating capillary tubes by insertion into a solid aluminium alloy heating block; they used this to study destruction of bacterial spores at 120°C, but the method is also suitable for studying destruction of Gram-negative bacteria at lower temperatures.

12.4.2 Using test-tubes to determine D values and the *z* value of an asporogenous culture

1. Dilute a 24-h broth culture (e.g. of *Escherichia*), to give a total count of about 3×10^8 per ml. (This can be determined by microscopic count, or by nephelometry or opacity tubes if these have been calibrated using microscopic counts.)
2. Put 10 ml of this dilution into 90 ml of sterile diluent (e.g. 0.1% peptone water) which contains a number of glass beads. Shake well to mix and to break up clumps. Distribute in 5-ml amounts into five sterile 150 × 16-mm test-tubes (preferably with polypropylene closures).
3. Put the tubes prepared in (2) into a water-bath at 58°C with a control test-tube containing a thermometer and 5 ml of diluent. When the temperature of the contents of the control tube reaches 1°C below the temperature of the water-bath, remove one of the tubes and simultaneously start timing from this moment (t_0).
4. Rapidly cool the removed tube in cold water. Prepare six decimal dilutions and perform a plate count using all of these dilutions.

5. Withdraw another tube and repeat step (4) after each of the following times: 2½, 5, 7½ and 10 min.
6. Incubate the plates and determine the number of survivors after the various heating periods. Draw a graph of $\log_{10}$ (number of survivors) against time. From this determine the $D_{58°C}$ (in minutes) as the reciprocal of the slope of the best straight line.
7. Repeat steps (1) to (6), using a water-bath set at each of the following temperatures and removing tubes after heating for each of the times shown:

   56°C for 0, 10, 20, 30, 40 and 50 min
   54°C for 0, 10, 20, 30, 40 and 50 min
   52°C for 0, 20, 35, 50, 65 and 80 min
   51°C for 0, 20, 40, 60, 80 and 100 min
   50°C for 0, 20, 40, 60, 80 and 100 min

8. Plot $\log_{10} D_T$ against T, to estimate the z value.

12.4.3 Using sealed tubes to determine the D values and *z* value of spores

The previous method represents the simplest method for determining D and *z* values. It has many sources of inaccuracy, amongst which are the long heat penetration times (for both heating and cooling), the temperature gradients that exist within the tubes, the likely presence of bacteria on (a) aerosol droplets in the contained air and (b) on the test-tube walls above the bulk liquid level (these bacteria can become reintroduced into the bulk liquid from time to time). These inaccuracies become more significant the shorter the heating times studied. The following method is described for use with spores of, for example, *Bacillus stearothermophilus*.

1. Cut glass tubing of 4 mm external diameter, 2 mm internal diameter, into 100-mm lengths. Using a suitable gas flame seal one end of each tube to give a hemispherical end. Plug with cottonwool and sterilize by autoclaving.
2. Prepare a spored culture of *B. stearothermophilus* by culturing on fortified nutrient agar (Gould, 1971) in large sterile medical flat bottles, incubated at 55°C for 6 days. The incubation temperature and period to provide a high ratio of spores to vegetative cells should be determined by experiment for the strain under investigation (the incubation conditions described were found to be satisfactory for *B. stearothermophilus* NCIB 8923 (ATCC 7954).
3. Rinse the spore crop from the slopes by the addition of a small amount of sterile phosphate buffer at pH 7.0, with a few glass beads to allow this to be effected using gentle agitation.
4. Add papain to the spore suspension to a concentration of 1 mg per ml, and incubate at 37°C for 48 h, to destroy vegetative cells.
5. Centrifuge the spore suspension, pour off the supernatant liquid and resuspend the pellet in phosphate buffer. Repeat this centrifuging and washing process six

times. (The use of sterile disposable centrifuge tubes with screw caps is recommended.)

6. Finally resuspend the spores in 0.85% sodium chloride to a concentration of about 10^8 spores per ml (using a counting chamber to determine the count microscopically) and store in a refrigerator.

7(a). Use 1 ml of this working spore suspension to prepare a dilution series and carry out a plate count to determine the viable spore population. Incubate the plates at 55°C with daily counting of colonies, avoiding contamination, until the count does not increase. This indicates the viable spore population *relative to the medium used.* (*Note.* Plates incubated at 55°C or above should be placed in sealed bags or other containers to minimize dehydration of the medium.)

7(b). Prepare a similar plate count after the suspended spores have been exposed to 80°C for 10 min. This brief heat treatment may be found to activate the spores and, by thus encouraging germination, to result in a higher count than that obtained on the unheated suspension.

8. This and following steps are carried out concurrently with step (7); that is, all the plate counts produced by steps (7) to (13) are incubated together.

   Introduce the well-mixed spore suspension into eight of the tubes prepared in step (1), filling each tube to approximately one-quarter of its length. The amount of spore suspension introduced into each tube can be predetermined by using a sterile microlitre syringe, or by weight if uncalibrated Pasteur pipettes are used.

   Rapidly seal the open end of each tube by rotation in a Bunsen flame, taking care as far as possible to protect the spore suspension from heating.

9. Use one of the tubes to ascertain the count at t_0. The surface of the tube should be swabbed with 70% ethanol and a sterile glass cutter used with aseptic precautions to weaken the glass at several points. Aseptically place the tube in a sterile Universal container with 10 ml of sterile phosphate buffer. Break the tube using a sterile metal rod. Mix the contents well and prepare a dilution series and a plate count in the usual way.

10. Immerse seven of the tubes in an oil bath set at 110°C. Remove one tube after each of the following heating periods: 1, 3, 5, 10, 15, 25 and 30 min.

11. Each tube should be cooled rapidly in a beaker of cold water, and a plate count made as described in step (9).

12. Incubate the plates at 55°C and determine the number of survivors per millilitre of suspension after the various heating periods. Draw a graph of $\log_{10}$ (number of survivors) against time, and ascertain the $D_{110°C}$ (see above).

13. Repeat the experiment using an oil bath set at each of the following temperatures and determine the number of survivors per millilitre of suspension after heating for each of the following periods:

    115°C for 1, 3, 5, 10, 15, 20 and 25 min
    118°C for 1, 3, 5, 10, 15, 20 and 25 min
    121°C for 1, 3, 5, 7, 10, 13 and 16 min

These times and temperatures have been found satisfactory in experiments involving *B. stearothermophilus* NCIB 8923 (ATCC 7954) (T. Player and N. Marsden, unpublished results, 1973).

14. Plot $\log_{10} D_T$; against T to estimate the z value.

12.4.4 Use of D and z values in process calculations

The D and z values for an organism may be used to determine the destructive effect of a heat treatment on that organism. For any heating and cooling curve there may be calculated a heating time at a given single temperature (the F value) (assuming instantaneous heating to and cooling from that temperature) which has an equivalent destructive effect on a population of the organisms (Stumbo, 1973).

It should be emphasized that the reliability of such calculations depends on the extent to which the z value *is* constant with temperature, i.e. whether there is a straight-line relationship of the form:

$$\log \mathrm{D} = a - \frac{T}{z}$$

If, however, the thermal destruction of organisms is described by an Arrhenius relationship (log D is proportional to the reciprocal of the temperature) then z will *vary* with temperature. In such a case, the z value determined experimentally will provide an estimate of the effect of temperature on the D value *only over the range of temperatures used in the experiment*. Other mathematical models for describing thermal destruction have also been proposed (see the references given in Section 12.4.1). The D and z values should not be used, therefore, for process calculations relating to temperatures very far outside the experimentally used range; for example, D and z values determined by heat resistance experiments performed at 80°, 90°, 100° and 110°C should not be used in process calculations relating to UHT processes at temperatures of around 130–140°C.

12.5 pH TOLERANCE

The effect of pH on growth rate needs to be examined by using liquid media adjusted over a specific range of pH. One of the problems in such a study is that the buffering efficiency of the medium around its initial pH will have a profound significance on the microorganisms as they produce metabolites.

Differences between microorganisms in respect of their pH tolerance can be screened for qualitatively by using pH gradient plates (Sacks, 1956; Huhtanen, 1975). This technique can establish the approximate limiting pH for growth, and thus reduce the range of pH over which liquid media need to be prepared for growth rate studies.

12.5.1 Procedure to study effects over pH 3.5–6.0

1. Use square plastic Petri dishes (e.g. 23-cm^2 dishes from Nunc, Kamstrup, Denmark). Determine the working volume of the medium (*w* ml) required for the size of dish actually used. The amount of each of the two media to be added will be one-half of this amount, ½*w* ml.
2. Mark one end of each dish, and draw lines 1 cm apart across the back of the dish parallel to the marked end.
3. Rest the marked end of the dishes on a piece of rectangular rod of a suitable diameter so that when ½*w* ml of molten medium is poured into the dish it just reaches the marked end of the dish when a working depth of medium is achieved at the lower end of the dish.
4. Pour into each dish ½*w* ml of an appropriate medium (e.g. tryptone soya agar) which has been adjusted to pH 6.0 with sterile hydrochloric acid just before pouring (the amount of acid can be predetermined on a bottle of medium using a pH meter as described in Section 7; this amount can then be used each time, and, as the actual pH in the dishes will be determined, any minor variations will be detected). Allow to solidify.
5. Place the dishes on a horizontal surface and pour into each ½*w* ml of molten tryptone soya agar adjusted to pH 3.5 with sterile acid immediately before pouring.
6. Allow the medium to solidify, and then refrigerate for 18–24 h to permit equilibration between the two layers.
7. Before inoculation, allow the dishes of medium to reach room temperature. Set aside one dish for pH determinations in (8) below. Streak cultures under investigation across the remainder of the dishes at right angles to the marked lines. A number of cultures can be streaked on to each plate, provided the streaks are separated by at least 2 cm.
8. Using a pH meter fitted with a flat-tipped combination electrode, at this time measure the pH in each 1-cm sector of the dish, taking measurements along the sector to determine whether there are any inconsistencies.
9. Incubate the plates. After incubation, determine the lowest pH at which each culture shows growth.

Notes

(a) This gradient plate procedure is described for the study of effects over a range of approximately pH 3.5–6.0, using mineral acid, with no specific additional buffering being used. Filter-sterilized lactic acid, citric acid or acetic acid solutions can be used to adjust the pH, so that the responses of the microorganisms to the organic acids at the measured pHs can be assessed.

(b) For example, if the lower pH be produced by adding filter-sterilized lactic acid to 0.34% into the upper layer of tryptone soya agar, this should give a range of pH across the plate of approximately pH 4.6–7.0.

(c) For phosphate-buffered gradients, prepare filter-sterilized 1.0-M solutions of potassium dihydrogen phosphate, dipotassium hydrogen phosphate and orthophosphoric acid. Incorporate these aseptically in the molten tryptone soya agar at 1: 10 just before pouring the plates. Using two of these three will provide different pH ranges. For example, with potassium dihydrogen phosphate in the upper layer and dipotassium hydrogen phosphate in the lower layer, the pH range will be from approximately 5.6 to 7.8. With potassium dihydrogen phosphate in the upper layer and orthophosphoric acid in the lower layer, the pH range will be from approximately 2 to 5.

13

Serological Methods

13.1 INTRODUCTION

The *in vitro* reactions between antigens and antibodies are of great value. Microbiological applications of these reactions are principally of the following three types.

13.1.1 Classification and identification of microorganisms

In this case the bacterium is unknown and the antibodies are known. This type of study is known as *antigenic analysis* and is used most often in epidemiological investigations and in systematic bacteriology. For example, in the investigation of an outbreak of food poisoning, *Salmonella* may have been isolated from a food that is under suspicion. The strain of *Salmonella* isolated from the food can be tested against a series of antisera containing antibodies against antigens of known serological types (serovars) of *Salmonella*. Cultures of *Salmonella* isolated from faeces of patients can be similarly tested, and the *Salmonella* in the food and faeces identified. Slide agglutination tests are used for this purpose (see Section 13.2).

13.1.2 Detection and isolation of microorganisms and their toxins

It is possible to use immunological techniques to detect target microorganisms in foods or in enrichment cultures, and to concentrate and isolate such organisms. To achieve this, the specific antibodies will normally be linked or conjugated to an enzyme label (enzyme-linked immunosorbent assay or ELISA; see Section 13.4) or to magnetic beads (see Section 13.5).

Another technology that can be used is affinity chromatography. In this, one component (e.g. an antibody) is immobilized on a stationary phase material such as agarose beads; this adsorbs the other reactant (in this case the specific antigen) from the liquid phase (e.g. dilution of food). This procedure is used in a number of commercial kits for detection of specific mycotoxins (e.g. aflatoxins) in food samples.

13.1.3 Detection and identification of antibodies in blood, milk and other body fluids

In this application, known bacterial cultures are used to detect the presence of specific antibodies against them. This type of serological investigation is used most frequently in clinical and veterinary microbiology. The more active an infection is, or the more recent, the greater the concentration of antibodies present in the patient's serum. To determine the concentration (or titre) of antibodies, a tube agglutination test is used in which a series of dilutions of serum is tested against the microbial suspension to determine the highest dilution of serum at which agglutination occurs.

In food microbiology, an example of a test for detection of antibodies is the *Brucella* milk ring test used in the diagnosis of brucellosis in cattle. A suspension of killed *Brucella* cells stained with, for example, haematoxylin is used to provide known antigen and tested against the milk from the animal. If antibodies are present in the milk there will be a (positive) reaction. (See Section 13.3.)

13.2 SLIDE AGGLUTINATION TESTS

Agglutination results from mixing together cellular antigen and homologous antiserum. As a result of the antigen–antibody reaction, the bacterial cells clump together and form flocculent masses or dense granules. In the case of bacteria of the Enterobacteriaceae, the reaction between 'H' (i.e. flagellar) antigen and its homologous antiserum results in flocculent clumping, whereas the reaction involving the 'O' (i.e. somatic) antigen results in a more dense and granular clumping.

Antisera commercially available are of two types: polyvalent antisera which react with organisms of a particular genus or with groups of serovars, and which are suitable for preliminary screening; and specific monoclonal antibodies, use of which allows identification of a particular serovar. Bottles of antisera should be stored in the dark at 4–7°C, and under these conditions they may remain usable for many months longer than the indicated shelf-life, although the titre (or concentration) of effective antibody should be checked periodically.

13.2.1 The slide agglutination test

Procedure

1. Using a clean, grease-free slide, mark it into three with a wax pencil labelling the three sections 'H', 'O' and 'C' (control).
2. Place one or two loopsful of physiological saline in section 'C'.
3. Place a drop of a formaldehyde-treated suspension of bacteria (see below) in the section marked 'H', and a drop of a heat-treated bacterial suspension (see below) in the section marked 'O'.

4. Mix one loopful of undiluted polyvalent 'H' antiserum with the bacterial suspension labelled 'H'.
5. Mix one loopful of undiluted polyvalent 'O' antiserum with the bacterial suspension labelled 'O'.
6. Rock the slide gently backwards and forwards and observe for 3 min over a dark background. If agglutination occurs there will be a clumping of the bacteria, usually within 30 s. There should be *no* change in the control suspension.

Note. It is important to flame the wire loop *and* allow it to cool between each transfer.

It is suggested that a presumptive *Salmonella* should be tested by slide agglutination using a polyvalent *Salmonella* 'O' antiserum. If negative, repeat the test using polyvalent 'H' phase 1 and 2 antisera. If both are negative, repeat using *S. typhi* 'Vi' antiserum. For further identification of agglutinating cultures, single factor sera can be used. Harvey and Price (1961, 1974) have described a simple method of inducing H-antigen phase change in *Salmonella* cultures, when this is necessary for identification of particular serovars.

13.2.2 Latex agglutination tests

A wide range of commercially available kits utilize latex beads (e.g. of polystyrene and of less than 1 μm in diameter) on to which the antibodies have been adsorbed. If the beads are white, the test is usually performed and observed against a black background. A more complex type of reagent uses a suspension of latex beads of more than one colour (e.g. red, blue and green). Each colour is associated with a different antibody or combination of antibodies. The agglutination is read against a white background. The colour of any flocculation indicates the presence of a particular serogroup. Examples of such coloured latex kits are Wellcolex Colour Kits for *Salmonella* and *Shigella* (obtainable from Murex Diagnostics, Dartford, UK).

Another type of product, the immunochromatographic kit, offers a one-use disposable plastic dip-slide in which coloured particles are used in conjunction with chromatographic transport, with a positive antigen–antibody reaction appearing as a coloured line situated in an observation window where immobilized antibody is located. Such disposable dip-slides are usually used to detect the target organisms directly in enrichment cultures, as they are not at present sufficiently sensitive to detect the low concentration of organisms normally present in food samples. Examples of such kits for *Salmonella*, *Escherichia coli* O157:H7 and *Listeria* are manufactured by Oxoid and Lumac.

13.2.3 Preparation of killed cell preparations of Enterobacteriaceae for agglutination tests

Procedure

1. The organism is subcultured on to a freshly made nutrient agar slope to examine reactions involving 'O' antigens, and into nutrient broth for 'H' antigens. Incubate for 4–18 h at 37°C.

2(a). For a preparation to contain 'H' antigens, a 5-ml broth culture should be treated by the addition of two or three drops of commercial formalin and diluted to an opacity equivalent to a concentration of about 5×10^8 organisms per ml.

2(b). For a preparation containing 'O' antigens, wash the growth from the slope with about 3 ml of saline. The bacterial suspension should be heated in a boiling water bath for 10 min to destroy the less heat-stable 'H' and 'Vi' antigens. The suspension should be quite turbid for slide agglutination tests.

Note. Disinfectants based on hypochlorite or other chlorine-containing compounds should *not* be used for discard jars, as this introduces a hazard of toxic fumes resulting from a reaction with formaldehyde.

13.3 THE MILK RING TEST FOR *BRUCELLA ABORTUS* AND *BR. MELITENSIS*

The milk ring test (MRT) is an agglutination reaction used to diagnose brucellosis in dairy herds *and* to detect from bulk milk supplies herds in which brucellosis is present.

It is essentially a test that demonstrates the presence of agglutinins (antibodies) in the milk, by the addition of stained bacteria to the milk sample followed by incubation for 30–60 min. The stained *Brucella* are clumped by any agglutinins present and, in cow's or sheep's milk, the stained agglutinin–antigen complex rises with the fat globules. This causes the cream layer at the top to become deeply coloured. (In the case of goat's milk, the agglutinated stained bacteria go to the bottom of the test-tube, owing to the different creaming properties of the milk.)

The milk ring test is extremely sensitive and is very suitable for testing cans or churns of milk because the milk of one infected cow can be detected when mixed with that of many uninfected cows.

Procedure

1. Add one drop of a commercial preparation of stained *Br. abortus* or *Br. melitensis* suspension (e.g. as obtainable from Murex Diagnostics, Dartford, UK) to a 25-mm

high column of a well-mixed milk sample in a small, narrow 75 × 10-mm test-tube. Mix thoroughly by shaking but *avoid frothing*.
2. Incubate the mixture at 37°C for 30–60 min and then examine.

Result

If the cream layer is deeply coloured and the milk beneath the cream layer is white or nearly so, the test is regarded as positive, indicating the presence of agglutinins in the milk. If the cream layer is white and the milk beneath deeply coloured, or the cream layer is the same colour as the milk layer, the test is recorded as negative.

13.4 ENZYME-LINKED IMMUNOSORBENT ASSAYS

This technique involves the use of antibodies to which enzymes have been bound covalently. The enzyme chosen is one that catalyses a reaction resulting in the release of a chromophore or fluorophore. The traditional ELISA procedure uses wells of a microtitre plate coated with an adsorbent material which are treated with antibody. The immobilized antibody can then capture the specific antigen.

In the *direct* ELISA technique, the antigen so trapped may be the target microorganism or target toxin in a food dilution or an enrichment broth. After carefully washing unbound material from the wells, a preparation consisting of antibody conjugated with enzyme is added. This conjugate will in turn be adsorbed on to the antigen. Once again, unbound material is carefully washed away, and the substrate for the enzyme added. Release of the chromophore or fluorophore by the enzymic activity can be measured: the intensity of colour is proportional to the amount of antigen present. This methodology is often termed the 'non-competitive double-antibody sandwich' technique.

Specific ELISA techniques are discussed in Parts II and III. Readers are also referred to Wreghitt and Morgan-Capner (1990) and Grange *et al.* (1987).

13.5 CONCENTRATION OF MICROORGANISMS BY IMMUNOCAPTURE

The most successful applications at present are in the detection of target microorganisms in pre-enrichment or selective enrichment broths, perhaps after a shorter-than-normal incubation period.

This can be achieved by stirring the liquid selective enrichment culture with an antibody-coated paddle (e.g. as used in combination with enzyme-catalysed generation of a chromophore in the TECRA® Unique™ procedure for *Salmonella*; see Section 21.3.2).

An alternative capture method uses magnetic beads coated with antibodies specific to the target microorganisms (e.g. obtainable as Dynabeads® from Dynal, Oslo, Norway). The magnetic beads can be stirred into an enrichment broth culture. After allowing time to permit binding of the target organisms to the beads, the beads are retained with a magnet placed against the side of the container whilst the liquid is removed. The beads are washed and can then be transferred to a solid selective medium to detect the presence of any microorganisms on the surface of the beads (see Section 21.4.1.2). Alternatively the bacteria-carrying beads can be examined by an impedance/conductance instrument, by use of polymerase chain reaction, or by ELISA (see also Kroll *et al.*, 1993).

13.6 THE PRECIPITIN TEST

Whereas, in the agglutination test, the antibody causes clumping of an already particulate antigen material (e.g. bacterial cells), in the precipitin test a *clear* solution of a protein or polysaccharide antigen extracted from bacterial cells is used. In the traditional test, the clear antigen solution when mixed with the appropriate clear antiserum will produce a mixture that first turns cloudy and then precipitates.

13.6.1 Preparation of the antigen solution for Lancefield's method of streptococcal grouping

1. Grow the strain of *Streptococcus* under investigation overnight in 50 ml of yeast glucose lemco broth.
2. Centrifuge the broth and discard the supernatant liquid.
3. Add 2 ml of 0.05 N hydrochloric acid containing 0.85% w/v sodium chloride to the sediment of bacteria and transfer the suspension to a small test-tube, using a pipette.
4. Place the tube in boiling water and leave for 10 min.
5. Cool the tube quickly under running water.
6. Add one drop of 0.04% phenol red solution. The liquid should turn the phenol red to yellow.
7. Slowly, *drop by drop*, add 0.3 N sodium hydroxide until the fluid is alkaline (red in colour). Then add, *drop by drop*, 0.1 N hydrochloric acid until the liquid is neutral (the colour should be salmon-pink).
8. Centrifuge and remove the *supernatant liquid* by means of a Pasteur pipette to another test-tube, and discard the sediment. The liquid constitutes the antigen solution, which is ready for use. If it is desired to keep the antigen extract, 0.5% phenol should be added, and the extract kept refrigerated.

13.6.2 The precipitin ring test

This was the most common method of performing a precipitin test and is the test usually employed in streptococcal grouping. It can be made roughly quantitative by using a series of tenfold dilutions of antigen ranging from $1:10$ to $1:10^7$.

Procedure

1. 0.1 ml of antiserum is placed in a 6×50-mm tube.
2. 0.1 ml of prepared antigen or of a serial dilution of antigen solution is layered over the antiserum. This must be done slowly and with care so that a sharp interface forms between the two solutions. A Pasteur pipette with a finely drawn end is used and the tip is placed against the inside of the tube just above the surface of the antiserum so that the antigen solution runs on to the antiserum.
3. Set up, in the same way, a saline–antiserum control and an antigen–normal serum control.

Result

Formation of a white precipitate at the interface of the reagents within 30 min indicates a positive reaction.

If economy of reagents is necessary, capillary tubes may be employed instead of 6×50-mm tubes. The capillary tubes can be held vertical in a block of 'Plasticine'.

When a precipitin test is set up using serial dilutions of antigens, one tube often shows a precipitate before the others. The ratio of dilution of antigen to dilution of antiserum in this tube is called the *optimal ratio*.

Note. Standard antisera are obtainable from Murex Diagnostics, Dartford, UK.

13.6.3 The latex agglutination test

A more rapid method is commercially available, employing latex beads coated with group-specific antibodies. The antigen in the extract is tested against the range of latex reagents. Some commercially available kits (e.g. that from Murex Diagnostics) provide reagents to permit rapid enzymic extraction of the antigen, in place of the more laborious acid extraction described above.

14

Moulds and Yeasts

This section describes very simple basic techniques that can be used in preliminary studies of the microfungi encountered by the food microbiologist. More detailed procedures, such as physiological tests for yeasts, and slide culture techniques will be found in Section 40.

14.1 GENERAL CONDITIONS FOR THE GROWTH OF MOULDS AND YEASTS

Mycological media differ from bacteriological media in a number of ways because of the differing requirements for growth. Most fungi have an optimum pH much lower than that of most bacteria. Also, fungi are more capable of growing on media of inorganic salts with the addition of carbohydrate as an energy source, although some fungi possess a growth requirement for B-group vitamins or other growth factors which may need to be fulfilled by adding 0.1% yeast extract. Almost all moulds, particularly the saprophytes, are obligate aerobes.

14.2 MEDIA FOR THE GROWTH OF MOULDS AND YEASTS

There are many media used for the culture of moulds and yeasts, the following being among the more useful.

1. *Malt extract agar* for the isolation, counting and cultivation of moulds and yeasts. The pH may be at 5.4 or 3.5 depending on the purpose of the medium.
2. *Czapek–Dox agar* and *potato dextrose agar* are general purpose media for the cultivation of both moulds and yeasts.
3. *Davis's yeast salt agar* (Davis, 1958) for counting moulds and yeasts. The pH of this medium can be lowered to pH 3.5 to inhibit bacterial growth.

4. *Orange serum agar* contains clarified orange juice, yeast extract and glucose, and is a medium suitable for culture and counting of microfungi and aciduric bacteria.
5. *Dichloran rose bengal chloramphenicol agar (DRBCA)* (Pitt and Hocking, 1985) for the isolation of moulds from samples containing large numbers of bacteria.
6. *DG18 agar (dichloran 18% glycerol agar)* (Pitt and Hocking, 1985; Pitt *et al.*, 1992) is suitable for the culture of osmophilic (xerophilic) fungi. However, malt yeast 50% glucose agar has been recommended for the culture of extreme xerophiles such as *Xeromyces bisporus* and *Eremascus*.
7. *Osmophilic agar* (Scarr, 1959; Beech and Davenport, 1969) is a wort agar with a high concentration of sucrose and glucose, for the growth of osmophilic and osmotolerant organisms.

Most of the above media are readily available in dehydrated form.

As an alternative to acidification, media such as Davis's yeast salt agar, malt extract agar or orange serum agar can be rendered selective against bacteria either by the addition of chloramphenicol (as used in DRBCA) or, alternatively, by the addition immediately before pouring the plates of sterile penicillin and streptomycin solutions to give final concentrations of 20 and 40 units per ml respectively (Buckley *et al.*, 1969).

For a comprehensive general survey of media used to culture yeasts and moulds see Booth (1971).

14.3 EXAMINATION OF MOULDS

1. Record the colonial characteristics and examine the colonies under the ×10 (low-power) objective of the microscope.
2. Prepare slides of the mould growth for microscopic examination in the following way. Pick off a portion of the growth with a needle and tease it out in a drop of lactophenol–picric acid or lactophenol–cotton blue placed on a microscope slide. Cover with a clean coverslip, taking care to exclude air bubbles.

   Examine the prepared slide under the microscope, first using the low-power objective, and then using the ×40 (high-power dry) objective for a closer examination of a selected field.

   Moulds are examined microscopically in *wet* preparations as described above. Yeasts and some moulds (e.g. *Fusarium*, *Mucor*, *Rhizopus*) can be studied in aqueous mounts, but many moulds are inadequately wetted by water and become enclosed in air bubbles. Consequently a mountant such as lactophenol is employed, usually containing a stain such as picric acid or cotton blue. Lactophenol also has the advantage that wet preparations do not spoil by rapid evaporation as is the case with water-mounted preparations, although a gradual deterioration will occur from other causes.
3. Moulds can also be grown and examined by the use of slide cultures (see p. 359–361).

14.4 EXAMINATION OF YEASTS

1. Colonial characteristics should be noted.
2. Prepare wet mounts by suspending a portion of culture in a drop of water. Add a small drop of Gram's iodine and cover with a coverslip. Observe with the ×40 (high-power dry) objective. Loeffler's methylene blue can be used instead of iodine.
3. Prepare a heat-fixed smear in the usual way and stain by Gram's method.

14.5 IDENTIFICATION OF MOULDS AND YEASTS

Moulds are identified on the basis of morphological and cultural characteristics including:

(a) The colonial characteristics: size, surface, appearance, texture and colour of the colony.
(b) The vegetative mycelium: presence or absence of cross-walls, and diameter of hyphae.
(c) The asexual and sexual reproductive structures: (e.g. sporangia, conidial heads, zygospores, arthrospores).

Yeasts may reproduce by budding, binary fission, ascospores or, less commonly, by other methods. Identification is more difficult than in the case of multicellular moulds, partly because yeasts rarely produce ascospores on ordinary media. Sporogenous yeasts can be induced to form ascospores by subculturing twice on a nutrient agar containing 5% glucose and 0.5% tartaric acid, followed by subculture on to an agar medium containing 0.04% glucose and 0.14% anhydrous sodium acetate only. Other media are described in Section 40.

PART II

Techniques for the Microbiological Examination of Foods

15 Introduction

There are a number of aspects to the role of the food microbiologist in the food industry. These include quality assurance/control on incoming raw materials, quality assurance/control of production, hygiene training for production staff, the development of suitable codes of practice for hygienic food production, assessment of detergent–disinfectants and the establishment of efficient cleaning regimes, the examination of samples of finished products, and the investigation of customer complaints of a microbiological nature. Often, undue emphasis is placed on the microbiological examination of many and frequent samples of the finished product only, with over-simple tests being performed which provide very little real information about the microbiological status of the food. The International Commission on Microbiological Specifications for Foods (ICMSF, 1974, 1986) drew attention to the unsatisfactory nature of casual sampling. It cautioned against deriving an unjustified sense of security from the interpretation of results obtained in unsatisfactory sampling plans. It recommended that sampling procedures should be statistically based: improved sampling plans of the type suggested by the ICMSF are able to provide a basis for statistically valid conclusions about the microbiological quality of the foods tested. Readers involved in the design of sampling plans and in the choice of microbiological standards are strongly recommended to consult ICMSF (1986).

It may be possible for a manufacturer or caterer to require that the suppliers of 'raw' materials meet certain specifications for the supplies to be accepted. For example, in the production of many low-acid canned foods, the heat treatment given needs to be far in excess of any legally required 'botulinum cook', in order to achieve an acceptably low level of spoilage in the finished product. This is because the spore-forming spoilage bacteria will be present in the foodstuff before heat treatment in much greater numbers than will *Clostridium botulinum* spores, and very many of such spores will have much higher D-values than will the *Cl. botulinum* spores. If the raw materials (e.g. rice, spices) that are used in the manufacture of the canned foods, and which are known to contribute the majority of the bacterial spores, are required to meet a specification relating to a low spore count, a less severe heat treatment will be needed to obtain an acceptably low level of spoilage, with a consequent improvement in organoleptic quality.

Quality assurance on production is likely to operate at three levels: hygiene control on the processing (and this includes cleaning regimes for equipment, etc.); detection of possible hazards from pathogenic organisms in the product; and the assessment of the potential shelf-life (or storage-life) of the product. The application of a repressive system of quality control based only on the sampling of finished products and rejection of batches that fail the standard is unlikely to succeed in its aim since microbiological testing is destructive, and in consequence only a relatively small sample of the entire batch can be taken. The passing of the standard by the samples tested can provide no absolute assurance that the rest of the batch would also pass if similarly tested (see also Ingram and Kitchell, 1970; ICMSF, 1986). An example of a repressive system that can provide a good measure of assurance (except in the case of deliberate sabotage) is the in-line detection of foods that contain adventitious metal particles, which by means of electronic metal detectors can be applied non-destructively to every item in a production run for a very wide range of food products. There is no way in which a food company can protect itself against a statutory zero-tolerance on pathogens by microbiological testing of samples of the finished product. The only possibility of protection would come from extremely rigorous on-line control of the hygienic precautions taken. For this to be successful the cooperation of food handlers and production staff is necessary.

Thus a more positive approach to quality assurance is one based on the provision of training courses in hygiene for production staff, and the application of agreed codes of practice which, although they may be based on officially published Codes of Practice, will be more helpful and easier to understand if they are written for the specific production situation. In this case results obtained on samples of the finished products are used as an indicator of the success of the hygiene codes. In factories with fairly stable operating conditions in which a processing line produces a few lines only, it may be possible to use control chart procedures to provide early warning of emerging problems. Interested readers are referred to the useful discussions by Steiner (1984) and Wetherill (1977) of control chart procedures in quality control.

Detailed microbiological examinations of samples taken during and after the production and processing of food products *can* provide useful information, allowing an assessment of the probable shelf-life, a check of the process in order to correct or anticipate any deterioration in the production methods, and some indication of potential public health hazard. However, the choice of the methods of examination for a particular food product requires, amongst other things, an intimate knowledge of the preparation, storage and distribution of the raw constituents. The ICMSF (1974) suggested that, in deciding whether to test a food sample for a given pathogen, the known food-borne disease record of that food should be considered. However, the lack of a record of implication in food poisoning may be merely due to the non-recognition of the existence of the hazard. Space only allows two examples to be given here to indicate the possibilities for misjudgement, and which caused substantial outbreaks 10 years or more before the recommendations of ICMSF (1974). Other examples are mentioned in Parts II and III of this book. Until the late 1950s many cakes and confectionery products incorporated in their external decoration raw shredded desiccated coconut. Such products were frequently decorated after cooking so that the coconut received no heat treatment. Outbreaks of salmonellosis occurred, in which uncooked desiccated coconut was the common epidemiological factor. It was discovered that the desiccated coconut was peculiarly liable to contamination by *Salmonella* because of the method of preparation and the drying process involved (Wilson and MacKenzie, 1955). Until the outbreaks of salmonellosis occurred in which the coconut was implicated, few people would have considered routine examination for *Salmonella* to be a test that had any significance in the microbiological analysis of desiccated coconut. Yet this *Salmonella* hazard may have existed for some time and may have resulted in a number of outbreaks of food poisoning. Without knowledge of the hazard, the questions asked of patients concerning their food consumption histories could easily have been insufficient in these hypothetical earlier outbreaks to detect the fact that desiccated coconut was the common factor.

In the second example, until the laboratory investigations that resulted from the Aberdeen typhoid fever outbreak in 1964 (Report, 1964; Howie, 1968) it had been considered that unspoiled cans of heat-processed meats such as corned beef could present no significant hazard as a vehicle for heat-sensitive pathogens such as *Salmonella*. On investigation it was found that when gas-producing coliform organisms were introduced into a can of corned beef together with *S. typhi*, the latter could easily outgrow the coliforms and prevent their producing gas. It was demonstrated *in vitro* that the anaerobic growth of *S. typhi* was enhanced by the presence of nitrate in concentrations in the medium equivalent to those found in corned beef (Meers and Goode, 1965). The Aberdeen typhoid outbreak in 1964 caused the Committee of Enquiry (Report, 1964) to review an outbreak of typhoid fever which occurred in Oswestry in 1948. When the Oswestry outbreak had been first investigated it had been decided that circumstantial evidence pointed to milk as the vehicle but later investigations showed this to be unlikely. The re-examination of the results of the epidemiological investigations (Report, 1964) strongly suggested the possibility of corned beef being the vehicle. The *S. typhi* strain involved in the Oswestry outbreak was then retyped and found to be phage-type 34 – the same type as that responsible for the Aberdeen outbreak.

Thus, in placing reliance on known records of involvement when trying to decide which tests should be performed on a food, it should be recognized that these records are in themselves imperfect indicators of hazards. The ICMSF (1974) pointed out that 'most food control efforts should be directed to the areas of greatest risk'. Nevertheless, whilst *priority* in testing should be given to known hazards, when time, expense and other considerations allow, it may be salutary to use tests given a lower priority.

The two examples just given concern contaminated materials which themselves caused the outbreaks of disease. However, the fact that a contaminated raw ingredient is to be incorporated in a product receiving a heat treatment sufficient to destroy the pathogens is not necessarily a sufficient safeguard, as in most factories, restaurants and shops (and indeed homes) cross-contamination can occur all too easily. For example, outbreaks of paratyphoid fever in which cakes filled or decorated with imitation cream were implicated were found by Newell (1955) to be due to contaminated frozen egg. The frozen egg was being incorporated into the cake or pastry. While the baking may be expected to have destroyed the salmonellae in the cake mixture, the imitation cream, which did not contain frozen egg as an ingredient, was liable to contamination with *Salmonella* because in each of the bakeries involved the same mixing machine was used for mixing both cake and filling. In some cases the bowls and machines were not even rinsed between mixing the cake mix and the imitation cream.

The investigation of complex food products and their constituents for the presence of potential spoilage or food poisoning organisms also requires care. For example, certain spices may give very high viable counts. Black pepper normally contains a large number of aerobic spore-bearers. The introduction of pepper into a food product towards the end of its preparation (e.g. in *pommes de terre duchesse*) may therefore introduce significant numbers of *Bacillus*, which may, under certain circumstances, be able to multiply sufficiently to cause spoilage or *B. cereus* food poisoning. In the preparation of the product already mentioned, namely *pommes de terre duchesse*, cooked sieved potato is mixed with butter, egg, pepper and nutmeg, and piped into attractive shapes which are then flashed under a grill, both to brown the outside and to reheat to serving temperature. If this were being prepared in a large restaurant, the situation might arise where large amounts of the product were made at one time and kept until required. Immediately before serving, it may or may not be warmed under a grill or in an oven. Such a situation could result in numbers of *B. cereus* sufficient to cause food poisoning. *B. cereus* food poisoning was very common in Hungary, and Ormay and Novotny (1969) suggested that this may have been due to the use of large amounts of spices customary in the production of traditional Hungarian meat and vegetable dishes. The outbreaks of *B. cereus* food poisoning caused by fried rice dishes (Vernon and Tillett, 1972) provided another example of the potential impact of modifications to kitchen practice caused by problems resulting from scaling up production in catering establishments (see Gilbert *et al.*, 1974).

Thus, the more complex a food product, the more care is required in the correct selection of samples to be tested, in the correct selection of tests and counts to be made, and in the interpretation of the results obtained.

16

Methods of Sampling and Investigation

It is necessary to take large and representative samples of foods. In the case of a packaged food product, the package should be opened only in the laboratory and sampling must be performed aseptically. All apparatus used for sampling should have been previously wrapped and sterilized. Batches of foods that are in packets, cans, bags or other containers should be examined by individually testing a number of units selected at random. The International Commission on Microbiological Specifications for Foods (ICMSF, 1986) discusses in detail the criteria for determining the stringency of sampling that may be required for any given combination of food, type of processing, method of storage and type of microorganisms under consideration.

16.1 LIQUID SAMPLES

It is usually relatively easy to obtain representative samples of liquids. Frequently the liquid (e.g. milk, ice-cream mix, sugar syrups) will be held in a vat and will be subject to continuous or periodic mixing; otherwise the liquid should be mixed thoroughly up and down (e.g. using a sterile ladle) before the sample is taken. A large sample (e.g. 100–500 ml) should be withdrawn into a sterile container for transport to the laboratory. In the laboratory the liquid should be mixed thoroughly once again before pipetting the amounts required for investigation.

16.2 SOLID SAMPLES

Sampling of solids may be performed using sterile scalpels, spoons or cork-borers depending on the nature of the material to be sampled. A particulate food such as flour or dried milk is capable of sufficient mixing to enable a single, fairly small sample (e.g. 100 g) to be taken. If, however, the product is in bulk, larger samples from more than one location should be taken. The large samples should then be treated separately, each being mixed thoroughly in the laboratory before smaller samples are removed for testing.

Meat, fish and similar foods may be examined by taking deep samples as well as surface samples. Deep samples should be taken with care to minimize contamination from superficial levels. Some foodstuffs such as fresh and cooked meats can be sampled using sterile carving knife, scalpel and forceps. In the case of frozen foods a cork-borer or even an electric drill fitted with a bore-extracting bit can be used to obtain deep samples without the need for thawing.

16.3 SAMPLING OF SURFACES

Surface slices may be removed, or alternatively surfaces may be examined by transferring the microorganisms from sample to microbiological medium with the aid of a supposedly inert carrier that neither causes death nor allows multiplication of the microorganisms. Such carriers include rinses, swabs and adhesive tape. As the continued long-term viability of the microorganisms on the carrier without death or multiplication is difficult, if not impossible, to achieve, the microorganisms must be inoculated into suitable media at the earliest opportunity. The longer the delay before inoculation, the less reliable the quantitative assessments.

A further method is to transfer the microorganisms directly from the sample surface to the surface of the medium by impressing the one upon the other. The two principal impression techniques that achieve this are the agar sausage (Ten Cate, 1965) and the impression plate. Plates with a suitable profile are available commercially – they have a cross-section similar to that in Fig. 16.1. They are listed in manufacturers' catalogues as 'contact plates' or 'RODAC plates' (Replicate Organism Direct Agar Contact).

Impression techniques and adhesive tape transfer do not allow dilution series to be prepared, and therefore colony counting is possible only when the microbial load is small. A great advantage of impression techniques, adhesive tape, contact slides and swabs is that they allow a non-destructive examination of the food samples.

The seven methods described below are not quantitatively equivalent and sometimes may not even rank samples in the same order according to their apparent bacterial load.

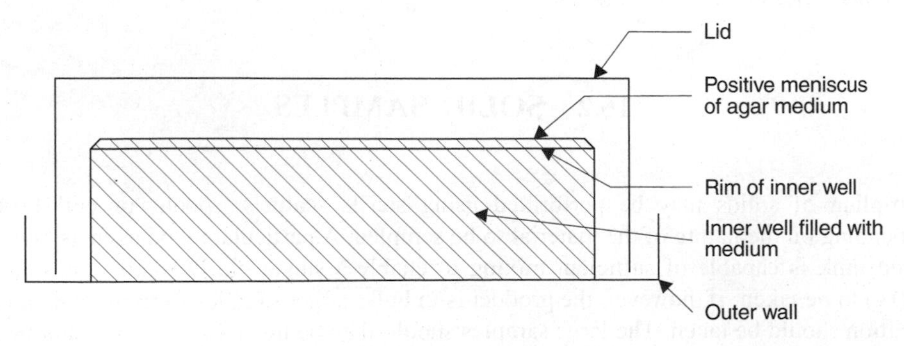

FIG. 16.1 Cross-section of an impression plate.

16.3.1 Surface slices

Remove very thin slices of the superficial layers of the food using sterile scalpels and forceps. In the case of table poultry, for example, the skin provides suitable material for sampling. Homogenize the slices in a suitable diluent to obtain an initial 10^{-1} dilution.

16.3.2 Rinses and washes

Rinse or wash the food (one part by weight) in sterile diluent (ten parts by weight) and then consider the washings to be the initial 10^{-1} dilution. This procedure is applicable to foods such as sausages, dried fruits, vegetables and salad vegetables. It should be borne in mind that frequently the microorganisms on the surface of the food will not be detached merely by agitation in the diluent so it is advisable also to obtain samples from which dilutions can be prepared by comminution. The extent of agitation and consequent removal of microorganisms should be standardized. In reporting the results, it must be recorded that the count represents bacteria on the surface only.

16.3.3 Swabs

When quantitative results are required, the area to be examined should be defined by the use of a previously sterilized template.

16.3.3.1 Cotton-wool swabs

Cotton-wool swabs are prepared from non-absorbent cotton-wool wound to a length of 4 cm and a thickness of 1–1.5 cm on wooden sticks or stiff stainless-steel wire. They should be placed in alloy tubes which are then plugged and sterilized.

Procedure

Moisten the swab with sterile quarter-strength Ringer's solution and rub firmly over the surface being examined, using parallel strokes with slow rotation of the swab. Swab the surface a second time, using parallel strokes at right angles to the first set. Care must be taken that the whole of the predetermined area is swabbed. Replace the swab in the tube. To prepare counts, add 10 ml of quarter-strength Ringer's solution. Agitate the swab up and down in the tube ten times to assist the rinsing of the bacteria from the surface of the swab. Prepare plate counts from 1-ml amounts of the swab washings and from dilutions prepared as required. Counts should be recorded as the number per square centimetre of surface swabbed.

16.3.3.2 Alginate swabs

Alginate swabs are an alternative type of swab (Higgins, 1950), and are prepared from calcium alginate wool. It is recommended that the amount of alginate wool used for each swab should not exceed 50 mg. The alginate wool is wound on a 1.5-mm diameter wooden stick to give a swab 1–1.5 cm long by 7 mm in diameter, moistened *very slightly* in quarter-strength Ringer's solution, placed in a 10×1-cm metal tube and sterilized by autoclaving at 121°C for 15 min. It is convenient to close the tube with a cork instead of cottonwool, and to mount the wooden stick in the cork. The cork may then be used as a handle.

Procedure

Swab a predetermined area of the surface to be examined by rubbing firmly over the surface in parallel strokes, with slow rotation of the swab. Since alginate wool is very smooth, it is necessary for the rotation of the swab to be in the direction that will prevent the swab from unwinding. Then swab the same surface a second time, using parallel strokes at right angles to the first set. After swabbing the surface to be examined, and when it is required to carry out the laboratory examination, break off the swab aseptically into a screw-capped bottle containing 10 ml of a sterile 1% solution of Calgon (sodium hexametaphosphate) in quarter-strength Ringer's solution. Replace the cap and shake the bottle vigorously. This causes the alginate wool to disperse and dissolve, giving a suspension of all the bacteria present on the swab. Prepare plate counts from 1- and 0.1-ml amounts (or further dilutions as required) using appropriate media. Trypticase soya agar (or an equivalent such as tryptone soya agar or plate count agar) in place of nutrient agar has been recommended by Post and Krishnamurty (1964), as it appears to nullify to some extent the slightly bacteriostatic nature of the Calgon (Post *et al.*, 1963). To reduce the final concentration of Calgon to a less inhibitory level, 15–20 ml of medium should be added to each plate. Counts should be recorded as the number per square centimetre of surface swabbed.

16.3.4 Adhesive tape

This method of surface sampling involves the use of a suitable self-adhesive tape (e.g. 'Sellotape', 'Scotch Tape') or of self-adhesive labels. Self-adhesive labels have the advantage that the sampling details can be written on the back of the label and that they are already attached to a non-adhesive mount; self-adhesive tape needs to be transferred to a suitable sterile mount. Thomas (1961) described the use of self-adhesive tape and labels for investigations on the microbial flora of human skin, but food surfaces and equipment surfaces also can be examined in this way. Tapes and labels are frequently self-sterilizing for a short time after manufacture, by the action of volatile solvents, etc. Trials should be run on any tape intended for use in this way, to obtain assurance on the sterility of the tape and on the absence of residual bactericidal activity.

The adhesive strip or label should be turned back on itself at one end for about 1 cm to form a tab for holding it. The strip is removed from its mount, pressed against the surface to be examined, pulled off immediately and replaced on its mount. In the laboratory, the strip is removed from the mount, pressed against the surface of an appropriate culture medium, and then removed and discarded.

16.3.5 Agar sausages

The agar sausage consists of sterile agar medium solidified inside a sterile cylindrical plastic casing (Ten Cate, 1965). The sausages may be made in the laboratory, or may be obtainable ready-made in some countries. To use, the end of the agar and casing is cut off aseptically and the exposed sterile agar surface pressed against the sample surface. A slice of agar is then removed using a sterile scalpel, placed aseptically into a Petri dish, impressed side uppermost, and incubated.

16.3.6 Impression plates

Impression plates are plates of the form shown in Fig. 16.1; they are available in a sterile disposable plastic form from a number of manufacturers.

The centre well is filled with sufficient of the required agar medium to produce a convex meniscus. When set, the agar surface can be pressed against the surface to be sampled.

16.3.7 Contact slides (Thomas, 1966)

Press a sterile glass slide against the food sample to be examined. Transport the slide back to the laboratory. It may then be examined microscopically after fixing and staining (e.g. by Gram's method). Alternatively, the slides may be impressed on to poured plates of media, to achieve some transfer of the organisms to the agar surface. After removal of the slides (using sterile forceps) the plates may then be incubated. Although this method does not allow quantitative estimates to be made, it is useful for rapidly determining the main types that comprise the dominant microflora, especially on such foods as raw meat, poultry and soft cheeses.

16.4 SAMPLING FOR ANAEROBIC BACTERIA

If an examination for anaerobic bacteria is to be undertaken it is important that food samples likely to contain little free oxygen, for example deep tissues of meat, should not be exposed to normal atmospheric concentrations of oxygen as would occur if small samples were taken. When small samples are unavoidable, and also when swabs are used, a suitable transport medium (e.g. Stuart's transport medium) which is capable of maintaining reduced conditions must be employed. Thus, if an alginate wool swab is used, it should not be reinserted into its original tube but placed in a bottle of Stuart's transport medium. The swabs may also be moistened with reinforced clostridial medium before use.

16.5 ATTRIBUTES SAMPLING PLANS

16.5.1 Sampling to detect and count low concentrations of organisms

When it is desired to count concentrations of viable organisms in foodstuffs that are below the limit of sensitivity of colony count techniques, it is possible to do so using a multiple tube technique, employing successive dilutions as described in Section 5.3. If necessary, large amounts (e.g. 100, 10 and 1 g) of food can be used as the inocula into suitable large containers holding appropriate amounts of medium.

An alternative approach, known as an attributes sampling scheme, is to take a number of samples of the same size, and to determine whether any contain the organisms in question. If the organisms are not detected it is then possible to calculate the maximum proportion of units of that size that will, with a given probability, contain at least one of the organisms in question.

If the foodstuff is distributed already in discrete units (as is the case with canned foods), it is preferable to adopt this unit as the basis for each sample. Thus such attributes sampling schemes are readily applied to problems such as the determination of the proportion of cans of food, or packets of milk, that contain spoilage organisms. It is not so obvious that attributes sampling schemes can be applied to foodstuffs that are not distributed in a large number of discrete units (e.g. a vat of ice-cream mix), and also to foods in which the unit size is inconveniently large (e.g. a large can of frozen egg albumen).

In such cases, it is assumed that the distribution of the organisms is random throughout the batch, and that the batch of food is made up of units of a given size (e.g. 10 or 25 g). A number, n, of such units is examined for the presence of the organisms being sought. Then the proportion, in the entire batch, of positive units (i.e. units containing at least one organism) that may be detected with a given probability by at least one positive unit being found amongst the n units examined is given by:

$$d = 100\,(1 - \sqrt[n]{1-p})^{*}$$

*This is derived from the formula $p = 1 - (1 - d/100)^n$, which shows the probability of at least one defective unit being in the sample of n units taken, when d is the percentage defective in the entire batch. This relationship is true provided that n is a small fraction (less than one-quarter) of the entire batch.

For example, suppose that we take six units each of 25 g and examine them for the presence of bacteria. Then the percentage of positive units in the whole batch that will be detected with a 95% probability, by at least one of the six units being found positive, is given by:

$$d = 100\,(1 - \sqrt[6]{1 - 0.95})$$

Hence

$$\frac{d}{100} = 1 - \sqrt[6]{0.05}$$

$$1 - \frac{d}{100} = \sqrt[6]{0.05}$$

$$\log\left(1 - \frac{d}{100}\right) = \frac{1}{6}(\log 0.05) = \frac{1}{6}(\bar{2}.6990) = \frac{1}{6}(-1.3010)$$

$$1 - \frac{d}{100} = \text{antilog}(-0.2168) = \text{antilog}(\bar{1}.7832) = 0.6070$$

$$\frac{d}{100} = 1 - 0.6070 = 0.393$$

Therefore, $d = 39.3\%$.

That is, if all the units are negative then, with $p = 0.95$, the number of microorganisms in the batch is such as to give less than 39.3% of 25-g units positive (in other words there are fewer than 40 bacteria per 2.5 kg in the long run).

If we wish to find the number of units that needs to be sampled to detect a given proportion of defective units, we can do so from:

$$n = \frac{\log(1 - p)}{\log\left(1 - \dfrac{d}{100}\right)}$$

For example, suppose that we wish to apply a standard for a food such that, with a 95% probability, the foodstuff contains less than one *Salmonella* per 500 g. How many 25-g units must be sampled, and found to be positive, in order to detect a level of contamination of one organism per 500 g? There are 20 units of 25 g in 500 g. Thus the standard we wish to apply is equivalent to there being less than 5% of defective 25 g units (i.e. less than one unit in any 20 units being positive in the batch as a whole), with a 95% probability. The number of units, n, that needs to be taken is given by:

$$n = \frac{\log(1 - 0.95)}{\log\left(1 - \dfrac{5}{100}\right)} = \frac{\log 0.05}{\log 0.95} = \frac{\bar{2}.6990}{\bar{1}.9777} = \frac{-1.3010}{-0.0223} = 58.3$$

That is, we need to sample 59 units, each of 25 g, all of which must be negative, in order to ensure that in the long run the food contains (with a 95% probability) less than one *Salmonella* per 500 g.

16.5.2 Use of attributes sampling plans in conjunction with colony or most probable number (MPN) counts

In the above section we considered the use of attributes sampling plans for detecting small numbers of organisms. The ICMSF (1974, 1986) recommended the application of three-class attributes plans to microbial counts in general, and used such plans as the basis for its recommended microbiological standards for a wide range of foodstuffs. In a three-class plan, instead of the yes–no decision of the two-class plan discussed above, a further class is distinguished, the three classes being: acceptable, marginally acceptable, and wholly unacceptable. The two counts relating to the two thresholds are conventionally indicated as m and M respectively. Whereas any count above the limit for unacceptability will lead to rejection of the batch, a certain proportion, c, of counts in the 'marginally acceptable' range is allowable. Thus the use of a three-class attributes plan in conjunction with counts such as general viable counts, coliform counts, etc. will require the testing of multiple samples taken at random. One of the stated advantages of three-class plans is that the probability of acceptance or rejection of a batch is less affected by the nature of the distribution of the organisms within the batch than is the case with two-class plans. For further details see ICMSF (1986) and Harrigan and Park (1991).

16.6 CHOICE OF SAMPLES ON A NON-RANDOM BASIS

Usually attempts are made to take samples randomly. The samples to be taken from a production line or from a store may, for example, be chosen using a table of random numbers to identify the time of sampling or the location of samples. However, a conveyor line for a cooked food may show a gradual build-up in microbial numbers over the period of a run. On the other hand, a pipeline system for food distribution, if inadequately cleansed and if at least a rinse is not performed immediately before starting a production run, may cause higher counts to be obtained at the beginning of the run. If the food is then distributed through the pipeline at a temperature within microbial growth range, a build-up of microorganisms may occur towards the end of the run.

In the case of a batch of food in store, taking samples on a random basis may be justified if the storage conditions are the same for the entire batch. Often, however, temperature and other gradients will occur across the stack of food. Obviously in such cases most information would be obtained by the choice of samples from specific locations in the stack. The microbiological effects of such gradients can then be determined, particularly if there is opportunity for recording the environmental parameters involved (e.g. by use of a multichannel recorder and a number of thermocouples to detect temperature gradients in batches of food kept in cold stores).

In a microbiological examination of a food production to determine the effects of the elements of that process (e.g. as would be required in setting up and verifying a Hazard Analysis Critical Control Point (HACCP) protocol for a specific production line), the samples taken will be carefully chosen in order to verify the correct identification of hazards, their location points, and the appropriateness of the critical control points.

16.7 TRANSPORT AND STORAGE OF SAMPLES

Whenever possible, the original state of the sample should be maintained until the laboratory tests are carried out. For example, frozen foods should be kept frozen by using solid carbon dioxide (with precautions to protect the frozen food samples against exposure to the gaseous carbon dioxide) or 'cold packs', in conjunction with insulated containers for transport to the laboratory, followed by storage in a deep freezer until tested.

Perishable but unfrozen food samples should not be frozen before testing (unless the testing *must* be long delayed) but, if storage for a short period is unavoidable, the samples should be chilled and kept refrigerated at about 4°C until tested. It should be remembered, however, that refrigeration of samples for periods of 3 days or more will result in the multiplication of any psychrotrophic microorganisms present in the food, and may cause the death of some mesophiles or thermophiles.

The history of sampling, transport and storage should be recorded for all samples.

17

Preparation of Dilutions

Throughout the methods for food sampling it is suggested that initial 1 : 10 dilutions of food samples be prepared from 10-g amounts of the food, as frequently this is the largest sample that can conveniently be handled in many laboratories. Nevertheless, it is recommended that, if possible, 1 : 10 dilutions are prepared by suspending or homogenizing 50 g, as this gives much more reliable and representative results, whilst remaining a quantity reasonably easy to handle.

It is preferable to prepare the initial 1 : 10 dilution by using a sterile bottle marked at 100 ml (for a 10-g sample) or at 500 ml (for a 50-g sample), the homogenate being made up to volume with diluent. Subsequent dilutions can then be prepared volumetrically by pipette. Calculations of counts can correctly be expressed per gram of sample. The more usual method of preparing a 10^{-1} dilution by adding 90 ml of diluent to 10 g of food introduces an uncertainty into the calculation as the extent of solubility of the food and the food volume are unknown.

It should be noted that the use of refrigerated diluent when the sample is not at refrigerator temperature may cause a drop in count as the result of cold-shock (Meynell, 1958; Gorrill and McNeil, 1960; Strange and Dark, 1962). Most workers have been in agreement that it is usually Gram-negative bacteria that show cold-shock, that exponential-phase cells are most susceptible (stationary-phase cells being most resistant) and that the effect is more marked in quarter-strength Ringer's solution than in sucrose solutions or complex diluents. However, there has been at least one report (MacKelvie *et al.*, 1968) of a *Pseudomonas* that became more susceptible to cold-shock as the cells aged.

17.1 CHOICE OF DILUENT

17.1.1 General-purpose diluents

For many purposes sterile 0.1% peptone water at pH 6.8–7.0 (Straka and Stokes, 1957), phosphate buffer or quarter-strength Ringer's solution are satisfactory diluents; 0.1% peptone water appears to have a greater protective effect than phosphate buffer or quarter-strength Ringer's solution, and has probably become the most commonly used diluent. However, ISO standard methods in recent years have started to specify 0.1% peptone plus 0.85% sodium chloride as the normal general purpose diluent (see, for example, ISO 6887:1983).

The possible modification of the nature of the diluent by the sample at the lowest dilutions should not be overlooked. In particular, if the food sample contains a high proportion of undissolved water-soluble material, what will be the effect of its solution in the diluent? Will the pH or available water (a_w) be greatly affected? If in doubt, the pH of the first dilution should be checked, and a_w can be determined by use of a SINA® equihygroscope or similar instrument (some indication of a possible change in a_w may be obtained by simple determination of the soluble solids). It may be necessary to add phosphate buffer to the diluent to counter the pH drift. For example, Dixon and Wilson (1960) succeeded in isolating *Salmonella* from superphosphate-containing horticultural fertilizers by suspending the samples in 0.5 M phosphate buffer at pH 7.0–7.2. Such fertilizers had been thought previously to be free from *Salmonella*.

If dilutions of a highly soluble dry specimen (e.g. milk powder, baby food preparations) are being prepared, so that a low a_w occurs at the lowest dilution, the preferred diluent may even be distilled water (Silverstolpe *et al.*, 1961); the most suitable diluent should be chosen by testing a range of diluents and the one that gives the highest recovery rates used subsequently.

Inoculation of media should be carried out within 15–30 min of the preparation of the dilutions.

17.1.2 Diluents for anaerobes

When examining a sample for anaerobic bacteria either qualitatively or quantitatively, reinforced clostridial medium or a similar formulation must be used as the diluent in order to maintain a low oxidation–reduction (redox) potential. Dispersion of the sample in the diluent should be achieved by a method that provides least opportunity for introducing oxygen into the mixture. For example, the Colworth Stomacher (see below) is better than a top-driven homogenizer or a bottom-driven blender unless the sample container is flushed with oxygen-free nitrogen before blending.

If an examination is being undertaken for very oxygen-sensitive anaerobes, as well as using appropriate diluents, special precautions need to be taken such as using the Hungate technique (Hungate, 1969) or the use of an anaerobic workstation.

17.1.3 Diluents for osmophiles and halophiles

A suitable diluent for osmophilic counts is sterile 20% sucrose solution. For investigations of halophilic organisms (e.g. in samples of curing brines), sterile 15% sodium chloride can be used as the diluent.

17.2 LIQUID SAMPLES

Dilutions of liquids can be prepared in a manner similar to that described in Part I, Section 5.1. Pipette aseptically 10 ml of the thoroughly mixed sample into a sterile glass bottle with a ground glass stopper and make up to 100 ml with sterile diluent to give a 1:10 dilution v/v. Alternatively, weigh, with aseptic precautions, 10 g of the thoroughly mixed sample into the bottle and make up to 100 ml with sterile diluent to give a 1:10 dilution w/v, which, for all practical purposes, is equivalent to a 1:10 w/w. Prepare further decimal dilutions as necessary in the usual way. Inoculation of media should be carried out within 15–30 min of the preparation of the dilutions.

17.3 FINE PARTICULATE SOLID SAMPLES

Preparation of the initial dilution of fine particulate solid samples such as flour and milk powder is accomplished easily. Weigh aseptically 10 g of sample into a sterile glass bottle which is marked at 100-ml capacity and fitted with a ground glass stopper. Add sterile diluent to the 100-ml mark to give a 1 : 10 dilution w/v. Shake the suspension 25 times with an excursion of 30 cm. Prepare further dilutions as necessary in the usual way. Care should be taken with highly soluble samples that the counts obtained by distribution of measured volumes can be related accurately to the original sample (expressed as the count per gram). In order to achieve this it is necessary to confirm the volume of the first dilution as 100 ml. The possibility of alteration of pH and a_w must also be considered as already mentioned above as 100 ml. Inoculation of media should be carried out within 30 min of the preparation of the dilutions.

17.4 OTHER SOLID SAMPLES

17.4.1 Comminution

Tests of foods that may contain microorganisms below the surface should be carried out by blending at least 10 g of the food in the appropriate amount of sterile diluent using suitable apparatus. The preferred method of blending is by use of the Stomacher.

The Colworth Stomacher was developed at the Unilever Research Laboratories (Sharpe and Jackson, 1972). This type of apparatus is now available from a large number of manufacturers and suppliers of laboratory equipment.

The sample plus diluent are placed inside a sterile, disposable, thin and flexible polythene bag. The bag is placed into the blender chamber with a few centimetres of bag projecting above the top of the door of the chamber; closing the door firmly seals the bag while blending occurs. Starting the Stomacher causes two large flat stainless steel paddles alternately to compress the bag plus its contents against the flat inner surface of the door. Operation for 30 s is sufficient for dispersing most samples, although samples containing high concentrations of fat should be processed for 90 s.

Amongst the advantages of this apparatus are that the samples do not come into contact with the blender but are contained in cheap disposable bags; practically no temperature rise occurs; good dispersal is obtained even with deep-frozen samples; in the case of many foods very low dilutions can be prepared if low counts are suspected. It is also quiet compared with other forms of homogenizer. The apparatus is highly recommended. Tuttlebee (1975) in a comparative study of the use of Stomacher and Atomix concluded that in all respects the Stomacher was to be preferred.

17.4.2 Examination of surfaces

The methods of examining surfaces of food samples have been described in Section 16.

18

General Viable Counts

General viable counts are determined usually by colony counting methods (see Section 5.1), although the multiple tube technique may be used if low concentrations of bacteria are expected. The choice of medium and incubation conditions is difficult when general viable counts are attempted on the mixed microflora usually found in foods. Frequently viable counts are required of populations for which there is little knowledge of the types of organisms present, and in these circumstances, because of the variety of nutritional and physical requirements represented, it is impossible to obtain counts that truly indicate the number of viable organisms present.

A sample may contain some organisms that are obligate aerobes, some that are obligate anaerobes and some that are facultative with respect to oxygen; some of the organisms may be capable of good growth at 37°C but not at 20°C, some may grow at 20°C but not at 37°C, and others may grow at both temperatures. Even if four sets of plates were to be incubated – one set aerobically at 20°C, another aerobically at 37°C, a third anaerobically at 20°C, and the fourth anaerobically at 37°C – it would not be possible to determine the *total* viable population of these organisms, because an unknown proportion of the population would grow and be counted on more than one set of plates. Microbiologists should not be tempted to report an 'aerobic mesophilic count' as a 'total viable count' – it is no such thing.

Similar problems occur when attempting to choose media. Colony counts and multiple tube counts assess viability as the proportion of the population capable of continued multiplication (i.e. clonal viability) in the incubation conditions provided. In other words, the development of a colony (on or in an agar medium) or of turbidity indicates clonal viability in the incubation environment and *not* clonal viability in the environmental conditions found either in the normal post-processing state of the food or in the consumer. This distinction is especially important when stressed or damaged organisms are encountered in, for example, processed or preserved foods. Thus if a food sample be examined using both total (microscopic) count and viable count, a high total count in conjunction with a low viable count does not necessarily indicate that the majority of the organisms observed microscopically are dead, but perhaps merely that they are unable to multiply in the particular incubation environment.

The food microbiologist has to choose one or more sets of incubating conditions on empirical grounds, and as a general rule incubation conditions are chosen that have been found to give the highest counts in the largest number of samples. In most test situations more than one incubation temperature is likely to be employed:

0–10°C for counts of psychrotrophs and psychrophiles
20–32°C for counts of saprophytic mesophiles
35–37 (or 45)°C for counts of parasites and commensals of homoiothermic animals
55–63°C or higher for counts of thermophiles

Psychrotrophic and psychrophilic counts are best determined using surface count techniques so that the bacteria are not exposed to molten agar. True psychrophiles, which are probably rare in foods except those of marine origin, may be killed by exposure to ambient temperatures and, if their presence is suspected, samples, diluents and plates must be kept at temperatures below 10°C (Farrell and Rose, 1967).

The more complex the medium, the greater the range of types *likely* to be capable of growth (but see Section 35.1). Thus, glucose tryptone yeast agar (plate count agar) allows the growth of more types than does nutrient agar. The author obtained a viable count of less than 10 000 per gram of a frozen precooked meat dish when nutrient agar was used, whereas the same sample gave a count of several million per gram when plate count agar was used as the plating medium. The discrepancy occurred because the predominant component of the microflora of the food was *Enterococcus*, which required glucose for growth. Supplementation by the addition of serum, egg yolk or milk may provide even greater counts with certain types of sample, but such supplementation is not used routinely because of the extra cost and the increased complexity of medium preparation (serum and egg yolk need to be added immediately before pouring the plates).

If the interest is in the spoilage potential of the microflora then the medium should be at least partially habitat-simulating. For example, plate count agar plus 0.1% skimmed milk is frequently specified for the examination of dairy products (see ISO 6610:1992). In addition, the incubation temperature(s) should be chosen to reflect the storage temperatures to which the food is exposed. In attempting to assess the efficacy of the incubation conditions, the use of direct microscopic examination of the samples and dilutions can be useful. If a particular morphological type of organism is seen microscopically but is not recovered on isolation media, there is a strong possibility that the isolation environment is proving unsuitable for the growth of that microbial type. It is not uncommon for a significant component of the microflora of a food or other habitat to be unable to grow unless media are supplemented with an extract of the food or other material comprising the habitat. A good example of the usefulness of comparing microscopic examinations of samples with the types of microorganisms recovered by plating was provided by Varnam and Grainger (1973). Gram-negative rod-shaped bacteria were observed by microscopic examination to predominate in Wiltshire bacon-curing brines. However, to recover many of these organisms, Varnam and Grainger found it necessary to use media that included 10% and 20% sodium chloride and 15–25% filter-sterilized unheated aqueous extract of pork. They subsequently concluded (Varnam and Grainger, 1975) that the extract was not providing an essential nutrient, but rather acted by removing peroxide from the medium, as the pork extract could be satisfactorily substituted by manganese dioxide as suggested by Meynell and Meynell (1970).

Many marine bacteria are unable to grow unless at least part of the distilled water in the medium is replaced by sea water.

When attempting to study the constituent microflora of a habitat, the effects of similar supplementations should always be tested. However, care must be taken in interpreting and comparing microscopic and cultural examinations as some organisms (e.g. coryneform bacteria) may exhibit different morphological appearances in the sample and on the medium.

Bacterial counts in the presence of moulds and yeasts may be determined by using media incorporating the antifungal antibiotic cycloheximide to 10 p.p.m.

18.1 AEROBIC MESOPHILIC COUNTS

In some cases general viable counts may be used as indicators of the standard of hygiene or of plant sanitation (incubation temperatures of 25–30°C probably being best for this purpose), or as indicators of a potential health hazard (incubation temperatures of 30–37°C probably being best). Usually in these cases glucose tryptone yeast extract agar (plate count agar) is specified, usually incubated at 30° or 32°C (e.g. see ISO 4833:1991); this count is known as an 'aerobic mesophilic count' (AMC). However, it is likely that counts of more restricted groups of organisms will serve as better indicators in the case of most foods.

18.2 YEAST AND MOULD COUNTS

Yeast and mould counts were previously obtained by employing media such as orange serum agar, Davis's yeast salt agar or malt extract agar, in which the acidity was adjusted to around pH 3.5 by use of sterile lactic acid or citric acid immediately before pouring the plates. However, mould counts are now usually obtained by the use of dichloran rose bengal chloramphenicol agar (DRBCA). It must be realized, however, that, whereas counts of yeasts present similar problems to bacterial counts (yeasts being unicellular organisms), mould counts may be difficult or impossible to interpret. As most moulds will grow in a mycelial form in foods, the number of colonies obtained on a plate will depend on the degree of homogenization and the extent of the consequent fragmentation of the hyphae. If the mould growth has resulted in formation of conidia or other spores, vast numbers of these may be present, each potentially capable of forming a colony. Thus the number of colonies obtained in a plate count does not reflect the amount of mould growth in the foodstuff.

19

Psychrotrophic, Psychrophilic and Thermophilic Counts

As mentioned earlier (Section 12.3), if interest is centred on spoilage potential at refrigeration temperatures then surface count techniques should be adopted for counts of psychrotrophs and psychrophiles, as such organisms may be killed quickly when exposed to molten media at 45°C. There is still discussion about the best standard method (incubation temperature and time of incubation) for psychrotrophic counts, and some workers use consensus seems to be emerging for incubation at 6.5°C for 10 days. However, if the definitions in Table 12.2 (see Section 12.3) were followed, an incubation temperature of, for example, 4.5°C would be required (with an incubation period of, say, 14 days), and this is the temperature-time combination suggested in this book.

Thermophilic counts can be performed at 55°C incubated for 3 days. Plates or tubes of media should be enclosed (e.g. in large plastic bags) to minimize dehydration of the media.

20

Detection and Enumeration of Indicator and Index Organisms

The term 'indicator organisms' can be applied to any taxonomic, physiological or ecological group of organisms whose presence or absence provides indirect evidence concerning either a particular feature in the past history of the sample or the contemporary presence of a feature not directly investigated. It is often associated with organisms of intestinal origin but other groups may act as indicators for other situations. For example, the presence of members of the set 'all Gram-negative bacteria' in heat-treated foodstuffs is indicative of inadequate heat treatment (relative to the initial numbers of these organisms) or of contamination subsequent to heating. Coliform counts, as the coliforms represent only a subset of 'all Gram-negative bacteria', provide a much less sensitive indicator of problems associated with heat treatment, but are still frequently used in the examination of heat-treated foodstuffs.

The term 'index organism' was suggested by Ingram in 1977 (Mossel, 1981, 1982) for a marker whose presence indicated the possible presence of an ecologically similar pathogen, retaining the term 'indicator' organism as defined in the paragraph above.

The use of marker organisms in quality assessments originated in microbiological examinations of drinking water supplies, in which intestinal commensals such as *Escherichia coli* were considered to be indicative of a potential hazard from the possible presence of intestinal pathogens. For *E. coli* to act satisfactorily as an indicator (or 'index organism' as we would term it today) in this situation:

(a) it must be present whenever the pathogens concerned are present;
(b) it must occur in much greater numbers than the pathogens; and
(c) it must not be less resistant than the pathogen to the environment, or to any process or manipulation imposed.

For many years it appeared that the indicator bacteria used satisfied all these criteria; for example, *Salmonella* is affected by chlorination in a manner similar to the indicator bacteria. However, there is evidence that certain pathogenic viruses may be inactivated more slowly than the indicator bacteria are destroyed (Bonde, 1977) and the cysts of *Cryptosporidium* are also very resistant to destruction by chlorination.

If we consider the use of *E. coli* as an index of the possible presence of intestinal pathogens in foodstuffs, there have been many examples of foodborne diseases in which at least one the above criteria has not been satisfied. In spite of the evidence of the shortcomings, absence of *E. coli* is still frequently and rather indiscriminately used as an indication of 'safety' of the food in question.

If we attempt to list all the characteristics that an ideal generalized indicator or index organism should possess in food microbiology, they will include the following:

(a) The indicator should indicate in a positive way the conditions being indirectly assessed.
(b) It should preferably be correlated in a quantitative manner with the condition being assessed, and this quantitative correlation should itself be assessable.
(c) It should yield characteristic and simple reactions enabling, as far as is possible, an unambiguous identification of the group or species.
(d) Its growth in artificial media must be largely independent of any other organism present.
(e) It should preferably be randomly distributed in samples, unless the nature of the distribution coincides with the distribution of the aspect being assessed.

The extent to which the commonly used indicator and index organisms satisfy these requirements will be discussed in the individual cases listed below.

Although most counts of indicator and index bacteria are based on the use of selective media, it is possible that higher recovery rates may be obtained with a non-selective resuscitation stage preceding the use of the selective medium (see Section 21.1).

20.1 'TOTAL ENTEROBACTERIACEAE', COLIFORM ORGANISMS AND *ESCHERICHIA COLI*

In water analysis, the presence of *E. coli* indicates faecal pollution of the water, there being a positive correlation between the concentration of the organisms and the amount and/or recency of the pollution (see Section 35). Thus the presence of *E. coli* implies that the water may contain pathogens of intestinal origin. *Escherichia coli* may be used with care as an indicator or index in food microbiology. In addition, certain serovars of *E. coli* are enteropathogenic. They are a significant cause of infantile gastroenteritis, but may also cause outbreaks of food poisoning amongst the adult population (ICMSF, 1996; Rabinowitz and Donnenberg, 1996).

Raw food or foods containing uncooked ingredients will frequently contain coliform organisms including *E. coli*, but multiplication of the organisms may have occurred in the food, so that there may be no correlation between numbers and the level of initial contamination. However, whereas the presence of *E. coli* and other enteric commensals in foodstuffs may indicate the possible presence of enteric pathogens, the *absence* of the commensals cannot be taken as indicating the *absence* of such pathogens. For example, raw food materials of animal origin (milk, eggs, meat), may contain *Salmonella* in the absence of *E. coli* if the animal had been suffering from the relevant *Salmonella* infection (mastitis, oviduct infection or bacteraemia). Therefore, care must be taken in the interpretation of counts.

A further problem is that *E. coli* may be less resistant than *Salmonella* in certain foods (e.g. salt foods containing nitrite, see Section 15) or to certain processes. So even in those situations in which *E. coli* derives from faecal contamination (e.g. handling of a food by a worker with faecally contaminated hands) only examination *immediately after* the contamination will detect *E. coli* in numbers proportional to the amount of faecal contamination.

If, however, sampling and examination do not take place immediately after the contamination then, depending on the nature of the food and the conditions of storage, one of the following courses of events may ensue:

(a) The *E. coli* may die off so that the occurrence of the initial faecal contamination is undetected. In this case we may be incorrect in assuming that other enteric pathogens will have died at the same rate.
(b) The numbers of *E. coli* remain approximately the same. Again, this cannot be taken to indicate that other enteric organisms will react in the same way to the environments in which they find themselves.
(c) *E. coli* grows and increases in number. In this case we must assume that the conditions that were favourable for the growth and multiplication of *E. coli* probably also favoured the growth and multiplication of other bacteria of enteric origin.

Thus, although large numbers of *E. coli* do not indicate recent or heavy faecal pollution, they must be taken to indicate a possible hazard from enteric pathogens.

Nevertheless the ability of *E. coli* to survive and/or multiply cannot be used as a reliable indicator of the ability of pathogens such as *Salmonella* to survive and/or multiply (Anderson and Hobbs, 1973; Akman and Park, 1974). In consequence, it has been proposed by Mossel and colleagues that a better measure of the hygienic quality of foods would be a 'total Enterobacteriaceae count' (Mossel, 1957, 1967; Mossel *et al.*, 1962). Such counts are easily made by adding glucose to the selective coliform counting medium. If violet red bile agar is used then glucose should be added to the recipe to a final concentration of 1%. Excellent reviews of the rationale of the use of the 'total Enterobacteriaceae count' have been made by Mossel and colleagues (Mossel, 1967; Mossel *et al.*, 1995).

Throughout this book the following definitions relating to bacteria in this group have been assumed (note that each group constitutes a subset of the previously defined group):

1. *Total Enterobacteriaceae*. Bacteria that, in the presence of bile salts, will grow and produce acid from glucose (as determined by use of violet red bile glucose agar).
2. *Coli–aerogenes bacteria*. Bacteria that, in the presence of bile salts or other equivalent selective agents, can grow and produce acid *and* gas from lactose when incubated at 30°C.

3. *Coliform bacteria*. Bacteria that, in the presence of bile salts or other equivalent selective agents, can grow and produce acid *and* gas from lactose when incubated at 35 or 37°C.
4. *Faecal coliform bacteria*. Bacteria that, in the presence of bile salts or other equivalent selective agents, can grow and produce acid *and* gas from lactose when incubated at 44–45.5°C. Note that the incubation temperature is critical, and a water-bath should always be used for this test. The choice of temperature within the indicated range depends on the choice of medium, if equivalence of counts is used as the criterion, but it does not necessarily follow that when the same counts are obtained by two different procedures the same bacterial types are being detected.
5. *Escherichia coli*. Bacteria that, in addition to showing the above characteristics, are also methyl red positive, Voges–Proskauer negative, and cannot utilize citrate as a sole carbon source. Indole-positive strains are termed *E. coli* type 1, and are presumed to have the intestine as their primary natural habitat.

However, it should be noted that the verocytotoxigenic *E. coli* O157:H7 grows very poorly at 44°C (see Section 21.4). In addition, many taxonomists consider *Shigella* to be taxonomically indistinguishable as a genus from *Escherichia*, with some of the enteroinvasive serovars of *E. coli* (EIEC) being equivalent to certain shigellae. The EIEC ferment lactose slowly or late, or not at all, and are non-motile (ICMSF, 1996). Consequently verocytotoxigenic *E. coli* (VTEC) and EIEC will not be detected by isolation techniques based on the definitions given here; they are discussed in Section 21.

The presence of *any* of the above groups in heat-treated foods is indicative of one or more of the following:

(a) the initial concentration of the bacteria was so high that the heat treatment was inadequate to reduce the concentration to an undetectably low level;
(b) post-heating conditions allowed multiplication of survivors until their numbers became detectable;
(c) contamination occurred subsequent to the heat treatment.

Rarely will survival be due to the presence of relatively heat-resistant strains, and unless isolates are tested for their heat resistance by D-value determinations unusual heat resistance should not be hypothesized in preference to one of the three explanations given above.

The counting procedures described below can be modified by use of glucose or choice of incubation temperature as indicated to count any of the categories defined above. In general, the more inclusive categories should be used in examining heat-treated or shelf-stable foods. The less inclusive ones are used in the examination of raw meats and similar foods in which faecal contamination is likely to have resulted in the presence of enteric commensals or when it is appropriate to use the counts to assess levels of innate contamination, rather than their being used to assess efficacy of processing and preservation procedures.

20.1.1 Total Enterobacteriaceae

The International Commission on Microbiological Specifications for Foods (ICMSF, 1978) recommended that, to detect low concentrations, liquid enrichment is carried out in phosphate-buffered peptone water or tryptone soya peptone broth followed by selective liquid enrichment in Enterobacteriaceae Enrichment Broth (containing oxgall and brilliant green), and plating on violet red bile glucose agar. Glucose-fermenting colonies are checked to be oxidase negative. This procedure has been specified in ISO 8523:1991. Similar procedures, suitable for counting higher population concentrations and therefore omitting non-selective enrichment, have been described for both most probable number (MPN) and colony counts (ISO 7402:1993).

Usually 'total Enterobacteriaceae counts' that search for counts of 10–100 per gram or more will be obtained by preparing a decimal dilution series and producing pour plate counts on violet red bile glucose agar (VRBGA). To each plate, 15 ml of VRBGA (melted and tempered to 45°C) are added, mixed well with the inocula, and allowed to set. The set medium should then be overlaid with another 5 ml of molten VRBGA at 45°C to suppress surface colonies. After leaving to solidify, incubate the plates for 24 h at the chosen temperature. The ISO standards specify an incubation temperature of 35° or 37°C. However, some members of this family grow poorly (and are less biochemically active) at 35°C and above (Holt *et al.*, 1994), so that for a truly 'total Enterobacteriaceae' count, an incubation temperature of 30°C would be preferable as it may result in higher counts. If it is required to follow an ISO standard, then there may be no choice (unless counts at 35° or 37°C *and* 30°C are run in parallel). Otherwise, it is suggested that the differences in counts at 30° and 35°C (or 37°C) on the sample types usually being examined should be studied over a period of time and analysed statistically, to determine which incubation temperature should be chosen, with due consideration to the reasons for performing the test.

20.1.2 Counts of coliforms, faecal coliforms and *E. coli* by the multiple tube technique

The presence of coliform organisms in media described below is detected by the production of acid and gas from lactose. As many foods contain significant amounts of other carbohydrates (e.g. sucrose, glucose, fructose) it is especially important that *presumptive* positive results are confirmed by subculture as described when high concentrations of the food sample have been incorporated into the medium.

The incubation temperature is a matter of debate and agreement, but temperatures between 30° and 37°C can be used. Obviously, the lower the incubation temperature in this range, the greater the count is likely to be.

20.1.2.1 Lauryl sulphate tryptose broth at 30°, 35° or 37°C (ISO 4831:1991)

Pipette aseptically 1 ml of each of the prepared dilutions of the food into each of three or five tubes of lauryl sulphate tryptose broth. If very low concentrations of organisms are expected, 10-ml or even 100-ml amounts of the lowest dilution may be added to equal volumes of double-strength medium, three or five bottles being prepared at each dilution.

Incubate at 30°, 35° or 37°C and examine after 24 and 48 h for growth accompanied by the production of gas. (The production of acid may be detected by the addition of a pH indicator to the tubes after incubation.) A tube showing the production of acid together with sufficient gas to fill the concave of the Durham tube (i.e. the meniscus should be at a point where the sides of the Durham tube are parallel) is recorded as being *presumptively positive*. A presumptive positive result is also recorded if the Durham tube contains less than the stated amount of gas, but effervescence occurs when the side of the test-tube is tapped. In routine testing, gas production is usually regarded as indicating a presumptive positive result without the need for testing for acid production.

Presumptive positive tubes should be confirmed by subculturing two or three loopsful from each tube into a tube of brilliant green lactose bile broth. These are incubated at 35°C for 48 h, gas production (judged on the criteria described above) being taken as indicating a *confirmed positive* result. By comparing the combination of positive and negative tubes obtained at each stage with probability tables (see Appendix 2), the most probable number of 'presumptive coliforms' and of 'confirmed coliforms' can be determined.

Faecal coliforms are detected by subculturing from presumptive positive tubes into EC Medium incubated at 45.5 ± 0.2°C (but see Section 21.14).

20.1.2.2 Brilliant green lactose bile broth

An alternative multiple tube technique which may be used involves the incubation at 30° (BS 4285: 3.7: 1987), 35° or 37°C of inoculated tubes of brilliant green lactose bile broth to establish presumptive positive results (by gas production). Confirmation of presumptive positive results is by streaking a loopful from each such tube across a plate of Endo agar, violet red bile lactose agar or MacConkey's agar.

20.1.2.3 MacConkey's broth

An alternative procedure involves the use of MacConkey's broth with incubation for 48 h at 30°C (for 'coli–aerogenes' counts) or at 37°C (for coliform counts). Presumptive positive results (acid and gas produced) should be confirmed as above by subculture into brilliant green lactose bile broth or by streaking across MacConkey's agar plates. Counts are determined by reference to probability tables (see Appendix 2).

Faecal coliforms are detected by subculturing from presumptive positive tubes into brilliant green lactose bile broth incubated at 44 ± 0.1°C (but see Section 21.4).

20.1.2.4 Counts of faecal coliforms and *E. coli* by the multiple tube technique

An extension of the coliform count made at the time of subculturing for confirmation of positives can provide the additional information needed for counting faecal coliform organisms that are able to produce acid *and* gas from lactose at 44–45°C (and which, for most purposes, can be regarded as equivalent to *E. coli*). The test for this characteristic is known as the modified Eijkman test.

Presumptive positive tubes from the foregoing tests are subcultured into either:

(a) *EC medium*, incubated at 44.5±0.2°C (AOAC, 1995) for 24 h; or
(b) *Brilliant green lactose bile broth*, incubated at 44.0 ± 0.1°C for 24 h.

In Sections 20.1.2.1 and 20.1.2.3, the usual combination for the modified Eijkman test has been indicated. It is advisable to prewarm the tubes to the incubation temperature before inoculation. (The use of MacConkey's broth at 44–45°C for this test has been largely superseded by the media listed above.)

Escherichia coli 'type 1' is indole positive at 44–45°C and therefore the Eijkman test can be supplemented by inoculating a peptone water and incubating at the raised temperature for 24 h and then testing for indole production (see Section 11.1.4). The confirmation of *E. coli* by IMViC tests (indole production, methyl red test, Voges–Proskauer test and citrate utilization) is time consuming: because cultures for these tests are incubated at 35–37°C, it is first necessary to obtain pure cultures by streaking the appropriate broth cultures on to a solid medium (preferably one that is differential but non-selective).

A more rapid count of *E. coli* may be attempted by inoculating the original samples and dilutions into tubes of media which are incubated for 4 h at 30°C, and then transferred to a water-bath at 44.0 ± 0.2°C for 18–20 h.

Note. As already pointed out, a number of the enteropathogenic strains of *E. coli* (especially the EIEC and VTEC strains) will record negative results to these tests.

20.1.3 Colony counts of coliform bacteria and *E. coli*

20.1.3.1 Petri dish counting method

The usual medium employed is violet red bile lactose agar (ISO 4832:1991). Prepare duplicate sets of plates using 1-ml amounts of the chosen range of dilutions of the food. Add to each plate 15 ml of violet red bile agar, melted and cooled to 45°C, mix well and allow to set. Finally overlay with another 5 ml of violet red bile agar. After allowing to solidify, invert the plates and incubate at 35°C for 24 h for coliform counts. After incubation, count the number of dark red colonies. Normally typical coliform colonies will be 0.5 mm or more in diameter and show evidence of the precipitation of bile salts in the medium immediately surrounding the colonies, but of course overcrowding of the colonies will cause a reduction in size. In the case of foods containing carbohydrates other than lactose, media to which low dilutions (i.e. high concentrations) of the food have been added may give anomalous results as already mentioned. With plates of 10^{-2} or lower dilutions, it is therefore necessary to confirm presumptive coliform colonies either as coliforms by picking them off and inoculating lactose broth or MacConkey's broth and incubating at 35°C or as *E. coli* by the Eijkman test (see above).

20.1.3.2 Use of dry rehydratable films

A useful Petrifilm (3M, St Paul, Minnesota, USA) product (see Section 5.1.5) is available for counting *E. coli*. The medium is based on violet red bile medium, but it also contains 5-bromo-4-chloro-3-indolyl-β-glucoronide to detect glucuronidase activity. Colonies of *E. coli* are blue with an adjacent bubble of gas. The author has found this product both easy to use, and easy to read. It is an AOAC Official Method (AOAC, 1995).

20.1.3.3 Membrane filtration

Counts of coliforms and *E. coli* in appropriate samples may also be determined by using membrane filtration (MF) in combination with MF modifications of relevant selective media. This technique is particularly applied to analysis of drinking water.

20.2 *ENTEROCOCCUS* ('FAECAL STREPTOCOCCI')

The streptococci (faecal streptococci) that normally inhabit the alimentary tract of homoiothermic animals mostly belong to the antigenic Lancefield group D (see Part IV, p. 344). They include the species *Enterococcus faecalis*, found in human beings and in cattle and other ruminants; *E. faecium*, found in human beings, cattle and other ruminants, poultry and pigs; *Streptococcus bovis*, found in cattle and other ruminants, pigs and horses; and *Strep. equinus*, found in horses and pigs.

These organisms (and especially *E. faecalis* and *E. faecium*) are useful indicator and index organisms. Compared with *E. coli*, they are more resistant to freezing, low pH and moderate heat treatment. Thus the enterococci may frequently be detected in frozen foods, fruit juices or foods that have received a cursory heat treatment, even when *E. coli* has been killed by the inimical conditions. For example, large-size catering pack canned hams may receive a moderate heat treatment, rather than a full 'botulinum cook', and enterococci can often be detected in such products. Enterococci may help in assessing the standard of hygiene in factories (Slanetz *et al.*, 1963).

For reviews of the taxonomy of the genus *Enterococcus*, see Schleifer and Kilpper-Bälz (1984, 1987) and Devriese *et al.* (1993).

The group is, for the food microbiologist, operationally defined in terms of a presumptive positive reaction on a selective diagnostic medium.

These organisms have been detected by plate counts using either Packer's crystal violet azide blood agar or maltose azide agar (KF streptococcus agar). These methods were described by the ICMSF (1978). However, Mossel (1978) proposed kanamycin aesculin azide agar, and this is currently recommended by the International Committee on Food

Microbiology and Hygiene (ICFMH) (Corry *et al.*, 1995a).

Alternatively, for smaller concentrations of organisms, an MPN technique can be used employing maltose azide broth (KF streptococcus broth) or glucose azide broth.

In fact, the media used tend also to permit the growth of a number of enterococci and group D streptococci that are not necessarily of faecal origin.

The presence of azide in the media helps to inhibit staphylococci, as a result of its inhibitory action against the cytochromes. As with all selective techniques, a compromise must be reached between selectivity and sensitivity. In order to have the maximum sensitivity against the widest range of strains there may be a reduction in the selectivity against organisms not belonging to the enterococci; time-consuming confirmatory tests have great importance in such a case – in just those circumstances when many colonies develop that need to be screened, yet which prove not to be of the types that it is desired to enumerate.

20.2.1 Glucose azide broth

This procedure employs the multiple tube technique, so that small numbers of enterococci can be detected.

Pipette aseptically 1 ml of each of the prepared dilutions of the sample into each of three or five tubes of glucose azide broth. If very low concentrations of organisms are expected, 10-ml or even 100-ml amounts of the lowest dilution may be added to equal volumes of double-strength medium, three or five bottles being prepared at each dilution. Incubate the inoculated glucose azide broths at 35°C for 72 h and examine for the production of acid. Record those tubes that are positive (i.e. in which acid is produced) and subculture a loopful from each positive tube into a fresh single-strength glucose azide broth (5 ml per tube) and incubate at 45°C in a water-bath for 48 h, examining the tubes after 18 and 48 h. The production of acid at 45°C within 18 h indicates enterococci. If acid is produced after 18 h but within 48 h, it is presumptive evidence of enterococci, and can be confirmed rapidly by microscopic examination for the presence of short-chained streptococci. The most probable number of enterococci can be determined using probability tables (see Appendix 2).

20.2.2 Kanamycin aesculin azide (KAA) agar

Surface inoculate plates of well-dried medium by spreading 0.1-ml amounts of dilutions over the medium. Incubate at 37°C or 42°C for 24 h.

This medium is made selective by the use of azide and kanamycin; incubating at 42°C increases its selectivity.

Colonies of enterococci are small (up to 2 mm in diameter), white or grey, and surrounded by large black haloes (up to 1 cm or so in diameter). Some strains of mesophilic lactobacilli may grow on this medium, and some of these are capable of splitting aesculin and therefore producing black haloes. These can be distinguished by microscopic examination.

20.2.3 Packer's crystal violet azide blood agar

This medium is used with a pour plate technique. Prepare duplicate sets of plates using 1-ml amounts of the chosen range of dilutions of the sample. Add to each plate 15 ml of Packer's crystal violet azide blood agar, melted and cooled to 45°C, mix well and allow to set. Incubate at 35°C for 3 days. After incubation count all small violet-coloured colonies as presumptive enterococci.

Representative colonies of presumptive enterococci should be subcultured by streaking across poured plates of Barnes' thallium acetate tetrazolium glucose agar (Barnes, 1956). Incubate at 35°C for 24 h. Typical colony forms are: *E. faecalis* – colonies with a red centre and with or without a white periphery; other group D streptococci – white or pink colonies. Morphology and Gram-staining reaction can be determined at this stage to provide reasonable confirmation.

20.2.4 Maltose azide media

20.2.4.1 Pour plate counts

Pipette 1-ml amounts of the chosen range of dilutions into a set of Petri dishes. Add to each plate 15 ml of molten maltose azide tetrazolium agar at 45°C, mix thoroughly with the inoculum and allow to set. Incubate at 35°C for 48 h. After incubation, count the number of colonies that are dark red or have a red or pink central area.

20.2.4.2 Multiple tube technique

If very small numbers of enterococci are expected to be present in a food sample, use a multiple tube count with maltose azide broth as the selective medium. Pipette aseptically 1 ml of each of the prepared dilutions into each of five tubes of maltose azide broth. Incubate at 35°C for 48 h. After incubation examine the tubes for acid production, which is indicative of enterococci. Calculate the most probable number of streptococci using probability tables (see Appendix 2).

20.3 *BACILLUS* AS INDICATOR ORGANISMS

In the case of foods that (1) have received a pasteurization heat treatment and (2) have not permitted subsequent germination of surviving spores, post-processing contamination may be detected by determining the proportion of viable bacteria present which are spore-forming.

The procedure is to set up an aerobic mesophilic count on a food sample, and compare this with an aerobic mesophilic count after a laboratory pasteurization of a 10^{-1} dilution of the food.

Procedure

1. Prepare a 10^{-1} dilution of the food in 0.1% peptone.
2. Use this dilution to set up a decimal dilution series in 0.1% peptone, and inoculate plates with 1-ml amounts. Prepare pour plates with plate count agar.
3. Without delay after preparing the 10^{-1} dilution, transfer 10 ml of this 10^{-1} dilution to each of two sterile test-tubes. Insert a thermometer into one of the test-tubes.
4. Place both tubes into a stirred water-bath set at 81°C, so that the water in the water-bath is a little above the level of the dilutions in the test-tubes. Heat the tubes for 1 min after the temperature indicated by the in-tube thermometer reaches 80°C (Mossel *et al.*, 1973). Remove the tube not fitted with the thermometer and cool rapidly under running water or in ice-water.
5. Use the heated 10^{-1} dilution to prepare another decimal dilution series in 0.1% peptone, and inoculate sterile Petri dishes with 1-ml amounts. Prepare pour plates with plate count agar. This series of plates will provide an 'aerobic mesophilic spore count'.
6. Incubate both sets of plates at 30°C for 2 days, and then determine the aerobic mesophilic count and aerobic mesophilic spore count.

If there has been no significant post-processing contamination, the two colony counts should be the same (within experimental error).

The first category of foods for which this procedure can be used to determine the adequacy of the heat treatment and the absence of post-processing contamination consists of foods in which surviving spores have not been able, or have not had the opportunity, to germinate. This includes dried foods and foods with a low a_w, and heat-treated foods immediately subjected to freezing, at all stages of storage and distribution.

The procedure could also be applied to samples of other types of foods taken off-line in the factory immediately after the pasteurization process (e.g. pasteurized milk *immediately* after the cooling and packaging stage). However, in these cases the laboratory examination would need to be performed without delay, to avoid the opportunity for surviving spores to start to germinate (and thereby become heat sensitive). In this second category, if it is considered that psychrotrophic contaminants may be potentially important (e.g. in the case of pasteurized milk packaged into reused and washed glass bottles), surface plating (Miles and Misra drop counts, spread plates, or spiral plating; see Section 5.1) should be used rather than pour plates.

21

Detection and Enumeration of Pathogenic and Toxigenic Organisms

21.1 INTRODUCTION

In addition to the microorganisms that may be present in food and cause 'food poisoning', a number of diseases other than so-called food poisoning may be spread via food (see Harrigan and Park, 1991; Mossel *et al.*, 1995).

The possibility of foods playing an important part in the spread of some viral diseases is being given an increasing amount of attention. Among viral diseases that may be foodborne are infectious hepatitis A, poliomyelitis, and enteritis caused by a range of viruses including rotavirus and Norwalk virus (small round structured viruses, or SRSVs) (Kapikian, 1994). One reason for the inadequacy of the information on this topic is the complex and expensive procedure necessary for the detection of viruses in foods. In the present author's opinion the current evidence is such as to indicate that provided sufficient care is taken over bacteriological examinations and the interpretation of results, and appropriate action taken, there is little justification at present for the examination for viruses being attempted in a quality assurance laboratory. A combination of immunocapture with polymerase chain reaction (PCR) amplification is likely eventually to provide an opportunity for the detection of small numbers of specific virus particles in samples such as foods.

'Food poisoning' can be caused either by toxins produced by bacteria which may be non-viable when the food is consumed, or by the multiplication of bacteria in the intestine after consumption. In the former case, when the organisms cause an intoxination, a normal condition for toxin production is the multiplication of the organism in the food resulting in a large total count (not necessarily a large viable count) at the time of testing, and this implies that the food is, or has been, in a state such as to allow the growth of bacteria. In the latter case, where the organisms cause an infection, they need be present in only relatively small numbers in the food but they must be viable. Food that does not support the growth and multiplication of the organisms concerned may none the less constitute a health hazard if it is not of a nature that causes the death of the organisms, as the food may be used in a situation that allows transfer of even very small numbers of the organisms to another food that *will* support growth. An example of such a hazard was provided by the outbreaks of salmonellosis caused by contaminated desiccated coconut (Wilson and MacKenzie, 1955).

If a food poisoning syndrome is caused by ingestion of viable organisms (as in the case of *Salmonella* enteritis or *Clostridium perfringens* food poisoning), the laboratory procedures for assessing the hazard will depend on the use of highly selective media to detect the viable organisms. Very often assessment of a hazard from toxigenic organisms (e.g. *Staphylococcus aureus*, *Bacillus cereus*) is based similarly on the use of selective media to detect the viable organisms, although the syndrome is caused by ingestion of a preformed toxin which may be present in the food even when a process has subsequently rendered the causative organisms non-viable. In the case of *Staph. aureus* enterotoxin, aflatoxin and some other mycotoxins, the toxins are sufficiently well characterized that chemical or serological detection techniques can be used. Some toxins, such as those of *B. cereus*, have not been sufficiently well characterized that detection techniques have entered routine use.

In the case of mycotoxins, although bioassay techniques are used to identify previously unknown toxins, chemical detection (by thin-layer chromatography or high-pressure liquid chromatography) is the preferred method of routine hazard assessment rather than cultural examination for given fungi. Consequently mycotoxins will not be considered in detail in this book. Reviews of chemical and bioassay techniques have been made by Moreau (1979) and Moss (1980).

The selective and differential media discussed so far in Section 20.1. are normally employed in circumstances in which the organisms sought are expected to constitute a relatively large proportion of that section of the microflora which shows physiological and/or ecological similarities. If this assumption be untrue then an unacceptably high level of false-positive (or false-negative) results may be obtained; for example, anaerobic spore-bearing organisms may cause misleading results in coliform tests performed on samples of chlorinated water (Report, 1969).

In the case of pathogenic and toxigenic organisms it is necessary to use selective and differential media and methods that can detect the organisms when they are in the presence of perhaps very large numbers of physiologically and/or ecologically similar organisms. Most selective pressures are only *relatively* selective and usually selectivity is improved only at the expense of sensitivity (also known as 'productivity'). This problem is exacerbated in food microbiology by the likelihood of the organisms having suffered some metabolic injury from a food-processing procedure so that the organisms are inhibited by the very media that are expected to select for them. Yet because *Salmonella* in a frozen or dried food cannot recover and grow in selenite broth, for example, it does not follow that the organisms could not recover in a non-selective environment (e.g. trifle). Thus selective procedures are relatively complicated and lengthy: liquid enrichment usually precedes the use of solid selective media, and non-selective resuscitation may precede the liquid enrichment stage.

The composition of the non-selective resuscitation medium will depend on the type of damage most likely to have been suffered (Tomlins and Ordal, 1976). If organisms in lag or late stationary phase are frozen or dried, then the metabolic injury is likely to result in a temporary nutritional requirement for substances to be supplied exogenously that normally can be synthesized endogenously, so that higher recoveries are likely in rich media (see, for example, Straka and Stokes, 1959; Fisher, 1963). On the other hand, if logarithmic-phase cells are heated the damage suffered is likely to include nucleic acid damage; such organisms may recover in a minimal medium but not in a rich medium (Gomez *et al.*, 1973; Wilson and Davies, 1976). It has been suggested (Mackey, 1984) that minimal medium resuscitation is usually associated with cells *grown* in a minimal nutrient environment which are then injured whilst in the exponential phase. If this be the case, then minimal medium resuscitation is not likely to give higher recovery rates than 'rich medium' resuscitation when applied to microorganisms in processed foods. However, when trying to determine the optimal isolation procedure, it is advisable to use *both* rich medium resuscitation *and* minimal medium resuscitation in order to ascertain which gives the best recovery rates from a particular food-processing–storage system.

Confirmatory and identification tests that follow the use of selective media are usually specially designed and offer confirmation only in the context of the prior selection of presumptive positive results. It is essential to ensure that the organisms submitted to confirmatory tests are obtained as pure cultures. For example, consider the case of a mixed microflora containing some coliform organisms and some salmonellae cultured on to desoxycholate citrate agar, a medium selective for *Salmonella*, which differentiates on the basis of lactose fermentation. Coliform organisms grow only reluctantly on this medium and, if in association with a developing colony of *Salmonella*, they may not show their lactose-fermenting ability. The presumptive *Salmonella* colony transferred directly to confirmatory physiological test media is very likely to show positive results characteristic of both the *Salmonella* and the coliform organism. The purification of the *Salmonella* culture on a non-selective but differential medium would have demonstrated the presence of the coliform organisms.

Therefore, the stages in isolation and identification are likely to be:

(a) pretreatment of the sample if necessary (see Section 17.1.1);
(b) resuscitation in a 'non-selective' non-differential medium;
(c) enrichment in a liquid selective medium;
(d) detection on solid selective and/or differential media;
(e) purification of cultures on a differential non-selective medium;
(f) confirmatory physiological tests; and
(g) serological tests and possibly phage typing.

In microbiological analysis associated with the epidemiological investigation of outbreaks, it is often necessary to obtain a more detailed identification of the organisms involved than can be provided by serotyping. Phage typing is commonly employed in investigations involving, for example, strains of *Salmonella enteritidis*, *S. typhimurium* and *S. typhi*. Subdivision of strains of a given serovar can also be made by utilizing pulsed-field gel electrophoresis (PFGE) (Thomson-Carter *et al.*, 1993). For example, Bender and co-workers (1997) have described the use of PFGE in studying 344 cases of *Escherichia coli* O157:H7 infection. It would at first sight appear that PFGE provides too fine a subdivision, since 143 different PFGE patterns were obtained. However, four general outbreaks were detected solely as a result of the PFGE subtyping, three of these outbreaks being associated with the consumption of hamburgers obtained from specific retail outlets. Lee and colleagues (1998) used PFGE similarly in the study of a *Salmonella javiana* outbreak.

A procedure involving stages (a) to (g) may occupy 2 weeks, so that the incentive to introduce rapid alternative procedures will be readily appreciated. Conventional methodology is still used widely in the detection of pathogens in foods. These procedures are discussed first in the section on each pathogen. Many different approaches to rapid methods are being developed, which may permit the detection time to be reduced to 2 days or less. The procedures and systems detailed in the following pages (e.g. immunocapture) are those that the author has used. Procedures that have not been used by the author (e.g. nucleic acid probe techniques) are mentioned with some key references so that the interested reader can gain an entry to the literature on the topic.

Some of these techniques attempt to modify or short-cut the conventional method, for example by detecting the pathogens in the enrichment broth. One such procedure, investigated during the 1960s, involved the use of fluorescent antibody techniques to detect the organisms by direct microscopic examination of the liquid enrichment cultures obtained at stage (c), using a fluorescence microscope. In the case of this particular application, detection of fluorescing cells can be used to hold up the distribution of the food, but as non-specific reactions can occur this presumptive positive result must be confirmed by following and completing the normal procedure.

A screening technique that may give fewer non-specific reactions would be to use immunomagnetic separation from the enrichment broth (e.g by using Dyna-Beads), followed perhaps by enzyme-linked immunosorbent assay (ELISA) detection of the separated cells. A cheaper method of immunocapture involves the use of antibody immobilized on a surface on a dipstick. The dipstick is immersed in the enrichment culture, and captured pathogens subsequently detected by an ELISA stage. Examples of this approach are the TECRA® Visual Immunoassay, the Oxoid 'Clearview'™ Immunoassay and the Lumac® Path-Stik.

At present such techniques are often used in conjunction with the subsequent conventional stages, for example permitting early detection by immunocapture to provide rapid presumptive positive results for first action, but confirming presumptive positives in the usual way.

An interesting alternative approach to modifying conventional methodology to provide rapid results has been the development of the Oxoid Salmonella Rapid Test, which is described in Section 21.3.3.

A completely different approach is to use electrometry (impedance/conductance measurements; see Section 5.5) in conjunction with selective media (see Stannard *et al.*, 1989).

Unfortunately standardization of isolation procedures irrespective of the foodstuffs to be examined may be impracticable in the resuscitation and enrichment steps: there is increasing evidence that the effectiveness of a given enrichment technique (in terms of selectivity and sensitivity) may depend as much on the unintentional modification of the medium caused by the addition of (usually) appreciable amounts of food as it does upon the recipe used to prepare the medium. Obviously different foods will have different effects. A possible approach to the standardization of detection methods which has much to commend it is to specify for each pathogen the resuscitation stage and the first liquid enrichment stage for the particular type of food being examined, with later stages being common to the examination of all categories of sample. In the present state of knowledge, specifying the most effective first steps is possible with only a very few foods. Consequently generalized procedures must be recommended with the important proviso that such procedures should not be applied uncritically but should be regarded as open to modification to suit particular circumstances.

21.2 QUANTIFICATION OF SELECTIVE ISOLATION TECHNIQUES

If the procedure does *not* involve a liquid enrichment technique, counts may be performed on the solid selective media by the usual methods outlined in Section 5.1 (e.g. Miles and Misra surface drop count). Procedures that incorporate liquid resuscitation or liquid enrichment stages can be made quantitative by a modification of the multiple tube count technique, using the most probable number (MPN) tables in Appendix 2. This necessitates the use of replicate food samples at three consecutive 'dilution levels' (e.g. 100, 10 and 1 g of food) during the *first* culture stage.

An alternative method of rendering the isolation procedure quantitative whilst using liquid enrichment is to follow the suggestion made by the Committee on Salmonella (1969) and use an attribute sampling plan (see Section 16.5). The Committee proposed the use of a unit size of 25 g, and also suggested the following simple approximation to the formula given on p. 153:

$$n = (-2.3/d)\log(1 - p)$$

where n is the number of units that, if examined and all found to be negative, will imply with confidence p that the population of positive units in the whole lot averages less than d.

For 95% confidence in the result:

$$n \approx 3/d$$

It has been determined (Gabis and Silliker, 1974) that comparable reliability can be obtained by compositing the samples taken. The units, n, are chosen randomly from the batch, but in the laboratory they may be composited before examination. For example, for $n = 30$, and a unit size of 25 g, the units may be combined so that three 250-g amounts are examined; although the same total weight is examined, savings are made in time and media for isolation and identification.

A number of studies have also been made of the use of hydrophobic grid membrane filters (HGMF) in the selective isolation of pathogens, permitting a single-dilution MPN count (see Section 5.3.2). These techniques can incorporate resuscitation and primary enrichment, by transferring a membrane from one medium to the next. However, Blackburn and Patel (1991) found that the HGMF technique applied to detection of

Salmonella gave a high level (31%) of false-negative results, and this, together with a lack of significant time saving, led them not to recommend the technique for detection and enumeration of salmonellae.

21.3 *SALMONELLA* AND *SHIGELLA*

21.3.1 Conventional methodology

The ISO standard method for isolation of *Salmonella* is illustrated in Fig. 21.1. The notes that follow suggest possible modifications or extensions to the ISO method that (i) may improve sensitivity, and (ii) may increase the possibility of detecting *Shigella*.

As already pointed out in Section 20.1, *Shigella* is more closely related genetically to *Escherichia coli* than it is to *Salmonella*. Nevertheless, the role of clinical microbiology in developing selective media to isolate the causative organisms of enteric infections led to the search for media that would permit the growth of *Shigella* as well as *Salmonella*. It was soon appreciated that *Shigella* was as sensitive as *E. coli* to brilliant green, so that the solid media of choice for clinical investigation (e.g. of faecal specimens) for shigellae became desoxycholate citrate agar and *Salmonella–Shigella* agar.

Shigella dysentery may be spread by food as a result of contamination of the food by food-handlers, etc. Unfortunately the minimum infective dose is very low, and completely satisfactory enrichment methods have yet to be described; in consequence the real significance of foods as vehicles of infection is unknown, as the difficulties experienced in recovering shigellae from suspect foods mean that most reported outbreaks rely on epidemiological and statistical correlative evidence (see, for example, CDSC, 1992). However, it would seem that, of the liquid media devised for enrichment of enterobacteria, Hajna's GN broth (Hajna, 1955; Taylor and Harris, 1965; Taylor and Schelhart, 1967) is still the best for this purpose. Isolation of shigellae on solid media is probably best accomplished with Taylor's xylose lysine desoxycholate (XLD) agar (Taylor, 1965; Taylor and Schelhart, 1971) or Hektoen Enteric agar, with *Salmonella–Shigella* agar as a slightly less effective alternative.

21.3.1.1 Non-selective resuscitation

One of the drawbacks of the ISO horizontal method is that only a single resuscitation (pre-enrichment) medium is specified. It has, however, been recognized that one of a range of different resuscitation media may be the most successful for different foods. The International Commission on Microbiological Specifications for Foods (ICMSF, 1978) listed seven possible resuscitation media.

(a) In cases in which any salmonellae present may have suffered metabolic damage when in lag or stationary phase (e.g. dried foods, frozen foods) add aseptically

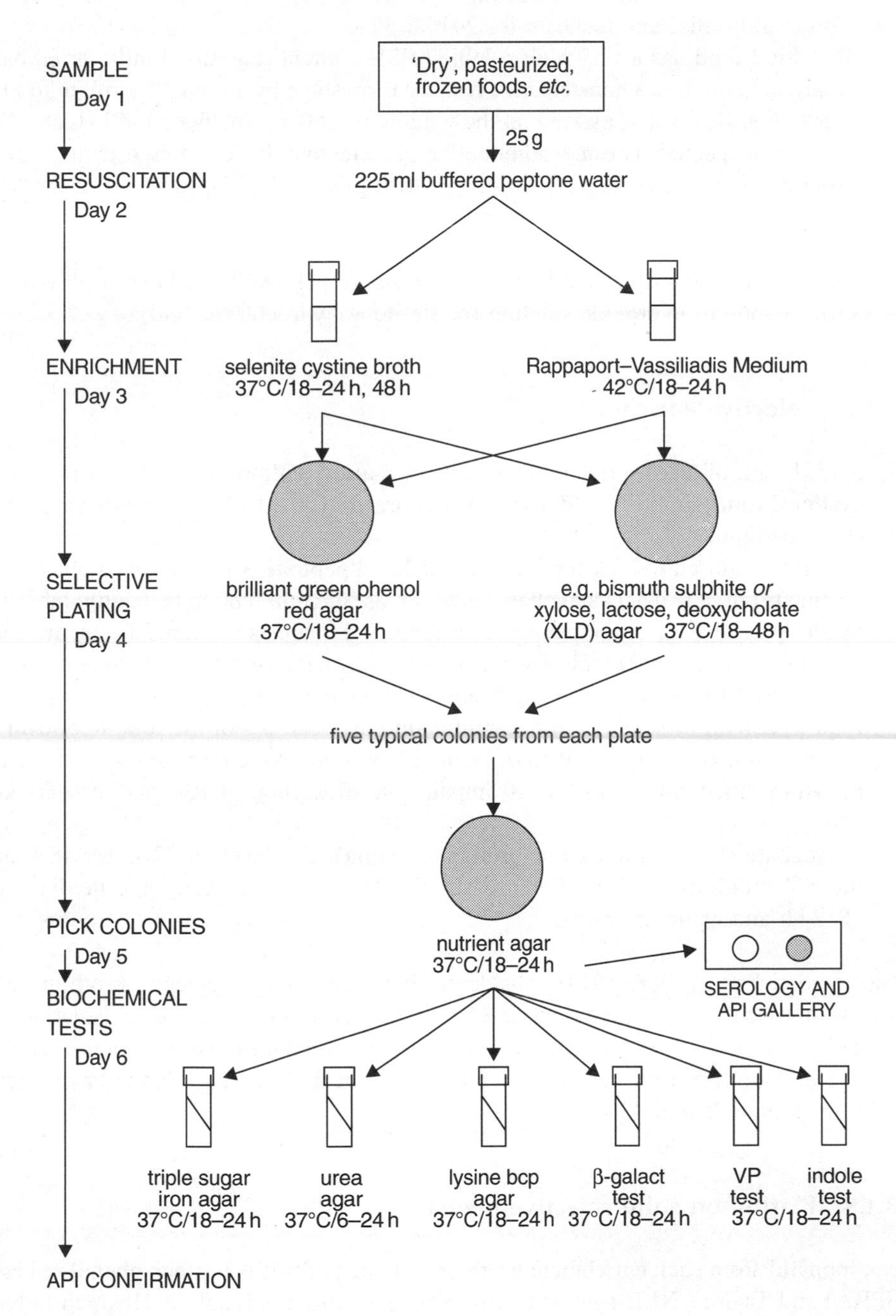

FIG. 21.1 Detection of salmonellae (ISO 6579:1993).

25 g of the food product to 225 ml of buffered peptone water (ISO 6579:1993). Thoroughly mix, and incubate for 24 h at 37°C.

(b) If a dried food has a very high soluble solids content (e.g. dried milk, dried baby food), also prepare a non-selective resuscitation stage by adding 25 g of the food to sterile distilled water, making up the volume to 250 ml. Incubate for 24 h at 37°C.

(c) If it be suspected that any salmonellae present may have suffered damage (e.g. from heating or freezing) whilst in a growth phase add 25 g (or 10 g) of the food product to 250 ml (or 100 ml) of minimal nutrients recovery medium.

Note. If the pH of the suspension is outside the range pH 6.8–7.0, adjust to pH 6.8–7.0 with sterile N sodium hydroxide solution (or sterile N hydrochloric acid).

21.3.1.2 Selective enrichment

1. Add 10 ml of each pre-enrichment resuscitation culture to 100 ml of selenite cystine broth; similarly add 10 ml of culture to 100 ml of Rappaport–Vassiliadis (RV) medium.

   If it is required to look for *Shigella*, buffered peptone water followed by selenite cystine broth *may* detect shigellae. However, as *Shigella* tend to be readily inhibited by the presence of a competing microflora, it is suggested that 25 g of the food sample be added to 225 ml of Hajna's GN broth *without* a pre-enrichment stage.

2. In the case of foods not requiring a non-selective resuscitation stage add 25 g of the food to 250 ml of selenite cystine broth; similarly add 5 g of the food to 500 ml of RV medium, and 25 g to 250 ml of Hajna's GN broth. Note that the use of the usual inoculum : medium ratio of 1 : 10 impairs the efficiency of RV medium (Fricker, 1987).

   Incubate the selenite cystine broth and Hajna's GN broth at 37°C for 48 h, and the RV medium at 42°C. Prepare streak plates on solid selective media after 18–24 h and again after 48 h.

Harvey and Price (1968, 1974) considered that incubation of selenite broth at 43°C following non-selective resuscitation at 37°C provided a better chance of isolating the salmonellae, but if 43°C is chosen as the incubation temperature for the selenite cystine broth, the medium must be sterilized by filtration as heat-sterilized selenite broth shows toxicity to salmonellae at 43°C.

21.3.1.3 Plating on solid selective media

Streak loopsful from each enrichment broth on to plates of brilliant green phenol red agar (BGPRA) and Taylor's XLD agar. If a third plating medium is required, Hektoen Enteric agar, *Salmonella–Shigella* agar or Wilson and Blair's bismuth sulphite agar may be used. Incubate the plates for 24 and 48 h at 37°C and examine for the presence of typical colonies. Typical *Salmonella* colonies (including S. *arizonae*) on Taylor's XLD agar are red with black centres; colonies of *Shigella* typically are uniformly red. (*Note*. Some strains of *Pseudomonas* and *Proteus* also produce uniformly red colonies on this medium.) Typical *Salmonella* colonies on BGPRA are pink or red (occasionally colourless) surrounded by a zone of bright red medium; shigellae are likely to be inhibited on this medium. However, in the author's experience, BGPRA has a lower selectivity against unwanted organisms than the other media named. Because of this, *Salmonella* colonies on brilliant green agar may be rendered atypical by the presence of lactose-fermenting organisms.

21.3.1.4 Subculture of presumptive salmonellae and shigellae

Pick off suspect colonies (the ISO standard specifies at least five from each plate) and streak on to nutrient agar or a non-selective lactose agar (neutral red chalk lactose agar may be used) to confirm the purity of the cultures. As already explained (p. 177), this step is most important to prevent physiological tests being performed on mixed cultures.

It is worth remarking that outbreaks of enteritis caused by lactose-fermenting variants of *Salmonella* have been reported (Poelma, 1968; CDC, 1993) and that S. *arizonae* and some strains of *Shigella* will slowly attack lactose (see p. 325) so that the purification stage should not be employed for *rejecting* cultures, especially if these have been derived as typical colonies on isolation media *not* containing lactose (e.g. Wilson and Blair's bismuth sulphite agar).

Harvey and Price (1967) found that salmonellae could be recovered more frequently if an additional stage was introduced, which they termed a selective motility technique. The method (Harvey and Price, 1967, 1974) consists of removing the total growth from a selective agar medium with a short swab. The swab is placed in a short length of glass tubing inside a screw-capped bottle containing a semi-solid nutrient agar, the level of which is sufficient to immerse the swab. The swab is prevented from touching the base of the bottle by a short piece of glass rod placed within the glass tube (Fig. 21.2). The whole is incubated at 37°C for 24 h. The growth from the surface of the agar *outside* the inner

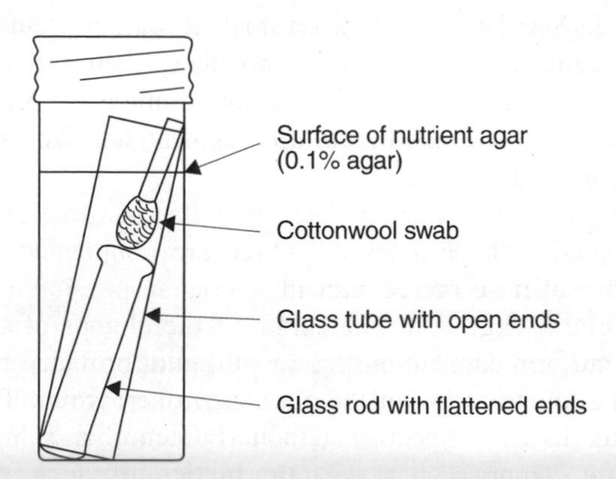

FIG. 21.2 Secondary enrichment by selective motility technique (Harvey and Price, 1967).

tube is subcultured on to brilliant green MacConkey's agar which is then incubated at 37°C for 24 h. Presumptive *Salmonella* and *Shigella* are picked off, purified, and subjected to physiological and serological tests.

21.3.1.5 Physiological test media for identification of salmonellae and shigellae

The ISO standard specifies confirmation of *Salmonella* by the following biochemical tests: TSI agar; urea agar; L-lysine decarboxylase medium; detection of β-galactosidase; and VP test.

Alternatively the culture can be picked off into a proprietary rapid multi-media testing system such as the API 20E (bioMérieux). Results can be checked against a comprehensive computer database available from bioMérieux, which provides a probability with the indicated identification, and suggestions for any secondary confirmatory tests. The author has found the API20E very convenient and easy to use, as is the Enterotube (Roche). A number of the well established commercial systems have been evaluated by Holmes and Costas (1992), who found that, of the seven systems tested, these two gave the highest number of correct identifications.

It is important to take care in the interpretation of biochemical results. In the USA, food-poisoning outbreaks caused by lactose-fermenting salmonellae have been reported, and sucrose-fermenting salmonellae have also been reported. As has been pointed out by Harrigan and Park (1991), a procedure that discards lactose-positive organisms as being 'not *Salmonella*' is unable to determine whether lactose-fermenting salmonellae are common, rare or non-existent. Although this is self-evident, unfortunately many microbiologists fall into this trap.

21.3.1.6 Serological differentiation of *Salmonella*

The identity of a presumptive *Salmonella* culture may be confirmed by a slide agglutination test (see Section 13.2). Full serological typing is outside the scope of the ordinary quality assurance or investigational laboratory, requiring as it does a wide range of expensive antisera of limited shelf-life, including single-factor antisera. However, a useful intermediate level of identification is provided by multi-coloured latex agglutination kits (Hadfield *et al.*, 1987).

The reagents consist of a mixture of latex particles of three different colours, each colour being associated with particles on which are conjugated a different range of antibodies. The latex reagent is mixed with a bacterial suspension on a disposable 'slide' which provides a white background. The colour of the clumps of agglutinated particles against a particular uniform background colour of liquid (provided by the unagglutinated particles) permits the identification of the *Salmonella* serogroup. The two makes of kit that the author has used – Spectate (Rhône-Poulenc) and Wellcolex (Wellcome Diagnostics) – both comprise two separate bottles of such mixtures, permitting identification within the subgroups A–G, and also detection of Vi antigen. Wellcome

Diagnostics has a similar Wellcolex kit for identification and differentiation of the four species of *Shigella*. It should be noted that it is important to use a *dense* suspension to obtain adequate colour development. Also, whilst it is possible to mix the bacterial suspensions and latex reagents on the slide by rocking the slides manually, these reagents do require a very thorough mixing, which is best achieved by means of an electrical rotator designed for the purpose.

21.3.2 Immunocapture

There are now many commercially available systems, of which the author has used the two described here.

The TECRA® 'Unique'™ Salmonella Immunocapture technique operates at the pre-enrichment stage (i.e. after 21.3.1.1). A sample from the buffered peptone water pre-enrichment culture is placed in a tube and a plastic paddle dipstick inserted. The dipstick is coated with antibodies and, during agitation in the culture medium, picks up organisms bearing the relevant antigens. The paddle dipstick is then washed and transferred to another tube containing an enrichment broth (TECRA M-broth). This is incubated for 4 h to permit multiplication of the captured salmonellae. The larger number of salmonellae present on the dipstick is detected using an ELISA conjugate and chromogenic substrate. The presence of salmonellae causes the flat blades of the paddle to become purple. The author has found the TECRA® 'Unique'™ system extremely easy to use and to read.

The Lumac® Path-Stik is a rather similar ELISA dipstick, but is applied at the selective enrichment stage (i.e. after 21.3.1.2). One millilitre of an incubated Rappaport–Vassiliadis (RV) medium culture is transferred to 10 ml of buffered peptone water (the 'post-enrichment' culture). This is incubated for 6–8 h at 35° or 37°C. The Path-Stik dipstick is then dipped into the post-enrichment culture, and the result read after 10 min. The presence of *Salmonella* is shown by the development of a coloured line in a test window.

21.3.3 The Oxoid Salmonella Rapid Test (SRT)

The Oxoid SRT is designed to combine liquid enrichment stages with liquid indicator media, motile salmonellae being self-selecting by migration through the selective medium under the influence of chemoattractants (Holbrook *et al.*, 1989a,b). The test is performed in a disposable culture vessel with screw cap in which there are two tubes with porous bases. Each of the tubes contains two dehydrated media separated by a porous disc. The media in the bottom halves of the tubes are liquid selective media (RV medium, and lysine iron desoxycholate medium respectively), and the media in the upper halves of the tubes are indicator media (lysine iron cystine neutral red medium, and modified brilliant green lactose medium respectively). Presumptive identification of *Salmonella* is based on hydrogen sulphide production and lack of lactose fermentation.

To prepare the culture vessel and tubes, firstly the dehydrated media are rehydrated. Then an elective medium and novobiocin are added to the culture vessel. The elective medium is inoculated with 1 ml of a pre-enrichment culture of a sample. The vessel is incubated at 41°C for 24 h. The elective medium is designed to allow rapid growth of *Salmonella* whilst inhibiting growth of *Proteus*, *Edwardsiella* and Gram-positive bacteria. During incubation the motile salmonellae migrate into the tubes, and act on the media there. The colour changes occurring in the upper tubes are noted. A loopful of any presumptive *Salmonella* culture is removed from the top of a positive tube and tested against latex agglutination antisera.

The principal advantages of the Oxoid SRT are that the combining of media into one vessel reduces hands-on time, and incubator space, and that only 24 h of incubation is required after the pre-enrichment culture. However, it should be noted that the isolates obtained from the tops of the inner tubes are not pure clones (not derived from single cells) so that any investigations subsequent to the initial agglutination test will require the cultures to be purified properly. At the time of writing, although the culture vessels are supplied in impermeable plastic laminate packaging with desiccant sachets, shelf-life is limited by the hygroscopic and caking characteristics of two of the dehydrated media.

The author has found the Oxoid SRT to be easy to use and to read.

21.3.4 Nucleic acid probe techniques

Amongst the commercially available systems for detecting *Salmonella* by use of a nucleic acid probe is Gene-Trak (see also Section 5.6). The Gene-Trak System (Gene-TrakSystems, Hopkinton, Massachusetts, USA) is an AOAC Official Method (AOAC, 1995) for detecting *Salmonella* in selective enrichment broths. At the time of writing it is still necessary to increase the numbers of organisms from the food sample by a selective enrichment stage. An assessment of the performance of Gene-Trak (D'Aoûst *et al.*, 1995) concluded that direct probing of selective enrichment cultures yielded poor results, and RV broth in particular adversely affected probe sensitivity even when it was diluted. The study showed that best results were to be obtained by incubating selective enrichment media for 24 h, followed by a 6-h post-enrichment culture in GN broth, this GN broth culture then being subjected to nucleic acid probing.

21.4 PATHOGENIC *ESCHERICHIA COLI*

Six clinically distinct categories (pathovars) of disease-causing *E. coli* have been described (Smith and Scotland, 1994; Rabinowitz and Donnenberg, 1996). The different virulence and infective characteristics derive from a combination of the factors determining surface adsorption and adherence of the bacterial cells to the intestinal epithelium and the toxins that may or may not be produced.

Most of these organisms are not distinguishable from one another by simple biochemical or physiological tests. As can be imagined, the processes of surface adsorption, adherence and colonization of the intestinal epithelium are substantially affected by the bacterial cell wall antigens, and therefore particular serovars tend to correlate with particular pathovars (Levine, 1987). Consequently identification of most pathogenic *E. coli* is dependent on serotyping. However, because of the developing problem presented by verocytotoxigenic *E. coli* (VTEC), especially the serovar O157:H7, a range of procedures for detecting and isolating these organisms has been developed.

The six types are:

1. *Enteropathogenic* E. coli (*EPEC*) do not produce high levels of Shiga-like toxins, but attach to, and efface, microvilli. Serovars include O55, O86, O111, O119, O125, O126, O127, O128ab, O142 and O158.
2. *Enterotoxigenic* E. coli (*ETEC*) possess particular colonization factor antigens and also produce heat-labile (LT) toxin and heat-stable (ST) toxin. Serovars include O6, O8, O15, O20, O25, O27, O63, O73, O78, O80, O85, O114, O115, O128ac, O139, O148, O149, O153, O159, O166, O167 and O169.
3. *Enteroinvasive* E. coli (*EIEC*) and *Shigella* are genetically nearly identical and have a number of antigens in common. In fact, in *Bergey's Manual* (Krieg *et al.*, 1984) it is suggested that *E. coli* and the four species of *Shigella* form a single species on the basis of DNA relatednesss, and shigellae are merely metabolically inactive biogroups of *E. coli.* EIEC serovars include O28ac, O29, O112, O124 (equivalent to *Sh. dysenteriae* 3), O136, O143 (equivalent to *Sh. boydii* 8), O144, O152 (equivalent to *Sh. dysenteriae* 12), O164 and O167.
4. *Verocytotoxin-producing* E. coli (*VTEC*) *or enterohaemorrhagic* E. coli (*EHEC*) are strains of *E. coli* that produce a cytotoxin against Vero cells (a tissue culture cell line from African green monkey kidney). They were first described in 1977 (Konowalchuk *et al.*, 1977). (See Section 21.4.1 below.)
5. *Enteroaggregative* E. coli (*EaggEC*) are self-adherent, tending to autoagglutinate, giving the appearance under the microscope of a stack of bricks. These organisms, their serological characterization, and the clinical disease they cause have yet to be fully described.
6. *Diffusively adherent* E. coli (*DAEC*) are even less well characterized than EaggEC.

21.4.1 Detection of *E. coli* O157:H7 (VTEC)

Since VTEC were first reported as human pathogens (Riley *et al.*, 1983), there has been a great increase in many countries in the incidence of foodborne infections caused by VTEC. This increase has not been just because of heightened awareness, as the clinical diseases can and do frequently extend beyond mere diarrhoeal disease to haemorrhagic colitis (Riley *et al.*, 1983), haemolytic uraemic syndrome (Karmali *et al.*, 1983) and thrombotic thrombocytopenic purpura (Moake, 1994; Neild, 1994), with significant case fatality rates. A number of serovars (e.g. O26, O48, O111, O113, O128, O145, O157, O163) may produce verocytotoxins, but the serovar most commonly incriminated is O157:H7. This organism is spreading through herds of beef and dairy cattle in a number of countries, and also is found in sheep (Kudva *et al.*, 1996) and goats (Bielaszewska *et al.*, 1997). Consequently the outbreaks have been associated mostly with undercooked beef or raw milk, or dairy products prepared from raw milk (ACMSF, 1995). However, other food and drink products have been involved, for example unpasteurized apple cider (Besser *et al.*, 1993). Contamination of foods and beverages can occur at any stage in the food chain. Because of the very low infective dose (perhaps even a single organism), a hazard is presented in any situation where survival can occur – multiplication is not necessary. Thus, in the case of the apple cider outbreak reported by Besser and co-workers, 90% of the apples used were 'drops' and collected from the ground under the trees; they were not washed before being pressed, and the cider was not pasteurized. It was possible that animal manures used beneath the fruit trees was the source of contamination: the acidity of the apple juice might retard or prevent multiplication, but this would not in itself confer safety.

Because of the serious nature of the diseases caused, specific screening for *E. coli* O157:H7 is being regarded as important in the relevant sectors of the food industry. A whole range of techniques has been and still is being developed. A few are described below. Nevertheless, it is most important to realize that the very low infectious dose makes this organism a particular hazard for workers in the microbiological laboratory, and laboratory-acquired infections have been reported (Gopal Rao *et al.*, 1996).

Unfortunately, of course, *any* procedure for isolating coliforms that involves incubation temperatures up to 37°C may lead to isolation and growth of VTEC, so if VTEC continues to become more prevalent, all food microbiologists should be aware of the possible presence of this organism even in circumstances in which it is not being investigated actively. The safety precautions described in Section 2.1 should be followed carefully, and if it is suspected that VTEC is likely to be detected and cultured, a local risk assessment of the laboratory facility should be performed to determine whether further containment conditions need to be applied; see the discussion of the interim advice of the UK's Health and Safety Executive (Anonymous, 1997).

E. coli O157:H7 grows poorly at 44°C, so the normal extension tests described in Sections 20.1.2 and 20.1.3 will fail to detect this organism. This organism is also normally sorbitol negative, whereas most *E. coli* are sorbitol positive (but see below). In addition, it is negative to the MUG test (i.e. it does not hydrolyse 4-methylumbelliferyl-β-D-glucuronide).

21.4.1.1 Plating media

Isolations can be made on a MacConkey's agar containing 1% sorbitol instead of lactose (SMAC). This medium can be made more selective by adding cefixime (0.05 mg l^{-1}) and potassium tellurite (2.5 mg l^{-1}) (CT-SMAC). Plates should be incubated at 37°C, and examined for non-fermenting colonies. *Hafnia* and *Escherichia hermanii* give similar colonies on SMAC (Feng, 1995) so presumptive *E. coli* O157 need to be confirmed serologically. It should be noted that sorbitol-positive mutants of *E. coli* O157 have been selected in *in vitro* experiments (Fratamico *et al.*, 1993). At the time of writing it is extremely difficult to design a protocol to detect sorbitol-positive *E. coli* O157 in the environment; the continued development of immunocapture techniques combined with appropriate differential media may permit the detection of such mutants. Bolton and colleagues (1995) investigated the use of selective enrichment followed by plating on CT-SMAC for food samples, and the use of immunomagnetic capture to detect the organism in clinical samples.

21.4.1.2 Liquid enrichment media

To detect small numbers, a 10^{-1} homogenate of the food should be prepared in buffered 1% peptone water containing cefixime ($0.05\,mg\,l^{-1}$), cefsulodin ($10\,mg\,l^{-1}$) and vancomycin ($8\,mg\,l^{-1}$). This should be incubated at 37°C for 20–24 h, followed by subculture on to CT-SMAC (Roberts *et al.*, 1995).

Greater sensitivity can be obtained by using immunomagnetic separation from the liquid enrichment medium, followed by plating the magnetic beads on to CT-SMAC (Bolton *et al.*, 1996). This technique uses magnetic beads coated with antibodies specific for the target organism. Beads are available commercially for detection of *E. coli* O157 (e.g. Dynabeads® made by Dynal). After incubation of the enrichment medium, the beads are mixed into the enrichment medium and then collected magnetically. The bacteria-carrying beads are then washed, resuspended, and plated on to CT-SMAC, which is then incubated as normal.

21.4.1.3 Confirmation

Escherichia coli O157 can be confirmed by using a latex agglutination kit (available from a number of manufacturers).

21.5 *YERSINIA ENTEROCOLITICA*

This Gram-negative bacterium can cause an enteric infection, resulting in diarrhoea and abdominal pain (Kapperud, 1991). Complications seen can include reactive arthritis. It seems probable that human infections are contracted by consumption of foods derived from animal sources or infected from animal sources (including pigs, cattle and sheep). The bacteria have been isolated from ice cream samples, and milk (Christensen, 1981; Vidon and Delmas, 1981), and dairy products could offer a potential hazard, particularly as the bacteria can grow at 4°C.

Various liquid enrichment media have been evaluated (see, for example, Park *et al.*, 1981; Vidon and Delmas, 1981). However, cold enrichment can be achieved in homogenates prepared in non-inhibitory Tris-buffered or phosphate-buffered saline incubated for up to 21 days at 4°C, or 14 days at 9°C. The major problem with cold enrichment is that the extended incubation period makes the method generally unsuitable for food quality assurance applications. Treatment of enrichment broths with alkali after incubation increases the chances of recovery (Aulisio *et al.*, 1980). The combination of a low temperature of incubation with non-toxic mixtures, followed by alkaline treatment, seems to give the best recovery rates. ISO 10273:1994 specifies enrichment in a selective irgasan–ticarcillin–potassium chlorate broth (ITPC), or in a slightly less selective peptone–sorbitol–bile salts broth (PSB), with the ITPC and the PSB being subcultured without alkali treatment, and the PSB also being subcultured after alkali treatment.

A selective solid medium has been developed, containing desoxycholate, cefsulodin, irgasan and novobiocin (CIN agar) (Schiemann, 1979, 1982; see also the review by Swaminathan *et al.*, 1982). This medium is available from manufacturers of dehydrated media (such as Oxoid). To achieve selectivity the PSB is subcultured on to CIN agar and the ITPC on to a high bile salts medium (ISO 10273:1994 specifies a modified *Salmonella–Shigella* (MSS) agar).

The organism is lactose negative, motile at 25°C, produces acid from glucose but no gas, and is urease positive.

21.5.1 Liquid enrichment

Add 25 g of food to 225 ml of PSB broth and mix using a Stomacher. Similarly mix 25 g of food into 225 ml of ITPC broth. Incubate the PSB either as a shaken culture at 22–25°C for 48–72 h (or as an unshaken culture for 5 days), and the ITPC at 25°C for 48 h.

21.5.2 Plating

1. Streak a loopful of PSB culture on to CIN agar.
2. Streak a loopful of ITPC culture on to MSS agar.
3. Add 0.5 ml of PSB culture to 4.5 ml of potassium hydroxide saline solution, and mix. After 30 s (20 s specified by ISO 10273:1994), streak a loopful on to CIN agar.

Incubate plates aerobically at 30°C for 24 h and 48 h.

After incubation for 24 h, examine plates for presumptive *Y. enterocolitica* colonies. If no characteristic colonies are seen, incubate for a further 24 h. Characteristic colonies on CIN agar are small (less than 2 mm in diameter), smooth with a red centre and translucent edge, surrounded by a zone of precipitated bile. On MSS agar, characteristic colonies are small (less than 2 mm in diameter) and grey.

21.5.3 Confirmation

Select colonies of presumptive *Y. enterocolitica* and streak on to plates of nutrient agar to ensure purity of the cultures, and incubate at 25°C for 24 h.

Perform an oxidase test on the isolates.

Also inoculate the following (or alternatively use a biochemical multi-test kit):

(a) Christensen's urea agar;
(b) Kligler's iron agar;
(c) Simmon's citrate agar;
(d) Carbohydrate fermentation tests with sucrose, rhamnose, melibiose, raffinose;
(e) Møller's lysine decarboxylase medium and Møller's ornithine decarboxylase medium.

Incubate at 25°C for 24–48 h.

Y. enterocolitica is urease positive. On Kligler's iron agar it produces an acid butt (yellow), no gas, and no hydrogen sulphide (no blackening). It cannot utilize citrate, and it cannot ferment rhamnose, melibiose or raffinose. It cannot decarboxylate lysine. Biotypes 1–4 (the human pathogens) ferment sucrose and can decarboxylate ornithine. *Y. enterocolitica* is motile when incubated at 25°C, but non-motile when grown at 35–37°C.

21.6 *VIBRIO*

Vibrio is a genus of predominantly oxidase-positive, fermentative, motile Gram-negative straight or curved rods. Thus a key characteristic differentiating them from the Enterobacteriaceae is their positive reaction to the oxidase test, although *V. metschnikovii* is oxidase negative.

The principal species of *Vibrio* involved in foodborne infections are *V. cholerae*, *V. parahaemolyticus* and *V. vulnificus*.

V. cholerae, the causative organism of cholera, is found in small numbers in freshwater and seawater environments. However, disease outbreaks are caused predominantly by contamination of water by faeces from infected persons, which leads to an increase in bacterial concentration in the water to above the minimum infective dose. Outbreaks can be caused by bottled waters (Blake *et al.*, 1977). Cholera can also be spread by foods that are contaminated by *V. cholerae*-containing faeces or by water itself contaminated with *V. cholerae*-containing faeces. Such foods can include salad vegetables and also shellfish.

V. parahaemolyticus enteritis, an infection-type food poisoning, is common in countries with warm coastal waters, including the USA but especially Japan, with raw or undercooked seafood and shellfish being implicated. In the UK and other countries with cooler coastal seawater temperatures, it is usually associated with frozen precooked seafoods imported from countries with warm coastal waters.

V. vulnificus has caused enteritis after consumption of raw or undercooked shellfish such as oysters (Blake *et al.*, 1980), once again being associated with warm coastal waters such as those found off the USA. Oliver *et al.* (1992) found that colistin–polymyxin B–cellobiose agar was superior to TCBS agar in the detection of *V. vulnificus*.

Five other species have been reported as causing human enteritis: *V. mimicus* (Davis *et al.*, 1981; Shandera *et al.*, 1983), *V. fluvialis* (Huq *et al.*, 1980; Lee *et al.*, 1981), *V. furnissii* (Brenner *et al.*, 1983), *V. metschnikovii* (Lee *et al.*, 1978) and *V. hollisae* (Hickman *et al.*, 1982). *V. hollisae* will not grow on Thiosulphate Citrate Bile Salts (TCBS) agar or the other media commonly used for the isolation of the pathogenic vibrios, and a satisfactory selective isolation medium has yet to be described; consequently *V. hollisae* will not be considered further here.

Described below are a method for detection of *V. parahaemolyticus*, and a method for the detection of pathogenic *Vibrio* species in general.

21.6.1 Method for the detection and isolation of *Vibrio parahaemolyticus*

This method is adapted from ISO 8914:1990.

1. Add 25 g of food sample to 225 ml of alkaline saline peptone water (ASPW). Mix, using a Stomacher, to give a 10^{-1} suspension. Similarly prepare a 10^{-1} suspension in salt polymyxin B broth (SPB).

   Incubate both enrichment broths at 35° or 37°C for 8 and 24 h.

2. At 8 and 24 h, inoculate a loopful of each enrichment broth (*without* first shaking the flasks) on to plates of TCBS agar.

   Incubate the TCBS at 35° or 37°C for 18–24 h.

3. Examine plates for typical *V. parahaemolyticus* colonies, which are smooth, 2–4 mm in diameter and either blue-green or colourless with a green centre (i.e. not sucrose fermenting). (Sucrose-fermenting organisms form yellow or yellow-brown colonies.)
4. Pick off and subculture typical *V. parahaemolyticus* colonies, streaking on to plates of salt (3% w/v) nutrient agar. Incubate at 35° or 37°C for 16–24 h.
5. Confirm identity as described below.

21.6.2 Method for the detection and isolation of pathogenic *Vibrio* species

The use of ASPW and TCBS will, in principle, permit the isolation of many *Vibrio* species in addition to *V. parahaemolyticus*. However, the less halophilic vibrios (e.g. *V. cholerae*) will tend to be overgrown in ASPW by the more halophilic vibrios, so Roberts *et al.* (1995) suggested using a modified ASPW (MASPW) by reducing sodium chloride to 1%, and adding also 0.4% magnesium chloride and 0.4% potassium chloride.

1. Add 25 g of food sample to 225 g of MASPW. Mix with a Stomacher. Incubate at 35° or 37°C for 8 and 24 h.

2. At 8 and 24 h inoculate a loopful (*without* shaking the flask) on to TCBS agar and on to modified cellobiose polymyxin B colistin (MCPC) agar (Massad and Oliver, 1987; FDA, 1992).

   Incubate the TCBS at 35° or 37°C, and the MCPC at 40°C, for 18–24 h.

3. On TCBS agar, *V. cholerae*, *V. fluvialis*, *V. furnissii* and *V. metschnikovii* are sucrose positive (yellow colonies), whereas *V. vulnificus* and *V. parahaemolyticus* are sucrose negative (blue-green or green colonies).

   On MCPC agar, colonies of *V. cholerae* are purple, whereas the cellobiose-fermenting *V. vulnificus* gives flat yellow colonies. Other species of *Vibrio* grow poorly on MCPC agar.

21.6.3 Identification of cultures from isolation methods 1 and 2

1. Pick off and subculture selected colonies on to plates of salt (2.5% w/v) nutrient agar.

   Incubate at 35° or 37°C for 16–24 h.

2. Identify purified cultures by:
   (a) oxidase test
   (b) β-galactosidase test
   (c) motility test

   and *either* inoculation of API20E and/or API20NE (using sterile 0.85% or 3% sodium chloride solution as the suspending liquid), *or* the following:

   (d) triple sugar iron (TSI) agar containing 2.5% (w/v) sodium chloride, using a straight wire to stab the butt and streak the slope
   (e) tryptone water containing 2.5% (w/v) sodium chloride, for the indole test
   (f) salt (2.5%) nutrient agar, densely surface inoculated, on which place, with aseptic precautions, a 10-μg O129 disc and a 150-μg O129 disc. (Discs of the 'O129 vibriostat' 2,4-diamino-6,7-di-iso-propylpteridine phosphate are available commercially from many suppliers of microbiological media and reagents.)

   Incubate at 35° or 37°C for 18–24 h.

Results

The reactions on API20E and API20NE are provided by bioMérieux on their API database, and are also listed by Austin and Lee (1992).

On TSI agar, *Vibrio* produces no gas and no blackening (hydrogen sulphide production). *V. parahaemolyticus* on TSI agar produces a yellow (acid) butt and red (neutral or alkaline) slant. Other *Vibrio* species also produce a yellow butt, but some will in addition produce a yellow (acid) slant because of the metabolism of sucrose. The reactions differentiating the relevant species are shown in Table 21.1.

TABLE 21.1 Reactions of the foodborne pathogenic species of *Vibrio*

	Acid from sucrose	Acid from cellobiose on MCPC*	Oxidase	β-Galactosidase	Indole	O129 10 μg	O129 150 μg
V. cholerae	+	–	+	+	+	S	S
V. parahaemolyticus	–	(V)	+	–	+	R	S
V. vulnificus	–	+	+	+	+	S	S
V. mimicus	–	–	+	+	+	S	S
V. fluvialis	+	(+)	+	+	+	R	S
V. furnissii	+	–	+	+	V	R	S
V. metschnikovii	+	–	–	V	V	S	S
Aeromonas	V	(+)	+	V	V	R	R

*The response shown here is for cellobiose utilization on the selective medium MCPC agar. This medium tends to select for *V. cholerae* and *V. vulnificus*. The reactions shown in parentheses are the responses to cellobiose that may be seen in non-selective biochemical tests.

+, Indicates that 80% or more of strains are positive; –, indicates that 20% or less of strains are negative; V, indicates that between 21 and 79% of strains are positive.

21.7 *CLOSTRIDIUM PERFRINGENS*

Food poisoning caused by *Clostridium perfringens* is an infection rather than an intoxination, if by intoxination we mean symptoms caused by the ingestion of preformed toxin. The symptoms are produced as the result of formation in the gut of toxin during sporulation of a large number of ingested vegetative cells (Duncan, 1973). *Cl. perfringens* rarely, if ever, sporulates in foods, but it sporulates readily in the intestine. Large numbers of viable vegetative cells must be present in the food to cause symptoms. Food poisoning caused by *Cl. perfringens* therefore normally requires conditions favourable for the multiplication of the bacteria in the food. Large numbers of *Cl. perfringens* may be expected in a food that has caused food poisoning, whereas in routine microbiological quality assurance the emphasis is on detecting the presence of *Cl. perfringens* (both spores and vegetative cells) in small numbers if, due to subsequent storage and treatment of the food, bacterial multiplication is possible.

21.7.1 Colony count method

Prepare serial dilutions of the food to be examined, using reinforced clostridial medium as the diluent. Carry out plate counts on the dilutions using egg yolk-free tryptose sulphite iron citrate cycloserine agar (SICA) (ISO 7937:1977), Shahidi and Ferguson's polymyxin kanamycin sulphite agar (SFA) (ICMSF, 1978) or neomycin blood agar (Roberts *et al*., 1995). SICA and SFA can be inoculated either as pour plates using 1 ml amounts of a dilution (in which case overlay the plates with a further 10 ml of the relevant medium) or by surface inoculation of predried, prepoured plates. Neomycin blood agar plates should be prepoured and predried, and surface inoculated. Incubate the plates anaerobically (see Section 9.5) at 35° or 37°C for 24 h.

Presumptive *Cl. perfringens* colonies on SICA and SFA are coloured black (due to the reduction of sulphite causing a precipitation of iron sulphide). On neomycin blood (bovine, equine or human blood) agar, colonies of *Cl. perfringens* produce a narrow zone of complete haemolysis (due to the theta toxin) and a further surrounding narrow zone of incomplete haemolysis (due to the alpha toxin).

Pick off colonies of presumptive *Cl. perfringens* and stab into tubes of nitrate peptone water plus 0.3% agar and also streak on plates of Willis and Hobbs's lactose egg-yolk milk agar (Willis, 1962), having previously spread half of each plate with a few drops of *Cl. perfringens* antitoxin and having allowed it to dry. Streak the organism under examination over the whole plate. Also streak a plate of blood agar.

Incubate the tubes of nitrate agar and the plates of Willis and Hobbs's medium anaerobically for 24 h at 37°C. Incubate the blood agar plates *aerobically* for 24 h at 37°C. After incubation, examine the nitrate agar for the type of growth (indicating motility or lack of it) and test for nitrate reduction (see Section 11.1.9). Examine a Gram-stained smear prepared from the growth on Willis and Hobbs's medium. Gram-positive non-motile rods, capable of reducing nitrate, are *Cl. perfringens*. This is confirmed by: (a) the reaction on Willis and Hobbs's medium – *Cl. perfringens* gives cloudy zones around the colonies on the section of the plate without antitoxin, but no zones on the section with antitoxin, and also ferments lactose (see also Willis, 1965); and (b) no growth on the aerobically incubated blood agar plates.

21.7.2 Multiple tube count method

If very small numbers of clostridia are expected to be present in a food sample, a multiple tube count may be carried out using differential reinforced clostridial medium (DRCM) as the selective medium. Pipette aseptically 1 ml of each of the chosen dilutions (prepared using reinforced clostridial medium as the diluent) into each of five (or three) bottles of DRCM. A duplicate set of tubes can be prepared and 'laboratory-pasteurized' at 63±2°C for 15 min. This will destroy vegetative cells, but will heat-activate and encourage germination of spores within the incubation period. Consequently spores and vegetative cells can be differentiated by the two MPN counts. Incubate at 35° or 37°C for 5 days. Tubes showing blackening should be confirmed for *Cl. perfringens* as described under method 1. Calculate the most probable number of *Cl. perfringens* using probability tables (see Appendix 2).

21.8 *CAMPYLOBACTER JEJUNI*

21.8.1 Introduction

In 1977, Skirrow described a form of enteritis caused by certain strains of *Campylobacter jejuni* and *C. coli*. Other species, including *C. lari* and possibly *C. upsaliensis*, may also be involved in human enteritis. The closely related *Helicobacter pylori* is considered to be involved in causing gastric and duodenal ulcers, and possibly implicated in stomach cancer. However, at present it is thought that food does not act as a vehicle for *H. pylori* but that it is spread from person to person.

Since 1977 the number of outbreaks of *Campylobacter* enteritis identified and investigated in many countries, including the UK, indicates that *Campylobacter* enteritis is one of the commonest causes of enteritis. Vehicles of infection are raw or ineffectively pasteurized milk, inadequately cooked poultry, and cross-contamination from raw poultry. In rural communities, direct human infection from farm and domestic animals is a highly important route. Occasionally outbreaks of campylobacteriosis may be waterborne, for example where well-water has become contaminated by faecal material from farm animals. In the UK, where doorstep delivery of bottled milk is commonplace, there have been family outbreaks and point source outbreaks as a result of contamination of bottled pasteurized milk by birds pecking through the bottle top seals to drink some of the milk (Riordan *et al.*, 1993). It has been estimated (Palmer and McGuirk, 1995) that bird-pecked milk bottle tops can explain around 17% of cases occurring in the April–June peak seen every year in the UK.

Without doubt, *Campylobacter* was a very common cause of enteritis long before 1977, but because of its fastidious nutritional and gaseous requirements it will not be detected unless specifically sought. One of the currently used procedures consists of using a blood agar (10% sheep blood or 5–7% laked horse blood) containing a selective combination of three to five antibiotics. A large number of such media has been described. Widely used media include a modified version of the medium of Butzler (Lauwers *et al.*, 1978), the Preston medium of Bolton and Robertson (1982), the Exeter medium of Humphrey (1986) and the Preston blood-free cefoperazone medium of Hutchinson and Bolton (1983).

To detect small numbers, a liquid enrichment culture stage can be used, followed by subculturing on to solid media. Exeter liquid medium incubated at 37°C has been recommended for maximum recovery of sublethally injured cells (Roberts *et al.*, 1995). *Campylobacter* species that are human pathogens are referred to as 'thermotolerant', being able to grow at 42–43°C. This property is certainly valuable when isolating *Campylobacter* from faecal specimens or other samples in which any *Campylobacter* organisms are likely *not* to be injured. However, in examining food samples, swabs of equipment surfaces, and many other samples from the food processing environment, any viable *Campylobacter* present are likely to have suffered sublethal damage from heating, refrigeration, freezing or exposure to bacteriostatic or bactericidal substances. Such sublethally injured organisms may well be inhibited at the higher temperature of 42° or 43°C by the combination of antibiotics found in the selective media. Consequently, either the entire incubation period for the initial culture should be at 37°C, or an initial incubation at 37°C for 4–6 h should precede transfer to 42° or 43°C.

Park and Sanders (1991) developed a selective medium for isolating injured *Campylobacter*. It is their medium that is recommended in ISO 10272:1995; however, the ISO method (as currently published at 1997) and the British Standard 1996 modified version both describe inappropriate methods for preparing the antibiotic mixture, so for the time being the procedure of Roberts *et al.* (1995) should be followed, using Exeter medium. A detailed discussion of the rationales and relative merits of the various enrichment and isolation media have been presented by Corry *et al.* (1995b).

Because of the complex formulations of the media, their performance is very sensitive to minor deviations during preparation, so it is suggested that commercially available dehydrated media and ready prepared antibiotic supplements (e.g. as obtainable from Oxoid or Difco Laboratories) be used. Accordingly, recipes for these media are not given in Appendix 1.

Plates should be incubated in a microaerobic gaseous atmosphere of 5–10% oxygen, 10% carbon dioxide and 85–90% nitrogen approximately. This mixture is most easily obtained by using an anaerobic jar *without* a catalyst, partially evacuating the jar and refilling to atmospheric pressure with a nitrogen–carbon dioxide mixture; using gas-absorbing/generating sachets designed for jars without catalysts (e.g. Campygen from Oxoid); or using gas-generating sachets for use with catalyst-containing jars (e.g Oxoid *Campylobacter* Gas Generating Kits – BR56 for use in conjunction with their 3.4-litre catalyst-containing anaerobic jars, or BR60 for use with 2.5-litre jars). For further details see Section 10.1.

21.8.2 Procedures

21.8.2.1 Direct isolation from raw poultry, etc.

Swab either the chicken skin (especially near the vent area) or the abdominal cavity. Streak the swab over the surface of a plate of Preston blood agar, Exeter agar, or Preston blood-free cefoperazone agar. Incubate in a microaerobic atmosphere (see above): Preston blood agar or Exeter agar for 4 h at 37°C, followed by 44–68 h at 42°C; Preston blood-free cefoperazone agar at 37°C for 48 h.

21.8.2.2 Enrichment from frozen foods, etc.

1. Mix 25 g of food in 225 ml of Exeter liquid medium using a Stomacher. Transfer to a screw-capped jar or bottle of a size that leaves very little headspace (so that the necessary microaerobic environment will develop), and tightly fit the screw cap. Incubate at 37°C for 18–48 h.
2. Subculture on to a selective agar medium (e.g. Preston blood-free cefoperazone agar, Exeter agar or Preston blood agar).

21.8.2.3 Confirmation

Confirmation of identity is on the basis of microscopic examination (motile Gram-negative non-sporing curved S-shaped rods), and the isolates being positive to catalase and oxidase tests. For a discussion of the differentiation of *Campylobacter* species, see Corry *et al.* (1995).

21.9 *LISTERIA MONOCYTOGENES*

21.9.1 Introduction

There has been increasing interest in the presence (or absence) of *Listeria monocytogenes* in foods, as the result of some substantial outbreaks of foodborne listeriosis in North America and Europe, with case fatality rates as high as 30%. Counts of up to 10^6 per gram have been found in soft cheeses and meat pâtés. Other foods that may contain substantial numbers are chilled ready-prepared salads. A large proportion of foods are commonly found to contain small numbers of *L. monocytogenes*, although some microbiological specifications stipulate the absence of *L. monocytogenes* from 25 g of food.

Thus detection of *L. monocytogenes* in some circumstances will require liquid enrichment; in other circumstances it will require a pre-enrichment resuscitation stage, and at other times it can be performed by direct plating. The US Food and Drug Administration developed methods that were then adopted by the International Dairy Federation, and these formed the basis for ISO 10560:1993 for detection of *L. monocytogenes* in milk and milk products. This was based on a single-stage enrichment in an acriflavine nalidixic acid broth, followed by plating on Oxford agar.

Since then, the International Organization for Standards has produced horizontal methods for both detection (ISO 11290-1:1997) and enumeration (ISO 11290-2:1997) of *L. monocytogenes*. For detection, a two-stage enrichment is used, with inoculation of a food sample into so-called Half Fraser broth, followed by subculture into Fraser broth as secondary enrichment medium. This secondary enrichment medium is then subcultured on to Oxford agar and polymyxin–acriflavine–lithium chloride–ceftazidime–aesculin–mannitol (PALCAM) agar. For enumeration, a 10^{-1} dilution of food sample is prepared using as diluent either buffered peptone water or Fraser broth base (i.e. Fraser broth without the selective supplements). A dilution series can then be prepared in the usual way, and plated on to PALCAM agar. The advantage of using Fraser broth base as the initial diluent is that, once the dilution series and plate counts have been set up, the appropriate supplements can be added to the inoculated Fraser broth base to produce the Half Fraser broth primary enrichment for detection. Thus both enumeration of large numbers (more than 10 per g) and detection of smaller numbers (1 per 10 g or 1 per 25 g) can be performed. The procedure described below follows this double approach.

21.9.2 Procedure for detection and enumeration of *L. monocytogenes*

21.9.2.1 Enumeration

1. Weigh out 25 g of food sample, add 225 ml of Fraser broth base to form the 10^{-1} dilution, and mix with a Stomacher.
2. Prepare a decimal dilution series in 0.1% peptone in the usual way.

3. Surface inoculate 0.1-ml amounts of dilutions on to prepoured, predried plates of PALCAM and Oxford agar.
   Note. Retain the 10^{-1} dilution for the detection procedure.
4. Incubate the plates at 37°C for 48 h (the Oxford agar plates aerobically, and the PALCAM agar plates under the microaerobic conditions obtained as described in Section 10.1).

21.9.2.2 Detection by enrichment of small numbers of *Listeria*

1. To the 225 ml of 10^{-1} dilution in Fraser broth base, add 2.25 ml of a sterile solution of lithium chloride, 0.23 ml of a sterile solution of nalidixic acid sodium salt, 1.12 ml of a sterile solution of acriflavine, and 2.25 ml of a sterile solution of ammonium iron citrate. (See Appendix 1 for details of the medium and the sterile supplements.) Mix to form the primary enrichment in Half Fraser broth. Incubate at 30°C for 24 h.

2(a). Streak a loopful of the primary enrichment culture on to a plate of Oxford agar and another loopful on to a plate of PALCAM agar. Incubate the plates at 35° or 37°C for 48 h (the Oxford agar plates aerobically, and the PALCAM agar plates under the microaerobic conditions obtained as described in Section 10.1).

2(b). Transfer 0.1 ml of the primary enrichment broth to a tube containing 10 ml of Fraser broth, to give a secondary enrichment. Incubate at 35° or 37°C for 48 h.

3. Using the secondary enrichment medium, repeat the plating step described in 2(a) above.

21.9.3 Rapid screening test for *Listeria*

To provide a rapid screening test, the primary or secondary enrichment culture described in Section 21.9.2.2 can be examined using an appropriate ELISA kit (e.g. TECRA Listeria Visual Immunoassay).

21.9.4 Identification and confirmation of *Listeria*

1. *Listeria* hydrolyses aesculin, and all species of *Listeria* except *L. grayi* cannot produce acid from mannitol. Colonies of *Listeria* on Oxford agar are 2–3 mm in diameter with dark brown or black haloes (due to aesculin hydrolysis); the colonies often have sunken centres. Colonies of *Listeria* on PALCAM agar are 1.5–2 mm in diameter, green and have black haloes (they may also have black centres).
2. To confirm *Listeria*, pick off five typical colonies and subculture each on to tryptone soya yeast extract agar (TSYEA), streaking to obtain separate colonies to ensure purity of culture. Incubate aerobically at 35° or 37°C for 24–48 h.
3. For each isolate, examine the colonies that have developed on TSYEA, using Henry illumination (see Section 3.4.1). *Listeria* produces thin semi-transparent

colonies which, with Henry illumination, appear blue, blue-green or blue-grey with a finely granular surface resembling coarsely ground glass.

Select an isolated colony on TSYEA, and inoculate a tube of tryptone soya yeast extract broth (TSYEB) and a slope of TSYEA. Incubate at 25°C for 18–24 h.

4(a). Perform a motility test and a Gram stain on the TSYEB. *Listeria* shows a tumbling motility; this can be distinguished from a rapid, directional motility by using a stock culture of a non-pathogenic strain of *Listeria*. *Listeria* produces slim, short, Gram-positive, non-sporing rods.

4(b). Perform a catalase test and oxidase test on growth from the TSYEA slope. *Listeria* is catalase positive and oxidase negative.

4(c). Use the TSYEB culture to set up confirmatory and differential tests for any presumptive *Listeria* by inoculating:

(i) D-glucose, D-salicin, L-rhamnose, and D-xylose fermentation broths (with bromcresol purple; see Appendix 1, p. 401). Incubate at 35° or 37°C for 7 days. Examine daily for acid production (purple → yellow).

(ii) Nitrate peptone water for nitrate reduction (see Section 11.1.9). Incubate at 35° or 37°C for 5 days.

(iii) A CAMP test using *Staphylococcus aureus* and *Rhodococcus equi* (see Section 11.5.3). Incubate at 35° or 37°C for 24 h.

Alternatively the API Listeria kit (bioMérieux) may be used. This has been found by McLauchlin (1997) to give good identification and differentiation of *Listeria* species.

Results

Bacteria in the genus *Listeria* are Gram-positive, small slender rods, catalase positive, oxidase negative, produce acid from D-glucose, acid from D-salicin, and are motile with a tumbling motility. The reactions differentiating the relevant species are shown in Table 21.2.

TABLE 21.2 Differentiation of *Listeria* species

	Enhanced haemolysis with *Staph. aureus*	Enhanced haemolysis with *Rhodo. equi*	Acid from D-mannitol	Acid from L-rhamnose	Acid from D-xylose	Nitrate reduction
L. monocytogenes	+	–	–	+	–	–
L. ivanovii	–	+	–	–	+	–
L. seeligeri	(+)	–	–	–	+	–
L. grayi	–	–	+	–	–	–
L. grayi subsp. *murrayi*	–	–	+	V	–	+
L. welshimeri	–	–	–	V	+	–
L. innocua	–	–	–	V	–	–

+, 90% or more of strains are positive; –, 90% or more of strains are negative; V, variable, 11–89% of strains are positive.

21.10 *STAPHYLOCOCCUS AUREUS*

21.10.1 Introduction

Staphylococcal food poisoning is an intoxination that depends on the ability of the food concerned to support the growth of the staphylococci which produce the toxin. The staphylococcal toxin can withstand heating at 100°C for 30 min and therefore the absence of viable organisms in the food is not proof of safety if there is a high total (microscopic) count of cocci. *Staph. aureus* may grow and multiply in pickled and cured foods, although there is evidence that toxin production is inhibited by salt concentrations that allow the growth of the organisms. However, this appears to depend on the composition of the food in which the organisms are growing, and there have been reports of toxin production in habitats containing salt concentrations typical of many cured foods.

The two methods most appropriate for the examination of food products are direct plating on Baird-Parker's medium or liquid enrichment in Giolitti and Cantoni's tellurite mannitol glycine broth followed by streaking on milk salt agar (ICMSF, 1978). The method described in ISO/DIS6888-1 employs Baird-Parker's medium followed by confirmation using tests for the production of coagulase and thermostable nuclease. The method in ISO/DIS6888-2 examines for coagulase-positive staphylococci directly on the initial isolation medium by using instead of Baird-Parker's medium, a rabbit plasma fibrinogen agar. This is stated to have the advantage of detecting the reportedly enterotoxigenic *S. hyicus*. This species produces atypical colonies on Baird-Parker's medium and does not produce a rapid reaction to the coagulase test, requiring the test to be incubated for 24 h; it will therefore be missed by the method using Baird-Parker's medium followed by confirmation with the coagulase test.

Counts of low concentrations of *Staph. aureus* can be undertaken by the MPN technique in conjunction with Giolitti and Cantoni's medium, with all blackened tubes being streaked.

21.10.2 Procedure

1. Prepare serial dilutions of the food and transfer 0.1-ml amounts to duplicate plates of poured, well dried, Baird-Parker's medium. Spread the inoculum evenly over the surface with a sterile, bent glass rod. Incubate for 24 h at 37°C. If no colonies have developed, reincubate for a further 24 h. *Staph. aureus* typically forms colonies that are 1.0–1.5 mm in diameter, black, shiny, convex, with a narrow white entire margin and surrounded by clear zones extending 2–5 mm into the opaque medium. Pick off a number, or all, of the colonies that are typical of *Staph. aureus* and test for coagulase production (see Section 11.4.3) and either thermostable nuclease production or DNase production. The absence of coagulase should not automatically exclude a diagnosis of staphylococcal food poisoning as there have been reports of enterotoxin production by coagulase-negative strains (e.g. Breckinridge and Bergdoll, 1971).

2. If low concentrations of *Staph. aureus* are suspected, a liquid enrichment stage can be included, adding 10-g amounts of food to 100 ml of Giolitti and Cantoni's tellurite mannitol glycine broth. Incubate at 37°C for 24 h and then streak loopsful on to milk salt agar. The enrichment technique can be rendered quantitative as described in Section 21.2. Note that a salt enrichment broth (recommended by AOAC (1995) and FDA (1992)) may be insufficiently selective when applied to high-salt food samples (e.g. cured or pickled foods), and also that in refrigerated foods *Staph. aureus* may have become salt sensitive as a result of the refrigeration. For the enrichment of *Staph. aureus* from high-salt foods Giolitti and Cantoni's medium (Giolitti and Cantoni, 1966) may prove more effective. To obtain a cross-selectivity in the two stages of isolation, milk salt agar is recommended for plating rather than the tellurite-containing Baird-Parker's medium.

21.10.3 Staphylococcal enterotoxins

The ICMSF (1978) recommended two methods of extracting and concentrating enterotoxin in foods, followed by detection and identification of the enterotoxin by the microslide gel diffusion test (see also AOAC, 1995; FDA, 1992). The sensitivity of these techniques is barely adequate to detect the minimum concentration of enterotoxin in foods that can cause the clinical syndrome when consumed. Concentration techniques using, for example, affinity chromatography columns may help, but sometimes significant loss of toxin can occur during the procedure. More sensitive methods include the ELISA technique (Saunders and Bartlett, 1977; Stiffler-Rosenberg and Fey, 1978; Koper *et al.*, 1980). The ELISA method is suitable for use in routine laboratories, and a number of manufacturers (e.g. TECRA, and R-Biopharm, Darmstadt, Germany) produce suitable diagnostic kits.

21.11 *BACILLUS CEREUS*

21.11.1 Introduction

Bacillus cereus food poisoning (see Hauge, 1955; Davies and Wilkinson 1973; and also pp. 145–6) is caused by consuming foods containing very large numbers of bacteria (usually of the order of 10^7 per gram) (Goepfert *et al.*, 1973). Therefore, when examining suspected foodstuffs during epidemiological investigations of outbreaks of food poisoning, a selective medium is not necessary. A count of aerobic phospholipase producers can be obtained by the surface inoculation of egg-yolk agar (Section 11.3.4), followed by microscopic confirmation of Gram-positive rods and biochemical tests for *B. cereus* (see below). *B. cereus* and the insect pathogen *B. thuringiensis* are very closely related; the principal difference is that *B. thuringiensis* has a plasmid-encoded insect-pathogenic δ-endotoxin; δ-endotoxin is deposited as a parasporal crystalline inclusion body which can be detected by microscopic examination. Damgaard *et al.* (1996) have found that most strains of *B. thuringiensis* display cytotoxigenicity similar to *B. cereus*; Jackson *et al.* (1995) described the isolation of a cytotoxigenic *B. thuringiensis* strain from a food poisoning outbreak. As *B. thuringiensis* is used as an agricultural pesticide and may also be a natural inhabitant of soil it may be found in many foods. It is important that an isolate is not rejected as of public health significance merely because it is identified as *B. thuringiensis* because of the microscopical observation of parasporal crystals.

Other species of *Bacillus*, for example *B. subtilis* and *B. licheniformis*, have been reported as capable of causing food poisoning (Kramer *et al.*, 1982).

Many foods can be expected to contain small numbers of *B. cereus*, because it is such a common environmental contaminant. In consequence selective enrichment techniques will not usually provide information of much value. At intermediate population concentrations, in which a foodstuff may represent a potential hazard because some proliferation of *B. cereus* has occurred, it may be useful to employ a selective and differential medium such as mannitol egg-yolk phenol red polymyxin agar (MEPPA) (Mossel *et al.*, 1967) with direct plate counts. This medium is specified by ISO 7932:1993.

However, MEPPA is readily overgrown and has poor buffering power against acid produced by mannitol fermenters (*B. cereus* does not ferment mannitol). It is suggested that parallel counts be set up on the polymyxin pyruvate egg-yolk mannitol bromthymol blue agar (PPEMBA) of Holbrook and Anderson (1980). This medium is available as a dehydrated medium from Oxoid. Lecithinase activity and spore production are reported to be enhanced on this medium by the inclusion of the pyruvate and peptone Unfortunately, this medium also has rather poor buffering power, so plates should be examined after 24 h of incubation. Presumptive *B. cereus* may be confirmed by: Hugh and Leifson's medium to test for anaerobic dissimilation of glucose; hydrolysis of gelatin; reduction of nitrate; ability to grow on Knisely's chloral hydrate agar; inability to attack mannitol or xylose.

Foods can also be examined directly for the presence of *B. cereus* enterotoxin. For example, an ELISA kit is available from TECRA, and a reversed passive latex agglutination kit is available from Oxoid (see also Rusul and Yaacob, 1995).

21.11.2 Procedure

Prepare serial dilutions of the food (up to 10^{-4}) and transfer 0.1-ml amounts to duplicate plates of poured, well dried MEPPA and PPEMBA. Spread the inoculum evenly over the surface of each plate with a sterile, bent glass rod, starting with the plates at the highest dilution first, to avoid the necessity of sterilizing the rod between plates. Incubate for 24 and 48 h at 30°C.

Results

On MEPPA, presumptive *B. cereus* colonies are rough and dry with a violet-red background and a halo of dense white precipitate; on PPEMBA the colonies will be turquoise to peacock blue, and surrounded by a precipitate. These reactions result from absence of attack on mannitol, and the presence of lecithinase activity against the egg yolk.

In contrast, colonies of *B. subtilis* and *B. licheniformis* will be yellow and/or surrounded by a yellow zone (because of acid produced from mannitol) and show no zone of precipitation (because of lack of a lecithinase).

21.11.3 Confirmation

To confirm colonies as *B. cereus* purify by streaking on egg-yolk agar plates, and prepare pure cultures on nutrient agar slopes. Examine microscopically for the presence of Gram-positive rods, which will usually be capable of sporulation, the spores being ellipsoidal and not large enough to cause swelling of the vegetative cells.

Biochemical reactions typical of *B. cereus* that may be tested for include:

(a) anaerobic dissimilation of glucose (Hugh and Leifson's medium, Section 11.2.3);
(b) *inability* to attack mannitol or xylose (Section 11.2.2);
(c) hydrolysis of gelatin (Section 11.1.1);
(d) reduction of nitrate (Section 11.1.9);
(e) ability to grow on Knisely's chloral hydrate agar.

In addition *most* strains of *B. cereus* can hydrolyse starch (Section 11.2.1), and *most* are Voges–Proskauer positive (Section 11.2.6). (See also Part IV, p. 337.) The presence of a parasporal body detected by microscopic examination should *not* be considered to exclude the possibility that the isolate is enterotoxigenic (see above).

Lecithinase-negative strains of *B. cereus* are known to occur. Their significance in food poisoning is still in doubt so that in epidemiological work associated with outbreaks it may be considered desirable to search for lecithinase-negative strains in addition to the more typical lecithinase-positive strains.

If it is desired to identify in addition *B. subtilis* and *B. licheniformis*, other differential tests will be necessary. It is suggested that in this case a commercial diagnostic kit be used (e.g. API 50CHB from bioMérieux), to distinguish all three species.

21.12 *CLOSTRIDIUM BOTULINUM*

Whereas food poisoning caused by *Cl. perfringens* is an infection, botulism, which is caused by *Cl. botulinum*, is an intoxination. The bacteria multiply in the food before it is consumed, producing an extremely potent exotoxin. Most outbreaks of botulism have been caused by smoked, cured, pickled or canned food, which in Europe is usually meat or prepared meat products (e.g. sausage, pâté, etc.) which may or may not be canned, and in America is usually canned fruit or vegetable. Vacuum-packed foods (including smoked fish) have also been implicated in outbreaks of botulism, even when such foods are refrigerated. *Cl. botulinum* type E is capable of toxin production in foods when stored at refrigeration temperatures down to 3°C (Hobbs, 1981).

The spores of *Cl. botulinum* are relatively heat resistant but are more sensitive to heat at low pH. As canned foods present suitably anaerobic conditions for the growth of *Cl. botulinum* if the pH is above 4.5, heat processing is designed to destroy the *Cl. botulinum* spores. In national statutory regulations, canned food is defined as food in a hermetically sealed container which has been sufficiently heat processed to destroy any *Clostridium botulinum* in that food or container or which has a pH of less than 4.5.

The detection or enumeration of *Cl. botulinum* in foods is difficult and dangerous, and not to be undertaken in normal quality assurance laboratories. Details of the techniques that may be used have been described by the AOAC (1995) and the FDA (1992), but work must be undertaken only by adequately trained staff (in whom high antitoxin titres are maintained) in specially equipped laboratories. In addition an examination of the suspected food for the toxin may be made. This involves immunodiffusion tests or the injection of mice, and will usually be carried out by the public health authority or a similar body in the event of an outbreak of botulism.

21.13 OTHER PATHOGENIC AND TOXIGENIC BACTERIA

21.13.1 *Aeromonas*

These bacteria are associated particularly with aquatic habitats and may be found in untreated water, raw shellfish and on foods washed in untreated water. *Aeromonas* species, including *A. hydrophila*, *A. sobria* and *A. caviae*, have been reported to be implicated in human intestinal disease (Champsaur *et al.*, 1982; Altwegg and Geiss, 1989; Wadström and Ljungh, 1991; Kirov, 1993; Merino *et al.*, 1995), although some doubt has been cast on their role as causative agents (Morgan and Wood, 1988). Nevertheless in some outbreaks the bacteria have been isolated both from the patients and from the implicated food (e.g. CDSC, 1993).

A modification of XLD agar (Ryan's XLD agar) designed to be selective for *Aeromonas* is available commercially from Oxoid (Pin *et al.*, 1994). Isolates can be identified using the API 20NE test kit (bioMérieux). (See also Jeppesen, 1995.)

Aeromonas can be distinguished from *Vibrio* in that it is resistant to vibriostatic reagent 0/129 (2,4-diamino-6,7-di-isopropylpteridine) and does not require added sodium chloride in the medium. Differentiation of the three species on the basis of biochemical tests is shown in Table 21.3. In addition, the three species show different reactions on blood agar.

Aeromonas hydrophila and *A. sobria* typically are both haemolytic on equine blood agar, producing wide zones of complete haemolysis. In the case of *A. hydrophila*, however, the haemolytic zone is surrounded by a zone (halo) of precipitation; also, in areas of heavy confluent growth of *A. hydrophila* the medium typically shows a green coloration.

Table 21.3 Differentiation of the three foodborne human pathogenic species of *Aeromonas*

	Aesculin hydrolysis	Gas from glucose	Acid from salicin	VP reaction
A. hydrophila	+	+	+	+
A. sobria	−	+	−	d
A. caviae	+	−	+	−

Isolated colonies of *A. caviae* are non-haemolytic but show a zone of precipitation. Haemolysis may be seen in areas of heavy confluent growth.

21.13.2 *Pseudomonas*

Cephaloridine–fucidin–cetrimide agar (CFCA) has been recommended as a selective medium for pigmented and non-pigmented *Pseudomonas* (Mead, 1985). Most *Pseudomonas* spp. can grow up to 30°C, but *Ps. aeruginosa* and a few other species are characterized by an ability to grow up to about 41–42°C.

The principal food-poisoning pseudomonad is *Ps. cocovenenans*, which causes 'bongkrek' poisoning in S.E. Asia by producing toxins (toxoflavin and bongkrekic acid) during saprophytic growth on the fermented coconut product tempeh bongkrek (Steinkraus *et al.*, 1983; Buckle and Kartadarma, 1989). This organism has been transferred to the genus *Burkholderia* as *B. cocovenenans* (Euzeby, 1997). At present there is no information on the ability of this bacterium to grow on CFCA or other selective media.

Pseudomonas aeruginosa is an organism of concern in drinking water (especially in bottled natural mineral waters).

21.13.2.1 Procedure

1. Prepare a decimal dilution series of the sample in 0.1% peptone.
2. Surface inoculate prepoured, predried plates of CFCA with 0.1-ml amounts of dilution. Spread with a sterile glass spreader, starting with the highest dilution. Incubate at 25°C for 48 h. An incubation temperature of 41 ± 1°C can be used if seeking *Ps. aeruginosa*, preceded by a resuscitation stage at a temperature of 30°C for 4 h, after which the plates are transferred to 41 ± 1°C.

21.13.2.2 Confirmation

Colonies producing yellow-green fluorescent diffusible pigment can be assumed to be *Pseudomonas*. *Pseudomonas* species not capable of producing fluorescent pigment will produce smooth translucent white or cream colonies.

Pseudomonas is oxidase positive, catalase positive, motile with polar flagella, and oxidatively metabolizes glucose (determined with Hugh and Leifson's medium).

21.14 TOXIGENIC MICROFUNGI AND THEIR MYCOTOXINS

The earliest firm description of food intoxination caused by a microfungus is ergotism caused by *Claviceps purpurea*, which has been known about for hundreds of years. Now many toxigenic genera have been listed, with a very large number of toxic secondary metabolites (mycotoxins) having been identified for each genus (see Moreau, translated by Moss, 1979).

Amongst the more important toxigenic genera are:

Aspergillus mycotoxins include aflatoxins, aspergillic acid, citreoviridin, kojic acid, ochratoxins, patulin, penicillic acid, sterigmatocystin and versicolorins
Penicillium mycotoxins include citreoviridin, griseofulvin, islandicin, luteoskyrin, ochratoxin A, patulin, penicillic acid and penitrems
Fusarium mycotoxins include fumonisins, moniliformin, trichothecenes including T-2 toxin and deoxynivalenol, and zearalenone

Other toxigenic genera include *Alternaria*, *Stachybotrys*, *Wallemia* and, of course, *Claviceps purpurea.*

Many mycotoxins have been identified by their toxic effects on animals other than humans, especially farm animals. Others have been identified by *in vitro* cytotoxic effects on tissue cultures. The involvement of most mycotoxins in human disease remains in doubt. However, there is no doubt that it is important to minimize the risks from food containing aflatoxins, ochratoxins, trichothecenes, fumonisins, zearalenone and patulin.

The principal approach must be to use Hazard Analysis Critical Control Point (HACCP) principles to control both contamination of food by microfungi and growth of microfungi in food. Secondly, the detection and identification of microfungi in food is important both in monitoring efficacy of HACCP procedures and in quality assessments particularly of food materials moving in international trade (see Section 14 and Section 40.2 for methods for isolation and identification).

The third approach is to examine foods directly for the presence of specific mycotoxins, an activity that once again is of particular importance for foods moving in international trade. Such analytical procedures are outside the scope of this book, and so details are not given here.

Extreme care must be taken whilst following such procedures, because of the potential carcinogenic or teratogenic hazard. Reagents are commercially available for immuno-affinity concentration and immunodetection (e.g. by ELISA) of a few mycotoxins, especially aflatoxins, ochratoxins, fumonisins and zearalenone. Examples of such kits are those manufactured by Vicam (Somerville, Massachusetts, USA), R-Biopharm (Darmstadt, Germany) and Rhône Diagnostics (Glasgow, UK).

Other analytical procedures are essentially of chemical analyses (see WHO, 1990) and identification by thin-layer chromatography and high-pressure liquid chromatography (see AOAC, 1995) or mass spectroscopy and nuclear magnetic resonance spectroscopy (see, for example, Miller and Trenholm, 1994).

21.15 PROTOZOAN AND HELMINTHIC PARASITES

These are of great importance worldwide. There are many routes of infection but in this book we are obviously concerned with parasites that infect by the oral route, with water and food being the vehicles. With some parasites (e.g. *Toxocara canis*) the normal mechanism for transmission from soil, vegetation or other host animal (the dog in the case of *T. canis*) to the mouth is with the fingers. However, with these parasites, food contaminated by soil or directly by the relevant animal faeces can also act as a vehicle.

Some parasites (e.g. *T. canis*, *Ascaris lumbricoides*) are widely distributed across many continents and climatic environments. Some (e.g. *Giardia lamblia*, *Cryptosporidium*) have been showing a tendency to become a problem in an increasing range of countries.

Of concern to the food scientist and technologist must be the increasing international trade in foods that are consumed without cooking. For example, salad vegetables are now air-freighted worldwide from countries in tropical and subtropical regions to free such fresh, perishable commodities from seasonality. This means that infection by soil-transmitted parasites (or parasites transferred to foods by infected food handlers) can now occur in any country involved in such imports. Unfortunately washing the food before consumption may only decrease rather than eliminate the risk of transmission, as many fruits and vegetables are fragile and have many crevices. In a series of outbreaks of *Cyclospora* infection in the USA caused by fresh raspberries imported from Guatemala (CDC, 1997b, 1997c) the raspberries had reportedly been washed in 10 of 14 outbreaks.

The amoebae are, of course, microorganisms and fall within the area of responsibility of the microbiologist. Some large parasites, easily visible to the naked eye, are transmitted by eggs which are microscopic (e.g. *A. lumbricoides*) or by other life-cycle stages which are microscopic or near-microscopic (e.g. cercariae of *Clonorchis sinensis*). In any case, most establishments (factories, quality assurance laboratories, etc.) are not large enough to employ a specialist parasitologist, so the microbiologist will find that any work on parasites is included in their duties.

As in the case of pathogenic bacteria, laboratory examinations for parasites cannot provide a reliable way of protecting the consumer. In fact, as parasites cannot be cultivated *in vitro* (with the exception of free-living pathogenic amoebae), laboratory examinations have until recently been restricted to direct visual examination (with or without the microscope), so the sensitivity of such methods has been extremely poor.

Laboratory examinations may be useful not only in epidemiological studies of outbreaks of disease, but also in monitoring the efficacy of quality management techniques, and this will be especially the case if concentration and amplification methods are developed for such examinations. In the case of macroscopic parasites, such as the nematode *Anisakis simplex* found in fish, direct naked-eye observation of the larvae may be possible using appropriate transmitted light through the food together with good incident light (see, for example, AOAC, 1995). However, the parasites can easily be missed. Elution and concentration methods, often employed in clinical parasitology laboratories in the examination of faecal specimens, have not been developed adequately for the detection of most parasites in foods, although Jackson *et al.* (1981) described a method for eluting nematodes from fish.

In the case of parasites ingested in microscopic or near-microscopic forms (e.g. cysts of amoebae such as *Entamoeba histolytica* and *Cryptosporidium parvum*; eggs of *A. lumbricoides*; metacercariae of *Fasciola hepatica*; cercariae of *Clonorchis sinensis*), attempts at detection in foods by microscopy will rarely be successful. For details of the morphology of the various stages in the life cycles of parasites, see Manson-Bahr and Bell (1995).

For many parasitic diseases, ELISA is used in clinical diagnosis to detect antibodies circulating in the serum of an infected person. However, the methodologies and reagents have yet to be developed to permit immunoconcentration (e.g. by Dynabeads®), ELISA or PCR to be used to detect infective stages of parasites in foods.

The best approach to these problems at present must be to employ good husbandry practices in agriculture, horticulture, fish-farming, etc., using a HACCP system to identify the range of hazards and to define the control points.

In general, parasites are heat sensitive. Also, most of the parasites are sensitive to freezing, being destroyed by freezing to −18°C for 1 week. This can be of use in disinfecting fish to be eaten raw or used raw in marinaded products (see Section 27).

Of course, heating or freezing is not appropriate for tackling the problem of parasites in fresh vegetables to be used for salads, or in fruit, so that a Hazard Analysis Critical Control Point (HACCP) evaluation leading to a Code of Good Agricultural Production Practice is required on the farm to reduce the chances of faecal contamination of such crops.

21.16 *BRUCELLA*

In many countries, for example in northern Europe, infection of farm animals by *Brucella* has been eradicated. In very many countries, however, human infection, particularly by *Br. abortus* and *Br. melitensis*, is still a hazard. Ovine and caprine infection by *Br. melitensis* is found in the Iberian peninsula and the Mediterranean countries; both species are widespread in many Asian countries; and elk and bison are reported to act as a reservoir of infection in some parts of the USA. The organism can be transmitted by raw milk and by dairy products made from raw milk that have no pasteurization step incorporated in the process. *Brucella* can survive quite well in such products (e.g. for up to 6 months in hard cheese and 4 months in butter) (see Brinley Morgan and Corbel, 1990; Garin-Bastuji and Verger, 1994).

Because of the high infectivity of the organisms, there is a high risk of laboratory-acquired infections for all laboratory staff working with *Brucella*. In countries where *Brucella* is enzootic, the *Brucella* milk ring test can be used safely to identify milk samples that contain antibodies to the organism and which therefore possibly originate from infected animals (see Section 13.3). Food microbiologists in food quality assurance laboratories should never attempt the *isolation* of *Brucella* from samples of raw milk or dairy products. Such investigations should be restricted to official specialist public health and clinical pathology laboratories.

21.17 *MYCOBACTERIUM BOVIS*

As in the case of *Brucella* infections, many countries in the world, for example the UK and many other northern European countries, have more or less eradicated *Mycobacterium bovis* from dairy herds, with just occasional sporadic cases occurring in dairy cattle. One problem in achieving complete eradication is that many wild animals can act as natural reservoirs of infection. In countries where *M. bovis* is still enzootic in dairy herds, raw milk and dairy products made from raw milk that do not incorporate a pasteurization step in the process will offer a hazard of non-pulmonary tuberculosis (often manifesting as cervical lymphadenitis). It has been estimated that udder involvement leading to milk containing *M. bovis* occurs in about 1% of infected cows (Collins and Grange, 1983).

Food microbiologists in food quality assurance laboratories should never attempt the isolation of *M. bovis* from samples of raw milk or dairy products. Such investigations should be restricted to official specialist public health and clinical pathology laboratories.

PART III

Microbiological Examination of Specific Foods

22

Introduction: Effect of the Food Environment on Constituent Microflora

As already mentioned (see Section 1) microbiological examinations of foods *may* assist in the assessment of hygienic precautions during production and of the efficacy of a preservation process, and *may* allow predictions of the potential shelf-life, and also the identification of potential health hazards by the use of suitable indicators or by direct detection of pathogens. However, it has already been pointed out that extreme care must be taken in the interpretation of results and in the conclusions drawn from them.

In Part III, outline procedures for the examination of food components and products on a commodity basis are provided. In many cases there is an appropriate ISO standard method, and this will be indicated. The methods described in Section 21 mostly follow the protocols for the ISO standard methods.

Often a food manufacturer or a retailer will obtain food materials from sources that cannot be inspected or can be inspected only infrequently. In such situations, quality assurance may have to rely rather heavily on monitoring the microbiological quality of the materials. For example, imported vegetables to be used in the preparation of prepacked delicatessen salad mixtures may very likely have been derived from fields fertilized with animal manures. Washing with water containing even quite a high residual chlorine concentration has been shown to be relatively ineffective in reducing the numbers of surface contaminants. An appropriate microbiological examination in this example would include a count of coliforms or *Escherichia coli* as an index of the presence of potential pathogens.

23

Raw Meat and Raw Meat Products

23.1 FRESH, CHILLED AND FROZEN RAW MEATS

If red meat is cooked when still in rigor it will be unacceptably tough. Consequently, after slaughter of the animal, the carcass requires ageing until rigor has disappeared – the muscle tissue relaxes once again, and in this state the meat is more tender. Thus tenderization may occur more rapidly if the carcass is held at a higher temperature, but of course the higher the temperature, the greater the chance of undesirable microbial growth and spoilage. At 0.5°C beef carcasses require weeks before they are tender, at 13°C around 5 days are required, at 18°C only 2 days, and at 29°C only a few hours (see also Ranken and Kill, 1993). Meat, and particularly lamb, can suffer significant toughening by being chilled too soon after slaughter. However, rather than too-rapid chilling, it is more common for the problem to be that chilling has not been applied soon enough, so that microbial deterioration has started to occur.

The muscle tissue of healthy animals contains few, if any, bacteria, but cut and exposed surfaces become easily contaminated after slaughter and during and after dressing and butchering. For example, the surface count on skin and hair may exceed 10^9 microorganisms per square centimetre. Microorganisms can multiply readily on the cut surfaces, although the microbial count of the interior of the meat usually remains much lower. Because of the respiration that occurs in the tissues after slaughter, the meat rapidly becomes anaerobic, except for a surface layer a few millimetres in thickness, after only a few hours. Thus obligate aerobes and facultative organisms may develop on the cut surfaces, but only facultative or obligately anaerobic organisms may develop in the depths of the meat. After the dressing procedures, the population on the cut surfaces will be 10^3 to 10^5 aerobic bacteria, of which fewer than 100 will be psychrotrophic, per square centimetre.

If the meat is kept at all at temperatures above 20°C, mesophiles will be able to multiply rapidly, including anaerobic clostridia. At lower temperatures, psychrotrophs become more important. When the carcasses are kept below 10°C in conditions of high humidity so that the surfaces remain moist, psychrotrophic bacteria such as *Pseudomonas*, *Acinetobacter* and *Moraxella* will be selected and will eventually predominate on the surface, causing surface slimes. When conditions of fairly low relative humidity prevail and when, for example, cut surfaces are exposed to the rapid movement of air, as occurs when there is forced air circulation in the chill room, the drying of the surface layers of the meat tends to prevent the growth of psychrotrophic bacteria, and psychrotrophic moulds or yeasts may develop instead. The genera that may be encountered under these circumstances include *Sporotrichum* (causing 'white spot'), *Cladosporium* (causing 'black spot'), *Thamnidium*, *Geotrichum* and *Trichosporon*. *Mucor*, *Rhizopus* and *Penicillium* may also be observed.

The microbial condition of the meat can be affected profoundly by the preslaughter condition of the animals. If there is a glycogen deficiency in the muscle tissue at the time of slaughter, little lactic acid is produced and the pH does not fall very far. A condition known as DFD (dark, firm, dry) meat results and, although such meat retains water well (of financial interest), the relatively high pH may lead to poor microbiological quality, with amino acids being broken down without any delay, to cause off-odours and off-flavours.

The storage and distribution of joints of meat or larger cuts in vacuum-packed gas-impermeable bags will usually lead to the *Pseudomonas–Acinetobacter–Moraxella* microflora being followed by *Brochothrix thermosphacta* (Gardner, 1981), and finally by lactic acid bacteria of the 'atypical meat streptobacteria' group of *Lactobacillus*.

23.1.1 Microbiological examination

The microbiological condition of the meat surface can be assessed most rapidly in a semi-quantitative manner by microscopic examination of contact slides stained by Gram's method. Cultural examinations can be performed on swabs, or on surface rinses of defined areas using sterile stainless steel cylinders to contain the rinse liquid whilst the surface is abraded, the rinse liquid then being removed by pipette. Alternatively, very thin superficial samples may be sliced away using sterile scalpels and forceps, these samples then being homogenized in diluent.

After a decimal dilution series is prepared to 10^{-5} (or more than one series if special diluents are needed (see Section 5)), amongst the examinations that may be made are as follows.

1. General viable (aerobic) counts on plate agar, incubated at 4.5°C, 20°C (or 25°C), 37°C (or 35°C) for 14, 3 and 2 days respectively (see Section 18). ISO 2293:1988 specifies 30°C.
2. Total Enterobacteriaceae, coliforms or *E. coli* counts using a violet red bile agar or multiple tube technique (the latter if small numbers are expected) (see Section 20).

   Some of the ISO vertical methods are currently being revised (e.g. ISO 3811, ISO 12074). It is suggested that, meanwhile, if wishing to follow an appropriate ISO standard method, the relevant horizontal methods are used as appropriate: ISO 7402 or 8523 for Enterobacteriaceae, ISO 4831 or 4832 for coliforms, ISO 7251 for *E. coli*. In addition *E. coli* O157 may be sought (see Section 21.4).

3. Colony count for *Brochothrix thermosphacta*, using Gardner's Agar (Gardner, 1966) (see ISO 13722:1996)
4. Examination for the presence of *Salmonella* (see Section 21.3). A revision of ISO 3565 is in preparation. It is suggested that meanwhile the horizontal method in ISO 6579 is used for guidance; the methods described in Section 21.3 broadly follow the protocol for ISO 6579.
5. Examination for the presence of *Yersinia enterocolitica* (see Section 21.5). ISO 10273:1994 is the appropriate horizontal method.
6. A count of *Campylobacter jejuni* (see Section 21.8). The horizontal method ISO 10272:1995 may be used.
7. A count of *Staphylococcus aureus* (see Section 21.10). A two-part revision of the horizontal method ISO 6888 is in preparation.
8. Viable counts of anaerobic bacteria using a dilution series prepared with reinforced clostridial medium as a diluent. Use plate count agar and blood agar; incubate anaerobically at 35–37°C for 2 days.
9. A count of *Clostridium perfringens* (see Section 21.7). The horizontal method ISO 7937:1997 may be used.

23.2 RAW SAUSAGES, BURGERS, GROUND BEEF AND OTHER COMMINUTED RAW MEAT PRODUCTS

Meat products in this group will often contain various combinations of cereal, rusk, salt and spices. In addition to refrigerated storage, in some countries certain products may be partially preserved by the addition of preservatives such as metabisulphite. This has a temporary preservative action, with the general viable count and total Enterobacteriaceae count falling during the first day after production. Thereafter the viable counts will rise again as the free sulphur dioxide content falls.

The initial microflora comprises microorganisms deriving from the raw meat, together with those deriving from the other components listed. The latter will be principally endosporogenous bacteria. However, the spoilage flora developing on refrigerated storage tends to reflect the modified environment presented by the comminuted product. Thus, after storage, *Brochothrix thermosphacta*, yeast, *Lactobacillus* and *Micrococcus* are frequently major components of the microflora (Dowdell and Board, 1968, 1971). As well as coliform organisms, faecal streptococci may also be found in significant numbers. *Pseudomonas*, *Acinetobacter* and *Kurthia zopfii* may occur as minor components. In some products *Pediococcus* or *Leuconostoc* can be found. As already discussed in Section 21.4 there is much concern about such products causing outbreaks of infections by *E. coli* O157:H7 (verocytotoxigenic *E. coli*). This strain seems to be becoming established in both beef and dairy herds. Many of these outbreaks in North America and Europe have involved fast-food outlets selling hamburgers. (See Roberts and Gross, 1990; Bolton *et al.*, 1995.)

23.2.1 Microbiological examination

Composite samples should be taken, and appropriate decimal dilution series prepared. Amongst the examinations that may be made are those described in Section 23.1.1 above, but in addition, the following may be made:

10. Acetate agar (or Rogosa agar) with cycloheximide (to inhibit microfungi) as double-layered pour plates to count *Lactobacillus*, with incubation at 30°C for 5 days. The vertical method ISO 13721:1995 may be used.
11. Counts of *Enterococcus*.
12. Counts of yeasts, using dichloran rose bengal chloramphenicol agar, incubated at 20–25°C for 5 days. (See Section 14.) The vertical method ISO 13681:1995 may be used.

23.3 MEATS AND MEAT PRODUCTS PRESERVED BY CURING AND PICKLING

This group of products will be taken to include those in which partial drying, smoking and/or fermentation are used. Often the food material is subjected to a combination of processes; for example, the use of brining, partial drying and smoking may be combined. The brine solutions used may contain sodium chloride alone, or sodium chloride together with sodium or potassium nitrite (sometimes with nitrate as well).

('Corned beef' and 'corned mutton' are products made by adding the dry salts (including nitrite and/or nitrate), followed usually by canning with a substantial heat treatment. These products have the microbiological characteristics of other canned foods processed for a 'botulinum cook', and therefore fall under Section 34.)

The possibly carcinogenic cornpounds resulting from the above processes – nitrosamines, and also various components of wood smokes – have led to modifications to the processes, which have reduced their antimicrobial effectiveness. In addition, consumers in many countries are seeking foods with lower sodium content, so that a reduction is also occurring in the salt content of many products. This is leading to problems of diminishing microbial inhibition and control. In consequence, many such products now rely more on the additive effects of a wider range of environmental parameters, any one of which is by itself insufficient to provide adequate control of microorganisms. Thus moderate sodium chloride content, low nitrite content, light or no smoking, may be combined with the use of preservatives such as sorbic acid, the use of humectants such as glycerol, sorbitol or other polyhydric alcohols, vacuum packaging, and application of refrigeration throughout the entire distribution chain to the point of consumption.

The pH of cured or pickled meats in which microorganisms have been allowed to contribute to flavour, etc., usually by natural development of the innate microflora that stabilize in the curing brine, will normally be in the range pH 5.5–6.5. Fermented sausages often incorporate additional fermentable carbohydrate, and starter organisms, such as *Lactobacillus plantarum*, *Pediococcus cerevisiae*, *Micrococcus* and *Debaryomyces*, may be added. After maturation at temperatures that may range from 10° to 30°C, the product will have a pH of below 5.5. With larger amounts of added fermentable carbohydrate, the pH obtained will be as low as 4.5.

Typical fermented sausage products of the type just described include the salamis of Eastern Europe, and such products as soudjouk, a Turkish product made from beef, water buffalo and/or mutton. An acidulant such as glucono-δ-lactone may be used to obtain a more rapid pH drop in the initial stages. In these products, partial dehydration is permitted during the maturation, so that the final product displays a substantial microbiological stability at ambient temperatures provided the ambient humidity is not too high. An outbreak of *E. coli* O157:H7 food poisoning in the USA in 1994 was found to be caused by Italian-type dry salami. As a result, the USFDA subsequently required all producers of fermented meat products to: (a) cook their products; or (b) validate their particular process by use of defined test procedures; or (c) hold and test each batch of product.

In some countries (e.g. Hungary, Bulgaria) a surface mould flora is encouraged to develop on salami. Some salami manufacturers prefer not to use starter cultures, and instead rely on the development of a 'house flora' of both lactic acid bacteria in the fermentation of the salami contents and of surface moulds in the maturation stage. In this case a range of genera and species will be found in both groups. Nevertheless, the predominant surface mould is almost certain to be *Penicillium*. Unfortunately, some of the moulds developing naturally in this way may be mycotoxinogenic. Other manufacturers use starter cultures of both the lactic acid bacteria and the surface moulds. Work is being done particularly in the selection of non-toxigenic strains of *Penicillium nalgiovense* (Cook, 1995).

The bacterial flora of cured or pickled meat products will include lactic acid bacteria, halotolerant micrococci and halotolerant yeasts such as *Debaryomyces*. Moulds may develop on the surfaces of dried products. Halophilic *Vibrio* may be found in meat-curing brines.

23.3.1 Microbiological examination

Composite samples should be taken, and appropriate decimal dilutions prepared. Surface counts may also be made; in the case of joints of meat these may be obtained as described in Section 23.1.1 above, but in the case of sausage products rinses should be used with either 0.1% peptone or 0.1% peptone plus 15% sodium chloride (the latter being used for halophilic/halotolerant counts). The 'atypical' meat lactobacilli should be detected using Tween 80 agar (i.e. acetate agar base without added acetate) (Varnam and Grainger, 1972, 1973).

General viable counts, especially aerobic mesophilic counts, are usually made in quality assurance microbiology to assess hygienic aspects of production. In this application it is appropriate to use the standard media such as plate count agar. However, in ecological studies of a meat microflora, especially in products such as these, it is important to remember that plate count agar provides an environment very different from the food, and plate count agar plus 15% sodium chloride gives an environment very different from that of a curing brine. The direct microscopic examination of samples and dilutions can help to assess the efficacy of an isolation medium in recovering the constituent microflora. If a particular morphological type of microorganism is seen microscopically but is not recovered on isolation media, there is a strong possibility that the isolation environment is proving unsuitable for the growth of that microbial type. If this is found to be the case, then one approach that can be tried is to add to the media a filter-sterilized extract of the raw meat (see, for example, Varnam and Grainger, 1973).

For quality assurance purposes, various of the following examinations may be performed (see also Section 23.1.1):

1. General viable counts on plate count agar, incubated at 4.5°, 20° (or 25°C), 37°C (or 35°C) for 14, 3 and 2 days respectively.
2. Halotolerant/halophilic counts on plate count agar plus 15% sodium chloride, incubated at 4.5° and 25°C for 7 and 3 days respectively. The diluent used for these counts should be 15% sodium chloride or 0.1% peptone plus 15% sodium chloride.
3. A count of *Staphylococcus aureus*.
4. A count of *Lactobacillus* on acetate agar base (i.e. with acetate buffer omitted), prepared as a surface colony count, incubated at 30°C for 5 days in an anaerobic atmosphere containing 5% carbon dioxide. The ISO standard, ISO 13721:1995 may be used.
5. A yeast and mould count using dichloran rose bengal chloramphenicol agar, incubated at 20–25°C for 5 days. If the presence of moulds is suspected, surface spread plates should be prepared. The vertical method ISO 13681:1995 may be used.

24

Cooked Meat Products

24.1 SLICED COOKED MEATS

Cooking will destroy a very large proportion of the microflora of the raw meat. Even thermoduric organisms such as *Clostridium perfringens* will usually be reduced to very small numbers, although improper storage after cooking can allow proliferation of such survivors. During slicing and serving operations, contamination of the food can occur from hands, slicing machines and other equipment. Thus inadequate attention to hygiene can lead to the meat being contaminated by a range of commensal and pathogenic microorganisms including enterobacteria and staphylococci.

24.1.1 Microbiological examination

Decimal dilutions should be prepared at least to 10^{-3}. Poor hygiene and high storage temperatures can lead to microbial counts occurring up to 10^{7} or more per gram. In such cases, the dilution series should be extended to 10^{-6}.

Perform general viable counts at 25° and 37°C, total Enterobacteriaceae counts, coliform or *Escherichia coli* counts, and counts of *Staphylococcus aureus* and *Clostridium perfringens*.

24.2 COOKED MEAT PIES, MEAT CASSEROLES, ETC.

Such products should have low viable counts immediately after cooking, any surviving microorganisms being most probably spores of *Bacillus* and *Clostridium*, and examination for both aerobic and anaerobic bacteria should therefore be performed. In the case of meat pies that have a jelly or stock added after cooking, the stock and meat should be sampled separately as well as together in a combined sample. In such products the stock is usually injected into the cooked pie, the stock being kept in a holding reservoir in which it is maintained at a temperature just below boiling point. Too low a holding temperature may lead to a multiplication of thermophilic or even mesophilic bacteria in the stock.

24.2.1 Microbiological examination

Amongst the microbial counts that should be performed are: aerobic plate counts on plate count agar at 25°, and 37° and 55°C, and anaerobic plate counts at 37°C; *Clostridium perfringens* count; and *Bacillus cereus* count.

25

Poultry and Poultry Products

Chicken and turkey meat have consistently been a major source of food poisoning and foodborne diseases, especially *Salmonella, Campylobacter* and *Clostridium perfringens*, and possibly also enteropathogenic *Escherichia coli* and *Yersinia enterocolitica.*

Certain *Salmonella* serovars are host adapted. The 1960s saw the establishment of *S. agona* in chickens and *S. hadar* in turkeys in the UK and certain other countries. Since the 1980s there has been in a number of countries, especially in Europe and North America, an increase in infection of poultry flocks by certain phage types of *S. enteritidis*. These have shown an unusual facility for being vertically transmitted by transovarian infection of the egg. In the UK and in some other countries including Spain, this has involved mainly phage type PT4. In North America, other phage types showing the same ecological traits have become established. At present it is unclear what has led to this change in the enzootic pattern. However, it has resulted in eggs and egg products and poultry and poultry products in these countries becoming the predominant foods implicated in *S. enteritidis* enteritis in human beings, and even in the relevant phage type(s) of this serovar becoming the principal salmonella to be isolated from human enteritis. It is not surprising that eggs and egg products have re-emerged as foods of major food poisoning significance, since, whereas poultry is always intended to be cooked thoroughly (even though it often is not), eggs and egg-containing foods often deliberately receive no heat treatment (e.g. as in mayonnaise, mousse) or a very light heat treatment (e.g. meringues, zabaglione). When an egg can be infected internally with salmonellae, it is difficult to achieve a hazard-free preparation of some of these recipes. In these countries, the catering and manufacturing industries are being encouraged or even compelled to use only pasteurized egg for these purposes.

25.1 REFRIGERATED POULTRY

Psychrotrophic microorganisms that cause spoilage of refrigerated raw poultry are *Pseudomonas* (both non-pigmented and pigmented species) and *Acinetobacter/Moraxella*. In addition, and to a lesser extent (except when pseudomonads fail to grow), *Aeromonas*, 'atypical' lactobacilli, *Brochothrix thermosphacta* and certain members of the

Enterobacteriaceae may be involved. These psychrotrophs, which are found on the exterior of the live bird, derive from the external environment and not from the intestinal contents, whereas the organisms of public health significance mentioned above will be harboured primarily in the intestinal contents of the poultry.

In the processing plant, a number of processes lead to the spread of microorganisms, with cross-contamination occurring from one bird to many others.

Cross-contamination can occur in scalding tanks if the water temperature is below 55°C. The 'soft scald' process, using temperatures of 50–52°C, tends to be used for poultry that are to be air-chilled and then sold as 'fresh' (ACMSF, 1996). Substantial contamination can occur during defeathering, with microbial build-up being possible on the flails, and during evisceration.

Before chilling, spray washing can be used in an attempt to reduce surface contamination but in fact this may result in a reduction in the aerobic mesophilic colony count only from 10^5 to 10^4 per cm^2 of skin surface, and even this level of reduction is difficult to achieve.

Chilling may be carried out by circulation of cold air (a procedure used in Europe especially on poultry to be sold without freezing) or by immersion in cold water. Static tank chilling is rarely used now, and usually the carcasses are agitated whilst being immersed in water flowing through the tanks. The chill water may flow in the same direction as the carcasses ('through-flow') or in the opposite direction to the carcasses ('counter-flow'). The latter is preferable because the carcasses just about to leave the tank are in contact with the cleanest water.

A better control of microorganisms *in the chill water* is achieved by having secondary chlorination of the process water, with residual chlorine levels of up to 200 p.p.m. in the chill water used for immersion chilling. However, because the chlorine is quickly inactivated on contact with the skin of the poultry, chlorination of the chill water will not significantly reduce the microbial concentrations on the skins of the carcasses (ACMSF, 1996).

25.2 'NEW YORK' DRESSED POULTRY AND GAME

These are uneviscerated and usually unplucked. In some countries, retail sale of poultry in this form is prohibited. When held refrigerated 'New York' dressed poultry and game have a much longer shelf-life than when eviscerated and plucked. Psychrotrophic spoilage bacteria cannot readily colonize the muscle tissue. Instead, greening eventually occurs as hydrogen sulphide produced in the intestine diffuses into the muscle tissue and results in sulphaemoglobin formation.

Game birds are usually hung at cool, rather than refrigeration, temperatures, preferably around 10°C. As with all game, much of the organoleptic appeal depends on the tenderization of the muscle tissue during the hanging, together with flavour which develops, probably at least partially, as a result of microbial metabolites diffusing from the intestinal lumen. Microbial development does not normally occur in the meat itself.

25.3 CURED AND SMOKED POULTRY PRODUCTS, POULTRY SAUSAGES, ETC.

These are produced in rather small amounts, in most countries fulfilling minority demands. The microbiology of such products should be approached in a similar manner to that of cured and smoked red meats, etc.

25.4 MICROBIOLOGICAL EXAMINATIONS

There are substantial differences between poultry meat and red meat as microbial habitats, because of differences in chemical composition and pH, for example, and this has to be taken into account in ecological studies. Nevertheless, procedures for microbiological examinations for quality assessment purposes can be the same as those described in Section 23 for the analogous red meat products.

Since the microbial flora of table poultry is confined largely to the skin surface, visceral cavity or cut surfaces, it follows that the methods most appropriate for the examination of table poultry are surface-sampling techniques. Some indication of the kinds of microorganisms and their relative numbers and disposition on the skin surface can be obtained by the use of contact slides or adhesive tape. These techniques are particularly useful in the examination of non-frozen specimens after keeping quality tests, and for the examination of frozen poultry after thawing.

For quantitative studies, swab procedures can be used with sterile templates to define the area to be swabbed, or the whole or part carcass may be rinsed in a known volume (e.g. 500 ml) of sterile diluent. Neck skin also provides convenient sampling material, particularly during in-line sampling, skin homogenates then being prepared in the laboratory (see also Barnes *et al.*, 1973). Dilution series can be prepared from swabs, rinses and skin homogenates, and cultural studies made as described for other raw meat products in Section 23.

Because of the potential importance of poultry as a vehicle for *Salmonella* and *Campylobacter*, specific examinations for these bacteria are likely to be important components of such quality assessments. However, it should be recognized that innate contamination of poultry by these two organisms is extremely likely, so that end-product specifications should be drafted and applied with care to avoid adopting unrealistic goals. Reduction in human salmonellosis and campylobacteriosis is better achieved by concentrating on reducing or eliminating contamination of the animal feedstuffs, applying good hygienic measures in processing plants, and ensuring that consumers, caterers, restaurateurs, etc. understand the need for both taking hygienic precautions to prevent cross-contamination of cooked foods from the raw poultry, and applying *adequate* cooking procedures for poultry.

26

Eggs and Egg Products

26.1 INTRODUCTION

Although *Salmonella* has been shown to be capable of infecting hen eggs as well as duck eggs *in vivo*, the contents of the great majority of new-laid hen eggs are sterile. The outside of the shell normally carries a large number of bacteria as a result of the contamination of the eggshell with faeces, dust, etc. These bacteria include Gram-positive bacteria of the genera *Micrococcus*, *Staphylococcus*, *Arthrobacter* and *Bacillus*, and Gram-negative bacteria of the genera *Pseudomonas*, *Escherichia*, *Enterobacter*, *Proteus*, *Serratia*, *Aeromonas*, and members of the *Acinetobacter–Alcaligenes–Flavobacterium* group (Board, 1966; Board and Fuller, 1994). Of these, it is the Gram-negative organisms that are most frequently isolated from egg contents. Rough handling or washing of the eggs facilitates penetration of the shell and membranes by the surface contaminants, whereas dry cleaning does not so readily assist penetration.

In the preparation of bulked egg materials, whether whole egg, egg albumen or egg yolk, there is opportunity for contamination of the contents of the eggs by microorganisms from the shell exteriors, and of course any microorganisms such as *Salmonella enteritidis* PT4 present in the *contents* of even one egg will contaminate the entire bulk. It was this characteristic, together with the ability of *S. enteritidis* PT4 to transmit vertically from chicken to chick by ovarian infection, that led to *S. enteritidis* PT4 being an important foodborne pathogen in the UK and much of the rest of Europe (ACMSF, 1993). Consequently trade in, and use of, unpasteurized bulked egg is to be discouraged. The bulked whole egg, egg albumen or egg yolk should be pasteurized before drying or freezing. A series of outbreaks in the USA caused by *S. enteritidis* PT4 was reported by Boyce *et al.* (1996).

26.2 SHELL EGGS

26.2.1 Examination of the contents of shell eggs

Scrub the eggs with warm soapy water and a stiff brush, then rinse well and drain. Immerse in alcohol for 10 min, then allow to drain well and flame quickly. Cut a hole in the end opposite the air sac (which is located at the blunt end) using a small carborundum disc on an electric drill, or using a sterile scalpel. Remove the contents aseptically and homogenize with a Stomacher. Counts can be carried out on single eggs or on the bulked contents of a number of eggs. Prepare serial decimal dilutions in the usual way. Carry out aerobic mesophilic counts on plate count agar at 25° and 37°C, total Enterobacteriaceae counts, coliform counts, and an examination for the presence of *Salmonella*.

26.3 FROZEN WHOLE EGG

Frozen whole egg is normally packed in metal containers with press-on or screw-on lids. The product may contain many contaminating bacteria derived from the processes involved in its preparation. The International Commission on Microbiological Specifications for Foods (ICMSF, 1974) recommended that pasteurized whole egg should have a direct microscopic count of less than 5×10^6 per gram, a general viable count of less than 10^6 per gram, and that *Salmonella* should be absent from 125 g.

Procedure

Sample the product while still in the frozen state. Clean the lid and top of the tin, swab with alcohol, flame and then remove the lid. With a sterile auger or other suitable instrument, remove two cores, one from the centre of the can and one at the edge, extending from the top surface to as deep a level as possible with the instrument used. Transfer these to a sterile container and examine as soon as possible. Allow the frozen samples to soften slightly and, while still very slightly frozen, blend thoroughly. Prepare serial decimal dilutions to 10^{-5}. Carry out:

1. Aerobic mesophilic counts on plate count agar at 25° and 37°C.
2. Total Enterobacteriaceae counts using VRBGA pour plates.
3. Presumptive coliform counts using either violet red bile lactose agar pour plates or the multiple tube technique.
4. A direct microscopic count by the Breed's smear method, on the 10^{-2} dilution.
5. Examination for the presence of *Salmonella*.

26.4 DRIED EGG

Dried egg may have a viable count ranging from a few hundred to several hundred millions of bacteria per gram. Consequently the dilutions used for viable counts should be chosen to cover this range. The procedure is similar to that used for frozen egg except that sampling is simplified because there is no question of thawing, and the use of a Stomacher is not necessary.

26.5 FROZEN, DRIED AND FLAKE ALBUMEN

The methods used are similar to those for frozen or dried whole egg, but some samples of dried or flake albumen may be found to have high total (microscopic) counts owing to the intentional use of a bacterial fermentation stage in the manufacturing process. The subsequent drying, in the case of dried albumen, may cause a great reduction in the viable count.

27

Fish, Shellfish and Crustacea

27.1 INTRODUCTION

In general, fish is more prone to microbial spoilage than is meat because the latter has a lower pH and is less moist, bacterial growth being to some extent inhibited. Because of the low ambient temperatures in marine and fresh waters, the autochthonous bacterial flora includes a greater proportion of psychrotrophs and fewer bacteria with an optimum growth temperature of 37°C compared with the usual flora of meat. The predominant types of bacteria are species of *Moraxella*, *Pseudomonas*, *Arthrobacter*, *Flavobacterium–Cytophaga*, *Acinetobacter* and *Micrococcus* (Shewan, 1971). The microbial spoilage of fish has been discussed by Hobbs (1983).

A number of parasites can be transmitted by the consumption of raw or undercooked fish, including:

(a) the tapeworm *Diphyllobothrium latum* (in trout, Pacific salmon, and perch) (see, for example, Ruttenber *et al.*, 1984);
(b) the nematodes *Anisakis simplex* (occurring worldwide in a wide range of marine fish) (see, for example, Kliks, 1983) and *Capillaria philippinensis* (in fish from fresh and brackish water in S.E. Asia);
(c) the trematodes *Clonarchis sinensis* (especially in carp in S.E. Asia), *Opisthorchis* (in a range of fish in Eastern Europe, India and S.E. Asia) and *Paragonimus* (in crabs and crayfish in Japan, China, S.E. Asia, Africa and South America).

As already mentioned in Section 21.15, laboratory examinations for such parasites do not provide a reliable way of protecting the consumer. However, such parasites are not only heat sensitive but also for the most part are killed readily by freezing to −18°C and below for about 1 week. As many parasites are not killed by marinades, fish for pickled and marinaded products should first be either heat-treated or frozen.

27.2 SALT-WATER FISH

Many of the bacteria found will be halophilic or at least salt-tolerant. However, many marine bacteria have a requirement for *low* concentrations of salt and will be inhibited on media containing 5% sodium chloride (MacLeod, 1965). Large numbers of bacteria are frequently found in the slime on the skin surface, which can support the growth of many bacteria which contaminate fish after catching.

Scombrotoxic food poisoning results from the consumption of scombroid fish (including tuna, bonito and mackerel) in which some bacterial spoilage has occurred, particularly by *Morganella*, *Proteus* and *Hafnia*, resulting in the metabolism of histidine to histamine or other metabolites with similar physiological action. The active compounds are heat stable, so that commercially canned fish may cause such poisoning if a spoilage microflora has been allowed to develop before canning. Specifications for such foods may include maximum levels for histamine, detected by chemical analytical techniques (for such analytical techniques see, for example, AOAC, 1995).

Coliform organisms and *Staphylococcus aureus* are good indicators of the standard of hygiene during handling. Shewan (1970) suggested that the following standards be adopted:

(a) an aerobic mesophilic count at 35–37°C of not more than 10^5 per gram;
(b) a coliform count of less than 200 per gram (*E. coli* less than 100 per gram); and
(c) *Staph. aureus* fewer than 100 per gram

If agar liquefiers are encountered frequently, silica gel may be used as the solidifying agent.

Procedure

Carry out surface counts and tissue counts in a manner similar to that described for meat, using:

1. Plate count agar for aerobic mesophilic counts, incubated at 4.5°, 25° and 37°C for 14, 3 and 2 days respectively.
2. Plate count agar plus 1% sodium chloride for counts of non-extreme halophiles, incubated at 4.5°, 25° and 37°C for 14, 3 and 2 days respectively.
3. Plate count agar plus 15% sodium chloride for counts of extreme halophiles and salt-tolerant organisms, incubated at 4.5°, 25° and 37°C for 14, 3 and 2 days respectively. In this case the diluent should be 15% sodium chloride solution.
4. The multiple tube count method for coliform counts.
5. Baird-Parker's medium for *Staph. aureus* (surface counts) at 37°C.
6. Media to detect *Vibrio parahaemolyticus* and other pathogenic vibrios. This is recommended especially in the case of fish from inshore waters.

27.3 FRESHWATER FISH

Many types of Gram-negative bacteria are commonly found on freshwater fish; Trust (1975) found that *Pseudomonas* and *Cytophaga* predominated on salmonid fish.

Coryneform organisms are the most common Gram-positive types. Many of the organisms will be psychrotrophic but not salt-tolerant or halophilic. The florae of healthy and diseased fish have been surveyed by Collins (1970). In addition, it is possible for freshwater fish to be contaminated with *Salmonella*, when the fish come from waters contaminated with faecal matter from humans or farm animals, as can occur particularly in the intensive fish-rearing ponds found in rural farming communities in some countries.

Procedure

The procedure is as described for salt-water fish but omitting counts of salt-tolerant organisms. Include examinations for *Salmonella* if appropriate.

27.4 SMOKED, MARINADED AND PICKLED FISH

Traditionally smoked fish is much less prone to microbial spoilage than raw fish, and moulds become more important as spoilage organisms, owing to the partial dehydration that occurs during smoking (Shewan and Hobbs, 1967). Some modern smoking methods, however, do not decrease the susceptibility of the fish to microbial spoilage, or remove the hazard from *Clostridium botulinum* or parasites. The possible survival of parasites in marinaded and pickled products has already been mentioned.

Procedure

This is as for raw fish, with the addition of a qualitative examination for moulds by streaking the 10^{-1} dilution (or a higher concentration) across the surface of Davis's yeast salt agar or Czapek–Dox agar.

27.5 SHELLFISH AND CRUSTACEA

In general, the bacteria responsible for spoilage are the same as those responsible for the spoilage of salt-water fish. Sewage pollution of the estuarine and inshore habitat of many shellfish may result in concentration by the animal of human intestinal organisms, which may include pathogens such as *Salmonella* and enteric viruses. Therefore, in addition to total and aerobic mesophilic counts, the presence or absence of *E. coli* (and possibly also enterococci and *Clostridium perfringens*) should always be determined as an indication of faecal pollution. Information should be sought on whether the samples have been taken before or after keeping the shellfish in chlorinated water to allow self-cleansing to take place. Pasteurized and/or pickled shellfish products should be free from all enterobacteria.

Another type of hazard of microbial origin is the possible presence of paralytic shellfish poison (PSP) and diarrhoetic shellfish poison (DSP) as the result of the shellfish growing in waters in which certain dinoflagellates (e.g. *Gonyaulax* and *Dinophysis*) are found. These phytoplankton occasionally multiply in coastal waters in large numbers, resulting in the appearance of so-called 'red blooms'. Avoidance of outbreaks of this type of poisoning is achieved primarily by identifying the proliferation of the causative phytoplankton and by issuing orders that prohibit the harvesting of shellfish from the relevant areas whilst the phytoplankton persist. Examination of samples for the presence of PSP and DSP has traditionally been by biological testing using, for example, mice – a technique not suitable for routine food quality assurance laboratories. However, detection of PSP and DSP can also be achieved by chemical analytical techniques (e.g. Botana *et al.*, 1996). In either case these analyses are not microbiological in nature, and are therefore not considered further here. A RIDASCREEN® ELISA kit (R-Biopharm, Darmstadt, Germany) for PSP is available (Kasuga *et al.*, 1996).

Procedure for the examination of shellfish

Samples should be examined within 24 h of collection. They should be transported and stored dry at between 5° and 10°C until examined.

Scrub the shells of several shellfish (100 ml of body material will be required) thoroughly under running water. Finally wash in sterile water and then place one in the middle of a wad of six sterile, large filter papers. In the case of oysters hold the oyster in the filter paper with the flat side uppermost and open with a sterile oyster knife by cutting through the muscle which holds the two valves of the shell together. Remove the flat shell, taking care not to spill the liquid in the concave shell. Other shellfish can be opened in a similar way by cutting through the adductor muscles. Remove the liquid to a sterile wide-mouthed glass bottle using a sterile pipette, and transfer the body of the shellfish to the bottle using sterile forceps. Cut the body into small pieces with sterile scissors and repeat the process with more shellfish until there is about 100 ml of mixed liquor and finely chopped pieces. Transfer aseptically exactly 50 ml to a sterile glass bottle with ground glass stopper, containing a small number of glass beads. Shake vigorously, add 50 ml of sterile quarter-strength Ringer's solution, and shake again. Add 20 ml of this 1 : 2 dilution to 80 ml of diluent to give a 10^{-1} dilution. Prepare further dilutions in the usual way.

The following tests may be performed:

1. Aerobic mesophilic counts on plate count agar at 25° and 37°C.
2. Presumptive coliform counts by the multiple tube technique, inoculating each of three or five tubes with 2 ml of the 1 : 2 dilution (giving a 10^0 dilution), three or five tubes with 1 ml of the 10^{-1} dilution, and three or five tubes with 1 ml of the 10^{-2} dilution. Confirm presumptive coliform organisms as *E. coli* by the Eijkman test.
3. Counts of salt-tolerant organisms on plate count agar plus 15% sodium chloride at 25° and 37°C, having used 15% sodium chloride solution as the diluent.
4. A count of enterococci by the multiple tube method using three or five tubes with 2 ml of the 1 : 2 dilution, three or five tubes with 1 ml of the 10^{-1} dilution, and three or five tubes with 1 ml of the 10^{-2} dilution.
5. Examination for *Vibrio parahaemolyticus* and other pathogenic vibrios.

27.6 RECOMMENDED SPECIFICATIONS

(1) The consumer has to recognize that there will always be some risk associated with eating raw fish, raw shellfish or raw crustacea.

Good-quality cooked fish products (including frozen precooked crustacea) should have aerobic mesophilic counts of less than 5×10^5 per g, fewer than 10 *E. coli* per g, fewer than 10 *V. parahaemolyticus* per g, and *Salmonella* not detected in five 25-g samples.

Precooked battered and/or breaded fish products may present a hazard from *Staph. aureus*, so this organism should be tested for, and counts of fewer than 100 per g should be readily achievable by good manufacturing practice.

The ICMSF (1986) has suggested detailed specifications for six fish product groups based on aerobic mesophilic counts, counts of *E. coli* and *Staph. aureus*, and, where appropriate, *V. parahaemolyticus* and *Salmonella*.

(2) There are European Union (EU) directives (EC, 1991) on the microbiological criteria to be applied to crustacea and molluscan shellfish, which have been implemented in national legislation (e.g. in the UK's *Fishery Products and Live Shellfish (Hygiene) Regulations 1998* (Anonymous, 1998)). The relevant microbiological specifications are as follows:

Live shellfish intended for immediate human consumption (from Class A production areas)

Faecal coliforms*	<300 per 100 g
*E. coli**†	<230 per 100 g
Salmonella	Absent in 25 g
Total PSP	Not more than 80 μg per 100 g
DSP	Not detected by biological test methods

Live bivalve molluscs for treatment either in a purification centre or by heat (from Class B production areas)

The following specification must be achieved by at least 90% of samples:

Faecal coliforms*	Not more than 6000 per 100 g
*E. coli**†	Not more than 4600 per 100 g

Live bivalve molluscs that require relaying for at least 2 months followed, where necessary, by treatment either in a purification centre or by heat (from Class C production areas)

Faecal coliforms*	Not more than 60 000 per 100 g
*E. coli**†	Not more than 46 000 per 100 g

*Determined using a five-tube, three-dilution MPN test or equivalent

†*E. coli* is functionally defined as faecal coliforms that produce indole from tryptophan at 44 ± 0.2°C within 24 h.

Fish

Must be inspected for parasites, and must be frozen to −20°C or lower for at least 24 h, if the fish and fish products are:

(a) consumed raw or almost raw;
(b) are herring, sprats, Atlantic salmon or Pacific salmon, cold-smoked at a temperature of less than 60°C;
(c) marinated and/or salted herring with insufficient processing to kill nematode larvae.

(3) The EU decided on microbiological specifications applicable to cooked crustacea and molluscan shellfish (EC, 1993). These were as follows:

'Thermotolerant coliforms' or *E. coli*	$m = 10$, $M = 10^2$, $n = 5$, $c = 2$
Staph. aureus	$m = 10^2$, $M = 10^3$, $n = 5$, $c = 2$
Salmonella	Absent in 25 g, $n = 5$, $c = 0$

with the following guidelines for aerobic mesophilic count (AMC) at 30°C:

Whole products	$m = 10^4$, $M = 10^5$, $n = 5$, $c = 2$
Shelled or shucked products (except crab meat)	$m = 5 \times 10^4$, $M = 5 \times 10^5$, $n = 5$, $c = 2$
Crab meat	$m = 10^5$, $M = 10^6$, $n = 5$, $c = 2$

28

Milk and Milk Products

Consumption of raw milk, and direct utilization of raw milk in food products (e.g. cheeses) that do not receive heat treatment, should be actively discouraged or, where appropriate, prohibited, as raw milk is so readily capable of acting as a vehicle for transmitting pathogens. In many countries, *Brucella* is still a substantial hazard in raw milk or dairy products made from raw milk (see, for example, CDSC, 1995). Even when the raw milk is obtained from herds certified as free of tuberculosis and brucellosis, the possibility still remains that the milk may contain other pathogens such as *Salmonella*, pathogenic *Escherichia coli*, *Campylobacter*, *Staphylococcus aureus*, *Streptococcus* species in Lancefield groups A, B and C, rickettsiae such as *Coxiella burnetii*, and some pathogenic viruses. Consequently in countries where raw milk is permitted to be sold for consumption, or raw milk is used in food products not subsequently heat treated, it is important that, in addition to the quality assessments discussed below, the cattle and/or raw milk be examined for at least the pathogenic bacteria listed above.

In regions where animal feeding stuffs may have been subject to mould attack, it may also be advisable to screen milk for aflatoxin M_1 either by using an enzyme-linked immunosorbent assay (ELISA) kit (see ISO/DIS14675) or by thin-layer chromatography (TLC) (see ISO/DIS14674).

28.1 LIQUID MILK

28.1.1 Liquid raw milk

The quality assessments discussed below are appropriate for examining the hygienic quality of raw milk intended to be heat treated.

Raw milk drawn aseptically from a healthy udder will contain only a few hundred to a few thousand bacteria per millilitre, mostly of the genus *Micrococcus* and the udder diphtheroid *Corynebacterium bovis*. In normal non-aseptic milking, many microorganisms are introduced into even high-quality milks from the external environment – from the udder surface, from bovine faeces, soil, bedding, feed and so on. After milking there will be a delay before the milk is heat treated, so there will be an opportunity for bacteria to grow and multiply. The nature and number of the microflora will be dependent on the storage temperature. If the milk has been chilled soon after milking and retained at refrigeration temperatures, there will be growth and multiplication of psychrotrophic mesophiles such as *Pseudomonas*, and also certain *Bacillus* species. If a delay occurs before chilling is applied, then the numbers of lactic acid bacteria may increase substantially.

Chilled raw milk sampled at the farm should have an aerobic mesophilic count (AMC) at 30°C of fewer than 10^4 per ml, and the bulked chilled raw milk at the factory or creamery should have an AMC at 30°C of fewer than 10^5 per ml. Many of the assessments of raw milk such as microscopic examinations relate to detection of mastitis, and are best performed on individually drawn samples obtained on the farm. Other quality assessments such as AMCs are concerned with determining the hygienic quality of production and storage.

28.1.1.1 Examination for mastitis

Many of the procedures used in the laboratory for the diagnosis of bovine mastitis (inflammation of the udder) are methods for detecting changes in the character of the milk which follow microbial infection of the udder. Isolation by cultural methods of the specific causative organism may also be attempted; this procedure is also a necessary step in the evaluation by sensitivity tests of the appropriate antibiotic treatment. It is important that milk samples for bacteriological examination should be taken only after careful washing and disinfection both of the hands of the milker and of the udder. The sample is obtained, after discarding first milk into a strip cup, by milking directly into a sterile sample bottle. For preliminary investigations and herd tests, mixed quarter samples from single cows may be suitable, but for detailed investigations individual quarter samples from each suspect animal are required. Laboratory investigation is particularly necessary in cases of subclinical mastitis when the cow and the milk show no abnormality detectable by eye.

Methods to detect subclinical mastitis include:

1. Microscopic examination of Breed's smear stained with Newman's stain as modified by Charlett (1954), to detect somatic cells and/or bacteria; alternatively a modified Newman's stain omitting basic fuchsin may be used (ISO 13366-1:1997).

   Two internationally recognized alternatives to the microscopic method are: (a) an electronic particle counter (e.g. Coulter counter) (ISO 13366-2:1997), and (b) ethidium bromide stain followed by automatic cell counting by fluorescence microscopy (e.g. using a Fossomatic) (ISO 13366-3:1997).
2. Streaking on blood agar to detect mastitis-causing pathogens. If mastitis-causing *Streptococcus* species are common, a medium such as Edward's aesculin crystal violet blood agar may be used to advantage.

   If mastitis-causing pathogens are detected and clinical or subclinical mastitis is diagnosed the pathogens can be examined to determine their sensitivity to antibiotics.

3. Rapid, indirect tests such as the Whiteside test or California test. These give some indication of large numbers of cells being present in a milk by an increase in viscosity caused by release of nucleic acids as a result of adding either alkali or surface-active agent. The advantage of these tests is that they can be performed on the farm.

28.1.1.1.1 Determination of cell count by Breed's smears

One of the most useful methods for the diagnosis of mastitis in the laboratory is determination of the total cell count using the Breed's smear technique (see Section 6.1) and Newman's or similar stain. Bovine milk, even from healthy animals, contains some cells, mainly epithelial cells and small numbers of leucocytes, but in mastitis milk the number of leucocytes is greatly increased as a result of udder inflammation. An increase in cell count also occurs in early- and late-lactation milk but in this case the increase is due to epithelial cells. These cells can be differentiated by means of a special staining technique as described by Blackburn and Macadam (1954) but, as subsequently shown by Blackburn and co-workers (1955) and Laing and Malcolm (1956), differential cell counts have no particular advantage over total cell counts in the diagnosis of mastitis except for late lactation milk. When quarter samples are being examined it is convenient to prepare the four smears from each animal on the same slide.

Procedure

In the examination of whole milk, it is necessary to defat the smear either before staining with methylene blue or, more usually, during staining with modified Newman's stain (ISO 13366-1:1997) or with Charlett's stain. If using Newman's stain the dried smears should be stained for 10 min, followed by drying, washing in water, and drying again. If using Charlett's stain, the dried smears should be stained for only 10–30 s and drying the stained slides before washing is not necessary. Both forms of the stain contain methylene blue as the cellular stain.

Both recipes also originally contained 1,1,2,2-tetrachloroethane as the defatting agent, although it is preferable to substitute the rather less toxic 1,1,1-trichloroethane. Take all possible precautions to avoid inhaling the harmful vapour by using a fume cupboard.

Somatic cells and microorganisms stain blue, but in Charlett's stain the basic fuchsin present causes the milk casein to present a pink background to the blue-staining cells. It is best to perform the drying stages in air, using either a levelled hot-plate slide-drier or a fan drier.

Standards suggested by Laing and Malcolm (1956) are as follows.

Individual quarter samples:	
Below 250 000 cells per ml	Negative
250 000–500 000 cells per ml	Suspect
More than 500 000 cells per ml	Positive
Mixed quarter/single cow samples:	
More than 200 000 cells per ml	Suspect

28.1.1.1.2 Whiteside test

This is a simple, rapid, indirect test for the diagnosis of mastitis and is basically an indirect method of assessing numbers of cells present in the milk. When these are sufficiently numerous, as in mastitis milk, the nucleic acid set free from the cells on titration of the milk with alkali is sufficient to produce pronounced viscosity.

Procedure

Mix one part of N sodium hydroxide with five parts of milk or udder secretion on a glass plate or dish and stir with a glass rod for 15 s.

Record the extent of the viscosity developed, if any. An increase in viscosity occurs with an increase in the number of cells. Absence of increased viscosity is recorded as negative and indicates normal milk.

28.1.1.1.3 Cultural examinations

Isolations of mastitis organisms are normally carried out from individual quarter samples using either streak or pour plate techniques. The media used should include one capable of supporting the growth of most known mastitis organisms, although selective media for particular groups may be used if required. The most satisfactory non-selective medium is 5% blood agar. It must be noted that the kind of blood used will affect the haemolysis produced (see Section 11.5.2).

Edward's aesculin crystal violet blood agar (Edwards, 1933) is a selective medium for the isolation of mastitis streptococci. The selective agent, crystal violet, is at such a concentration, 1 in 500 000, as to permit growth of streptococci but to inhibit that of any staphylococci. Aesculin-fermenting organisms, including *Strep. uberis*, produce black colonies, whereas non-aesculin-fermenting organisms (e.g. *Strep. agalactiae* and *Strep. dysgalactiae*) produce colourless colonies.

An alternative selective medium for the pyogenic streptococci in general is Petts colistin oxolinic acid blood agar (COBA) (Petts, 1984). The oxolinic acid, an antibacterial compound related to nalidixic acid, inhibits Gram-negative bacteria and staphylococci.

Isolation procedure

Use a 4-mm loop to smear 0.01 ml of milk on to the surface of a previously prepared and well dried blood agar plate. When four quarter samples from one animal are being examined, these may conveniently be streaked in separate sections of one plate. Incubate at 37°C for 48 h, examining after 24 and 48 h. As standard quantities have been used for inoculations, a comparison can be made of the numbers developing from individual quarter samples. The presence of any haemolytic colonies should also be recorded.

The predominant organisms developing on the isolation plates are of most significance and are presumed to be the causative organisms. Record the size and shape of representative colonies, haemolysis if any, pigmentation, Gram reaction and morphology. These observations should be sufficient to indicate the probable genus of the organism, but further confirmatory tests can be carried out if required. It is convenient at this stage to carry out a slide coagulase test (see Section 11.4.3) on suspected staphylococci; a positive reaction indicates the presence of *Staphylococcus aureus*. In the case of suspected streptococci, in addition to physiological tests, the precipitin test (see Section 13.6) may be used to determine the Lancefield group of the isolate (see Section 39).

If mastitis-causing pathogens are detected and clinical or subclinical mastitis is diagnosed, the pathogens can be examined further to determine their sensitivity to antibiotics.

Sensitivity tests

It is important in the treatment of mastitis to determine which antibiotics are most likely to be effective in treatment, by carrying out sensitivity tests on a pure culture of the isolate from the mastitis milk.

Subculture the isolate into a suitable broth medium (nutrient broth or yeast glucose lemco broth) and incubate at 37°C for 24 h. Pour 0.1 ml of culture on to the surface of a previously poured and well dried agar plate (nutrient agar for staphylococci and yeast glucose lemco agar for streptococci). Spread the culture over the plate and allow to dry. Using a disc dispenser place a series of appropriate antibiotic discs on to the surface of the agar. Alternatively the discs can be placed on the agar using flamed forceps with aseptic precautions. Discs and disc dispensers are available from a number of manufacturers (e.g. Oxoid, Difco).

Incubate at 37°C for 24 h and record the presence of zones of inhibition around the discs. Inhibition indicates sensitivity to a particular antibiotic, which is therefore potentially valuable for mastitis treatment. Note, however, that a large zone around one antibiotic disc does not necessarily mean that it is more effective than another antibiotic which produces a small zone, as many factors are involved in determining the zone size (Cooper, 1955; Linton, 1961). A simple method of subsequently determining the minimum inhibitory concentration for an antibiotic showing a zone of inhibition around it is the E-Test (Difco), which consists of an antibiotic gradient strip that can be placed on a seeded plate.

28.1.1.2 Hygienic quality of raw milk

Although resazurin tests or methylene blue tests have been used to assess the development of mesophiles in non-chilled raw milk, they are much less useful when milk is stored on the farm in refrigerated bulk tanks and delivered to the creamery or factory by refrigerated tankers.

At present, colony counts at 5° and 30°C, and thermoduric counts, are probably the best method of assessing microbiological quality in such situations. However, it is obviously not possible to subject raw milk to this microbiological testing for control purposes. Lück (1972) has made a critical appraisal of the tests that may be applied to refrigerated bulk milk.

28.1.1.2.1 Aerobic mesophilic counts

Media suitable for this purpose are plate count agar and yeast extract milk agar. For official tests on milk samples in the UK, yeast extract milk agar has been the approved medium and the prescribed incubation period is 3 days at 30°C. Psychrotrophic counts and thermophilic counts can be determined by incubating sets of plates at 4.5° and 55°C for 14 and 2 days respectively. If low numbers of thermophiles are expected, they may be counted by the multiple tube technique either by using tryptone glucose yeast extract broth, with microscopic examination after incubation, or (for milk-spoiling thermophiles only) by detecting changes in the milk or in inoculated litmus milks. In examining milk from refrigerated bulk tanks, and alternate-day collection, the psychrotrophic count becomes of great significance (Lück, 1972).

28.1.1.2.2 Presumptive coliform counts

Although, as already explained in Section 20.1, coliforms can proliferate in a food such as milk and can establish a secondary habitat on poorly cleansed equipment, tests for coliforms or Enterobacteriaceae are useful as an index of careless handling and production methods in the case of raw milk, and of post-heating contamination (or less frequently of inadequate processing) in the case of heated milks. Methods to be used are as described in Section 20.1, unless specified differently for statutory control purposes.

28.1.1.2.3 Dye reduction tests

Standard quantities of dye and milk, usually 1 ml of dye solution and 10 ml of milk, are mixed in sterile rubber-stoppered test-tubes and incubated at 37.5°C or other temperature as required for the purposes of the test. The dyes most widely used for milk testing are methylene blue and resazurin, and to a lesser extent 2,3,5-triphenyltetrazolium chloride (TTC). In the case of TTC, reduction of the colourless compound yields a red-coloured compound, formazan. It is possible for leucocyte activity to contribute to dye reduction. This is particularly likely to occur when mastitis milk is examined and large numbers of leucocytes derived from the udder are present in the milk. Resazurin may be partially reduced in this way, although methylene blue is not so affected.

Extensive experimental work by Wilson (1935) established that raw milk contains a natural reducing system operating independently of bacterial activity but that it is incapable of bringing about reduction of methylene blue under normal test conditions, and for practical purposes may be disregarded. Pasteurization of milk largely destroys this natural reducing system, which is generated by enzyme activity, but in autoclaved milk an inherent reducing capacity reappears, although of a different character from that in raw milk.

Methylene blue test

The method described is based on the work of Wilson (1935) and on the *Milk (Special Designation) Regulations 1963.*

1. Prepare a stock solution of methylene blue. Add one standard tablet (Merck) to 200 ml of cold, sterile, glass-distilled water in a sterile flask, shake until completely dissolved, and make up to 800 ml with more distilled water. This solution can be stored in a cool dark place, preferably a refrigerator, for up to 2 months. Transfer aseptically each day's requirement into a sterile container and discard at the end of the day.
2. Thoroughly mix the sample to be tested and pour the milk aseptically into a sterile test-tube up to the 10-ml mark.
3. Add 1 ml of methylene blue solution and, after a lapse of 3 s, blow out the remaining drops. The same pipette may be used for a series of tubes provided it does not contact the milk or the wetted side of the tube.
4. Close the test-tube with a sterile rubber stopper and invert the tube slowly twice to mix the contents.
5. Within 5 min, place the tube in a water-bath at 37.5 ± 0.5°C and note the time. The level of the water in the bath should be above that of the milk in the tubes, and the bath should be fitted with a lid to exclude light.
6. Set up a control tube with each batch of tubes similar in colour and fat content to the milk under test. Pour 10 ml of milk into a sterile stoppered test-tube, add 1 ml of tap water and place in boiling water for 3 min, then cool and place in the water bath. The control tube will help to determine when decolorization is complete.
7. Examine the tubes after half an hour. The milk is regarded as decolorized when the whole column of milk is completely decolorized or decolorized to within 5 mm of the surface. A trace of colour at the bottom of the tube may be ignored provided it does not extend upwards for more than 5 mm.
8. When the test is to proceed beyond the half-hour period, the tubes should be examined at half-hourly intervals for the duration of the test. Tubes that have decolorized should be removed from the water-bath; tubes in which decolorization has begun should remain in the bath without inversion until decolorization is complete. All other tubes in the water-bath should be inverted once to redistribute surface cream within the milk and replaced. Excessive inversion should be avoided because it results in reoxidation of the methylene blue and consequently will invalidate the test result.

Applications of the test

The half-hour methylene blue test may indicate potential keeping quality. Before testing, the milk is aged by storing at atmospheric shade temperature from the time of arrival in the laboratory until testing at 0930 hours on the following day. During the winter months storage overnight (1700 to 0930 hours) is at 65 ± 2°F (18.3 ± 1.1°C). Samples are regarded as satisfactory provided the methylene blue is not decolorized after 30 min at 37–38°C.

The methylene blue test may also be used as a reference method when ambiguous results have been obtained from the resazurin test, as methylene blue is not affected as is resazurin by cellular activity present in early and late lactation milk and in mastitis milk.

Resazurin test

The reduction of resazurin takes place in two stages, first irreversibly into resorufin through shades of blue and mauve to pink, and then reversibly from resorufin into the colourless dihydroresorufin. The colours produced can be matched in a Lovibond comparator (obtainable for example from Merck) and given a number ranging from 0 (colourless) through 1 (pink) to 6 (blue).

1. Prepare the resazurin solution by adding one standard tablet (Merck) to 50 ml of cold sterile glass-distilled water to give a 0.005% standard resazurin solution. Resazurin in solution is unstable in the light, so when not in use the solution should be kept in a cool dark place, preferably a refrigerator, and should be discarded when 8 h old.
2. Thoroughly mix the sample to be examined and add 10 ml of milk to a test-tube.
3. Add 1 ml of resazurin solution, and after a lapse of 3 s, blow out the remaining drops, without contacting the milk or wetted side of the tube.
4. Close the test-tube with a sterile rubber stopper and invert the tube slowly twice (in 4 s) to mix the dye and the milk.
5. Transfer the tube to a water-bath at 37.5 ± 0.5°C and note the time. When large numbers of samples are being set up, resazurin should not be added to more than ten tubes before placing in the water-bath and, in the case of the 10-min rejection test (see below), not more than five tubes. Light must be excluded by using a lid on the water-bath.
6. The length of the incubation period depends on the particular form of the test being carried out. At the end of the prescribed time, remove and examine each tube. Any tube showing complete reduction, appearing white, is recorded as 0. Any tube showing an extremely pale pink, or pink and white mottling, is recorded as ½. Other tubes are inverted and immediately matched in the comparator as follows:

   Place a 'blank' tube of mixed milk without dye in the left section of the comparator and the incubated tube in the right section. The comparator must face a good source of daylight (preferably a north window) or a standard source of artificial daylight. Place the comparator on a bench and look down on the two apertures. Revolve the disc until the colour of the incubated tube is matched, then note the disc reading. When the colour falls between two disc numbers record as the half value (e.g. record a reading between 3 and 4 as 3½).

Applications of the test

(a) *Ten-minute platform rejection test*. This test has been widely used to detect and segregate milk of unsatisfactory bacteriological quality arriving at creamery platforms in churns or cans. Results are obtained after a 10-min incubation period at 37 ± 0.5°C. Standards are 4–6, accepted; 3½ or less, rejected.

(*b*) *Hygiene test*. Samples of milk are maintained at or below 7°C (churn milk) or 10°C (bulk milk) until testing. At the laboratory subsamples are kept overnight at 0–5°C until between 0800 and 1000 hours on the day following collection of milk from the farm. After preparing tubes with resazurin as described above, the tubes are then incubated at 37 ± 0.5°C for 2½ h with examination and inversion every half hour. Those showing complete reduction (i.e. samples with disc readings of 0 or ½) are regarded as having failed the test.

28.1.1.2.4 Laboratory pasteurization test

This test consists of pasteurizing a small sample of milk under laboratory conditions. The time–temperature combinations most conveniently carried out are equivalent to those of batch (holder) pasteurization. Any microorganisms surviving such heat treatment are regarded as (presumptively) thermoduric and may be subsequently enumerated and isolated by plate counts. The thermoduric nature of the isolates may be confirmed by D-value determinations (see Section 12.4). Isolation of mesophilic thermoduric bacteria is normally carried out at 30°C as certain thermoduric bacteria (e.g. microbacteria) grow sparsely or not at all at 37°C.

The reduction in numbers following heat treatment may be calculated as a percentage of the original numbers (the percentage reduction). A milk sample with few heat-resistant microorganisms will show a high percentage reduction, whereas samples containing large numbers of thermoduric microorganisms will show a low percentage reduction. The method described is based on the method of Egdell *et al*. (1950).

Procedure

1. Mix the sample and place 10 ml into a sterile test-tube. Close firmly with a sterile rubber bung.
2. Invert and completely immerse the tube to rest on its bung in a stirred water-bath at 63.5 ± 0.5°C for 35 min. This time period includes an allowance of 5 min for the temperature of the milk to reach 63.5°C (the ability of the water-bath to return rapidly to 63.5°C should be checked). Alternatively, the tube may be held in the water-bath for 30 min after the pasteurizing temperature has been reached in a control tube.
3. Remove the tube from the water-bath and cool rapidly in iced water.
4. Invert the tube three times to mix and prepare decimal dilutions up to 10^{-3} or as required.

5. Plate out on yeast extract milk agar or plate count agar, and incubate at 30°C for 3–4 days. If required, the thermoduric isolates can be identified using the identification schemes in Part IV, and the heat resistance can be determined by the methods described in Section 12.4.

28.1.1.2.5 Enumeration and isolation of Bacillus

Aerobic spore-formers can be isolated amongst other heat-resistant bacteria by means of the laboratory pasteurization test. However, because *Bacillus* spp. possess relatively heat-resistant spores, it is possible by subjecting milk to a more severe heat treatment to arrive at a more selective method of enumeration and isolation. A 'total' spore count may be determined by heating for 10–15 min at 70–80°C. Some *Bacillus* spp. produce particularly heat-resistant spores, and these can be determined after heating for 30 min at 100°C (Franklin *et al.*, 1956). When low numbers of survivors are expected, a multiple tube counting method should be used. *Bacillus cereus* can be enumerated by the method given in Section 21.11.

Procedure

1. Pour 5–10 ml of well mixed milk sample into a sterile test-tube. Heat in a water-bath at 80°C for 10 min after this temperature has been reached in a control tube. Cool, and plate out on starch milk agar; the starch in the medium encourages the germination of spores by absorbing inhibitors such as unsaturated fatty acids which may be present in the medium. Incubate at 30° or 37°C for 3 days for mesophiles, and at 55°C for thermophiles. When considerable numbers of microbacteria are present in a sample, it is possible for some of these to survive this heat treatment, so colonies should be confirmed as *Bacillus*.
2. Alternatively, counts may be determined by the multiple tube technique, using sterile litmus milk. A multiple tube technique needs to be used for 'resistant' spore counts determined by heating at 100°C. When using this counting technique the milk can be distributed in the dilution series *before* the heat treatment, which is then given to the tubes of inoculated medium before incubation. Estimate numbers by reference to probability tables (see Appendix 2).
3. Enumerate *B. cereus* using MEPP agar (see Section 21.11).

28.1.1.3 Investigation of faults and taints in milk

Microbial faults and taints may develop in milk when particular microorganisms grow to such an extent that their metabolic products become discernible and distasteful. The nature of the fault or taint itself provides considerable information about the causative organism. The term 'fault' is usually applied when there is a change in the physical condition of the milk, whereas 'taint' is used when the physical condition is normal but the flavour or smell is objectionable.

28.1.1.3.1 Isolation of causative organisms

Examine the milk sample and pour a few millilitres into a Petri dish to examine the physical condition; determine the pH reaction using indicator papers. Examine microscopically, staining the smears either with Charlett's improved Newman's stain, or by Gram's method (in which case first defat the smear with xylene, which should be drained from the smear and then dried in air before staining). *Note that xylene is volatile and irritant and should be used only in a fume cupboard, as should Newman's stain.*

Streak a loopful of the milk on to the surface of a non-selective medium such as yeast extract milk agar and on to any suitable selective or differential medium appropriate to the suspected organism, as indicated in the examples below. Incubate at temperatures over the range 4.5–63°C, including the temperature at which the milk sample had been stored before detection of the fault or taint. After incubation select colonies for further study; restreak to purify the cultures, and identify them using the identification schemes in Part IV. Check the pure cultures for their role as causative organism by subculturing into whole sterile milk and incubating; the causative organism has been isolated when the original fault or taint is reproduced in the subculture.

28.1.1.3.2 Examples of faults in milk

1. *Sour milk*. This is milk that has coagulated as a result of lactic acid production by lactose-fermenting bacteria (lactic streptococci and the coli–aerogenes group). In freshly soured raw milk, *Lactococcus lactis* may comprise about 90% of the total flora. To isolate *Lactoc. lactis*, streak on to yeast extract milk agar, yeast glucose lemco agar or neutral red chalk lactose agar. It is an advantage if thallium acetate is incorporated in the medium before pouring, at a final concentration of 1 : 2000 to inhibit Gram-negative bacteria.
2. *Gassiness or frothiness*. Gas bubbles are produced on the surface of the milk and may be trapped in the cream. The most common causative organisms are coli–aerogenes bacteria, especially *Enterobacter aerogenes*, and lactose-fermenting yeasts. Violet red bile agar can be used to isolate the Enterobacteriaceae, and malt extract agar (pH 3.5) to isolate yeasts.
3. *Sweet clotting or sweet curdling*. This is milk that has coagulated at an approximately neutral pH because of the activity of proteolytic enzymes. The organisms responsible are predominantly *Bacillus* spp.; the fault frequently develops in heat-treated milks, perhaps due partly to the removal of competing bacteria, and partly to heat activation of the spores. For isolation methods see above.
4. *Ropiness or sliminess*. This milk is viscous and can be drawn out into threads with a wire loop. The fault is caused by the growth of capsulate organisms, frequently developing in refrigerated milk. Common causative organisms are capsulate strains of coli–aerogenes bacteria (e.g. *Klebsiella aerogenes*), *Alcaligenes viscolactis*, capsulate strains of *Micrococcus* and *Bacillus subtilis*. Colonies will also show a viscous consistency.
5. *'Broken'* cream *or bitty cream*. This is a condition in which the cream breaks up into separate particles on the milk surface and does not re-emulsify. It is

particularly evident when poured into hot tea or coffee. The fault is caused by lecithinase produced by *Bacillus cereus* and *B. cereus* var. *mycoides*. The bacteria may be isolated from unheated or heated samples (see p. 242), using egg yolk agar (p. 112), or MEPP agar (p. 202). If small numbers of the organisms are anticipated they may be enumerated by a multiple tube count enrichment technique, in which replicate quantities or dilutions are set up in litmus milks, incubated at 25°C, followed by streaking loopsful from each tube on to egg yolk agar plates to detect lecithinase producers (Billing and Cuthbert, 1958).

28.1.1.3.3 Taints in milk

These may be accentuated and so more easily recognized if the milk is warmed slightly. The *propagation test* can be used to determine whether the taint is microbial in origin.

Transfer about 5 ml of tainted milk to 50 ml of sterile whole milk in a glass-stoppered bottle, and incubate at 15–25°C or other appropriate temperature, and examine twice daily for the development of the original taint. This will occur (i.e. can be propagated) only when it is microbial in origin.

1. *Malty or caramel taint*. The causative organism is *Lactococcus lactis* var. *maltigenes*, which differs from 'non-malty' strains of *Lactoc. lactis* only in its ability to produce 3-methylbutanal from the leucine component of casein.
2. *Carbolic or phenolic taint*. This occurs most commonly in bottles of commercially sterilized milk, and is caused by certain phenol-producing strains of *Bacillus circulans*, the spores of which may survive the heating process.

28.1.1.4 Detection of antibiotics in milk

Antibiotics excreted in milk following treatment for bovine mastitis are undesirable for public health reasons; for example, certain individuals may show allergic reactions, and transfer of drug resistance may be encouraged. In addition, residual antibiotics may cause interference with lactic acid fermentation in the manufacture of dairy products dependent on this process; for example, fermented milks, cheese, and ripened cream for butter making. The problem arises largely from a failure to withhold milk for the specified time after the last treatment of the cow, usually 48–72 h, the concentration of antibiotic excreted gradually diminishing during this period.

Several microbiologically-based tests sufficiently sensitive to detect traces of antibiotics in milk have been developed. Most are based on the inhibitory effect of the residual antibiotics on the growth or activity of a chosen organism. Except in the case of the β-lactam antibiotics such as the penicillins – for which β-lactamases are available – it is not generally practicable to determine the nature of the inhibitory substance, which may be an antibiotic, detergent or disinfectant. The possibility of phage or other heat-sensitive natural inhibitor producing a false-positive result can be eliminated by the heat treatment of presumptive positive samples before carrying out confirmatory tests. In all cases where comparative results are required, it is essential that the procedure is carried out under standardized conditions using a specified strain of test organism and standardized conditions of cultivation. The sensitivity of the tests and the extent to which traces of antibiotic can be detected depend on the sensitivity of the selected organism under the particular test conditions. Two methods used extensively for routine assay of antibiotics in milk are disc assay and dye reduction, and examples of these procedures are described here. In addition, rapid field test methods are available which can be used either on the farm or as a platform rejection test. An example of such a rapid field test is the SNAP(TM) test for β-lactam antibiotics (IDEXX Laboratories, Maine, USA). The SNAP(TM) test, which can give a result in 10 min, was shown by Bell *et al.* (1995) to be a reliable field method.

28.1.1.4.1 Disc assay

In this method an agar medium heavily seeded with the test organism is poured into a Petri dish and a small filter paper disc soaked in milk is then placed on the surface of the agar. Incubation until growth appears is at a temperature appropriate to the test organism. Clear zones occur around the discs where antibiotic or other inhibitor has diffused into the medium. By comparison with known controls, provided that conditions are standardized carefully, the diameter of the zone of inhibition can be used to give a quantitative estimation of a known antibiotic present in the milk (Cooper, 1955; Linton, 1961).

Any sensitive organism may be used in this test, for example *Micrococcus luteus*, *Bacillus subtilis* or *B. stearothermophilus* (*B. calidolactis*). An advantage of the thermophile *B. stearothermophilus* as used by Galesloot and Hassing (1962), in the modification by Crawford and Galloway (1964), is that results may be obtained comparatively rapidly, after incubation at 55°C. Disc assay tests using *B. stearothermophilus* and *B. subtilis* are AOAC official methods for β-lactam antibiotics (AOAC, 1995).

The following procedure using penicillin serves to demonstrate the principle of the disc assay, but for assays for any particular authority, reference should be made to the appropriate standard procedure. The presence of β-lactam antibiotic can be confirmed by using β-lactamase to inactivate any β-lactam present in the milk. This can be achieved either by adding β-lactamase to a sample of the milk (see description under dye reduction test method, below) or, more easily, by using, in addition to a plain paper disc to pick up and test the milk, a paper disc containing β-lactamase to absorb a milk sample (β-lactamase-containing discs are available commercially, for example from Difco Laboratories).

Procedure

1. Prepare penicillin control milk at the required concentration by adding penicillin solution to antibiotic-free milk. The concentration of penicillin in the control milk should be equivalent to the level of tolerance for the test; this in turn will depend on the sensitivity of the test organism. For *B. subtilis* 0.05 units of penicillin per ml provides a suitable standard, but with *B. stearothermophilus* (*B. calidolactis*) a more sensitive test is obtained and a penicillin control milk containing 0.02 units of penicillin per ml can be used.

2. Inoculate 1 ml of an exponential-phase culture of the test organism in nutrient broth into 5 ml of molten nutrient agar at 45–50°C and pour into a Petri dish. Allow to set, on a level surface, and mark off the base of the plate as required.
3. Using clean dry forceps, pick up a sterile blank antibiotic assay disc (e.g. 13-mm diameter; Difco) and dip into the well-mixed milk sample. Drain off the excess liquid and then place the disc in the appropriate sector of the plate. Rinse and dry the forceps before testing the control and any other samples.
4. Incubate the plate at the optimum temperature for the test organism (i.e. 30°C for *B. subtilis*, or 55°C for *B. stearothermophilus*) until growth appears, or for the statutorily prescribed period.
5. Examine the plate and compare the diameter of the zone of inhibition of the control with that of the test sample. Samples of milk are regarded as satisfactory when the zone of inhibition produced is less than that of the control. When zones of inhibition are equivalent to, or greater than, that of the control, and zones around the β-lactamase-treated sample are smaller than those of the untreated sample, the milk sample contains a β-lactam antibiotic. When zones of inhibition are equivalent to, or greater than, that of the control, but β-lactamase treatment has not reduced the size of the zone, the milk sample should be retested after first heating the milk to destroy any natural inhibitory substances. (Alternatively unheated and heated samples may be examined concurrently.) A milk sample in which the presence of inhibitory substances was confirmed in a concentration equivalent to, or greater than, that of the control would fail the test.

28.1.1.4.2 Dye reduction

In this method, developed by Neal and Calbert (1955) and Wright and Tramer (1961), the dye used is triphenyltetrazolium chloride (TTC) and the assay organism is *Streptococcus thermophilus*. When TTC dye is added to antibiotic-free milk previously inoculated with an active culture of *Strep. thermophilus*, it is rapidly reduced to the red-coloured formazan, as can readily be observed in the milk sample. This reduction process is irreversible, and the formazan form of the dye is therefore not reoxidized by molecular oxygen as is the case with methylene blue. When the activity of the bacterial cells is completely inhibited by the presence of more than a certain minimum concentration of residual antibiotics or other inhibitory substances, conversion of the TTC dye into formazan does not occur and the milk therefore remains white. Partial inhibition of the test organism is indicated when intermediate shades of red colour are produced. Since, as pointed out by Neal and Calbert (1955), the sensitivity of the test is affected by the age of culture, the size of inoculum, and the duration of the incubation period before the addition of TTC, it is important that carefully standardized conditions be maintained when comparative results are required. The main features of the TTC test are demonstrated in the following simplified version of the standard method (see also Crawford and Galloway, 1964) but, for assays for any particular authority, reference should be made to the appropriate standard procedure.

Procedure

1. Pipette 5 ml of well-mixed milk sample into each of two sterile, marked test-tubes with rubber bungs and make up to 10 ml with sterile antibiotic-free milk.
2. To one tube add 0.2 ml of a solution of 1000 IU of penicillinase (Calbiochem) per ml.
3. Add 1 ml of *Strep. thermophilus* culture (prepared from equal quantities of an 18-h culture and sterile antibiotic-free milk to facilitate pipetting) to each tube.
4. Invert to mix and incubate in a water-bath at 44 ± 0.5°C for 1½ h.
5. Add 1 ml of a 1% solution of TTC in sterile distilled water to each tube, invert the tubes to mix, and reincubate at 44°C for a further 1 h.
6. Examine the sample for inhibition by comparison with the control tubes. A control tube set up in antibiotic-free milk should be a deep pink in colour. A second control tube in which distilled water has replaced TTC indicates the colour to be found when complete inhibition of the test culture has occurred. A change in colour in the tube containing penicillinase indicates the presence of penicillin.
7. For comparative purposes, and to check the sensitivity of the culture, tests may also be carried out using milk with added penicillin at known concentrations. (The *Strep. thermophilus* culture is sufficiently sensitive when a pink colour is obtained in milk containing 0.01 units of penicillin per ml, but complete inhibition (i.e. no pink colour) occurs in the presence of 0.02 units of penicillin per ml.) Amounts of adventitious penicillin may be determined quantitatively by using a dilution series of the milk samples, and comparing inhibitions obtained against inhibitions in a series of tubes with known amounts of penicillin added to antibiotic-free milk.
8. All presumptive positive samples should be retested within 36 h of the time of sampling. Milk samples that cause complete inhibition of the culture in confirmatory tests are regarded as having failed. A similar test series using a sample that has been heated to 95°C will differentiate antibiotics from heat-sensitive inhibitors, as explained on p. 247.

28.1.2 Thermized milk

Thermization is a heat treatment at about 60–65°C for around 15 s, designed to be applied to raw milk incoming to a factory in order to provide the flexibility offered by an extended life under refrigerated storage in the factory before further processing. It is particularly useful in reducing the numbers of heat-sensitive psychrotrophs such as *Pseudomonas*, which could otherwise grow during refrigeration of the raw milk, producing extracellular enzymes (e.g. proteases and lipases) that may persist after subsequent heat treatment to give rise to spoilage of dairy products such as UHT-treated milk (Griffiths *et al.*, 1981) and cheese.

28.1.3 Pasteurized milk

As conventional microbiological tests require 1–2 days before the result is obtained, it is not possible to subject this highly perishable product to microbiological control on an *a priori* basis. Quality assurance can be obtained by confirming, using the phosphatase test, that pasteurization has been applied properly.

However, a number of outbreaks of enteritis caused by *Salmonella* or *Campylobacter* in pasteurized milk have been caused as the result of post-pasteurization contamination. In one outbreak, the contamination was through a faulty flow diversion valve. In another, the contamination occurred through faulty valves on a pipe loop which was associated with the cleaning-in-place circuit. As the pathogens could be introduced by leakage of relatively small volumes of raw milk into the pasteurized milk, it is extremely unlikely that a phosphatase test would be able to detect such a fault. Another source of post-pasteurization contamination in the case of flavoured milk drinks can be the addition, after pasteurization of the milk, of heat-sensitive flavour ingredients. This was the cause of an outbreak of yersiniosis in the USA, in which *Yersinia enterocolitica* was introduced with chocolate syrup added after pasteurization of the milk.

Thus, in addition to the phosphatase test, which can be used in a quality control role, microbiological assessments can be used in a quality assurance role to determine the quality of product already produced, distributed and sold, so that a decision can be taken on whether or not to accept future batches of product from that source. Obviously control chart procedures can prove valuable in these circumstances.

Aerobic mesophilic counts at 30–32°C, and coliform or total Enterobacteriaceae counts, may be performed. After pasteurization the general viable count should be not more than 30 000 per ml (and counts of less than 5000 per ml on the freshly pasteurized milk should be readily attainable). Total Enterobacteriaceae (or coliforms) should not be detected in 1 ml of product (less than 1 per ml should be a readily attainable standard).

The UK standards for pasteurized milk (Anonymous, 1995) are:

Pathogens	absent in 25 ml (n = 5, c = 0)
Coliforms per ml	m = 0, M = 5, n = 5, c = 1
Aerobic mesophilic count (21°C)	$m = 5 \times 10^4$, $M = 5 \times 10^5$, n = 5, c = 1

Pasteurized milk should be stored at refrigeration temperatures until consumption, so that the aerobic mesophilic count at 30°C will increase (many of the psychrotrophic mesophiles being detectable in counts incubated at 30°C). However, coliforms and other Enterobacteriaceae should not multiply in a pasteurized milk properly stored, so there is no justification for increasing the permitted count of these organisms in any standard applied to milk sampled at retail outlets.

28.1.4 UHT-treated and sterilized milk

These ‘commercially sterile’ products may be examined, after appropriate pre-incubation, for:

(a) Thermoduric thermophiles (counts at 55°C), which may survive the heat treatment in detectable numbers if spores are present in the raw milk in excessive numbers.
(b) Mesophiles (counts at 30°C), which may be there as post-heating contaminants.

The UK standard (Anonymous, 1995) for UHT milk and cream is that, after incubation in a closed container at 30°C for 15 days and at 55°C for 7 days, the plate count at 30°C should be 100 per ml or less.

28.2 MILK POWDER

Important factors influencing the microflora of milk powder are the heat treatment given the milk before the drying process and the method of drying the milk. Where comparatively severe exposure to heat occurs (for example, in roller-drying) the resulting powder shows a more restricted flora than does one in which the temperatures involved are less extreme (for example, spray-dried powder). In spray-drying the amount of heating to which contaminating bacteria may be exposed is insufficient for the drying process to decontaminate the milk even in respect of relatively heat-sensitive pathogens such as *Salmonella*. Therefore, milk for spray-drying should be pasteurized first, and care should be taken that there is no possibility of recontamination between pasteurizer and drier. Further factors influencing the microflora of the powder are the extent of contamination from the milk-plant and the extent to which microbial multiplication can occur before the drying process. It is particularly important at this stage that numbers of coagulase-positive *Staphylococcus aureus* do not reach levels at which enterotoxin production creates a health hazard in the subsequent milk powder. The packaging process may also allow the introduction of contaminants, particularly atmospheric contaminants such as yeasts and moulds. During storage of dried milk, any surviving or contaminating bacteria will slowly die, but the lower the moisture content the better they will be able to survive. Furthermore storage should not be relied upon to decontaminate an unsatisfactory batch. When powdered milks are reconstituted, any surviving microorganisms are capable of growth, and milks should therefore not be kept for longer than fresh milk once reconstituted.

The need for careful application of a Hazard Analysis Critical Control Point (HACCP) plan in individual factories is shown by the continuing occasional outbreaks of *Salmonella* food poisoning caused by dried milk products, some outbreaks being international (CDSC, 1997a).

28.2.1 Sampling

A standard procedure for sampling milk powder requires that samples should be taken with a dry, sterile, metal spatula or spoon after mixing the top 150 mm (6 inches) of the contents and then transferring not less than 115 g (4 oz) to a sterile sample jar of sufficient size to allow mixing by shaking. It is more informative if surface and subsurface contents of the packaged powder are taken as separate samples (see also ISO 707:1997).

28.2.2 Preparation of dilutions

Weigh out aseptically 10 g of milk powder into a sterile wide-mouthed glass bottle etched at the 100-ml mark and, adding sterile distilled water at 50°C with gentle agitation, make up to the 100-ml mark. This provides the 10^{-1} dilution. Shake the bottle 25 times in 12 s with an excursion of 30 cm. Place the reconstituted milk in a water-bath at 50°C for 15 min, then invert the bottle several times and examine immediately. Prepare further dilutions in 9-ml amounts of 0.1% peptone water as required, or as indicated by examination of the Breed's smear preparation (see Section 28.2.3).

Note that some standard methods specify the use of 0.1% peptone or even 0.1% peptone with 0.85% sodium chloride for reconstitution of the initial dilution. However, unless it is necessary to follow such a standard method, it is recommended that, because of the high proportion of water-soluble compounds present, use of sterile distilled water is to be preferred for the initial reconstitution, to avoid microorganisms being metabolically damaged by exposure to hypertonic solutions in the 10^{-1} dilution.

28.2.3 Microscopic examination

Set up a Breed's smear preparation, using Charlett's improved Newman's stain. The total count gives an indication of the extent to which microorganisms may have proliferated in the milk before drying.

28.2.4 Cultural examinations

Plate out 1 ml of each dilution on suitable media for particular groups of organisms as follows:

1. *Aerobic mesophilic counts*. Use plate count agar or yeast extract milk agar, incubated at 30°, 37° and 55°C for 5, 3 and 3 days respectively.
2. *Coliform or 'total Enterobacteriaceae' count*. Use a multiple tube method, incubating at 30°C; for the 'total Enterobacteriaceae' count use similar media containing glucose. ISO 5541:1986 specifies lauryl sulphate tryptose broth in a multiple tube method at 30°C to test for the presence of coliform bacteria.
3. *Enterococci*. See Section 20.2.
4. *Yeasts and moulds*. See Section 14.
5. *Staphylococcus aureus*. Use Baird-Parker's medium (surface counts), incubated at 37°C for 24 h. To detect low concentrations of staphylococci, use a liquid enrichment technique preceding plating on Baird-Parker's medium (see Section 21.10). Note that the absence of viable staphylococci implies no guarantee of the absence of enterotoxin, so testing for enterotoxin by ELISA may be desirable, especially if a microscopic count reveals more than 5×10^5 cocci (see Section 21.10.3).
6. *Bacillus spores and B. cereus*. Use the procedures described in Sections 28.1.1.2.5 and 21.11.

7. *Clostridium spores*. Use plate count agar or blood agar (the latter as surface counts) incubated at 30°, 37° and 55°C anaerobically for 3–5 days. *Clostridium perfringens* can be enumerated by a multiple tube method as described in Section 21.7.
8. *Salmonella*. Follow the isolation procedures described in Section 21.3, including a non-selective resuscitation stage, in which the reconstituted milk (10^{-1} dilution) is incubated at 37°C for 24 h.

28.2.5 Results and suggested standards

Report results per gram of milk powder. Davis (1968) suggested that freshly manufactured spray-dried milk powder should have a direct microscopic count of fewer than 10^6 per gram, a general viable count of fewer than 10^4 per gram, and counts of *coli–aerogenes* bacteria, yeasts and moulds each of fewer than 10 per gram. Galesloot and Stadhouders (1968) proposed that the count of *Staphylococcus aureus* should be lower than 10 per gram, and that *Salmonella* should be absent from 100 g. For freshly manufactured roller-dried powder, Davis (1968) suggested that the general viable count should be lower than 10^3 per gram, and that counts of *coli–aerogenes* bacteria, and of yeasts and moulds, should each be lower than 10 per gram.

The UK standards (Anonymous, 1995) for dried milk and dried milk-based products are:

Salmonella	
dried milk	absent in 25 g, n = 10, c = 0
dried milk products	absent in 25 g, n = 5, c = 0
Listeria monocytogenes	absent in 1 g
Staphylococcus aureus (per gram)	m = 10, M = 100, n = 5, c = 2
Coliforms at 30°C (per gram)	m = 0, M = 10, n = 5, c = 2

28.3 CANNED, CONCENTRATED MILK

Concentrated milks are commonly available either in the unsweetened form as evaporated milk or, with added sucrose, as sweetened condensed milk.

28.3.1 Unsweetened evaporated milk

The keeping quality depends on efficiently sterilizing the product in the can and preventing post-heating contamination. This contamination is most likely to occur through the seams developing temporary leaks during the cooling stage, but access of potential pathogens and spoilage organisms can be prevented by chlorination of the cooling water as described by Bettes (1965). The microbiological aspects are exactly the same as those of all other hermetically sealed, appertized products (see also Section 34).

Procedure

1. *Preincubation of cans*. Incubate representative cans at 55°C for 7 days, 35–37°C for 14 days and 25–27°C for 1 month as recommended by Davis (1963). Examine the cans after the stipulated incubation period and report on their appearance. Cans containing viable gas-producing microorganisms may become swollen (i.e. blown) and are easily recognized.
2. *Investigation of the bacteriological condition of the milk*. Prepare the can carefully for opening aseptically as follows. Shake the can and its contents thoroughly. Remove paper labels and wash the outside of the can with warm water. Wipe the can dry with a clean paper towel, sponge with a suitable disinfectant, and dry. Swab the surface to be punctured with ethanol and flame. Make the opening with a sterile can opener (taking great care for blown cans) and cover the opened can with a sterile Petri dish lid.

   To carry out a microscopic examination, prepare a smear of the milk on a slide and stain by Gram's method, defatting first with xylene if necessary. (*Use a fume cupboard.*)

   To carry out a cultural examination, inoculate 1 ml quantities of the milk into Crossley's milk peptone medium, and into 10 ml of molten yeast glucose lemco agar for a shake tube. Also inoculate 1 ml into a Petri dish and pour with yeast extract milk agar or plate count agar. Incubate at the appropriate temperature (i.e. that used for preincubation of the can).

   Cans that show milk in a normal condition and no organisms in 1 ml may be assumed to have been sterile.

28.3.2 Sweetened condensed milk

In this product a high sugar content and consequent low available water (a_w) is relied on to restrict metabolism and multiplication of any microorganisms present. It is not a sterile product, and microbiological problems are likely to be related to spoilage caused by osmophilic and osmotolerant yeasts and moulds. Some moulds are sufficiently microaerophilic or facultatively anaerobic as to give problems even in a satisfactorily canned product, but mould problems usually arise when cans are under-filled with a large headspace, providing a source of oxygen for mould growing on the surface of the product, typically producing 'buttons' of compacted mycelium. Davis (1968) suggested that sweetened condensed milk should have an aerobic mesophilic count of less than 100 per gram, a lipolytic count of less than 10 per gram, and counts of yeasts, moulds and coliforms should each be less than 1 per gram.

Procedure

1. *Preparation of the can*. The product may be viscous and it is therefore advisable to warm the contents of the can in a water-bath at 45°C for no more than 15 min

before opening, in order to reduce viscosity. The can should then be opened aseptically as described above for evaporated milk.

2. *Cultural examination*. Using a sterile 10-ml pipette with a large orifice, weigh out 10 g of milk into a sterile sample jar. Add 90 ml of diluent at 37°C and shake 25 times. Prepare further dilutions up to 10^{-3} in the usual way. Plate out 1 ml of each dilution on to suitable media as follows:
   (a) *Aerobic mesophilic count*: use yeast extract milk agar or plate count agar and incubate at 30°C for 3 days.
   (b) *Lipolytic count*: use tributyrin agar or Victoria blue butterfat (or margarine) agar and incubate at 30°C for 3 days (see Section 11.3).
   (c) *Total Enterobacteriaceae or coliform counts*: see Section 20.1. Use violet red bile glucose agar or violet red bile lactose agar respectively and incubate at 30°C for 24 h; alternatively, to detect smaller numbers, use a multiple tube counting technique. Presumptive positive results from multiple tube MPN counts must be confirmed by inoculating loopsful from positive tubes into tubes of similar medium or by streaking on MacConkey's agar or similar medium.
   (d) Yeast and mould count: see Section 14.
3. *Results and recommended standards*. Report the results per gram of condensed milk. Davis (1968) suggested that sweetened condensed milk should have a general viable count of less than 100 per gram, and a lipolytic count of less than 10 per gram. Counts of yeasts, moulds and total Enterobacteriaceae or coliform bacteria should each be less than 1 per gram.

28.4 CREAM

The microbiology of cream is similar to that of milk in that microorganisms present in the original milk may also be present in the cream, and survivors of any subsequent heat treatment (e.g. sterilization or pasteurization) together with post-heating contaminants may have an adverse effect on the keeping quality, particularly if subsequent storage temperatures are insufficiently low to inhibit microbial growth. Consumption of raw cream should no more be encouraged than consumption of raw milk.

From a microbiological point of view it is preferable to apply homogenization before pasteurization. However, in this case problems of rancidity through the action of milk lipases may be greater than when homogenization occurs after pasteurization. The homogenizer can sometimes act as a source of microbiological contamination.

Sterilized cream in cans, bottles or cartons should contain few if any viable microorganisms, and appropriate methods of examination would therefore resemble the sterility tests as described for evaporated milk rather than tests for fresh cream described below.

The application of a dye reduction test to cream gives little information about potential public health hazard, and appropriate cultural tests are to be recommended. (Davis and

Wilbey (1990) presented an extensive review of the microbiology of cream and dairy desserts.)

Procedure

1. *Sampling.* Samples may consist either of individual cartons or packages or of a sample from bulk taken with a sterile dipper sufficient to fill a sterile sample jar. When samples cannot be delivered to the laboratory within 2 h, they should be packed in ice and delivered to the laboratory before 1700 hours on the day of sampling. On arrival at the laboratory, samples should be stored in the refrigerator until testing begins.
2. *Cultural examinations.* Prepare decimal dilutions in 9-ml amounts of general-purpose diluent up to 10^{-4} or as required. Perform the following cultural examinations:
   (a) Yeast extract milk agar or plate count agar for aerobic mesophilic counts incubated at 30°C for 3 days.
   (b) Violet red bile glucose agar and violet red bile lactose agar, incubated at 30°C for 24 h (or use a multiple tube technique), for total Enterobacteriaceae and coliform counts respectively (see Section 20.1).
   (c) Baird-Parker's medium for *Staphylococcus aureus* (surface counts) at 37°C.
   (d) Tributyrin agar incubated at 4.5°C and 30°C for 14 days and 3 days respectively for lipolytic psychrotrophs and for a general lipolytic count.
   (e) Examination for the presence of *Salmonella*; see Section 21.3.
   (f) Examination for the presence of *Listeria monocytogenes*; see Section 21.9.

Results and recommended standards

Report the counts per gram of cream. Retail pasteurized cream should have an aerobic mesophilic count less than 10^4 per gram. The UK standards (Anonymous, 1995) for pasteurized cream are:

Salmonella	absent in 25 ml (n = 5, c = 0),
Listeria monocytogenes	absent in 1 ml
Coliforms (per ml)	m = 0, M = 5, n = 5, c = 2

28.5 DAIRY STARTER CULTURES

The term 'starter culture' as used in the dairy industry refers to a culture of lactic-acid bacteria in milk which is used to induce a lactic acid fermentation in dairy products in which fermentation is an essential part of the manufacturing process (see also Mäyrä-

Mäkinen and Bigret, 1993). The satisfactory production of fermented dairy products in which starter cultures are used depends on the activity and purity of the starter cultures. Consequently microbiological quality assessment of the starter cultures must be undertaken. The particular lactic-acid bacteria required in any given starter culture depend on the purpose for which it is to be used. In starter cultures for cheese making, the active production of lactic acid is an essential requirement and cultures may consist of single strains of *Lactococcus lactis* (*Lactoc. lactis* subsp. *lactis* or *Lactoc. lactis* subsp. *cremoris*), or of combinations of both subspecies, with or without aroma-forming bacteria. Traditionally starter cultures have been maintained in the factory as liquid mother cultures to provide a feed culture subcultured again into a larger volume of milk in order to provide a sufficiently large inoculum to be used in production. These multi-stage procedures provide opportunities for contamination by unwanted bacteria or by bacteriophage, and opportunities for selection of mutant starter organisms with changed and perhaps undesirable properties. Consequently the use of the quality assurance checks listed below are recommended in such circumstances.

An activity test can be performed on a starter culture to determine that the lactic-acid bacteria are sufficiently active. A modification of the activity test can be used for the detection of bacteriophage in cheese whey. Propagation of phage in the dairy starter cultures can be minimized by the use of one of the calcium-deficient ‘phage inhibitory media’ which are either available commercially or can be produced by use of phosphate as a sequestering agent. For example, Hargrove (1959) found that the addition of 2% (w/v) sodium phosphate to skim milk inhibited the development of phage, whilst permitting normal development of the starter cultures. Another approach is to develop strains that show phage resistance (Klaenhammer and Fitzgerald, 1994).

For products in which development of aroma is a special requirement, for example in ripened cream for butter making, cultures usually consist of citrate-fermenting organisms capable of producing diacetyl (e.g. *Lactoc. lactis* subsp. *diacetylactis* and/or *Leuconostoc lactis* or *Leuco. mesenteroides* subsp. *cremoris*), in addition to *Lactoc. lactis* subsp. *lactis* and/or *Lactoc. lactis* subsp. *cremoris*. The ability of a culture to produce acetoin (acetylmethylcarbinol) and diacetyl can be checked by application of the Voges–Proskauer test.

However, quite often today a modern manufacturer does not maintain starter cultures continuously, but instead uses commercially supplied culture concentrates which may be freeze-dried or liquid nitrogen-frozen in ring-pull cans. These can be used for preparation of a bulk starter. Some of the more highly concentrated commercially available starter culture preparations can be used for direct inoculation into the cheese milk. Direct in-vat inoculation obviously minimizes the chances for contamination by other bacteria or bacteriophage.

28.5.1 Microscopic examination

Examine smears stained by Gram’s method. Defatting with xylene is not required for the examination of skim milk cultures. Note any differences in relative numbers of pairs of cocci and short or long chains of cocci. In starter cultures long chains of cocci are indicative of *Lactoc. cremoris*.

28.5.2 Activity test

Add 1 ml of starter culture to 100 ml of sterile skim milk and incubate at 30°C in a water-bath for 6 h. After incubation determine the acidity that has developed by adding 1 ml of 0.5% phenolphthalein solution as indicator to 10 ml of the culture and titrating with N/9 sodium hydroxide until a faint pink colour is obtained. Divide the number of millilitres of N/9 sodium hydroxide required by 10 to obtain the titratable acidity of the culture.

28.5.3 Modified activity test for the detection of phage

Inoculate 1 ml of starter into each of two flasks containing 100 ml of sterile separated milk. To one of the flasks add 0.1% (1 ml of 1 : 10 dilution) of whey; to the other add 1.0 ml of diluent to serve as a control. Incubate in a water-bath at 30°C for 6 h.

Determine the titratable acidity for each culture as described previously. If the activity of the culture with added whey is more than 10% below that of the control, it may be assumed that the whey contains phage specifically affecting that particular starter culture.

28.5.4 Biochemical tests

The general appearance of a starter culture is a good indication of its condition, but the application of simple biochemical tests can confirm that its condition is satisfactory.

28.5.4.1 Catalase test (see Section 11.4.1)

Add 1 ml of 10 vol. hydrogen peroxide to about 5 ml of starter culture in a test-tube. As lactic acid bacteria are negative in the test, a positive reaction indicates gross contamination.

28.5.4.2 Voges–Proskauer test (Barritt's modification) (see Section 11.2.6)

To 2.5 ml of starter culture add the Voges–Proskauer reagents and shake the tube well. A pink coloration developing within 30 min constitutes a positive reaction.

28.5.5 Cultural examination

28.5.5.1 Enumeration of viable starter bacteria

Prepare dilutions of the starter culture in diluent up to 10^{-8} or as indicated from the Breed's smear examination. Plate out 1 ml of each of the last three dilutions and pour with media as follows:

1. Aerobic mesophilic count: use yeast glucose lemco agar and incubate at 30°C for 3 days.
2. Citrate-fermenting organisms (*Leuconostoc*, *Lactoc. lactis* subsp. *diacetylactis*): use the tomato juice lactate agar of Skean and Overcast (1962). This is a simplified form of the medium described by Galesloot and co-workers (1961), in which the whey agar base of the original medium is replaced by tomato juice agar (Difco). A suspension of calcium citrate incorporated in the medium enables citrate-fermenting organisms to be recognized by the formation of clear zones around the colonies, although according to Waes (1968) certain lactic acid-producing organisms may produce clear zones even though not citrate-fermenting. Growth of *Lactoc. lactis* subsp. *lactis* and *Lactoc. lactis* subsp. *cremoris* is inhibited by the incorporation of 0.5% calcium lactate in the medium. Incubate at 21°C for 4 days.
3. Differentiation of *Lactoc. lactis* subsp. *lactis* and *Lactoc. lactis* subsp. *cremoris*: use arginine tetrazolium agar (Turner *et al.*, 1963). Incubate at 30°C for 24–48 h. Colonies of *Lactoc. lactis* subsp. *lactis* are red (arginine +), whereas colonies of *Lactoc. lactis* subsp. *cremoris* are white (arginine –).

The percentage viability can be determined by comparing the colony count with the Breed's smear count (in which a clump or chain is taken to be the equivalent of one colony-forming unit).

28.5.5.2 Detection of contaminants

In most cases sufficient monitoring of contaminating organisms can be achieved by inoculating suitable selective media with 0.1 ml of starter culture. If quantitative estimations are required, use 1 ml quantities of each of the decimal dilutions already prepared.

1. To detect yeasts and moulds use media as described in Section 14.
2. To detect the presence of bacteria other than lactic-acid bacteria use plates of nutrient agar (or a similar non-selective medium not containing fermentable carbohydrate) incubated aerobically and anaerobically at 30°C for 3 days.
3. To detect total Enterobacteriaceae or coliform organisms use VRBGA or VRBLA respectively, or use MPN counts as appropriate (see Section 20.1).
4. To detect citrate-fermenting organisms in cultures of *Lactoc. lactis* subsp. *lactis* and/or *Lactoc. lactis* subsp. *cremoris* use semi-solid citrate milk agar prepared as below, in which fermentation of citrate is indicated by gas production. Inoculate citrated milk (prepared by adding 0.5 ml of 10% sodium citrate solution to 10 ml of milk) with 1% (0.1 ml) of starter culture and add to 4 ml of molten 2% agar at 48°C. Mix by inversion and incubate at 30°C for 3 days. In certain circumstances, as shown by Crawford (1962), the presence of these organisms in starter cultures for cheese making may prove deleterious in that gas holes are produced in the cheese as a result of the carbon dioxide evolved during the fermentation of citrate.
5. Specific pathogens can also be investigated (e.g. *Staphylococcus aureus, Listeria monocytogenes*, *Salmonella*, etc.) as described in Section 21.

28.5.6 Isolation and maintenance of pure cultures of starter bacteria

Obtain isolated colonies of starter bacteria either by using the colony count method described above or, when qualitative information only is required, by streaking on to the surface of the particular agar medium.

Select suitable isolated colonies and examine Gram-stained smears. Subculture into yeast glucose lemco broth or litmus milk and incubate at 30°C for 3 days. For the subculture of *Leuconostoc* spp. the broth medium should be used, as these organisms produce little or no acid in litmus milk and successful subculture in this medium is not immediately apparent. The culture obtained should be re-examined and restreaked if required. Once the culture has been established, single colony isolation should *not* be used during routine maintenance subculturing.

Maintenance of the pure culture can be achieved by inoculation into yeast glucose chalk litmus milk and incubating at 30°C for 24 h or until a *slight* acidity develops. As an alternative, Robertson's cooked meat medium may be used. Store the culture at room temperature and subculture every 3–6 months. Stocks of the starter cultures can also be held as freeze-dried cultures (see Section 4.4).

28.5.7 Demonstration of the effect of penicillin on starter cultures

This may be demonstrated by inoculating the test culture into a series of litmus milks containing added penicillin at a range of concentrations. The appearance of the litmus milk cultures following incubation indicates whether or not inhibition has occurred.

Procedure

1. Prepare a stock solution of penicillin at 100 units per ml by adding one tablet containing 10 000 units of benzylpenicillin to 100 ml of sterile distilled water, allowing to stand for 15 min and then shaking to dissolve.

   Prepare working solutions of 10, 1 and 0.1 units of penicillin per ml by serial decimal dilution of the stock solution.
2. Use the prepared penicillin solutions at 100, 10, 1 and 0.1 units per ml, to inoculate a series of 9-ml amounts of sterile litmus milks to give penicillin concentrations ranging from 10 to 0.001 units per ml as indicated in Table 28.1.
3. Inoculate a loopful of the culture to be tested into each tube of the series. Incubate overnight at a suitable temperature: 30°C for lactic streptococci and starter cultures, and 37°C for *Strep. thermophilus* and yoghurt cultures. Also incubate an uninoculated tube of litmus milk for comparative purposes.
4. Report the results for the cultures tested. When the control tubes show acid production, the absence of acidity indicates inhibition by the penicillin at the concentration tested. *Strep. thermophilus* is one of the most sensitive organisms to

TABLE 28.1 Litmus milk series to determine sensitivity of a starter culture to penicillins

	Tube no.							
	1	2	3	4	5	6	7	8
Litmus milk (ml)	9	9	9	9	9	9	9	9
Millilitres of penicillin solution (units per ml)	1 of 100	1 of 10	0.5 of 10	1 of 1	0.5 of 1	0.1 of 1	0.1 of 0.1	0
Sterile distilled water (ml)	0	0	0.5	0	0.5	0.9	0.9	1
Final concentration of penicillin (units per ml)	10	1	0.5	0.1	0.05	0.01	0.001	Control

penicillin and may therefore be used in penicillin detection tests. As it is a constituent of yoghurt starters, cultures for this fermented milk product are very sensitive to residual antibiotics in milk.

5. Examine Gram-stained smears prepared from the control tube and from tubes showing some inhibition of acid production. Note any differences in morphology between the control and the inhibited cultures. There is a tendency for elongated bacillary forms to be produced by the cocci in the presence of penicillin. Report the concentration at which such abnormalities are observed.

28.6 FERMENTED MILKS

Fermented milks are produced by the growth in milk of sufficient numbers of lactic-acid bacteria to produce curdling or thickening of the milk and to give it a typical sour flavour. This can occur by the development of bacteria already present in the milk as in the natural souring of raw milk when lactic streptococci make up the predominant flora, but for satisfactory large-scale production fermentation is usually induced by added cultures.

When lactic-acid bacteria are the predominant organisms, an acid-fermented milk is obtained as, for example, in cultured buttermilk. So-called bioactive yoghurts are becoming of major interest in many countries. These can be produced either by incorporating appropriate bacteria (such as *Bifidobacterium* and *Lactobacillus acidophilus*) in addition to the normal *Lactobacillus bulgaricus* and *Streptococcus thermophilus*, or by using so-called therapeutic starter cultures such as Biogarde® (*Streptococcus thermophilus*, *Lactobacillus acidophilus* and *Bifidobacterium bifidum*). (See also Ballongue, 1993; Salminen *et al.*, 1993.)

Other fermented milk products include:

1. Ymer – a thick spooning fermented milk produced in Denmark using mesophilic starters, which is concentrated by removal of whey.

2. Acidophilus milk – produced in many countries, using a heavily heat-treated skim or whole milk fermented by the slow-growing *Lactobacillus acidophilus*. Because *Lb. acidophilus* grows very slowly in milk, and competes poorly against any contaminating microorganisms, it is important to minimize the numbers of other organisms in the milk by using a heat treatment greater than ordinary pasteurization, and to use fairly high concentrations of added starter (up to 5%).
3. Kefir – this product is popular in Eastern Europe. It is produced by using 'kefir grains' as the inoculum. Amongst the microorganisms found in the grains are *Lb. lactis*, *Lb. acidophilus*, *Lb. casei*, *Candida*, *Kluyveromyces* and *Saccharomyces*. This results in a product that is both sour and weakly alcoholic, and as final fermentation is carried out in closed containers the kefir will be slightly effervescent.

The ability of yeasts to thrive in fermented milks may present problems in the manufacture and distribution of yoghurt and buttermilk, as frothiness and/or off-flavours may develop from low initial numbers of contaminating yeasts.

28.6.1 Microscopic examination

Prepare smears of the fermented milk in a drop of water on a slide and allow to dry. If whole milk products are being examined, first defat the smear by flooding with xylene, draining and drying before staining. Stain the smears by Gram's method. Report the kinds and relative proportions of microorganisms present in each type of product being examined. For example, in yoghurt cocci and rods should be present in approximately equal numbers. Contaminants will be detected only when they are present in considerable numbers.

To detect small numbers of Enterobacteriaceae or other contaminants, cultural methods with appropriate selective techniques, as already described, will need to be used, if necessary employing a multiple tube technique.

28.6.2 Isolation of constituent flora

Many fermented milks are produced using single or mixed mesophilic species of lactic-acid bacteria such as *Lactococcus lactis*. The constituent microflora can be assessed on neutral red chalk lactose agar and *Lactobacillus* on MRS agar.

The constituent starter organisms in yoghurt can be assessed using L-S differential medium.

Lactoc. lactis subsp. *cremoris* and *Lactoc. lactis* subsp. *lactis* can be differentiated using arginine tetrazolium agar

Although not normally required for routine testing purposes, these procedures also include methods to indicate whether particular culture organisms still remain viable. For quantitative investigations prepare decimal dilutions and inoculate the specified media with 1-ml quantities in the usual way. For detection or isolation, streak or spread plates may be used.

28.6.2.1 *Lactococcus* and *Streptococcus*

Use neutral red chalk lactose agar. Incubate as necessary, for example at 30°C to isolate lactococci from sour milk and from cultured buttermilk, and at 37–43°C to isolate *Strep. thermophilus* from yoghurt (or use L-S differential medium for this latter purpose).

The differential medium may also be made selective for Gram-positive bacteria by the inclusion of thallium acetate at a final concentration of 1 : 2000. This is useful when suppression of coliforms is required, as, for example, in the isolation of streptococci from sour raw milk.

On neutral red chalk lactose agar, acid-producing colonies of lactococci and streptococci are small, deep red, and surrounded by a clear zone where the acidity developed has dissolved the chalk present in the medium.

Select isolated presumptive lactococci and streptococci for examination, subculturing with a straight wire into litmus milk. Also prepare a smear for microscopic examination and stain by Gram's method to confirm the presence of Gram-positive cocci. Incubate the litmus milk at 30° or 37°C until acid is produced, then re-examine. Obtain a pure culture by restreaking on a solid medium if required and confirm identity as indicated in Part IV.

28.6.2.2 *Lactobacillus* spp.

Use a suitable selective medium for lactobacilli, for example Rogosa agar, and then cover with a further 5 ml of the molten agar medium to form a layer plate. The temperature of incubation depends on the particular species of *Lactobacillus* to be isolated. Streptobacteria are lactobacilli (e.g. *Lb. casei*) having a low optimum temperature, and require incubation at 30°C. Thermobacteria are lactobacilli (e.g. *Lb. acidophilus* and *Lb. bulgaricus*) with a higher optimum temperature, and incubation at 37°C is necessary. L-S differential medium can be used to detect *Lb. bulgaricus* in yoghurt, and to differentiate this organism from *Streptococcus thermophilus*.

Select isolated colonies and examine microscopically. Subculture from colonies of Gram-positive rods (presumptive positive lactobacilli) into litmus milk and incubate as previously. Re-examine the culture when growth is obtained. If further identification (p. 347) is required, purification of the cultures on solid media is necessary.

28.6.2.3 Yeasts

Use a selective medium for yeasts which is sufficiently acid to inhibit the growth of most bacteria, for example malt extract agar at pH 3.5, or Davis's yeast salt agar at pH 3.5. Alternatively use the other media described in Section 14. Incubate at 22° or 25°C for 3–5 days. Isolated yeast colonies may be picked off to malt extract agar slopes for further study. In particular, it may be of interest to investigate the fermentative capacities of the isolates by inoculating into a series of fermentation media containing, for example, glucose, galactose and lactose. See Section 40.1 for details of procedures for identification of the cultures.

28.6.2.4 Detection of contaminants

Methods for detecting contaminants in starter cultures combining streak plates and selective media can be applied as appropriate, and purified isolates identified using the diagnostic keys in Part IV.

It should be noted that the coliform test is indicative of the standard of plant hygiene only when performed on samples within 24 h of manufacture, as the coliform bacteria will die on storage. Yeasts may be of significance as potential spoilage organisms, and small numbers may be detected by a multiple tube count with Davis's yeast salt broth at pH 3.5 as the selective medium (yeasts must be confirmed microscopically in tubes showing growth with or without gas production).

28.7 CHEESE

Fermented foods such as cheese may be expected to contain large numbers of lactic-acid bacteria, the microorganisms responsible for the fermentation process. For a short period after manufacture, the predominant lactic-acid bacteria are those derived from the starter culture, the species composition of which largely determines the microbial flora of unripened soft cheese (e.g. cottage cheese and immature cheese of other varieties). In cheese of more prolonged ripening periods, the starter-derived organisms are gradually displaced by a population of more acid-tolerant bacteria, the lactobacilli, and in many hard-pressed varieties of cheese (e.g. Cheddar) these may be largely responsible for the satisfactory ripening of the cheese. In other varieties of cheese, the particular character of the final product may also depend on the growth and metabolic activity of other microorganisms, for example *Penicillium* spp. in mould-ripened cheese and propionibacteria in Swiss cheese.

When considering microbiological quality in respect of spoilage and public health hazard, two classes of cheese can be distinguished: the 'soft' and 'fresh' cheeses in which the a_w and pH are not sufficiently low to prevent development of spoilage unless kept refrigerated, and the 'hard' and 'semi-hard' cheeses such as Cheddar, Provolone and Edam, which will have a reasonable shelf-life at cool ambient temperatures up to 20°C. All types of cheese, even hard cheeses, have been implicated in outbreaks of *Salmonella* enteritis (see, for example, CDSC, 1996a, 1997b). Such outbreaks have usually involved the use of unpasteurized milk for cheese production, and, although *Salmonella* will gradually die in the low pH environment, significant numbers may survive well in excess of 2 months.

It has been demonstrated conclusively that soft and semi-soft cheese may present public health hazards in respect of staphylococcal food poisoning, enteropathogenic *Escherichia coli* O157 and *Listeria monocytogenes*. Hard cheeses, in offering a low a_w environment, can permit prolonged viability of any pathogens that have survived up to the stages of removal of whey and pressing. Even fastidious pathogens such as *Brucella abortus* have been shown to be capable of surviving many months in hard cheese.

Quality assessments of cheeses should therefore include an examination for coliforms, faecal coliforms or *E. coli*, as well as an examination for *Staph. aureus*, and any other pathogens thought to be of significance. All forms of cheese can be involved in staphylococcal intoxination-type food poisoning. The staphylococci may originate from a human carrier or from a cow with mastitis. The development of staphylococcal enterotoxin usually occurs because of partial starter failure resulting in slow acid production. Since numbers of *Staph. aureus* decline on ripening, any count in excess of 10^3 per gram must be regarded as indicative of the possible presence of hazardous concentrations of enterotoxin. However, the apparent absence of *Staph. aureus* in the product does not mean that the cheese is safe: the bacteria may have died during ripening, whilst leaving the active enterotoxin in the cheese.

28.7.1 Sampling

Sampling of cheese for bacteriological purposes should be carried out using sterile sampling equipment and the sample transferred aseptically to a sterile sample jar. Precise details of sampling technique will vary according to the type of cheese to be examined.

The weight of the sample should not be less than 50 g.

For small soft cheese and small packets of wrapped cheese, an entire cheese or packet may be taken as a sample. For other soft cheese and semi-hard cheese where use of a trier is not practicable, the sample may consist of a wedge of cheese taken with a sterile knife by making two cuts radiating from the centre of the cheese. Any inedible surface layer should be removed before transfer to the sample jar.

For hard cheese of large size, the sample is most conveniently obtained by the use of a sterile cheese trier. This should be inserted obliquely towards the centre of the cheese on one of the flat surfaces, not less than 10–20 cm from the edge. The outer 2 cm of cheese containing the rind is cut off and replaced in the cheese.

Standard sampling procedures are described in ISO 707:1997.

28.7.2 Preparation of dilutions

Several methods are available for the preparation of the initial 10^{-1} cheese emulsion. Earlier methods requiring sterile sand are now seldom used. The diluent may be either quarter-strength Ringer's solution or 2% sodium citrate solution at 45°C (Naylor and Sharpe, 1958a). The latter diluent facilitates dispersal of the curd and consequent release of microorganisms. Emulsification may be achieved by shaking by hand or, more efficiently, by use of a macerator or the Colworth Stomacher. If shaking by hand is used, first grate or mince the sample aseptically, using a sterile mincer or grater (ordinary domestic apparatus is satisfactory, and may be sterilized in the autoclave).

Further dilutions can be prepared from the 10^{-1} dilution in 0.1% peptone water in the usual way. Since the counts obtained will be influenced by the method of preparing the initial emulsion, the method selected should be adhered to throughout any series of experiments.

28.7.3 Microscopic examination

1. *Qualitative*. Press a clean slide firmly over a freshly cut level surface of the cheese or cut a section of cheese and press between two slides, then separate and remove excess cheese with the edge of a slide. Defat with xylene for 1 min (*use a fume cupboard*) and dry in air. Stain by Gram's method and examine, then pour off. Report the kinds and *relative* numbers of microorganisms present. A number of fields should be scanned as in a mature cheese the bacteria are often localized in colonies rather than scattered uniformly throughout the cheese.
2. *Quantitative*. Examine dilutions of the cheese sample by Breed's smear preparations stained with Charlett's improved Newman's stain. The information so obtained may be used as a guide in the preparation of dilutions for cultural examination.

28.7.4 General cultural examinations

1. *Aerobic mesophilic count*. Use yeast glucose lemco agar. This non-selective medium will support the growth of most lactic-acid bacteria, but will not reveal the presence of lactobacilli when these are greatly outnumbered by lactococci. Incubate at 30°C for 3 days.
2. *Lactobacilli*. Use acetate agar, Rogosa agar or the modified Rogosa agar of Mabbitt and Zielinska (1956), and pour layer plates. Any of these similar media will partially select for lactobacilli, suppressing growth of lactococci, although some growth of leuconostocs and pediococci may occur. This type of medium is therefore particularly useful for the detection and enumeration of lactobacilli in the early stages of cheese ripening when lactococci are the predominant organisms. Incubate at 30°C for 3–5 days for the isolation of streptobacteria from Cheddar and similar cheeses.
3. *Total Enterobacteriaceae and coliform bacteria*. Use either VRBGA or VRBLA respectively or an appropriate MPN technique, incubating at 30°C. For methods to detect *E. coli* O157, see Section 21.4.
4. *Yeasts and moulds*. See Section 14.
5. *Staphylococcus aureus*. Use Baird-Parker's medium, surface inoculated, incubated at 37°C for 24 h.
6. *Clostridium perfringens*. See Section 21.7.
7. *Listeria monocytogenes*. See Section 21.9.

28.7.5 Isolation and maintenance of pure cultures of *Lactobacillus* spp.

Select isolated colonies from Rogosa agar or modified Rogosa agar. Examine Gram-stained smears and pick off presumptive *Lactobacillus* colonies into a suitable broth medium, such as the MRS broth of de Man *et al.* (1960). Incubate broths at 30°C for 2–3 days.

The resulting culture should be re-examined by Gram's method and restreaked on to a suitable agar medium, for example acetate agar, Rogosa agar or the agar medium (MRS agar) of de Man *et al.* (1960). Confirm that the isolate is a *Lactobacillus* sp. and identify as far as possible by methods described in Part IV.

Pure cultures of lactobacilli may be maintained in a similar manner to streptococci (see p. 259).

28.7.6 Isolation of propionibacteria from Swiss cheese

Propionibacteria may be isolated from Swiss cheese by means of a yeast extract lactate medium devised by van Niel (1928). This medium, which may be termed elective (see p. 84) because it satisfies the minimum nutritional requirements of the propionibacteria but of few others, can thus also be used for enumeration of propionibacteria if required. Fermentation of lactate in the medium results in the evolution of carbon dioxide and, if the original broth medium of van Niel is converted to the semi-solid form by the addition of agar, disruption of the medium is presumptive evidence of propionibacteria. Isolation of propionibacteria should, if possible, be carried out under anaerobic conditions in an atmosphere enriched with 5% carbon dioxide as provided by the GasPak® system (see Section 9.5.2).

Procedure

1. Prepare dilutions as in Section 28.7.2 and inoculate into molten yeast extract lactate semi-solid medium at 45°C. Mix well by rotating the tubes between the hands and allow to cool. Seal the surface with molten 2% agar at 45°C. Incubate at 30°C, preferably anaerobically in an atmosphere enriched with 5% carbon dioxide, for 7–10 days.
2. Tubes of medium showing gas fissures are presumed positive and can be confirmed by microscopic examination and subculture. Propionibacteria frequently appear as small Gram-positive coccobacilli which are non-motile and can thus readily be differentiated from coliform bacteria and clostridia, which may also be present in the cultures. Report the highest dilution at which gas production is observed and hence estimate numbers of propionibacteria per gram of cheese.
3. Streak from one of the presumptive positive tubes, in effect an enrichment culture, on to yeast extract lactate agar. Incubate at 30°C for 5–7 days, either anaerobically, in an atmosphere enriched with 5% carbon dioxide or aerobically, using layer plates, in a carbon dioxide-enriched atmosphere. Anaerobic incubation is to be preferred.
4. Select isolated colonies for examination and stain by Gram's method. Subculture presumptive *Propionibacterium* colonies into yeast extract lactate broth or yeast glucose lemco broth and incubate at 30°C, restreaking if required.
5. Pure cultures of isolates can be maintained by stab inoculation into yeast glucose lemco agar, incubating until growth is established and then covering with a layer of sterile liquid paraffin.

28.7.7 Isolation of moulds from mould-ripened cheese

Mould-ripened cheeses include semi-hard cheese (e.g. Stilton, Roquefort) in which the ripening agent, *Penicillium roqueforti*, grows in the interior of the cheese, and also soft cheese of comparatively small size or shallow depth (e.g. Camembert or Brie) in which the mould-ripening agent *P. camemberti* or more usually *P. caseicolum* (*P. candidum*) grows on the surface of the cheese. Isolation of a particular mould in pure culture from a mould-ripened cheese can be achieved by plating a fragment of cheese showing mould growth on appropriate selective media.

Procedure

1. Remove a fragment of mould growth from the required cheese with a sterile loop in about 2 ml of sterile quarter-strength Ringer's solution.
2. Use a dilution technique to obtain isolated colonies. Inoculate one loopful of the prepared suspension into 10 ml of a rose bengal medium (see Section 14), molten malt extract agar (pH 3.5) or Davis's yeast salt agar (pH 3.5) at 45°C. Mix by rotation between the hands and transfer a loopful of the agar suspension mixture to a second tube of molten medium. The first tube of medium is then poured into a Petri dish. The second tube is mixed by rotation and used to inoculate a third tube before itself being poured into a Petri dish. The process is repeated until four plates have been prepared. Incubate at 25°C for 3–5 days.
3. Examine the plates for isolated colonies typical of the required mould species and record details of colonial appearance. The identity of the isolate can be confirmed by microscopic examination as described in Section 40.2.
4. Subculture from a suitable isolated colony using a sterile straight wire on to a malt extract agar slope, and incubate at 25°C for 3–5 days. Pure cultures can be maintained on this medium or on potato dextrose agar and should be subcultured at intervals of 3–6 months.
5. Proteolytic activity of the isolates can be determined by inoculating on to 10% milk agar as described in Section 11.1.2. Incubate at 25°C for up to 14 days. Proteolysis is indicated by clear zones surrounding the growth.
6. The lipolytic activity of the isolates can be examined by subculture on to tributyrin agar, butter-fat agar or Victoria blue butter-fat agar as described in Section 11.3. Incubate at 25°C for up to 14 days.

28.7.8 Isolation of *Brevibacterium linens* from bacteria-ripened soft cheese

A bacteria-ripened soft cheese is one in which a surface smear or coat consisting of the ripening agent *Brevibacterium linens* develops on the surface of the cheese and imparts a characteristic butyrous texture and orange or orange-brown colour. In addition to occurring on typical bacteria-ripened soft cheeses, such as Limburger, *Bbm. linens* may also be isolated from orange or orange-brown spots on the surface of mature mould-ripened soft cheese.

Procedure

1. Prepare a suspension of the surface smear in about 2 ml of quarter-strength Ringer's solution and streak on to the surface of nutrient agar containing 5% sodium chloride. Alternatively, the cheese agar of Albert *et al.* (1944) may be used to enhance orange pigment production by *Bbm. linens*. Incubate at 25°C for 5 days.
2. Examine orange pigmented colonies by Gram's method. Subculture from isolated presumptive *Bbm. linens* colonies into nutrient broth. Incubate at 25°C and restreak on to nutrient agar. Confirm the identity of the isolate as indicated on pp. 341–342.
3. Pure cultures may be maintained on nutrient agar slopes by subcultures at intervals of 3–6 months.

28.7.9 Results and microbiological specifications

The current UK standards for cheeses (Anonymous, 1995) are:

(1) Hard cheese produced from raw or thermized milk:

Staph. aureus	$m = 10^3$, $M = 10^4$, $n = 5$, $c = 2$
E. coli	$m = 10^4$, $M = 10^5$, $n = 5$, $c = 2$
L. monocytogenes	Absent in 1 g
Salmonella	Absent in 25 g, $n = 5$, $c = 0$

(2) Hard cheese produced from pasteurized milk:

L. monocytogenes	Absent in 1 g
Salmonella	Absent in 25 g, $n = 5$, $c = 0$

(3) Soft cheese produced from raw or thermized milk:

Staph. aureus	$m = 10^3$, $M = 10^4$, $n = 5$, $c = 2$
E. coli	$m = 10^4$, $M = 10^5$, $n = 5$, $c = 2$
L. monocytogenes	Absent in 25 g, $n = 5$, $c = 0$
Salmonella	Absent in 25 g, $n = 5$, $c = 0$

(4) Soft cheese produced from pasteurized milk:

Staph. aureus	$m = 10^2$, $M = 10^3$, $n = 5$, $c = 2$
Coliforms	$m = 10^4$, $M = 10^5$, $n = 5$, $c = 2$
E. coli	$m = 10^2$, $M = 10^3$, $n = 5$, $c = 2$
L. monocytogenes	Absent in 25 g, $n = 5$, $c = 0$
Salmonella	Absent in 25 g, $n = 5$, $c = 0$

(5) Fresh cheese:

Staph. aureus	m = 10, M = 10^2, n = 5, c = 2
L. monocytogenes	Absent in 25 g, n = 5, c = 0
Salmonella	Absent in 25 g, n = 5, c = 0

28.8 BUTTER

As butter should be made from pasteurized cream, there should be little hazard of causing food poisoning or of transmitting disease. However, the ICMSF (1980b) has described two outbreaks of staphylococcal food poisoning found to be caused by whipped butter.

General hygienic quality of butter can be assessed by aerobic mesophilic counts at 4.5° and 30°C, yeast and mould count, lipolytic count, coliform count and halotolerant count. If required, counts of *Staph. aureus*, or ELISA examination for staphylococcal enterotoxin, may be performed.

Butter may be manufactured either from unripened cream to make sweet cream butter or from cream ripened by the addition of a starter containing citrate-fermenting organisms to make ripened cream butter. Many of the organisms present in the cream are removed in the buttermilk during the process of manufacture but some will survive to the finished product. Butter may be either unsalted or salted, the concentrations of sodium chloride varying from 0.5 to 2.5%. The moisture content of butter is usually restricted by legislation: in the UK it should not exceed 16%. The effective concentration of salt in the aqueous phase is considerably higher than the overall concentration and, for a 2% salt and a 16% moisture content, the salt concentration in the aqueous phase is 12.5%. This is a considerable deterrent to the growth of many microorganisms, and growth is further restricted when the water droplets are of small size and evenly distributed throughout the butter. Refrigeration and storage of the butter at low temperatures also retards microbial growth. Spoilage of butter is therefore most likely to arise from the activity of microorganisms capable of growing at low temperatures, particularly those capable of lipolysis, proteolysis or loss of flavour, or those causing discoloration.

28.8.1 Sampling

The sampling of butter for bacteriological purposes requires the use of sterile equipment and the aseptic transfer of samples to sterile sample jars. The butter should not come into contact with paper or any absorbent surface. ISO 707:1997 describes a range of procedures to sample butter from large retail packs and from bulk containers of various sizes. In general, firstly a surface layer of 5 mm is removed from the sampling area. Next, a sample is taken with a sterile trier, with the top 25 mm of the core being removed and discarded (or used as part of the surface sample; see below). A sample of at least 50 g should be taken.

It is obvious that surface contamination under the packaging may be important, so surface and core samples should both be examined. When the surface is to be examined separately, remove the surface butter to the required depth with a sterile knife or spatula, and transfer to a sterile sample jar.

Samples should be kept cool during transport to the laboratory and examined as soon as possible.

Analysis may be on either a volumetric or a gravimetric basis, based on a shaken melted sample which thus includes both the fat and aqueous fractions. Alternatively, because most of the organisms are present in the aqueous fraction, this aqueous fraction only may be examined (although in this case it may be difficult to calculate the counts per gram of product).

28.8.2 Procedure for cultural examinations

Melt the sample quickly at a temperature not exceeding 45°C by immersing the sample bottle for a short time in a water-bath at 45°C. Mix the melted sample by shaking 50 times with an excursion of 30 cm in 1 min.

For analysis on a volumetric (v/v) basis, transfer 10 ml of melted sample to 90 ml of diluent in a bottle of 200-ml capacity. The diluent should be 0.1% peptone with 0.1% agar to stabilize the emulsion. The pipettes and diluent should be at a temperature of 45°C. For analysis on a gravimetric (w/v) basis, the butter may be transferred to the diluent with a pipette but using a balance to weigh 10 g of sample to which are added 90 ml of diluent. Shake the diluent and butter 25 times with an excursion of 30 cm to give a homogeneous suspension, the 10^{-1} dilution.

Prepare subsequent decimal dilutions up to 10^{-4} or as required.

Carry out:

1. Aerobic mesophilic and psychrotrophic counts on plate count agar incubated at 30° and 4.5°C for 3 and 14 days respectively.
2. Counts of yeasts and moulds; see Section 14.
3. Counts of caseolytic microorganisms using the 30% milk agar of Smith *et al.*, (1952) as suggested by Druce and Thomas (1959); incubate at 30°C for 5 days. Caseolytic colonies are those surrounded by clear zones and may be confirmed by flooding the plates with a protein precipitant before counting the colonies (see also Section 11.1.2).
4. Lipolytic test media: a number of different media is available (see Section 11.3). The disadvantage of the simplest medium, tributyrin agar, is that many organisms clearing this medium are incapable of hydrolysing more complex fats (e.g. butter-fat or margarine). For this reason, media using complex fats may be preferred or used in addition to tributyrin agar. An alternative procedure is to use one medium as the sole isolation medium, but subsequently to test for lipolysis on other media. This may be done by the separate subculture of individual colonies, or more speedily, by use of the replicator technique (see Section 8). Incubate at 22°C for 5 days or 30°C for 3 days.

5. Violet-red bile glucose agar or VRBLA for total Enterobacteriaceae and coliform bacteria respectively, incubated at 30°C for 24 h; alternatively use a multiple tube technique (see Section 20.1).
6. Halophilic/salt-tolerant counts using 15% sodium chloride to prepare a dilution series, and media containing 15% sodium chloride.

28.8.3 Results and recommended standards

Record results per millilitre or per gram of butter tested according to method of analysis used.

Davis (1968) suggested that butter manufactured from unripened cream should have an aerobic mesophilic count of less than 10 000 per gram, counts of proteolytic, lipolytic and psychrotrophic microorganisms each of less than 1000 per gram, of yeasts and moulds less than 100 per gram and coliform bacteria of less than 10 per gram.

The current UK standard (Anonymous, 1995) is:

Salmonella	Absent in 25 g, n = 5, c = 0
L. monocytogenes	Absent in 1 g
Coliforms per gram	m = 0, M = 10, n = 5, c = 2

28.8.4 Isolation and further study of lipolytic microorganisms

Select isolated colonies showing lipolysis. Stain smears by Gram's method and subculture into semi-solid yeast glucose lemco agar. Incubate at 22° or 30°C until growth is established, then re-examine and restreak on solid media to obtain pure cultures.

Determine the lipolytic activity of each isolate by single streak inoculations on to the surfaces of a range of test media (Section 11.3). Several cultures may be tested on one plate and media not used for counting purposes may be examined in this way (e.g. Tween agar).

To determine the salt tolerance of isolates, use the semi-solid nutrient agar cultures above to prepare 24-h nutrient broth cultures of the lipolytic isolates. Inoculate a standard loopful of each broth culture into each of a series of tubes of nutrient broth containing sodium chloride at a range of concentrations (e.g. 5, 10, 15 and 20%). Include also nutrient broth (0.5% sodium chloride). Incubate at 22° or 30°C for 3–5 days and record the relative amount of growth in each tube. The salt tolerance of isolates obtained in 'halophilic' counts may be determined similarly.

29

Ice-cream and Frozen Desserts

Amongst the microbiological examinations that can be applied are aerobic mesophilic count, a total Enterobacteriaceae count (or at least a coliform count), *Staphylococcus aureus* count, an examination for *Salmonella*, etc., and the current standards in the UK (Anonymous, 1995) utilize specific microbiological counts (see below). In the following section the term 'ice-cream' has been used, but the same procedures are applicable to other frozen desserts.

Because factory-produced ice-cream and frozen desserts are kept in hardening rooms and cold stores for some time after production before being distributed and consumed, there is no problem in applying microbiological assessments that require 24–48 h before the result has been obtained.

29.1 COLLECTION OF SAMPLES

1. *Hardened ice-cream*. For packages and tubs, one or more unopened packages constitute the sample, which should be delivered intact to the laboratory in a sterile container. For multi-layered ice-cream, the sample should contain the same proportions of each layer as in the original ice-cream. In the case of ice-cream in bulk containers, first remove the surface layer with a sterile spatula or spoon and, with a second sterile spatula, take a sample of not less than 60 g (2 oz) into a sterile jar. For information on bulk ice-cream as served to the consumer, the sample is taken from the surface layer with the retailer's own server.
2. *Soft ice-cream*. This is freshly frozen ice-cream sold direct from the freezer. In this case, fill the sample jars directly from the freezer outlet; the sample should be a minimum of 60 g.
3. *Transport of samples*. Samples should be transported to the laboratory in a refrigerated container and maintained at not more than −18°C until examined in the laboratory. A more convenient procedure is to transport samples to the laboratory in an insulated container for delivery within 2 h of the time of sampling, but when or where this time limit is not practicable, samples should be packed in ice and should arrive at the laboratory within 6 h of sampling.

29.2 TREATMENT OF SAMPLES

1. If the sample is in the original retail package, transfer aseptically the whole, or a representative portion, to a sterile container of not less than 60-ml capacity.
2. Frozen samples should be left at room temperature for a maximum period of 1 h until melted. Alternatively, the sample may be liquefied by holding the sample jar in a water-bath at 42–45°C for no more than 15 min.
3. If the sample is unfrozen, examine immediately.

29.3 CULTURAL EXAMINATION

The analysis should be carried out on a gravimetric basis because the weight of 10 ml of ice-cream may range from 4.5 to 10.5 g (Patton, 1950).

Procedure

Invert the sample bottle three times to mix the sample. Using a sterile 10-ml pipette, weigh out 10 g of melted ice-cream into a sterile container. Add 90 ml of sterile diluent and invert three times. This constitutes the 10^{-1} dilution. Prepare further dilutions up to 10^{-4} or as required. Perform examinations as follows:

1. Plate count agar or yeast extract milk agar incubated at 4.5° and 30°C for 14 and 3 days respectively for psychrotrophs and mesophiles.
2. Violet-red bile glucose agar or violet-red bile lactose agar incubated at 30°C for 24 h, for total Enterobacteriaceae and coliform bacteria respectively. Alternatively use a multiple tube technique for the detection of smaller numbers (see Section 20.1). Positive cultures can be confirmed as *Escherichia coli* as described in Section 20.1.
3. Media for yeasts and moulds; see Section 14.
4. Baird-Parker's medium (surface counts) incubated at 37° for 24 h, for *Staph. aureus*.
5. Thermoduric bacteria can be counted using the 10^{-1} dilution in a laboratory pasteurization test as described for milk (see p. 242).
6. Detection of *Salmonella*; see Section 21.3.
7. Detection of *Listeria monocytogenes*; see Section 21.9.

Results and specifications

Simple ice-creams which are subjected totally to a pasteurization should in theory be free of Enterobacteriaceae. The specification recommended by the ICMSF (1986) is very lenient in the coliform count and aerobic mesophilic count, as is the UK standard for frozen milk-based products, including ice-cream. The UK standard (Anonymous, 1995) is:

Aerobic mesophilic count (30°C) per g	$m=10^5$, $M=5\times10^5$, $n=5$, $c=2$
Coliforms (30°C) per g	$m=10$, $M=100$, $n=5$, $c=2$
Staphylococcus aureus per g	$m=10$, $M=100$, $n=5$, $c=2$
Listeria monocytogenes	Absent in 1 g
Salmonella	Absent in 25 g, $n=5$, $c=0$

It should be noted that the aerobic mesophilic count and coliform counts are not obligatory. But any well run, large ice-cream factory is likely to achieve a microbiological quality far superior to this specification; aerobic mesophilic counts of fewer than 1000 per g at 30°C and coliform counts of less than 1 per g are common.

A major problem, however, can exist in the freezing of ice-cream mix at the point of retail sale. Soft-serve ice-cream dispensers in mobile vans and temporary stands may produce ice-cream of very poor microbiological quality. Contamination of the mix, temperature abuse of the mix, and inadequate cleaning and sterilization of the equipment can lead to aerobic mesophilic counts in excess of 10^6 per g. (The author has sampled a soft-serve ice-cream from a franchised mobile ice-cream van which proved to have a *coliform* count in excess of 5×10^5 per g.)

A third type of product is the complex ice-cream and frozen dairy dessert, in which after the heat treatment other ingredients such as biscuit, meringue, fruit, piped cream are added, perhaps as external decoration to the frozen product. In this case it is unreasonable to expect that these post-pasteurization manipulations will not lead to some increase in the population of viable microorganisms. Indeed some of these added ingredients, for example chopped nuts, can contribute significant numbers of microorganisms. It is therefore important that the microbiological quality of these added ingredients is monitored carefully. These problems are reflected in the somewhat more lenient specifications recommended by the ICMSF (1986) for such products.

30

Fruit, Nuts and Vegetables

30.1 INTRODUCTION

Botanically, a fruit is a matured ovary from a flower, which carries the seed (or seeds). Thus peas, capsicum (green and red peppers, etc.) aubergines, cucumbers, avocadoes and tomatoes botanically are fruits, but from a culinary point of view are commonly regarded as 'vegetables'. Fruits that are regarded by most people as 'fruit' (apples, pears, oranges, plums, cherries, etc.) are sweet and sour. Rhubarb, which is used in cooking as though it were fruit, is of course no such thing, but is the leaf-stalk. The issue is complicated further in that unripe fruits such as bananas (or plantains) and papayas are used in cooking as vegetables.

The microbiologist needs to think about fruit and vegetables in terms of the microbial habitat that they provide. The fruits most people regard as 'fruit' (apples, pears, oranges, plums, peaches, loquats, tamarinds, etc.) are both sweet and acid in flavour. Thus 'fruit' typically has a high sugar content (especially of glucose, fructose and sucrose) and a low pH (usually pH 4.5 or lower) as a result of accumulation of organic acids (especially citric, malic and quinic acids); these, and the typically low protein and lipid contents, have a profound effect on the types of microorganisms involved in spoilage. From a microbiologist's viewpoint, therefore, tomatoes (with a pH between 4.0 and 4.5) are more sensibly included with 'fruit' when thinking about the fruit or juice as a microbial habitat. The pH of 'vegetables' ranges from 5 to 7.

The microbiological effects of the organic acids in 'fruit' do not merely result from the low pH. Undissociated molecules of the organic acids also have particular inhibitory effects (see ICMSF, 1980a). In addition to the organic acids already mentioned, many other organic acids are also found, sometimes forming a high proportion of the total organic acids, for example tartaric acid in grapes and tamarinds, benzoic acid in cranberries.

Note that some fruits (e.g. papaya, melon) are less acid and have a higher pH; for example, ripe papaya typically has a pH of 5.5–5.9. Nuts are, botanically speaking, fruits, but as a microbial habitat mostly they present a particular set of features (low available water (a_w) and a high protein water and lipid content) and consequently need to be considered separately.

Fresh fruit and vegetables normally carry a surface flora of soil saprophytes and some plant parasites. Some of these microorganisms will play an important part in any subsequent spoilage. For example, ascospores of *Byssochlamys fulva*, deriving principally from contamination by soil, are sufficiently heat resistant to give rise to problems in canned fruit. Microbiological analysis of these foods is not usually carried out, but in the case of fruit and vegetables to be eaten raw (particularly when considerable handling is involved, for example prepared salads or when the source is dubious), an examination for the presence of food-poisoning pathogens, or for indicators of faecal contamination, may be advisable. Fruit and vegetables grown on land that has been fertilized with organic manures of animal origin (or with 'night soil' of human origin) should be examined similarly. As already pointed out, an outbreak of *Escherichia coli* O157 food poisoning was caused by cider made from apples of which a large proportion were 'drops' or 'windfalls' and which could therefore be contaminated by animal manures used to fertilize the orchards.

Fruit and vegetables that are harvested by hand or hand-sorted may become contaminated with pathogens from an infected food-handler. Such pathogens may persist in frozen foods, especially when a blanching stage is not possible. For example, frozen raspberries have caused outbreaks of hepatitis A infection (see, for example, Reid and Robinson, 1987).

Frozen fruit and vegetables have a microflora similar to that of fresh products, although the proportions of types and their absolute numbers will be somewhat different owing to effects of general hygienic measures in the freezing plant, blanching procedures, storage temperatures before and after blanching, etc. Freeze-dried products may also show a similar flora to the original, with the same provisos concerning proportions and absolute numbers.

30.2 VEGETABLES

These frequently can be affected by bacterial soft rots caused particularly by Gram-negative organisms of the genera *Erwinia* and *Pseudomonas*. The development of such soft rots is encouraged by prepacking vegetables in insufficiently ventilated water-impermeable wraps, and vegetables showing spoilage of this type can be observed on shelves of supermarkets that have inadequate stock control procedures.

Fungi, particularly *Botrytis*, *Fusarium*, *Rhizopus* and the parasitic fungus *Peronospora* (which causes downy mildews), may also cause spoilage on stored vegetables.

Cultural examinations may be made using the usual media for general viable counts, yeasts and mould counts, etc. Pectinolytic organisms may be counted using polypectate gel medium, which is inoculated as spread plates. After incubation at 25–30°C, a count is made of the colonies forming depressions in the surface of the gel. The medium may be made partially selective for bacteria by the addition of cycloheximide, and for Gram-negative bacteria by the addition of cycloheximide and crystal violet (Pérombelon and Burnett, 1991) (see p. 451).

30.2.1 Refrigerated prepared salads, etc.

In recent years there has been a great increase in the retail sale of ready-to-eat mixed salad: chopped, shredded, raw vegetables with or without added fruit, meat or fish. Even when such products include a salad dressing such as vinaigrette or mayonnaise, although the dressing itself may have a low pH (3–5.5), many microorganisms contaminating the raw vegetables may not experience a lethal or even inhibitory pH. A storage temperature between 0° and 5°C is recommended for such products, and they will have a short shelf-life of a very few days.

Spoilage organisms include bacteria (such as *Pseudomonas*, *Cytophaga*, *Erwinia* and lactic-acid bacteria) and yeasts, which form part of the natural microflora of the vegetables. Delicatessen salads, including beansprouts, have been involved in a range of food-poisoning outbreaks. Amongst the pathogens that may be of significance in such products are:

(a) *Listeria monocytogenes* (Sizmur and Walker, 1988; Salamina *et al.*, 1996);
(b) *Salmonella*, which may derive from animal manures used to fertilize land on which salad vegetables are grown (and, of course, mayonnaise has been incriminated particularly in respect of *S. enteritidis* PT4);
(c) *E. coli* O157 (see, for example, CDC, 1994);
(d) *Cryptosporidium* (see, for example, CDC, 1996);
(e) *Staphylococcus aureus* (CDSC, 1996b); and
(f) hepatitis A virus (Rosenblum *et al.*, 1990).

30.2.2 Microbiological examinations

As pointed out in Section 22, vegetables (especially imported supplies) may have been fertilized in the field with animal manures without the knowledge of the user (food manufacturer or retailer). If such fresh vegetables are to be consumed raw, it is advisable to use a count of *E. coli* as index organism. Note that *Enterobacter* and some other genera of the Enterobacteriaceae can be part of the natural microflora of vegetables, so it is difficult to interpret results of total Enterobacteriaceae counts on fresh vegetables, whereas a total Enterobacteriaceae count is the more appropriate count to be performed on blanched frozen vegetables, for example.

Perform any of the following examinations as appropriate:

1. Aerobic mesophilic count
2. Total Enterobacteriaceae count on VRBGA (see Section 20.1)
3. Coliform/*E. coli* count on VRBLA (see Section 20.1) and detection of *E. coli* O157 if the vegetables are sourced from a region where the organism is prevalent (see Section 21.4)
4. *Staph. aureus* count (see Section 21.10)
5. *Listeria monocytogenes* count (see Section 21.9)
6. Mould count (see Section 14)

30.2.3 Suggested specifications or standards

Fresh vegetables to be eaten raw (including salad vegetables, bean sprouts, prepared delicatessen salads):

Aerobic mesophilic count	$m = 10^5$, $M = 10^6$, $n = 5$, $c = 2$
E. coli	$m = 10$, $M = 10^3$, $n = 5$, $c = 2$
Salmonella	Absent in 25 g $n = 10$, $c = 0$
L. monocytogenes	Not detected in 1 g

Blanched frozen vegetables:

Aerobic mesophilic count	$m = 10^4$, $M = 10^6$, $n = 5$, $c = 2$
Coliforms (or total Enterobacteriaceae)	$m = 10$, $M = 10^3$, $n = 5$, $c = 2$

Dried vegetables:

Coliforms (or total Enterobacteriaceae)	$m = 10^2$, $M = 10^3$, $n = 5$, $c = 2$

30.3 NUTS

The low a_w inhibits the *growth* of bacteria, but can help the *survival* of any bacteria present (in the case of shelled nuts include contaminants deriving from hand-sorting).

As nuts offer a hazard in respect of mycotoxins, there should be a Hazard Analysis Critical Control Point (HACCP) system operating before and after harvest and for storage. Often, a quality assurance microbiologist will have no information on the care taken before, during and after harvest in the country of origin. In this case consideration should be given to an inspection of nuts in shell, with acceptance depending on the proportion of mouldy nuts detected.

Shelled nuts, even those shelled by machine, may be contaminated by food-handlers sorting after shelling. Coconut is often or usually shelled by hand, and coconut has been responsible for outbreaks of *Salmonella* enteritis and of typhoid fever (Wilson and MacKenzie, 1955) and of cholera (Taylor *et al*., 1993). As shelled nuts are often used in many food products in which they will receive no heat treatment (for example, being put on to frozen desserts, or as confectionery decoration) a total Enterobacteriaceae count or coliform count is recommended as a quality check.

30.3.1 Microbiological examinations

1. Total Enterobacteriaceae counts on VRBGA or coliform counts on VRBLA (see Section 20.1)
2. Mould counts (see Section 14)

30.3.2 Suggested specifications

Coliforms (or total Enterobacteriaceae)	$m = 10$, $M = 10^2$, $n = 5$, $c = 2$
Moulds	$m = 10^2$, $M = 10^4$, $n = 5$, $c = 2$

30.4 FRUIT

It is difficult to suggest microbiological examinations and specifications for fruits and fruit products. For example, although *E. coli* O157 infections have been caused by fruit juice and apple cider, its presence is such a rare event, and *E. coli* counts would be so unlikely to result in positive detections, that the procedure would be uneconomic unless there were *a priori* reasons for the examination. Examinations for particular mycotoxins (e.g. for patulin in apple juice) may be considered appropriate on occasion. Aflatoxin has been found in dried figs and fig paste at concentrations up to 40 µg per kg (Sharman *et al.*, 1991).

Examination of fruit for incipient or potential spoilage, or of already spoiled samples, may be required but the types of organisms sought depend on the type of fruit and on the method of storage. The fungi can be placed in two main groups: firstly those that are true parasites and that have been able to invade the host tissue through the lenticels (e.g. *Glœosporium*), and secondly those that invade only or mainly through wounds (e.g. *Monilia fructigena*), although some of this latter group can attack senescent fruit through the lenticels.

The main economically important fungi causing spoilage of pome fruits are *Botrytis cinerea*, *Glœosporium* (especially *G. album*), *Monilia fructigena* and *Penicillium expansum*. The most important fungi causing spoilage of citrus fruits are *Penicillium* species, especially *P. italicum* and *P. digitatum*, *Alternaria* and *Colletotrichum*.

A clue to the fungal type is given by the form of spoilage: for example, *Penicillium* will usually cause blue and green mould rots (although *P. expansum* causes brown rots of pome fruits, and the blue-green spots of conidia production develop only in the advanced stages of rotting), *Botrytis* causes fluffy grey mould growth, and *Glœosporium* and *Colletotrichum* cause anthracnose with small sunken spots being evident on the surface of the fruit. A comprehensive review of fungal spoilage of pome fruits is given by Fidler *et al.* (1973). Many moulds can be identified microscopically by reference to Part IV. *Colletotrichum* and *Glœosporium*, when growing on the host plant, produce clumps of conidiophores which emerge from breaks in the plant cuticle as cushion- or disc-shaped structures known as acervuli. The differentiation of *Colletotrichum* and *Glœosporium* requires the detection and microscopic examination of these acervuli. *Colletotrichum* typically produces dark-pigmented pointed spines or setæ amongst the conidiophores, whereas *Glœosporium* does not have these setæ. The conidia of both genera are colourless, sickle- or crescent-shaped, ovoid or cylindrical. The macroconidia of *G. album* tend to be curved with rounded ends, whereas those of *G. perennans* are straight and pointed at one end; *G. perennans* also frequently produces many spherical microconidia.

Prepacked, prepared, fresh fruit salads may provide opportunity for the spread of pathogens. For example, the relatively high pH of melon flesh (pH 5.9–6.7) permits the rapid growth of *Salmonella* at temperatures above refrigeration temperature (Golden *et al*., 1993), and prepared melon has caused large outbreaks of *Salmonella* enteritis in the USA (CDC, 1991; Madden, 1992).

30.4.1 Fruit pastes, purees and comminuted fruit

Frequently these products are used in food manufacture, and the specific methods of examination will depend on the nature of the food or the end-product. For example, the examination for coli–aerogenes organisms in fruit preparations being used in recipes for ice-cream and frozen dairy confections; the examination for thermoduric organisms, flat-sour spoilage organisms, etc. in fruit preparations being used in food products to be canned (especially when the product as a whole will not be very acidic). The types of organisms sought and the methods of examination chosen depend on the food product or production method involved (see the appropriate section).

A direct microscopic examination of the fruit preparation for fungal hyphae can give a very rough indication of the extent of fungal spoilage of the original fruit, and arbitrary limits may be set for acceptance purposes. An example of this procedure is the Howard mould count for tomato products (see Williams, 1968; Aldred *et al*., 1971; AOAC, 1995) which can be modified for other fruits. The standard Howard mould count makes use of a special Howard counting chamber or cell, but any suitable haemocytometer or counting chamber could be used in an agreed modified procedure, assessing the number of hyphae in a given volume and expressing the result as the number or aggregated length of hyphae per millilitre. In microscopic examinations, phase-contrast microscopy, dark-field microscopy or appropriate stains (e.g. erythrosin solution, ethanolic solutions of methylene blue or basic fuchsin) may be used to visualize the fungal hyphae, depending on the nature of the background formed by suspension of the product.

Outline procedure for Howard mould count on tomato products

1. Mix the sample. Add water to give a total solids content which, at 20°C, results in a refractometer reading of 45.0–48.7 or a refractive index of 1.3447–1.3460. Mix well.
2. Add four or five loopsful to the Howard cell and carefully place the coverslip on the cell, lowering from one side to exclude air bubbles.
3. Adjust the drawtube of the microscope to give a field of view with a diameter of 1.382 mm using the ×10 objective. The quantity of liquid in each field of view will then be 0.15 mm^3 when a Howard cell is used. If a microscope is being used that is incapable of such adjustment, determine the field diameter and calculate the required correction factor.
4. Examine at least 25 randomly selected fields on each of two or more slide preparations, using a mechanical stage to facilitate the choice of fields. The field of view should be selected whilst *not* looking down the microscope.

5. A 'positive' field is recorded when the aggregate length of not more than three mould hyphae exceeds one-sixth of the field diameter. Record the number of 'positive' and 'negative' fields. The most difficult task in performing a Howard mould count is the differentiation of fungal hyphae from plant tissue.
6. Various government and legislative control agencies have set maximum limits in the case of tomato products. For example, the Food and Drug Administration of the USA sets a maximum limit of 40% 'positive' fields for tomato paste, and 20% 'positive' fields for tomato juice. It is important to remember that the test procedure is artificial and arbitrary, that the definition of a 'positive' field is arbitrary, and also that there is not necessarily any good correlation with the amount of spoilage of the product as judged chemically or organoleptically (as the amount of spoilage caused by a given amount of mould growth will depend on the fungal species involved and its particular metabolic activity and physiological capability).

30.4.2 Fruit juices and squashes

Fruit received at a processing plant will often be contaminated with large numbers of yeasts, e.g. 10^3–10^4 per gram on apples, and up to 10^7 per gram on grapes (Splittstoesser, 1996).

Bottled fruit juices and squashes will frequently have been pasteurized or may contain preservative. In consequence the microflora will be considerably modified from the original microflora of the raw fruit. Hot-filling of fruit juice at a fill temperature of around 88°C can provide a shelf-life of many months in glass bottles. Ascospores of *Byssochlamys*, *Talaromyces* and *Neosartorya* may survive hot-filling and produce either mycelial masses or a generalized haze. At ready-to-drink concentrations, preservative-containing cold-filled fruit squashes and fruit drinks usually have a limited shelf-life with high bacterial counts often developing. The growth of acetic-acid bacteria can cause off-flavours, and loss of oxygen from the headspace can result in partial collapse of plastic containers.

Frozen fruit juice concentrates normally show very low microbial counts (usually below 10^3 per ml), but after storage of the reconstituted ready-to-drink product to the end of shelf-life very high bacterial counts may be found (up to 10^7 per ml). These may be identified by the procedures listed in Part IV. Bacterial spoilage of citrus juices caused by diacetyl-producing species of lactic acid bacteria can be assessed by a quantitative version of the Voges–Proskauer test (see Hill and Wenzel, 1957; Murdock, 1968); this test may be used in control situations in processing factories.

Viable counts should be carried out on an agar medium containing 2 or 20% sucrose or glucose. The lower concentrations of sugar are used for a general viable count of microorganisms, the higher sugar concentrations for viable counts of osmophiles (alternatively osmophilic agar may be used). Acetic-acid bacteria can be identified by the procedures listed in Part IV.

Procedures for microbiological examinations

1. A general viable count of yeasts and moulds on DRBCA or similar medium. If it is desired to identify yeasts, standard procedures may be used; however, Deak and

Beuchat (1993) reported that, although the API20C kit (bioMérieux) was designed particularly to identify yeasts of clinical importance, the kit correctly identified 86% of yeasts which they had isolated from fruit juices.

2. An aerobic mesophilic count of yeasts, moulds and bacteria on malt extract agar, pH 5.4.
3. An osmophilic count on osmophilic agar or on orange serum agar containing 20% sucrose adjusted to pH 5.4 or pH 3.5. Plates should be incubated at 25–30°C (for 7 days) and examined after 3 and 7 days.
4. *Spore count.* Transfer aseptically 10 ml of the sample to each of two sterile, plugged test-tubes and into one insert a thermometer through the cottonwool plug, so that the thermometer bulb is immersed completely in the sample. Place both tubes in a water-bath at 80°C and allow the tubes to remain in the bath for 15 min after the temperature in the control tube attains a maximum (usually just below the temperature of the water-bath). An alternative time–temperature combination that may select a slightly different thermoduric population is obtained by holding at 100°C for 5 min. Remove the tubes and cool quickly in cold water. Prepare decimal dilutions. Carry out pour plate counts on plate count agar, or multiple tube counts using the equivalent liquid medium (glucose tryptone yeast extract broth). Incubate duplicate sets at 30°C and 55°C, for the mesophilic spore count and the thermophilic spore count respectively.
5. *Ascospore count.* Use the procedure described in (4), but heat-treat for 30 min at 70°C, and then use a mould medium such as DRBCA (see Section 14).
6. *Microscopic count.* Carry out a direct microscopic count for yeasts by placing a known volume (1/100 ml) of the sample or a low dilution of the sample on a slide, adding one drop of 0.02% erythrosin solution and mixing well. Spread the mixture over 1 cm^2 of the slide and allow to dry. Alternatively the sample stain mixture may be examined as a wet preparation using a counting chamber or haemocytometer slide. The yeasts and mould fragments can be seen and counted but, owing to the high solids content of most fruit juices and squashes, bacteria cannot usually be distinguished. A modification of a Howard mould count may also be adopted using wet mounts with a suitable counting chamber or haemocytometer slide.

30.5 SUGARS AND SUGAR SYRUPS

Sugar syrup is used as a constituent in the preparation of many food products and the nature of the microbiological examination will depend on the type of end-product. For example, if the sugar is to be used in the preparation of a heat-treated canned food, samples should be examined for the presence of organisms able to cause flat-souring (*Bacillus stearothermophilus*), hydrogen sulphide-producing clostridia using shake tube counts in iron sulphite agar, and non-hydrogen sulphide-producing clostridia using liver broth (see Section 34). In general, plate counts for bacteria, yeasts and moulds should be carried out using orange serum agar, DRBCA, Davis's yeast salt agar, and osmophilic agar or orange serum agar with 20% sucrose. Incubate at 25–30°C for up to 1 week. The genera most frequently isolated are *Bacillus*, *Lactobacillus*, *Leuconostoc*, *Hansenula*, *Pichia*, *Saccharomyces* and *Torulopsis*. Mesophilic and thermophilic spore counts and microscopic counts can be carried out as described for fruit juices and squashes. For methods of examination specifically related to the type of end-product, refer to the appropriate section. In general, viable counts should not exceed 10 per gram (Muller, 1972).

30.6 CHOCOLATE AND CHOCOLATE CONFECTIONERY, SUGAR CONFECTIONERY

Spoilage may be caused by yeasts, resulting, for example, in gas production in chocolate-enrobed fondants leading to explosion of the chocolates.

Localized mould growths can occur on marzipan. Mould growth is more likely to occur on products enclosed in water-impermeable packages and in which moisture migration has resulted in condensation.

Chocolate and chocolate confectionery have been involved in outbreaks of *Salmonella* food poisoning (Craven *et al.*, 1975; Greenwood and Hooper, 1983). A collaborative study (DeSmedt *et al.*, 1994) has shown that motility enrichment on modified semi-solid Rappaport–Vassiliadis medium (following pre-enrichment) gave a good sensitivity for *Salmonella* in chocolate and cocoa. An alternative to this method could obviously be the use of the Oxoid Salmonella Rapid Test.

For a discussion of the HACCP evaluation of chocolate, see Mazigh (1994).

The low a_w found in confectionery means that any pathogens introduced into such products will survive for quite long periods and can lead to food-poisoning outbreaks. The use of raw egg white in marshmallow confectionery has caused an outbreak of *Salmonella enteritidis* PT4 food poisoning (Lewis *et al.*, 1996).

30.7 SALTED, PICKLED AND FERMENTED VEGETABLES

The microbial flora of salted and pickled vegetables may be very different from that of fermented vegetables. A microbiological examination of such products can be carried out by sampling the brine or liquor. In the case of the finished products, it should be noted that frequently pickled and fermented vegetables are pasteurized to prolong the shelf-life. Such pasteurized products are bottled or canned, and the presence of heat-sensitive organisms indicates either faulty pasteurization or post-pasteurization contamination (e.g. due to jar-seal or can leakage).

Non-pasteurized fermented vegetables must be expected to contain many organisms, usually lactic-acid bacteria, particularly *Lactobacillus* and *Leuconostoc* (which can cause bloating by gas production in sweet and sour cucumbers), but also *Pediococcus*. Yeasts and moulds, members of the Enterobacteriaceae, *Bacillus* and *Clostridium* may all cause spoilage of fermented products. However, some fermented vegetable products are produced as the result of action by mixtures of lactic-acid bacteria, yeasts and/or moulds.

30.7.1 Microbiological examinations

Prepare serial dilutions in the appropriate diluent from the liquor or brine. In an actively fermenting brine the following population levels per millilitre may be present: 10^7 to 10^9 lactic-acid bacteria, 10^6 to 10^8 yeasts, 10^6 to 10^9 salt-tolerant or halophilic organisms, and up to 10^7 coliform organisms. In the finished, pasteurized product the expected counts will be low, requiring plating of 10^0 to 10^{-2} dilutions. Carry out:

1. A direct microscopic count of the undiluted liquor or brine by a Breed's smear using Gram's staining method. It is advisable to heat-fix slightly more thoroughly than usual to help prevent the organisms being washed from the slide when the salt or sugar is dissolved. An examination for yeasts or moulds may also be carried out using erythrosin.
2. A viable count of yeasts and moulds on, for example, DRBCA.
3. A viable count of yeasts, moulds and bacteria on malt extract agar at pH 5.4, incubated at 25–30°C for up to 1 week.
4. A count of lactic-acid bacteria on Rogosa agar or acetate agar, prepared as layer plates, and incubated at 30°C for 5 days.
5. A count of *Staphylococcus aureus* (see Section 21.10).
6. Counts of salt-tolerant and halophilic organisms, in the case of brine-cured and salted products. Use 15% sodium chloride as diluent for these counts, and media incorporating 15% sodium chloride.
7. Coliform counts using VRBLA, or a multiple tube count for smaller numbers (see Section 20.1).

30.8 SPICES AND CONDIMENTS

Typically, spices have high counts of bacteria and moulds, predominantly as spores. *Bacillus* and *Clostridium* spore counts can be up to 10^7 per gram. Even though the amount of spices and of spice and herb mixtures used in food products may be small, the high spore counts mean that such ingredients contribute significantly to the bacterial spore load that has to be destroyed by the cooking and the heating process. (See Pivnick, 1980.)

31

Alcoholic Beverages

The microbiology of the *production* of beers, ciders and wines is outside the scope of this book (interested readers are referred to Findlay (1971) and to Reed and Peppler (1973)) but the food microbiologist may occasionally be called on to examine reputedly spoiled samples.

In addition, as has already been mentioned, *Escherichia coli* O157 infections have been caused by unpasteurized cider (Besser *et al.*, 1993; CDC, 1997a), and also cryptosporidiosis has been caused by unpasteurized cider (CDC, 1997a).

Wild yeasts (i.e. yeasts other than the *Saccharomyces cerevisiae*, *S. carlsbergensis* and *S. ellipsoideus* strains involved in the primary fermentation) frequently prove to be more resistant to the presence of cycloheximide in a medium than the species named. Thus the incorporation of cycloheximide to a final concentration of 10 parts per million will enable yeasts such as *Pichia*, *Brettanomyces* and *Torulopsis* to be detected. However, it should be remembered that the flor yeasts of sherry and the initiating yeasts of Bordeaux wines are types that may grow on media containing cycloheximide, or on the lysine agar described below. Obviously in such cases yeasts so detected may or may not be spoilage yeasts. Cycloheximide-containing media will also enable bacterial contaminants to be detected, as bacteria are also resistant to cycloheximide.

A suitable medium is plate count agar with the glucose content increased to 20 g per litre to which is added 3 g of malt extract per litre, one set of plates being prepared at pH 7.0 and one set of plates at pH 4.0. Cycloheximide may be added to the medium before sterilization or it may be added as a sterile solution just before pouring the plates. If the latter procedure is adopted, a 0.1% stock solution of cycloheximide can be prepared, and sterilized by filtration or by autoclaving; 1 ml of this solution is added to every 100-ml bottle of molten medium and mixed immediately before pouring the plates. Such a medium should provide for the nutritional requirements of most bacteria and yeasts likely to be found in these samples. However, some lactobacilli found as contaminants in breweries or distilleries may not grow in media such as MRS medium unless the medium is supplemented with either hopped beer (Kirsop and Dolezil, 1975) or filter-sterilized malt extract and yeast autolysate (Bryan-Jones, 1975).

Sets of plates should be incubated aerobically and anaerobically at 30°C. Possible spoilage bacteria include lactobacilli, pediococci, acetic-acid bacteria and *Zymomonas mobilis*. *Z. mobilis* is an important spoilage organism in the brewing industry, causing turbidity and off-flavours. Some strains of *Z. mobilis* are facultatively anaerobic, but other strains are obligately anaerobic. The commonest microbial spoilage of cider is acetification caused by acetic-acid bacteria.

However, certain traditional alcoholic beverages may typically display such organisms (Steinkraus *et al.,* 1983). For example, pulque and palm wine fermentations involve yeasts, lactic-acid bacteria and *Zymomonas*, and, in the case of palm wine, acetic-acid bacteria are also normally present.

If it is required to assess low levels of contamination, the liquid version of such media can be used as an enrichment stage with a multiple tube counting technique to obtain counts of the organisms found. Alternatively, membrane filtration can be used to remove the microorganisms from a relatively large amount of sample. Membrane filtration will, however, be practicable only with filtered or clarified products, but in such situations it may be possible to attempt detection of contaminants at concentrations as low as one cell in 10 litres.

Another medium suitable for the detection of many wild yeasts is lysine agar, a defined medium (Morris and Eddy, 1957). *Saccharomyces cerevisiae* and *S. carlsbergensis* are unable to utilize lysine as a sole nitrogen source, whereas many other yeasts can utilize this amino acid; lysine agar is thus an elective medium. It is available in dehydrated form from Oxoid. When lysine agar is used to detect wild yeasts it is important to wash and centrifuge the yeast suspension at least three times using sterile distilled water to ensure that there is no adventitious extracellular source of nitrogen added to the medium. The pellet is finally resuspended to a known volume using sterile distilled water. In consequence this medium is more suitable for the detection of wild yeasts in samples such as yeast concentrates. It is also important with this medium to use levels of inoculum on the plates which provide 10^4 to 10^6 yeast cells per plate.

32

Bread, Cakes and Bakery Goods

These food products usually do not support the growth of bacteria, but cream and similar fillings and toppings of cakes are highly favourable for bacterial multiplication. Very rarely, species of *Bacillus* may cause a defect in bread known as 'ropiness' due to the production of capsular material.

The methods used in the microbiological examination of cake fillings and toppings depend on the nature of the constituents and are given in the sections appropriate to the constituents.

In addition, consideration should also be given to the possible contamination of fillings or toppings and decorations by food-handlers working in the preparation of the bakery products. The most well known hazard is the contamination of whipped cream or other filling or topping capable of supporting bacterial growth by *Staphylococcus aureus* from a food-handler. However, it is also possible for a food-handler to contaminate after baking *any* part of the product with a pathogen having a low minimum infective dose, such as *Shigella* or hepatitis A virus. For example, there have been two reported incidents of hepatitis A in which a food-handler contaminated sugar glaze applied to the tops of bakery products (Weltman *et al.*, 1996).

Microbiological control of bakery products other than fillings is concerned mainly with the possibility of spoilage due to the growth of moulds. As the baking temperature is sufficient to kill fungal conidiospores, subsequent spoilage is usually caused by mould contamination from the atmosphere or from wrapping material. The use of preservatives such as sorbic acid or propionate has greatly extended the shelf-lives of many bakery products by inhibiting mould development.

It should be noted that some light sponge cakes and meringues receive insufficient heat treatment to provide sufficient decimal reductions of *Salmonella* in the event of large numbers of *Salmonella* being present in the egg. As *S. enteritidis* PT4, in particular, can be transmitted vertically inside the egg, large bacterial concentrations can sometimes be found. It should therefore be a rule in commercial bakeries to employ only pasteurized whole egg or pasteurized egg albumen, and in some countries legislation requires bulk egg supplies to be so pasteurized.

32.1 EXAMINATION FOR MOULDS

Once aerial mycelium has started to sporulate, fungi are usually obvious by visual inspection. Growth of *Wallemia sebi* on wholemeal or brown bakery products may be less easy to see because the aerial mycelium is compact, forming low, small cushions, and the colour (light tan to brown) can make it difficult to see the colonies, especially if the foodstuff has a rough-textured surface (e.g. granary bread with whole grains). *Wallemia sebi* has long been thought of as merely a spoilage mould, but Wood *et al.* (1990) reported the production of a toxin by this mould.

With some products the development of aerial mycelium may be prevented. For example, the author has observed mould growth in cup-cakes in paper cases with fondant icing topping. The substrate mycelium pervaded the crumb structure of the cake, but the mould was prevented from producing aerial mycelium and spores, so that only close inspection (using a large magnifying glass or stereoscopic microscope) of the teased-out cake crumb could detect the mould hyphae.

If an item is suspected of being spoiled by the growth of moulds, make a visual examination for the presence of mould mycelium by teasing out with needles small samples of the food in a Petri dish placed on a black background with oblique lighting illuminating the sample. Follow this by a microscopic examination. Tease out small portions (including any that appear to contain fungal hyphae) on a microscope slide in a drop of lactophenol–picric acid or lactophenol–cotton blue, and cover with a coverslip. Identify the mould by reference to Part IV if the sporing stages are present. If only vegetative mycelium is present, it is necessary to culture the mould on a range of media in order to obtain spore formation.

32.2 EXAMINATION OF COMPRESSED BAKERS' YEAST

This may be examined for viability, the presence and numbers of bacterial contaminants and the numbers of wild yeasts (see also Reed and Peppler, 1973).

1. Weigh out a 10-g sample and prepare serial dilutions to 10^{-9}.
2. Perform microscopic (total) counts either by haemocytometer or by Gram-stained Breed's smear, using an appropriate dilution to give not more than 30 organisms per field.
3. Set up a viable count of the yeast on the high dilutions using Davis's yeast salt agar, incubated at 30°C for 3–5 days. From the microscopic count and the viable count the percentage viability can be calculated.
4. Set up viable counts for bacterial contaminants on a wide range of dilutions (10^{-1} to 10^{-7}) using plate count agar with cycloheximide added to a final concentration of 10 p.p.m. (add 1 ml of 0.1% sterile stock solution of cycloheximide to each 100 ml of molten medium immediately before pouring the plates). Incubate at 30°C for 5 days.

5. Set up viable counts for lactobacilli similarly, but using Rogosa agar or acetate agar, containing cycloheximide, as layer plates and incubating at 30°C for 5 days.
6. Set up a count for wild yeasts on the low dilutions (10^{-1} to 10^{-3}) using Davis's yeast salt agar with 10 p.p.m. cycloheximide. Incubate at 30°C for 5 days.

32.3 EXAMINATION OF STORED CEREAL GRAINS

The examination of whole cereal grain does not often come within the purview of the food microbiologist. The microflora of cereal grains is extremely varied, with bacteria (including actinomycetes and streptomycetes) outnumbering the fungi. Usually counts performed at 20–25°C will exceed those obtained at 37°C, but when grain is malted not only will bacterial counts increase, but also the ratio of the count at 37°C to that at 20–25°C often will be found to increase.

The main concern of the microbiologist, however, is to assess the extent of fungal growth. In addition to gross spoilage rendering grain unacceptable to the consumer, fungal growth frequently may involve the production of potentially hazardous metabolites.

The toxicity of *Claviceps purpurea* is well known, as ergotism is a disease well described in many ancient accounts. Inspections of cereal grain, for example at ports of entry, can include a simple visual inspection to determine the presence and concentration of ergotized grain – the grain becomes replaced by the dark purple sclerotia of *C. purpurea*.

A large number of mycotoxins are produced by a wide range of fungi, including *Penicillium*, *Aspergillus*, *Fusarium* and *Alternaria*. These mycotoxins have a wide range of chemical structures. Thus it is difficult to screen foodstuffs either microbiologically for mycotoxigenic fungi in general, or chemically for mycotoxins in general. However, certain mycotoxins have assumed greater public health significance, and consequently may be subjected to specific screening programmes. Those particularly associated with cereals are:

(a) ochratoxin, produced by various species of *Aspergillus* and *Penicillium*, and almost certainly the cause of Balkan endemic nephropathy in Central Europe;
(b) trichothecenes, produced predominantly by *Fusarium*, probably the cause of alimentary toxic aleukia in Russia, and more recently causing disease in China and India;
(c) fumonisins, produced by *Fusarium* and *Alternaria*, which may be associated with oesophageal cancer (Chu and Li, 1994).

Mycotoxins (other than aflatoxin) in cereals produced by *Aspergillus*, *Penicillium* and *Alternaria* have been reviewed by Miller and Trenholm (1994). Ochratoxins, trichothecenes and ergot, and the analytical procedures for their detection, have been reviewed by the International Programme on Chemical Safety of the World Health Organization (IPCS, 1990).

It is obvious that in this context a viable mould count on cereal grains or on flour provides little useful information on the possible amount of metabolites: the mould colonies on the plates may have developed from spores, or from small or large hyphal fragments as already explained (Section 18.2). When a *specific* toxin is being sought, as in the case of aflatoxin, chemical methods of detection and assessment may be used. Any attempt at assessing a potential hazard from *unspecified* toxic metabolites must surely be based on a method of estimating the amount of mycelial growth present (e.g. by a modification of the Howard mould count; see Section 30.4.1). Any acceptability standards based on such microscopic counts will be arbitrary and empirical. It must be emphasized that the presence of a species known to be capable of producing toxin is not evidence *per se* of the presence of toxin in the foodstuff. Firstly, toxigenicity varies from strain to strain within a species, and secondly toxin production is substrate dependent and can also be affected by environmental factors such as temperature, pH, etc. Nevertheless, until much more is known about the toxicity of cereals and flours (and other foods) that have been subject to fungal attack, it is suggested that the presence of any fungal hyphae at a level detectable microscopically as the equivalent of 20% 'positive' fields determined in the Howard mould count must be regarded with suspicion.

Obviously, a more positive approach to the problem of mycotoxins in cereals, flours, and so on, is to prevent their occurrence, by applying Hazard Analysis Critical Control Point (HACCP) procedures, risk assessment and risk management at all stages from the farm to the consumer. Some of these quality management aspects have been addressed by Kuiper-Goodman (1994).

32.4 EXAMINATION OF FLOUR

The indigenous flora of grain includes coliform organisms; therefore coliform counts on flours may be advisable when these are being incorporated into food products on which coliform counts are normally conducted, although usually the flour incorporated into products will receive a heat treatment sufficient to kill these organisms.

Since flour is usually to be subjected to a heat treatment, the most significant microorganisms to be sought are species of *Bacillus* and *Clostridium*. As already mentioned, certain *Bacillus* spp. (especially *B. subtilis*) may cause ropiness in bakery products; the presence of *Clostridium* is of special significance if the flour is being used as a thickener in meat-containing products because good conditions for development will be provided.

32.4.1 Procedure

1. Prepare decimal dilutions in 0.1% peptone water for aerobic counts, and in reinforced clostridial medium for anaerobic counts (see Section 17.1.2). Use a Colworth 'Stomacher' to ensure adequate mixing.

2. Carry out the following counts:
   (a) Mesophilic and thermophilic viable counts on plate count agar incubated aerobically and anaerobically at 25°, 37° and 55°C for 3 days.
   (b) A total Enterobacteriaceae or coliform count at 30°C using VRBGA or VRBLA, or by a multiple tube technique (see Section 20.1).
3. Using the appropriate 10^{-1} dilutions perform both aerobic and anaerobic spore counts on plate count agar at 30° and 55°C (see p. 282).

32.4.2 Recommended standard

Flour to be used in soups and meat products should contain not more than 15 thermophilic spores per gram; not more than ten flat-sour spores per gram, and less than one *Clostridium* spore per gram (see also Amos, 1968).

32.5 PASTA

Wheat pasta (e.g. spaghetti, macaroni, etc.) is made from semolina or fine flour (preferably milled from durum wheat). A stiff dough is extruded, or rolled and cut. Factory-made pasta may be sold retail without drying (so-called 'fresh' pasta), or dried to give a product that is shelf stable at ambient temperature.

Since there is no cooking stage during the dough preparation, both heat-sensitive and heat-resistant microorganisms may be present in the dough, and these can grow rapidly during mixing, extrusion and drying.

'Fresh' pasta is a highly perishable product with a short shelf-life even at refrigeration temperatures.

Staphylococcus aureus enterotoxin represents a particular hazard, as the subsequent cooking of the pasta will not inactivate the heat-resistant enterotoxin (Woolaway *et al.*, 1986).

33

Convenience Meals

33.1 INTRODUCTION

In countries with a factory-based food industry, there has been a developing market for 'convenience' foods which are centrally produced, and then distributed and sold refrigerated or frozen. In addition to this retail market, there are also the catering markets for 'cook–chill' and 'cook–frozen' products; there is a centralized production of precooked, portioned meals, refrigerated or frozen, and distributed to satellite kitchens for reheating.

Another development has been in 'cuisine sous vide', where the food products are vacuum-packaged in plastic pouches, cooked, and then stored and distributed refrigerated. 'Retorted' flexible packaged products receiving a 'botulinum cook', which are capable of being stored at ambient temperatures, are equivalent to canned foods and are discussed in Section 34. 'Sous vide' products, however, are only pasteurized, and it is important that such products are maintained at low temperature. Since psychrotrophic *Clostridium botulinum* can grow and produce toxin down to 3.3°C, the storage temperature of 'sous vide' products should be between 0° and 3°C (IFST, 1992). However, it has been recognized (ACMSF, 1992) that it is not always feasible to ensure a temperature below 3°C at all times, in which case such foods should also be characterized by one or more environmental factors that will control the growth and toxigenesis of psychrotrophic *C. botulinum*.

In the satellite kitchens, the ready meals should be reheated to give a heating-centre temperature of 70°C for at least 2 min; the temperature should not subsequently fall below 63°C, and the food served within 15 min (Anonymous, 1989; IFST, 1992). The same cooking and serving procedures for retail products should also be adopted in domestic kitchens.

Thus, with these types of product, appropriate codes of good manufacturing practice (CoGMP) and of good catering practice (CoGCP), and Hazard Analysis Critical Control Point (HACCP) systems, are perhaps even more important than for other food products. Consequently the microbiological examinations, which can give information only in hindsight on perishable products with a short shelf-life, should be regarded as providing validation and verification of the CoGMP, CoGCP and HACCP systems in place (see Harrigan and Park, 1991).

33.2 MICROBIOLOGICAL EXAMINATIONS

Generally, the methods for microbiological examinations will be those used for the component parts of the meal, already discussed in previous sections. In addition, keeping-quality tests are advisable in which samples of the foods in their original packets are stored under the conditions advised on the packets. Following this storage, carry out viable, coliform and direct microscopic counts.

1. Prepare serial decimal dilutions in the usual way. In the case of a frozen food, remove it from its pack, with full aseptic precautions, into a sterile container; allow the frozen sample to soften *slightly* and then weigh aseptically and homogenize the weighed sample with an appropriate amount sterile diluent to give the initial 10^{-1} dilution, preferably using a Colworth 'Stomacher'. Carry out the following tests:
   (a) Psychrotrophic and mesophilic counts (both aerobic and anaerobic) on plate count agar, incubated at 4.5°, 25° and 37°C for 14, 3 and 2 days respectively.
   (b) A total Enterobacteriaceae count on VRBGA, a coliform count on VRBLA, or a count of *Escherichia coli*, as required. To determine low concentrations use a multiple tube count. (See Section 20.1.)
   (c) A direct microscopic count on the 10^{-1} dilution or, if that is not possible, the 10^{-2} dilution, staining the Breed's smear by Gram's method or with Loeffler's methylene blue as appropriate (see Section 6.1).
   (d) An examination for the presence of *Clostridium perfringens*, using a dilution series prepared in reinforced clostridial medium (see Section 21.7).
   (e) A count of *Staphylococcus aureus* on Baird-Parker's medium (surface counts), incubated at 37°C for 24 h (see Section 21.10).
   (f) Detection or enumeration of *Listeria monocytogenes* (see Section 21.9).
2. Store further packets of the food at the maximum storage temperature and for the maximum time recommended by the manufacturer and repeat the tests after storage.

33.3 MICROBIOLOGICAL SPECIFICATIONS

The following have been suggested as guidelines for cook–chill and cook–freeze products examined at the point of consumption, before reheating is applied (Anonymous, 1989):

Aerobic mesophilic count at 37°C	Less than 10^5 per g
E. coli	Less than 10 per g
Staph. aureus	Less than 100 per g
Cl. perfringens	Less than 100 per g
Salmonella	Absent from 25 g
L. monocytogenes	Absent from 25 g

34

Canned Foods

34.1 INTRODUCTION

Although most canned foods are processed for sterility, some are not. For example, foods with a pH below 4.5 may be given a heat treatment sufficient to kill yeasts, moulds and their spores, and the vegetative forms of bacteria, without killing bacterial spores, which cannot usually germinate and grow in an acid food. Canned food products with a pH above 4.5 are, with few exceptions, given sufficient heat treatment to destroy the spores of *Clostridium botulinum*. The chief exceptions are some canned cured meats. These products would suffer from a significant deterioration of quality if given the usual fairly rigorous heat treatment. However, they can be given a less severe heat treatment, as the presence of nitrite, nitrate and sodium chloride combined with refrigerated storage and a limited shelf-life prevents the growth of surviving organisms. Thus, canned hams and similar products may be found to contain a variety of spore-bearing bacteria, or even *Enterococcus*. However, Gram-negative bacteria should be absent, as they would be indicative of grossly inadequate processing or of post-processing contamination.

Since the spores of some of the non-pathogenic spoilage organisms commonly associated with canned foods (particularly 'flat-souring' organisms such as *Bacillus stearothermophilus*) are both more numerous and more heat resistant than *Cl. botulinum* spores, the examination of the food for the presence of non-pathogenic thermoduric bacteria can suffice in the determination of the efficiency of the heating process.

After heat treatment cans are usually cooled in water. Temporary leaks through the seams of the cans may occur as a result of the stresses introduced by sudden cooling. When such leaks occur, small amounts of cooling water will enter the can. Consequently, water used for cooling must be of better than potable quality and contain fewer than 100 bacteria per millilitre, to reduce to a minimum the possibility of post-processing contamination. Post-processing contamination may involve microorganisms of many types including *Flavobacterium* (Bean and Everton, 1969), cocci and other non-sporing bacteria.

Laboratory examination of a sample of cans from a batch is *not* an appropriate quality assurance procedure (see Harrigan and Park, 1991). The correct quality management approach is to follow a Code of Good Manufacturing Practice, to apply Hazard Analysis Critical Control Point (HACCP) procedures, and to adopt such monitoring (including laboratory analyses) as necessary to verify that the quality management procedures are working. In particular, the microbiological quality of the starting materials should be monitored, as a sudden reduction in microbiological quality here could lead to an unexpected unacceptable survival rate after thermal processing.

An examination for relevant pathogens, toxigenic organisms or their toxins will be required if the examination is linked to a suspected involvement of the batch with a food-poisoning outbreak. The methods of examination should include those used for the similar product when uncanned, *with the addition* of sterility tests both before and after storage. Pathogens – *Cl. botulinum*, *Cl. perfringens*, *Staphylococcus aureus*, *Salmonella* and *Shigella* – should be completely absent. A large microscopic count coupled with the absence of viable organisms indicates the possibility of spoilage of the product before canning.

34.2 PROCEDURE FOR MICROBIOLOGICAL EXAMINATION

Select a representative number of cans from each batch (see Section 16.5). Examine the cans for physical defects which include faulty side or end seams, perforations, rust or other corrosion, dents and bulging ends. Bulging at one or both ends may be due to bad denting or to the multiplication in the food of microorganisms that ferment sugars with the production of gas. If any of the cans selected show such defects, examine these separately and remove further cans from the batch to make up the number of 'normal' cans examined. Since most cans are partially evacuated, opening the can provides an opportunity for contamination of the contents with airborne organisms. Therefore, cans should be opened in an inoculating chamber which should preferably ensure a sterile atmosphere by the use of forced ventilation of sterile air, ultraviolet light sterilization or other means, or at the very least provide completely draught-free conditions.

1. Examine half of the 'normal' cans selected as follows:
   (b) Swab the top of a can with alcohol and then flame it. Open the can with a sterile can opener (sterilize this by swabbing with alcohol and flaming). If the food product is liquid, remove a sample with a sterile pipette to a sterile container. In the case of a solid sample, use a sterile cork-borer to remove a core of the food from the centre of the can. Also remove a core from the immediate vicinity of the side seam.
   (b) Carry out a microscopic examination of a smear of the food, stained by Gram's method if possible.
   (c) Examine for pathogens or for toxigenic organisms and toxins as appropriate.
   (d) Inoculate five tubes each of tryptone soya broth, and glucose tryptone broth with portions of the food, to examine for the presence of viable bacteria, and 'flat-sour' organisms respectively. Inoculate five tubes of *freshly prepared* liver broth (Oxoid) and incubate anaerobically either by sealing with 2% agar or by

placing in an anaerobic jar. Incubate these media at 25°C for 3 days. Similarly inoculate two further sets of media and incubate at 37° and 55°C for 3 days. After incubation, examine the tubes for the growth of microorganisms and prepare Gram-stained smears from tubes that show growth. In the case of the glucose tryptone broth, flat-sour organisms will produce acid from the glucose and cause the medium to change colour from purple to yellow.
2. Incubate the remainder of the normal cans for 1 week at 37°C and then examine them for evidence of blowing. Sample and examine the contents of the cans as described above.
3. Examine blown cans by opening the cans aseptically with precautions to prevent the high pressure in the cans causing the contents to be scattered. For example, after sterilizing the top of the can with alcohol and by flaming, invert a sterilized funnel over the top of the can. Insert a sterile metal punch through the hole in the funnel and puncture the top of the can. When the pressure has been released, open the can with a sterile can opener and examine the contents. As well as the tests indicated above, inoculate media as follows:
   (a) Lauryl tryptose broth or VRBGA, to detect coliforms;
   (b) Robertson's cooked meat medium or anaerobically incubated liver broth (Oxoid), to detect *Clostridium* (mesophilic or thermophilic);
   (c) Tryptone soya broth, to detect *Bacillus*.

34.3 DIRECT DETERMINATION OF *F* VALUES

If the heating and cooling curve has been obtained by the use of thermocouples inserted into cans of food (radiotelemetry being required in the case of hydrostatic cookers), then the microbiological effect can be determined mathematically. This can be achieved either by calculation, using the methods described by Stumbo (1973), provided that some D values and the z value of the reference organism are known, or the thermocouples can be linked directly to an automatic F_0 computer (e.g. as made by Ellab). (See also Section 12.4.)

Alternatively, the total effect of a thermal process can be determined experimentally either by the following procedure, or by the use of spores of the reference organisms immobilized in food/alginate particles (Brown *et al*., 1984).

1. The thermal destruction curve for the reference organism (usually *Bacillus stearothermophilus*) at 121°C is determined in the laboratory as described in Section 12.4.3, a graph of log survivors: time being plotted in the usual way.
2. Some more ampoules of the spore suspension prepared at the same time are placed without undue delay in cans of the food product to be processed. Each ampoule is located in place within a can at the 'heating centre' using a suitable cradle (e.g. of thin but rigid wire).
3. After processing, the ampoules are recovered and the counts of surviving spores determined, and the average count calculated.

4. The average count obtained in (3) is matched on the curve obtained in (1) to find the time of heating at 121°C that is *equivalent* to the entire heating and cooling of the cans during the heat processing under investigation.

It should be noted that the accuracy of this determination depends on the ampoule cradle providing an insignificant change in the heat transfer characteristics. In addition, as the effect of heat on bacteria may be modified by other environmental parameters (e.g. pH, concentration of protein, fat or sugar), the suspending liquid used in the ampoules should be chosen carefully to match as far as is practicable the foodstuff being canned. One of the advantages of this technique over inoculated pack studies is that the bacterial suspensions being examined in the laboratory after the two types of heat treatment are identical in all respects except for the heat treatments.

35
Water

35.1 INTRODUCTION

Most microbiological examinations of water samples are carried out on water supplies or proposed water supplies to test the potability of the water. The bacteria found in water are mainly of three types: (a) the autochthonous (natural aquatic) bacteria; (b) soil-dwelling organisms; and (c) organisms that normally inhabit the intestines of humans and other animals. Most of the bacteria that are normally aquatic are Gram negative (including *Pseudomonas*, *Flavobacterium*, *Cytophaga*, *Acinetobacter* and *Chromobacterium*), although a few Gram-positive bacteria (coryneform bacteria, *Micrococcus* and *Bacillus*) may be found. Although some aquatic bacteria are extremely difficult to cultivate, most will be capable of growing on very dilute media, for example CPS medium (Collins *et al.*, 1973). Many of these organisms will not grow on standard nutrient agar or plate count agar. Such bacteria usually have an optimum temperature for growth of 25°C or less, and plates should be incubated for 14 days. Counts on nutrient agar or plate count agar will often be only one-tenth of those on CPS medium. The soil-dwelling bacteria, which may be washed by rain into streams, ponds, etc., include species of *Bacillus*, *Streptomyces* and saprophytic members of the Enterobacteriaceae such as *Enterobacter*. Most of these organisms will have an optimum temperature for growth around 25°C and will be capable of growing on nutrient or plate count agar.

The expected viable counts on CPS medium depend on the nature of the water sample. Unpolluted rivers may show comparatively low counts of up to around 100 per ml, although after heavy rain the run-off entering the river will contain large numbers of soil-dwelling organisms, but counts will also depend on the trophic state of the river. Downstream of towns counts will be much larger, partly because of the discharge of sewage and industrial effluents into the rivers. Water taken from the river at or immediately downstream of the town may contain more than one million bacteria per ml, but more usual counts for polluted river water are between about 10 000 and 200 000 per ml. Lakes and reservoirs may exhibit counts of 100 per ml or less when unpolluted, although this depends on the trophic state. Unpolluted waters from deep wells in good condition normally have counts of 100 per ml or less, but if contamination is able to occur (e.g. by seepage through a cracked well-casing) the counts will be much higher, although still usually below 10 000 per ml. Generally counts on plate count agar will be about one-tenth of those given above, although not such a great difference in counts may be seen in highly polluted waters.

35.2 HAZARDS FROM WATER USED FOR CONSUMPTION OR FOR THE PREPARATION AND PRODUCTION OF FOOD

Although high viable counts are usually indicative of both contamination of the water and the presence of bacteria other than aquatic bacteria, they do not necessarily indicate pollution by faeces, untreated sewage, etc., and the bacteria may be mostly soil saprophytes. Thus, a high viable count alone is not evidence that a water supply is potentially dangerous due to the possible presence of intestinal pathogens, but water supplies with high viable counts are nevertheless undesirable as they may contribute to food spoilage problems.

Faecal pollution of water gives rise to the presence of organisms derived from the intestine, including *Escherichia coli*, *Enterococcus faecalis*, *Clostridium perfringens* and possibly intestinal pathogens such as *Salmonella*, *E. coli* O157:H7, *Campylobacter*, *Vibrio cholerae* and *Yersinia*. In certain circumstances enteric pathogens may survive in water for comparatively long periods (see Geldreich, 1972). Methods to detect bacterial pathogens are described in Section 21. Also, a number of viral diseases may be transmitted via water. Methods for the concentration and detection of viruses at present depend mainly on detecting pathological effect on tissue culture cell lines, so are beyond the capabilities of many quality assurance laboratories; interested readers are referred to the booklet produced by the Standing Committee of Analysts (DoE/SCA, 1995).

A number of protozoa also present a hazard of water-borne disease, including *Cryptosporidium parvum* (Guerrant, 1997), *Giardia lamblia* (Flanagan, 1992) and *Entamoeba histolytica*. These protozoa produce cysts which may resist inimical environmental factors such as the presence of chlorine in treated water. Cysts of *C. parvum* are especially chlorine resistant, and drinking water has been responsible for very large outbreaks; in one outbreak in Milwaukee, USA in 1993, an estimated 403 000 people were affected. It was suggested by the US Environmental Protection Agency that an appropriate limit for cysts of *C. parvum* and *G. lamblia* in drinking water is less than 1 per litre (CDC, 1995). Concentration methods for detecting *Cryptosporidium* and *Giardia* have been discussed by the UN World Health Organization (WHO, 1993). In the UK the preferred procedure (DoE/SCA, 1989) has been to pass the water sample (in the case of treated water, at least 100 litres) through a cartridge filter. The trapped material is then eluted, centrifuged, resuspended and subjected to flotation in sucrose solution. *Giardia* cysts are detected microscopically after staining with iodine solution; *Cryptosporidium* cysts are stained with fluorescent dye-labelled antibody. As concern about these protozoa continues, improved antibody-based detection systems are becoming available commercially. Use of immunomagnetic capture techniques is being investigated actively (Whitmore and Sidorowicz, 1995).

Other microbiological hazards offered by water include toxins produced by cyanobacteria (blue-green algae) such as *Anabaena*, *Oscillatoria*, *Aphanizomenon*,

Nodularia, *Microcystis*, *Nostoc* and *Cylindrospermum* (Codd *et al*., 1991; WHO, 1993). At the time of writing, analytical techniques are too inadequate either to determine accurately the occurrence of these toxins in routine samples of drinking water or to perform proper risk assessments.

An additional hazard is presented by *Pseudomonas aeruginosa*, an opportunistic pathogen that is both widespread as a saprophyte in the natural environment and able to survive and grow and multiply in many aquatic habitats. Most human illness caused by *Ps. aeruginosa* in water is not an enteritis caused by drinking the water, but a contact disease, for example causing wound infections (WHO, 1996).

Aeromonas is another organism that can occur naturally in the aquatic environment. In stagnant fresh waters, *A. sobria* usually predominates, and numbers of *Aeromonas* in surface waters can range up to 10^3 per ml (WHO, 1996). To detect this organism in bottled mineral waters, membrane filtration can be combined with the use of Modified XLD Medium (see Section 21.13.1). When examining mains-distributed household water samples for *Aeromonas*, it is necessary to add ethylene diamine tetra-acetic acid (EDTA)-sodium salt at a concentration of $50\,mg\,l^{-1}$ as a complexing agent because *Aeromonas* is very sensitive to traces of copper (WHO, 1996).

For the purpose of determining the potability of a water supply, it is necessary to establish that the water is not contaminated with pathogenic microorganisms such as those named. Because intestinal bacterial pathogens, if present, would be greatly outnumbered by normal intestinal commensals such as non-pathogenic *Escherichia coli*, *Enterococcus faecalis* and *Clostridium perfringens*, it is more satisfactory to examine the water for the presence of the latter index organisms. *Enterococcus faecalis* is not as efficient an indicator as *E. coli* because it is usually present in the intestine in smaller numbers than *E. coli* and dies as quickly as *E. coli* in water. Nevertheless, occasionally *Enterococcus faecalis* may be predominant, as in a few animals it may outnumber *E. coli*. *Clostridium perfringens* is capable of surviving in water for longer than *Enteroc. faecalis* and *E. coli* and, in the absence of these two organisms, serves as an indicator of remote faecal pollution. Unfortunately the absence of the normal bacterial index organisms (*E. coli*, etc.) does not ensure that *Cryptosporidium*, *Giardia* and viruses are also absent.

However, counts of *E. coli*, aerobic mesophiles and *Cl. perfringens* are still generally useful (Report, 1994). In addition, in the case of water being used by a food manufacturer, other microbiological analyses may be required; for example, the numbers and types of psychrotrophic bacteria present may be of great importance.

35.3 BOTTLED MINERAL WATERS

Although any contaminating allochthonous microflora (i.e. those not indigenous to the aquatic environment) can be expected not to multiply in this environment, a decrease in numbers during storage of still mineral waters at ambient temperatures can be very slow (i.e. reductions detectable after weeks, rather than hours or days). *Vibrio cholerae* is highly tolerant of the aquatic environment (indeed there has been much discussion as to whether *V. cholerae* should be considered as a member of the autochthonous flora rather than the allochthonous microflora). In one large outbreak of cholera in Portugal, bottled non-carbonated mineral water was implicated as one of the primary vehicles (Blake *et al.*, 1977).

Carbonation of a mineral water has a significant bactericidal effect, and allochthonous bacteria will die over a period of hours or a few days.

Baird-Parker and Kooiman (1980) and Stickler (1989) have provided good brief reviews of the microbiological aspects of these products.

35.4 SAMPLING PROCEDURE FOR WATER FROM TAPS, STAND-PIPES, ETC.

Water samples are best collected in sterile wide-mouthed bottles with dustproof ground-glass stoppers. In the case of chlorinated water samples, the sample bottles should contain 0.1 ml of a 2% solution of sterile sodium thiosulphate for each 100 ml of water sample to be collected.

Great care should be taken during the sampling procedure to prevent contamination of the sample. The sample bottle should be filled completely at the time of sampling. If the sample is to be taken from a tap, the outside and inside of the tap nozzle should be cleaned thoroughly, after which the tap is turned on for a few minutes. The tap should then be turned off and heat-sterilized using, for example, an alcohol lamp. The tap is cooled by allowing water to run to waste for a minute or two, after which the sample bottle is filled with water. It should be noted that this method of sampling ensures that the bacteriological quality of the water supply delivered to the tap is tested, but if the reason for the bacteriological examination is to trace the source of contaminating organisms it would be advisable in addition to take either an initial sample from the tap before the sterilization procedure or a swab of the inside and outside of the nozzle to determine the possibility of contamination of the tap itself. The results of a single sample of a water supply are of limited value as often contamination may be intermittent.

In collecting water samples from reservoirs, ponds, wells, rivers, etc., it is advisable to use some sterilized mechanical device for holding the bottle and removing the stopper, and, in the case of obtaining samples from moving water, the mouth of the bottle should be directed against the current. The forms of apparatus suitable for collecting deep water samples have been reviewed by Collins *et al.* (1973). National protocols for the collection of official samples in most countries specify the use of these sampling devices. For *unofficial* sampling purposes, if such a device is unobtainable and cannot be improvised, extreme care should be taken in the manual collection of a sample so that no water is collected that may have contacted the hands.

35.5 CULTURAL EXAMINATIONS

As the sample bottle will be completely full, the sample should be mixed by inverting the bottle 25 times with a rapid rotary motion, then about a quarter of the contents poured away and the bottle shaken vertically 25 times with an excursion of 30 cm. Prepare aseptically serial decimal dilutions up to 10^{-2} in sterile quarter-strength Ringer's solution.

Carry out the following counts.

35.5.1 Aerobic mesophilic and psychrotrophic counts

Use CPS medium and plate count agar, incubated at 4.5°, 25° and 37°C for 14, 5 and 2 days respectively.

35.5.2 Coliform counts by the multiple tube technique

Use five (or three) bottles or tubes at each of the dilutions 10^2, 10^1, 10^0 and 10^{-1}. Quantities of water greater than 1 ml are inoculated into *double-strength* medium equal in volume to the amount of water inoculated. If the water sample is suspected of being highly polluted, dilutions higher than 10^{-1} should be tested setting up five (or three) tubes at each dilution. The advantage of comparability of results obtained on water samples with those obtained on food samples is probably great enough for the microbiologist in the food industry to standardize on the media already described (see Section 20.1).

35.5.3 Faecal streptococci by the multiple tube technique

Use five (or three) bottles or tubes at each of the dilutions 10^2, 10^1, 10^0 and 10^{-1}. Quantities of water greater than 1 ml are inoculated into double-strength medium equal in volume to the amounts of water being inoculated. Incubate the inoculated glucose azide broths at 37°C for 72 h. Record tubes that are positive (i.e. those in which acid is produced), and confirm by subculturing a loopful from each positive tube into a fresh single-strength glucose azide broth and incubating at 45°C (see Section 20.2).

35.5.4 Detection or counting of sulphite-reducing clostridia and *Clostridium perfringens*

Since the ratio of *Cl. perfringens to E. coli* is very low in faeces, the detection of *Cl. perfringens* in a water sample is an insensitive method of examination for recent contamination compared with the presumptive coliform count. The value of tests for *Cl. perfringens* lies in the fact that the spores are able to survive in water for comparatively long periods, and therefore their presence in water free from *E. coli* can be taken as some indication of remote faecal pollution. Three methods of detection may be used:

1. *Litmus milk method for gas-producing, lactose-fermenting clostridia.* Add 200 ml of water sample to 400 ml of sterile litmus milk (heated to 100°C in a steamer to drive off dissolved air and then cooled, immediately before inoculation) in a container of about 1000-ml capacity. (A large container must be used to allow for blowing without risk of contaminating the incubators.) Heat in a water-bath to 80°C for 10 min. The timing of this heating is best established by placing in the water-bath an exactly similar container (with the same type of closure) in which is 600 ml of water and a thermometer inserted so that the bulb of the thermometer is not in contact with the container wall. The moment when the contents attain 80°C can thus be established. Cool, cover the surface of the milk with melted sterile 'vaspar', and incubate at 37°C for 5 days. Examine every 24 h for the production of acidity, clotting and gas – the reaction known as a 'stormy clot' – which indicates the presence of *Cl. perfringens*.

   This test can be made semi-quantitative by putting up a range of volumes of the water sample and estimating the most probable number, using probability tables. It should be noted, however, that the number of *Cl. perfringens* present will be very low so that normally *Cl. perfringens* would not be detectable in quantities of less than 100 ml. Therefore, the amounts to be tested should be, for example, 200, 100 and 50 ml; thus specially designed probability tables may be required.
2. *Multiple tube technique for sulphite-reducing clostridia.* Use differential reinforced clostridial medium, followed by confirmation for *Cl. perfringens* if required (see Section 21.7).
3. *Wilson and Blair's sulphite medium for sulphite-reducing clostridia.* Heat to boiling 20 ml of Wilson and Blair's sulphite medium (dispensed in a Miller–Prickett tube) and cool to 50°C. Add 20 ml of water sample (previously heated to 80°C for 10 min to destroy vegetative forms of bacteria), also at 50°C, mix well and allow to set. Incubate aerobically at 37°C for 24 h. After incubation, examine for the presence of black (i.e. sulphite-reducing) colonies. The two main disadvantages of this method are that comparatively small amounts of water are tested, and that it is difficult to remove inocula from sulphite-reducing colonies to confirm them as *Cl. perfringens*.

35.5.5 Examination for *Pseudomonas aeruginosa* (see Section 21.13.2)

Use either spread plate inoculation, or membrane filtration to detect smaller numbers (see below), in conjunction with cephaloridine–fucidin–cetrimide agar (CFCA).

35.5.6 Examination for *Aeromonas* (see Section 21.13.1)

Use either spread plate inoculation, or membrane filtration to detect smaller numbers (see below), in conjunction with modifed XLD agar (Oxoid). If the samples are from mains distribution systems (e.g. household water samples), incorporate EDTA-sodium salt at 50 mg l^{-1} to protect the organisms against any traces of copper.

35.6 MEMBRANE FILTRATION METHODS

Aerobic mesophilic, coliform and *Enterococcus faecalis* counts can be carried out by using membrane filtration. It has already been mentioned that one advantage of membrane filtration is that small numbers of organisms can be detected, because the amount of water passed through the membrane is restricted only by the amount of gross suspended matter present in the water. (See Section 5.2 and 5.3.2 for the basic methodology.)

Aerobic mesophilic counts can be carried out on a water sample by using membrane-type nutrient broth or membrane-type tryptone soya broth, and incubating the filters at 35–37°C for 18–24 h or at 25°C for 2 days. For counts of *E. coli* use Millipore type HC filters. Incubate the filters for 2 h at 37°C on pads saturated with membrane-type nutrient broth, and then transfer the filters to pads saturated with membrane-type MacConkey's broth and incubate for 18 h at 44°C (Taylor *et al.*, 1955). By this means, *E. coli* type 1 counts can be determined within 1 day. Similarly, *Enterococcus faecalis* counts should be determined with preselective incubation of the filters on nutrient broth for 2 h at 37°C, before transferring the filters to pads saturated with glucose azide broth and incubating at 45°C for 18 h.

35.7 MICROBIOLOGICAL SPECIFICATIONS

The EU (EC, 1980a) has set for tap-water maximum admissible concentrations (MACs) of total coliforms, faecal coliforms and faecal streptococci each of less than 1 per 100 ml, and of sulphite-reducing clostridia less than 1 per 20 ml. For bottled natural mineral waters and spring waters, the EU (EC 1980b, 1996) has set MACs for total coliforms, *E. coli* and faecal streptococci each of less than 1 per 250 ml, for sulphite-reducing clostridia less than 1 per 50 ml, and for *Pseudomonas aeruginosa* less than 1 per 250 ml.

35.8 SPECIAL REQUIREMENTS FOR FOOD MANUFACTURE

It has already been mentioned that the food manufacturer may have more rigorous microbiological requirements of a water supply than that it is potable. For example, the presence of saprophytic psychrotrophs and other aquatic organisms has significance for spoilage of refrigerated or other foods. To achieve sufficiently low counts, additional chlorination at the factory may be necessary, for example in the case of cooling water in canning factories (water chlorination in canneries is frequently obligatory because a potable water supply is often not used for cooling purposes).

Another type of microorganism that may occasionally be of significance is the *Sphaerotilus–Leptothrix–Gallionella* group of sheathed and appendaged organisms.

Some of these, in addition to some strains of *Pseudomonas* and *Acinetobacter*, may cause trouble by blocking water distribution systems within the factory, and some may even cause spoilage troubles such as haze formation and the development of sliminess in fruit-flavoured liquid products of low pH, such as those designed for domestic production of 'ice lollies'. For isolation and identification of these organisms refer to Bergey's Manuals (Staley *et al.*, 1989; Holt *et al.,* 1994), Collins (1964) and Mulder (1964).

36

Examination of Food Processing Plant

36.1 INTRODUCTION

The very first stage in ensuring good, hygienic manufacture and processing of food products lies with good sanitary (hygienic) design and construction of premises and proper layout of equipment, to maximize the efficacy of cleaning and disinfection regimes. These aspects are outside the scope of this book, but readers are referred to Shapton and Shapton (1991).

A preliminary inspection of the premises, preferably during processing, is an invaluable aid in assessing the microbiological significance of general organizational procedures, particular patterns of layout and the processing methods used. An opportunity should also be taken at this stage to note the general condition of the equipment and the presence of any residues, film or scale. These observations should be recorded so that they may be correlated later with the results of any subsequent tests.

Information on the microbiological condition of the equipment can also be obtained by comparing the results on samples (taken at suitable points) of the first product passing over, and in effect rinsing, the equipment since the last cleaning and sterilizing process. This procedure is applicable to food-processing equipment carrying liquids (e.g. pipelines carrying milk, soups, sugar syrups, etc.). The samples obtained should be examined for aerobic mesophilic count and coliform count, and the presence of other organisms should be tested for as appropriate (see the relevant section).

When detailed information is required on a particular utensil or reasonably small piece of equipment, the rinse method is suitable, for example in the examination of churns and cans, bottles and cartons. It may also be used for the examination of pipeline installations by drawing an appropriate volume of rinse through the system.

For the examination of defined areas of large pieces of equipment (e.g. vats) or for assessing the microbiological condition of equipment or parts of equipment where the rinse method is not applicable (e.g. conveyor belts and cutting blocks), the swab method is suitable. The principal methods of examining surfaces have been described in Section 16.

In all cases where chemical sterilizing agents are known to have been used, the quarter-strength Ringer's solution of the rinse or swab should contain a supplement of an appropriate inactivator. For hypochlorites and iodophors, sodium thiosulphate is used and incorporated in the rinse at 0.05% final concentration. For quaternary ammonium compounds a mixture of 4% lecithin and 6% Cirrasol ALN-WF in water is incorporated to give a final concentration of 1% of mixture (Cirrasol ALN-WF is obtainable from ICI).

The methods for determining the numbers and kinds of microorganisms collected in the rinse and swab diluents usually involve a colony count on a non-selective medium and tests for the presence of coliforms and other groups using appropriate selective media. It is also possible to determine numbers of thermoduric organisms by laboratory pasteurization of the diluent, but in this case it is advisable to add a supplement of sterile separated milk before heat treatment (e.g. 5 ml of sterile milk to 5 ml of diluent).

An alternative method of examining the rinse and swab diluents is to use membrane filtration (see Section 5.2). This has the advantage that larger volumes of diluent can be examined than by traditional methods.

Because the proportion of microorganisms recovered from surfaces is greatly influenced by the extent of the rinsing and swabbing process, it is necessary that procedures should be standardized carefully if comparative results are to be obtained.

36.2 EXAMINATION OF PROCESSING PLANT, EQUIPMENT, WORKING SURFACES, ETC.

36.2.1 By swabs

In testing any piece of equipment, more than one area should be swabbed, paying particular attention to points that are difficult to access for cleaning (e.g. valves and junctions of pipelines that have been cleaned in place).

36.2.1.1 Preparation of swab

Cotton-wool or alginate-wool swabs may be used (see Section 16.3.3), or alternatively larger swabs prepared from unmedicated ribbon gauze. These are prepared by winding a 15-cm length of 5-cm wide ribbon gauze on to a 30-cm long stainless steel wire support. The wire support is made from 35 cm of stainless steel wire of about 3 mm in diameter, formed into a loop at one end and notched at the other to hold the gauze without slipping. Cotton-wool or gauze swabs should be placed in alloy or stainless steel tubes containing a known volume of quarter-strength Ringer's solution (containing an inactivator if a chemical sterilant has been used on the equipment).

36.2.1.2 Swabbing

Whenever possible, a minimum of 100 cm^2 should be swabbed. Before swabbing, press the swab against the side of its container to express excess liquid. Swab the surface as described in Section 16.3.3, that is by rubbing firmly over the surface using parallel strokes with slow rotation of the swab, and swabbing a second time using parallel strokes at right angles to the first set. Return the swab to the tube.

36.2.1.3 Examination of swabs

This should be completed as soon as possible, and in any case within 6 h of sampling. Examine alginate swabs by using Calgon Ringer's solution as described on p. 150. In the case of cotton-wool or gauze swabs, after not less than 5-min contact between the swab and the Ringer's solution, mix by twirling the swab vigorously in the Ringer's solution six times. After thorough mixing, prepare any dilutions thought to be necessary. Set up aerobic mesophilic counts on plate count agar (or any other medium so specified by a standard procedure being followed) and incubate at 30°C for 3 days. Examine also for the presence of coliforms and any other groups of significance; for example, in the case of equipment carrying hot gelatin stock for meat pies, examine for the presence of thermophilic spore-bearers.

36.2.1.4 Interpretation of results (see Table 36.1)

Table 36.1

Aerobic mesophilic count per cm^2	Classification
Not more than 5	Satisfactory
5–25	Requires further investigation
More than 25	Highly unsatisfactory; requires immediate action

Coliform counts. Equipment used for carrying, dispensing or holding heat-treated foods should bear fewer than 10 coliform bacteria per 100 cm^2. A result of 'no coliform bacteria found on 100 cm^2' can be regarded as satisfactory.

36.2.2 By rinses

Reasonably small pieces of equipment (e.g. buckets) can be examined using 500 ml of quarter-strength Ringer's solution. When a chemical sterilant has been used, the rinse liquid should contain the correct inactivator (see p. 307). When cleaned-in-place pipeline systems are examined by a rinse method, much larger quantities of sterile rinse liquid are required, which are pumped round the system and then aliquot quantities taken for examination.

In farm dairies, milking machines can be examined by either static or pulsating rinse procedures. A static rinse is carried out by passing the rinse solution twice through the teat cups and long milk tube into the original container. A pulsating rinse is carried out in a similar way but with the machine pump operating so as to pulsate the teat cup liners.

The advantage of this latter method is that it is more closely related to actual milking conditions.

Rinses should be tested as soon as possible and in any case within 6 h of sampling. Transport temperatures and conditions should be regulated and recorded.

Examine rinses for general viable counts, coliform counts and any other appropriate selective counts. Counts on rinses of small pieces of equipment should be calculated per utensil. Standards to be set will obviously depend on the individual situation, type of equipment, etc., and it is recommended that standards should be determined by prior survey of counts attainable under model cleaning conditions.

36.2.3 By impression plates

See pp. 148, 151 for methods of use. These techniques have a relatively limited range of uses in the examination of equipment as they can be used only on flat or nearly flat surfaces (e.g. conveyor belts, inside surfaces of rectangular vats, etc.). They have the virtue of providing a record of the position of viable organisms relative to any observed surface defects, corrosion, etc. A further advantage is that, in surveys of processing plant remote from the laboratory, transport and storage problems are obviated.

36.3 EXAMINATION OF WASHED BOTTLES AND FOOD CONTAINERS

Bottles should be selected for examination immediately after washing, closed with a sterile rubber bung and examined as soon as possible, and in any case within 6 h. Other food containers should be closed or covered to prevent contamination; they may be examined by a procedure similar to that for bottles described below.

36.3.1 Examination by rinses

Add 20 ml of sterile quarter-strength Ringer's solution containing 0.05% sodium thiosulphate to the bottle and replace the bung. Hold the bottle horizontally in the hands and rotate gently 12 times in one direction so that the whole of the internal surface is wetted thoroughly. Allow the bottle to stand for not less than 15 min and not more than 30 min, and again gently rotate 12 times so as to wet the whole of the internal surface.

Prepare aerobic mesophilic counts using an appropriate medium; for example, in the case of milk bottles use plate count agar or yeast extract milk agar, in the case of bottles for fruit squashes or juices use orange serum agar, and so on. Also perform a coliform count by the multiple tube technique. Incubate plates at 25–30°C for 3 days. Record results as the colony count or coliform count per bottle.

36.3.2 *In situ* culture of contaminants

An alternative method of examining washed clear glass bottles for the presence of contaminants is to add molten agar medium to the bottle, in sufficient volume to form a thin (5-mm) layer over the inside surface. The bottle is rolled slowly horizontally until the agar sets on the inside surface (this can be hastened by rolling the bottle in a shallow dish of iced water). The bottle should be incubated, bung down, in the same way as a roll tube. *In situ* culture is also very successful for demonstrating the presence of microorganisms on cracked and chipped crockery, etc. (see also Angelotti and Foter, 1958; Favero *et al.*, 1968).

36.3.3 Examination of large containers

Larger food containers, such as churns and cans used for transport and storage of milk, cream, sugar syrups, fruit syrups, etc., may be examined by a rinse method employing 500 ml of sterile quarter-strength Ringer's solution.

36.3.3.1 Selection and visual examination

Cans and churns should be examined not less than half an hour and not more than 1 h after washing. First a visual inspection should be made and the general condition recorded, in particular: (a) bad dents, rusting, open seams or poor lids; (b) presence or absence of film, scale or food or milk solids; and (c) degree of wetness of the churn – recorded as dry, moist or wet (with an obvious pool of water at the bottom of the churn or can).

Cans and churns containing turbid water or easily removable food or milk solids, as distinct from film or scale, can be reported as unsatisfactory without testing.

36.3.3.2 Rinsing

Pour 500 ml of sterile quarter-strength Ringer's solution over the inside of the lid into the can. Replace the lid, lay the can on its side, and roll it to and fro through 12 complete revolutions. Allow the can to stand for 5 min, and repeat the rolling. Pour the rinse solution from the can into the lid and then into the original sterile container.

36.3.3.3 Testing the rinses

Rinses should be tested as soon as possible, and in any case within 6 h of sampling. Mix by inverting the container slowly three times. Pour 1- and 0.1-ml plates using plate count agar, yeast extract milk agar, or other appropriate medium, and incubate at 30°C for 3 days. Test also for the presence of coliforms by a multiple tube technique, using tubes at

10^1, 10^0 and 10^{-1} dilutions (double-strength media are used for inocula exceeding 1 ml). Yeast and mould counts should also be performed on rinses of churns and cans used for sugar syrups, fruit syrups, etc. Record the results per can.

36.3.4 Interpretation of results (see Table 36.2)

Table 36.2

Viable count per container	Classification
Bottles and small containers	
Not more than 200	Satisfactory
200–1000	Cleaning procedure needs improvement
Over 1000	Unsatisfactory – take immediate action
Churns, cans and other large containers	
Not more 10 000	Satisfactory
10 000–100 000	Cleaning procedure needs improvement
More than 100 000	Unsatisfactory – take immediate action

Note. A wet container should be degraded to the next class below as it is unlikely to remain in a satisfactory condition.

36.4 AIR

Occasionally it may be useful to examine the air in food factories, dairies, etc. for the presence of specific organisms. This is particularly so in the case of mould contamination problems, as fungal spores are readily distributed through the air. If the source of a known and identified contaminant is being sought, it may be sufficient to use simple air exposure plates. Plates of already poured and set media are exposed in a variety of locations for 15 min. The number of colonies developing on one plate following a 15-min exposure period represents the number of particles carrying that type of organism settling on approximately 0.1 m^2 per min. The variation in counts of the organisms being traced is likely to indicate the focal points of the contamination. A second visit to the factory may then be sufficient to locate the source and to suggest possible remedial measures.

It is not often necessary in this type of work to require the use of a slit sampler or similar sampling device, although they are obviously needed to count absolute concentrations of microorganisms in the air. Membrane filtration may also be used; air sampling apparatus is available from the manufacturers of membrane filters.

A further environmental sampling procedure that can be used conveniently following the air sampling is the sweep plate technique as described by Cruickshank *et al*. (1975). Plates of culture media similar to those used for air sampling are exposed face down on the test surface (e.g. work surface) and are rubbed to and fro ten times over a distance of 30 cm. The lid is then replaced and the plate incubated with the air exposure plates. Although sweep plates obviously yield essentially qualitative information, the technique is particularly suitable for demonstrating reservoirs of organisms on infrequently cleaned surfaces and in demonstrating improvements following a satisfactory cleaning routine.

36.5 DETECTION OF *SALMONELLA* IN PROCESSING PLANT EFFLUENTS

In investigations of food processing plant or premises where a particular hazard of *Salmonella* contamination exists (e.g. poultry or meat processing plants), suitable samples can be obtained by the use of sewer swabs installed at convenient points in drains and gulleys (Harvey and Phillips, 1961; Georgala and Boothroyd, 1969; Harvey and Price, 1970, 1974). A 2-m length of string is attached to the swab (which may be folded gauze or a sanitary towel) and the whole wrapped in paper and sterilized in the autoclave. On site, the swab is suspended in the chosen drain or outlet, secured with the string and left in position for 1–7 days. Alternatively the swab may be used to wipe the drain or gulley surface and immediately removed to the laboratory. Extending the exposure time may increase the possibility of picking up salmonellae, but it will also increase the extent of contamination by other organisms, and in general the longer the exposure time, the longer the incubation time required in the liquid enrichment stage of isolation (and in the case of heavily soiled swabs a two-stage liquid enrichment may be required before plating).

After transporting the swab to the laboratory, liquid enrichment cultures are set up using selenite broth, etc. In the method of Patterson (1969) the swabs are transferred to a sterile plastic bag, a corner of which is cut off and the absorbed fluid in the swab then expressed and collected in a sterile container. (Sterile sodium thiosulphate solution can be added to neutralize any residual chlorine.) Ten millilitres of the fluid are added to each of 90 ml of selenite broth, tetrathionate broth and Hajna's GN broth (see Section 21.3). An alternative procedure recommended by Harvey and Price (1974) is to place the entire swab in a wide-mouthed jar (or to leave it in the jar that is used for transport if this is suitable) and to add 750 ml of selenite broth. Although this does not allow the use of more than one enrichment medium per swab, Harvey (1965) considered that culture from the entire swab increased the likelihood of isolating *Salmonella*.

The liquid enrichment cultures should be incubated and then used for isolation and identification procedures as already described (see Section 21.3).

The use of sewer swabs in this way enables premises to be screened for the possibility of the existence of a *Salmonella* problem more quickly and more easily than by the examination of many end-of-line samples. In the event of salmonellae being found, a return visit can be paid and detailed samples taken to attempt to detect the source of the contamination.

PART IV

Schemes for the Identification of Microorganisms

37

Introduction

In the schemes that follow, methods are given that should prove successful for the identification of the microorganisms most commonly isolated by, or of importance to, microbiologists in the food industry. However, if difficulty occurs in identification, readers are advised to refer to Holt *et al.* (1994) for bacterial identification, to Samson and van Reenen-Hoekstra (1988) and Pitt and Hocking (1997) for identification of moulds, and to Barnett and co-workers (1990) and Kreger-van Rij (1984) for identification of yeasts.

The diagnostic keys are separated into: Section 38 – a scheme for the identification of Gram-negative bacteria; Section 39 – a scheme for the identification of Gram-positive bacteria; and Section 40 – a scheme for the identification of yeasts and moulds. Certain fastidious and/or anaerobic animal pathogens have not been considered; for further details of such organisms see *Bergey's Manual of Systematic Bacteriology, Vol. 1 (BMSB1)* (Krieg *et al.*, 1984), *Bergey's Manual of Systematic Bacteriology, Vol. 2* (*BMSB2*) (Sneath *et al.*, 1986) and *Bergey's Manual of Determinative Bacteriology* (*BMDB*) (Holt *et al.*, 1994).

For the sake of brevity, the complete taxonomic descriptors (Latin binomial followed by authority of the name or combination of names) have not been given, and instead, as throughout this book, only the Latin binomial is used. However, the generic and specific names conform to current practice (see Euzeby, 1997), and the full taxonomic status can be found by accessing the 'List of bacterial names with standing in nomenclature' on the Internet at URL/:ftp://ftp.cict.fr/pub/bacterio/ (see Euzeby, 1997).

The methods for identification in these schemes have been put forward on the assumption that the use of selective and diagnostic media has not already enabled a tentative identification; in these cases, fewer tests may be required. For example, the screening with Kohn's two-tube media for presumptive *Salmonella* obtained on media such as desoxycholate citrate agar has already been described. Sequential identification methods have been used whenever possible, because they employ a minimum of tests. However, as already appreciated by all who have attempted to use a sequential key for 'identifying' a flowering plant whose identity is already known, and as reiterated by Sneath (1974), such sequential keys are very sensitive to error. Therefore, when the identity has been determined, the general description of the genus or species should be checked against the observed characteristics of the isolate. Some of the organisms described in Section 39 require special media for their isolation, and in such cases it is obvious that choice of primary isolation media may allow a restriction in the identification procedures to those employed for the group likely to be isolated, but the possible hazards of wrong identification should be fully realized.

Difficulties in identification may also be caused by the Gram-staining reaction; for example, the 'Gram-negative' *Acinetobacter* may retain the crystal violet–iodine complex and appear purple, whereas old cultures of *Bacillus* or young cultures of *Arthrobacter* may be Gram variable or Gram negative. In addition, many biochemical tests used for classification and identification show poor reproducibility within a laboratory or between different laboratories (see, for example, Sneath and Collins, 1974). Statistical analyses of interlaboratory tests, such as those organized by the International Commission on Microbiological Specifications for Foods (ICMSF) or the ISO, lead to continuing improvement of identification methods, by selection and use of the most reliable test methods both for classification and for identification.

The symbols used in the diagnostic tables (unless otherwise indicated) are as follows:

+	90% or more of strains give a positive reaction (usually within 1 or 2 days, unless otherwise stated)
–	90% of strains show no reaction or no growth
D	The reaction varies within the genus, depending on species, or within the family, depending on genus, as appropriate
V	Variable in reaction, 11–89% of strains are positive
V(+)	Variable, most (around 50–89%) are positive
V(–)	Variable, most (around 50–89%) are negative
(+)	Slowly developing positive
×	Late and irregularly positive
±	Slightly or weakly positive
.	Test not applicable or significant, or reaction not fully determined

38

A scheme for the identification of Gram-negative bacteria

As there are relatively few differences in cell morphology amongst the Gram-negative bacteria capable of being discerned with the optical microscope, the identification scheme that follows is based primarily on biochemical characteristics:

(a) Bacteria capable of growth on nutrient agar .... (Section 38.1)
(b) Bacteria isolated on glucose yeast extract agar, wort agar, malt extract agar or similar media, and giving poor or no growth on nutrient agar.. (Section 38.2)
(c) Bacteria isolated on media containing 12–20% sodium chloride, and incapable of growing on nutrient agar..................................... (Section 38.3)

38.1 GRAM-NEGATIVE BACTERIA THAT CAN GROW ON NUTRIENT AGAR OR PLATE COUNT AGAR

(after Krieg *et al*., 1984; Holt *et al*., 1994)

The organism to be identified should be examined for its ability to produce a pigment when grown on nutrient agar. It should also be tested to determine its mode of utilization of glucose. When these two characters have been determined, appropriate tests can be selected to complete the identification of the organism to generic level. In the case of tests not already described in the Manual, methods are given at the end of this section in the order in which they occur in the diagnostic scheme.

Production of pigment on nutrient agar

Streak the test organism on a nutrient agar plate. Incubate for 3 days at 25–30°C and examine for evidence of pigment production.

Mode of utilization of glucose

Inoculate the test organism, by stabbing with a straight wire, into Hugh and Leifson's medium with and without glucose (see Section 11.2.3). In some instances the modified medium described by Board and Holding (1960) may be more suitable. The original method of the Hugh and Leifson test requires two tubes of media plus glucose to be inoculated and one to be sealed with sterile liquid paraffin or vaspar, but the latter tube may not be necessary provided the tube is examined frequently (e.g. daily for 7 days), especially if the double-indicator version of the medium is used (see p. 424). The result obtained will be one of the following:

(a) Oxidation of the glucose, resulting in acid production at the surface of the medium, although, on prolonged incubation, the acid reaction may later spread downwards through the tube.
(b) Fermentation of the glucose, resulting in acid production uniformly throughout the medium along the entire length of the stab, with or without gas production. The reaction is first seen in the medium immediately surrounding the stab, and spreads rapidly throughout the tube. On prolonged incubation pH reversal may occur.
(c) Inability to utilize glucose with either no change in the colour of the medium or the production of alkali at the surface of, or throughout, the tube.

38.1.1 Interpretation of results of the two tests and further tests required

38.1.1.1 Division 1: Bacteria producing an intracellular (non-diffusible) purple pigment

Chromobacterium, *Iodobacter* and *Janthinobacterium* (see Table 38.1) produce a pigment of this type; tryptophan is required for pigment production. These organisms are found in soil and natural waters, and can therefore also be found as contaminants on, for example, raw vegetables. The violet pigment, violacein, is soluble in ethanol but insoluble in water; it becomes green in a 10% ethanolic solution of sulphuric acid. These genera are catalase positive, indole negative and Voges–Proskauer negative, and can reduce nitrate.

See also Logan and Moss (1992).

TABLE 38.1 Differentiation of bacteria in Division 1

	Chromobacterium	*Iodobacter*	*Janthinobacterium*
Growth at 4°C	−	+	+
Growth at 37°C	+	−	−
Growth on nutrient agar incubated anaerobically	+	+	−
Acid from L-arabinose	−	−	+

38.1.1.2 Division 2: Bacteria producing an intracellular (non-diffusible) yellow or orange pigment

Genera include *Cytophaga*, *Xanthomonas*, *Flexibacter*, *Flavobacterium*, *Flavomonas* and *Chryseomonas*. Some *Pseudomonas* species and some members of the Enterobacteriaceae may also produce yellow pigments. *Cytophaga* and *Flexibacter* consist of long, thin, often curved, rods. Organisms from actively growing cultures may show flexing movement when examined in hanging drop preparations. On nutrient agar, many strains form spreading or swarming colonies due to gliding motility (see Weibull, 1960; Perry, 1973). Some strains show motility better on a less rich medium such as Hayes' medium. Most strains of *Cytophaga* hydrolyse starch, which is a characteristic relatively uncommon in the other soil bacteria that produce orange and yellow pigments. Many strains of *Xanthomonas* and some strains of *Flavobacterium* may also hydrolyse starch, however. *Flavobacterium*, *Xanthomonas* and the yellow- or orange-pigmented enterobacteria will not show marked spreading on nutrient agar or Hayes' medium, nor display gliding motility.

Differentiation of this diverse range of yellow-pigmented bacteria is best achieved using API20NE kits (bioMérieux). If it is found that the isolate is a member of the Enterobacteriaceae, an API20E kit can then be used.

Certain other intracellularly yellow-pigmented Gram-negative bacteria, e.g. the methylotrophs *Blastobacter* and *Xanthobacter*, and *Alteromonas luteoviolacea* (the latter having a complex nutritional requirement for both components of sea-water and various organic growth factors) are unlikely to be isolated in routine food microbiology.

(See also McMeekin *et al*., 1971, 1972; McMeekin and Shewan, 1978.)

38.1.1.3 Division 3: Bacteria producing an intracellular (non-diffusible) red or pink pigment

Two types of Gram-negative bacteria which can grow on nutrient agar produce a pink or red intracellular pigment. They are *Methylobacterium*, which is strictly aerobic and does not ferment glucose, and *Serratia* species, which are facultatively anaerobic and do ferment glucose. These two genera are therefore differentiated with respect to the type of reaction in Hugh and Leifson's medium. In addition, flagella staining can be used as *Pseudomonas* possesses polar flagella, whereas *Serratia* possesses peritrichous flagella. (The red-pigmented *Alteromonas denitrificans* and *A. rubra* have a complex nutritional requirement for both components of sea-water and various organic growth factors, so they are unlikely to be isolated in routine food microbiology.)

38.1.1.4 Division 4: Non-pigmented fermentative bacteria

Bacteria of this type can be differentiated using the following tests (see Table 38.2).

For further differentiation of the Enterobacteriaceae, see Section 38.1.2.

TABLE 38.2 Differentiation of non-pigmented fermentative bacteria

	Group or genus				
	*Vibrio**	*Aeromonas*	*Plesiomonas*	*Photobacterium**	Enterobacteriaceae
0/129 sensitivity	V(+)	−	V(+)	+	−
Curved, S-shaped and spiral cells. Many spherical cells in 7-day cultures	V(+)	−	−	−	−
Oxidase test	+	+	+	V	−
Motility	D	V	+	+	D
Polar flagella	D	V	+	+	−
Thornley's arginine test	D	+	V	−	−
Gelatin hydrolysis	D	+	−	D	V
Starch hydrolysis	D	+	−	−	V
Luminescence	D	−	−	+	−

* Sodium ions are required by *Photobacterium* and most *Vibrio* species, and stimulate growth of all *Vibrio* spp. However, most do not require high concentrations of sodium, and can be isolated on plates incorporating low dilutions (high concentrations) of samples of marine origin, or on a medium that contains a sodium salt (e.g. nutrient agar).

Morphology

Most *Vibrio* cultures at 18–24 h on nutrient agar or in nutrient broth consist of slender organisms which are curved, S shaped or spiral. After 7 days these cultures contain many coccal forms. For photographs illustrating the morphology of *Vibrio* see Krieg *et al.* (1984), and Baker and Park (1975). Neither *Aeromonas* nor enterobacteria display this characteristic morphology, although some of the organisms in a culture may be slightly curved.

Thornley's arginine (Section 11.1.5) test gives positive results with all *Aeromonas* strains. Some *Plesiomonas* strains are positive (Eddy and Carpenter, 1964) but positive results are very rare in other fermentative Gram-negative bacteria.

Non-motile strains that are oxidase positive are identified as *Aeromonas* or *Vibrio*, whereas those that are oxidase negative are identified as enterobacteria.

Additional tests

The decarboxylase tests of Møller (1955) can be used for further confirmation of a culture as *Vibrio*, *Aeromonas* or *Plesiomonas* (see Table 38.3).

TABLE 38.3 Differentiation of *Vibrio*, *Aeromonas* and *Plesiomonas*

	Lysine decarboxylase	Ornithine decarboxylase	Arginine decarboxylase
Vibrio	+	+	D
Aeromonas	V	−	+
Plesiomonas	+	+	+

For differentiation of the foodborne pathogenic species of *Vibrio*, see Section 21.6. For differentiation of the foodborne pathogenic species of *Aeromonas*, see Section 21.13.1.

38.1.1.5 Division 5: Bacteria producing a yellow, green-yellow or green pigment which diffuses into the medium, and non-pigmented bacteria that do not ferment glucose (see Table 38.4)

Production of yellow, green-yellow or green pigments that diffuse into the medium is a property of many, but not all, *Pseudomonas* spp.

Most *Agrobacterium* are plant pathogens, causing crown gall, cane gall or hairy root disease. The genus *Acinetobacter* consists of coccobacilli, some of which may retain the crystal violet–iodine complex of the Gram strain, and so may resemble Gram-positive cocci except that division occurs in only one plane so that groups of organisms do not occur, and occasional short rods can be seen in the cultures. In early published work, these organisms were sometimes called *Achromobacter*. *Acinetobacter* and *Alcaligenes* occur naturally in soil and fresh waters.

Species identification within *Pseudomonas* can be achieved by tests for utilization of single sources of carbon, and both *BMSB1* and *BMDB* (Krieg *et al*., 1984; Holt *et al*., 1994) tabulate substrate utilization patterns for the species of *Pseudomonas*. Rosenthal (1974) has described a simple disc assay method for performing these tests.

TABLE 38.4 Differentiation within Division 5

	Group or genus				
	Agrobacterium	*Pseudomonas*	*Acinetobacter*	*Moraxella*[†]	*Alcaligenes*
Glucose oxidized	+	V	+	–	–
3-Ketolactose produced	D				
Inorganic N can be used as sole N source	+	V(+)	V(+)	V(–)	V(+)
Coccal morphology	–	–	+	+/–[†]	V
Tendency to retain Gram's stain	–	–	±	±	–
Thornley's arginine test	–	V(+)*	V	.	V(+)
Fluorescent diffusible pigment	–	V	–	–	–
Motility in a hanging drop	+	+	–	–	+
Polar flagellation	–	+	–	–	–[‡]
Penicillin sensitivity	–	–	–	+	–
Oxidase test	+	V(+)	–	+	+

*, Plant pathogenic pseudomonads are often arginine negative, as also are *Alteromonas*.
[†], The rods are considered to be in the subgenus *Moraxella* and the cocci in the subgenus *Branhamella*.
[‡], *Alcaligenes* is degenerately peritrichous, with only one to eight flagella, and will thus often appear to be polarly flagellate.

38.1.2 Differentiation of members of the Enterobacteriaceae

The members of this family are catalase-positive, oxidase-negative, rod-shaped bacteria, motile with peritrichous flagella or non-motile. Facultatively anaerobic, they ferment glucose rapidly with or without gas production. With the exception of certain strains of *Erwinia* and *Yersinia*, they reduce nitrate to nitrite. Brenner has provided a comprehensive discussion of this family in *BMSB1*.

Identification can be achieved using the API20E kit (bioMérieux) in conjunction with the API database for PCs.

In *BMSB1* (Krieg *et al*., 1984), 14 principal genera were listed in the family, with a further six less well known genera; in *BMDB* (Holt *et al*., 1994), 30 genera were listed, but some of these have been isolated very rarely. The principal genera likely to be encountered in food microbiology are:

Citrobacter	an environmental contaminant in water and food; occurs in the intestine and faeces; an opportunistic pathogen; verocytotoxigenic *C. freundii* have been reported as causing foodborne gastroenteritis and haemolytic uraemic syndrome (Tschäpe *et al*., 1995)
Enterobacter	widely distributed in water, soil, on plant surfaces; occurs in the intestine and faeces; an opportunistic pathogen
Erwinia	plant pathogen
Escherichia	intestinal commensal in homoiothermic animals; also an important intestinal pathogen and an opportunistic pathogen in other sites
Hafnia	occurs in soil, water and animal faeces; has been found in dairy products; an opportunistic pathogen
Klebsiella	occurs in soil, water and human faeces, and found on fruits and vegetables; some species are important opportunistic pathogens
Pantoea	plant pathogen
Proteus	found in soil, polluted waters and animal intestines; an opportunistic pathogen
Rahnella	an aquatic organism, but has been reported in milk processing environments (Cox *et al*., 1988)
Salmonella	an important pathogen of homoiothermic animals (including human beings), and of poikilothermic animals
Serratia	occurs in soil, water and on plant surfaces; an important opportunistic pathogen for human beings and other animals (including being a cause of bovine mastitis)
Shigella	intestinal pathogen of human beings and other primates
Yersinia	occurs widely in soil, water, animals and foods; variously pathogenic (*Y. enterocolitica* causes human enteritis, *Y. pestis* causes plague and *Y. ruckeri* causes disease in fish)

The diagnostic key that follows is based on the descriptions given in *BMSB1* and *BMDB* (Krieg *et al.*, 1984; Holt *et al.*, 1994). It is designed to enable further characterization of members of the Enterobacteriaceae likely to be encountered by food microbiologists which have been identified as such within Divisions 2 and 4. Organisms whose isolation appears to be restricted to association with opportunistic nosocomial infections have been omitted. Pink or red pigmented strains of *Serratia* have already been distinguished within Division 3. Rapid tests for the characterization of *Salmonella* and *Shigella* isolated on selective media have been described in Section 21.3, but such tests are not adequate for Gram-negative rod-shaped isolates obtained from media not selective for *Salmonella* or *Shigella*.

The tests in this section for differentiating the Enterobacteriaceae – with the exception of the test for the liquefaction of gelatin which is carried out at 20°C – should be carried out at an incubation temperature of 35–37°C unless otherwise indicated.

1. Isolated from diseased plant tissue .... Division 6
   Not isolated from diseased plant tissue *2*
2. Phenylalanine deaminase positive .... Division 7
   Phenylalanine deaminase negative... *3*
3. Growth in Simmon's citrate agar at 37°C, 22°C or both temperatures.............. *4*
   No growth in Simmon's citrate agar .... Division 8
4. Methyl red positive.. Division 9
   Methyl red negative..Division 10

38.1.2.1 Division 6: Plant pathogenic bacteria

This division comprises the genus *Erwinia* (which now includes the organisms of the previously recognized *Pectobacterium*) and members of the genus *Pantoea* (which includes certain organisms previously included in *Erwinia* and *Enterobacter*). They are distinguished from certain other members of the Enterobacteriaceae primarily on the basis of plant pathogenicity. The bacteria are motile; they may or may not reduce nitrate to nitrite; may or may not liquefy gelatin; and normally do not require organic nitrogen for growth. They frequently possess a cream, yellow or orange intracellular pigment. Some cause dry necroses, galls or wilts; others produce a pectinase and cause soft rots. For species differentiation see Holt *et al.* (1994), and for an explanation of current taxonomy refer to Euzeby (1997).

38.1.2.2 Division 7: Bacteria positive to the phenylalanine deaminase test (see Table 38.5)

Organisms in this division typically are also motile when grown at 20°C (but may be nonmotile when grown at 37°C), methyl red positive, and ferment neither lactose nor dulcitol (*Rahnella aquatilis* is non-motile and ferments lactose and dulcitol).

TABLE 38.5 Differentiation of phenylalanine deaminase-positive Enterobacteriaceae

	Organism					
	Proteus vulgaris	*Pr. mirabilis*	*Morganella morganii*	*Providencia* spp.	*Pantoea agglomerans*	*Rahnella aquatilis*
Acid from D-xylose	+	+	–	–	+	+
Acid from D-mannose	–	–	+	+	+	+
Acid from maltose	+	–	–	V(–)	+	+
Acid from melibiose	–	–	–	–	–	+
Indole production	+	–	+	D	–	–
H_2S production	+	+	–	–	–	–
Gelatin hydrolysis	+	+	–	–	–	–
Urease production	+	+	+	D	–	–
Ornithine decarboxylase	–	+	+	–	V	–
Lipase production	V(+)	+	–	–	–	–
DNase production	V(+)	V(+)	–	–	–	–

Proteus is capable of swarming to produce concentric zones on the surface of agar media. Swarming can be inhibited in a variety of ways (see, for example, Smith, 1975).

38.1.2.3 Division 8: Bacteria that are phenylalanine deaminase negative and Simmon's citrate agar negative (at 37°C) (see Table 38.6)

Apart from *Klebsiella*, whose reactions vary in these tests, organisms in this division are also methyl red positive; they do not ferment adonitol; do not produce urease; and most cannot liquefy gelatin or utilize malonate.

Cultures suspected of being *Salmonella* are best confirmed by serological methods using agglutination tests.

The species of *Yersinia* encountered in food microbiology can be differentiated as shown in Table 38.7. Note that all tests should be incubated at 25–28°C (Kapperud and Bergan, 1984).

38.1.2.4 Division 9: Bacteria that are phenylalanine deaminase negative, Simmon's citrate positive and methyl red positive (see Table 38.8)

Cultures suspected of being *Salmonella* are best confirmed by serological methods using agglutination tests (see Sections 13.2 and 21.3).

The three species of *Citrobacter* can be differentiated as shown in Table 38.9.

38.1.2.5 Division 10: Bacteria that are phenylalanine deaminase negative, Simmon's citrate positive and methyl red negative (see Table 38.10)

These organisms do not produce hydrogen sulphide. They are usually: indole negative; Voges–Proskauer positive; they usually ferment sucrose and mannitol; all possess β-galactosidase; and all can utilize gluconate.

TABLE 38.6 Differentiation of bacteria in Division 8

	Organism						
	Shigella	*Escherichia*	*Salmonella*[5]	*Edwardsiella*	*Klebsiella*	*Hafnia*	*Yersinia*
Gas from glucose	–[1]	+[4]	V[6]	+	V	+	V(–)
Acid from lactose	–[2]	+/×[4]	–	–	V	–	V(–)
Acid from salicin	–	V(+)	–	–	+	V(–)	D
Acid from myo-inositol	–	–	V(–)	–	+	–	V(–)
Acid from sucrose	V(–)	V	–[7]	D	+	–	D
H_2S production	–	–	V(+)[8]	+	–	–	–
Motility	–	+[4]	+	+	–	V(+)	–
Voges–Proskauer test	–	–	–	–	V(+)	V(+)	–
Lysine decarboxylase	–	+	+[9]	+	V(+)	+	–
Ornithine decarboxylase	–[3]	V	+[10]	V(+)	–	V(+)	+

[1] Some strains of *Shigella flexneri* type 6 produce small amounts of gas.
[2] *Shig. sonnei* ferments lactose very slowly.
[3] *Shig. sonnei* is positive.
[4] Certain strains of *Escherichia*, once known as the Alkalescens–Dispar group, do not produce gas from glucose, ferment lactose slowly or not at all, and are non-motile.
[5] *S. typhi*, some strains of *S. paratyphi* and a few other salmonellae.
[6] *S. typhi* does not produce gas.
[7] Sucrose-fermenting salmonellae have been described, but are probably rare.
[8] *S. paratyphi* and certain rare types do not produce hydrogen sulphide.
[9] *S. paratyphi* A is negative.
[10] *S. typhi* is negative.

TABLE 38.7 Differentiation of *Yersinia*

	Species						
	Y. aldovae	*Y. bercovieri*	*Y. enterocolitica*	*Y. frederiksenii*	*Y. intermedia*	*Y. kristensenii*	*Y. mollaretii*
Acid from L-rhamnose	+	–	–	+	+	–	–
Acid from L-sorbose	–	–	V	+	+	+	+
Acid from sucrose	–	+	+	+	+	–	+
Acid from melibiose	–	–	–	–	+	–	–
Acid from raffinose	–	–	–	–	+	–	–
Indole production	–	–	V	+	+	V	–
Voges–Proskauer test	+	–	V(+)	V(+)	+	–	–
Citrate utilization at 25–28°C (Simmon's citrate agar)	V	–	–	V	+	–	–

TABLE 38.8 Differentiation of Division 9 bacteria

	Genus						
	Salmonella	*Citrobacter*	*Klebsiella*	*Kluyvera*	*Serratia*	*Ewingella*	*Enterobacter*
Acid from lactose	–[1]	V	+	+	D	V	D
Acid from salicin	–	D	+	+	+	V(+)	+
Acid from myo-inositol	–[2]	–	V(+)	–	V(+)	–	–
H_2S production	+	D	–	–	–	–	–
Møller's lysine decarboxylase	+	–	+	D	+[3]	–	–
Møller's ornithine decarboxylase	+	D(+)	–	+	+[3]	–	+
Indole production	–	D	–	+	V(–)	–	–
Urea hydrolysis	–	D	D	–	–	–	D

[1] Some *Salmonella choleraesuis* in the subspecies *arizonae*, *diarizonae* and *indica* ferment lactose.
[2] Some salmonellae in *S. choleraesuis* subsp. *choleraesuis* ferment myo-inositol.
[3] *Serratia plymuthica* gives negative reactions in the lysine and ornithine decarboxylase tests.

TABLE 38.9 Differentiation of *Citrobacter* species

	C. freundii	*C. diversus*	*C. amalonaticus*
Acid from salicin	–	V(–)	V
Acid from D-adonitol	–	+	–
H_2S production	V(+)	–	–
Ornithine decarboxylase	V(–)	+	+
Indole production	–	+	+
Malonate utilization	V(–)	+	–

TABLE 38.10 Differentiation of bacteria in Division 10

	Genus			
	Klebsiella	*Enterobacter*	*Serratia*	*Pantoea*
Motility	–	+	+	+
Acid from lactose	+	V	D	V(–)
Gelatin liquefaction at 22°C	–	–	+	–
Malonate utilization	+	+	–*	D
Møller's ornithine decarboxylase	–	+	D	V
DNase production	–	–	+	–

* *Serratia rubidea* utilizes malonate.

38.1.3 Tests used: description of methods

Flagellation

Staining of flagella is not difficult provided that care is taken at every stage, following the staining technique described in Section 3.2.4.

0/129 sensitivity

Sensitivity to compound 0/129 (2,4-diamino-6,7-di-isopropylpteridine, obtainable from many laboratory chemical suppliers) can be determined by spotting one or two crystals of the compound on a plate streaked uniformly with the test organism. In practice, contamination is not a problem.

Luminescence

Most luminous bacteria have been isolated from marine sources, and may have a requirement for constituents of sea water. Organisms are likely to have been isolated on sea-water-containing media, or alternatively on low dilution plates (incorporating high concentrations of samples) in counts on samples of marine origin. Luminescence should be checked on the original plates, and, in the case of isolates, after 2–3 days of incubation on the sea-water yeast peptone agar of Hendrie *et al*., (1970).

Cultures should be examined in a dark room, after allowing 10 min for the eyes to accommodate.

Oxidase test

See Section 11.4.2.

Thornley's arginine test

See Section 11.1.5(c).

Møller's decarboxylase tests (see Section 11.1.6)

Lightly inoculate the organism, preferably with a straight wire, into a series of four tubes of Møller's decarboxylase medium. The series consists of a control (no added amino acid), medium plus lysine, medium plus ornithine, and medium plus arginine. Ensure that the wire penetrates beneath the layer of liquid paraffin. Incubate at 25°C and examine daily for up to 7 days. A positive result is indicated by the medium colour changing to violet after an initial change to yellow. Controls and negative reactions are yellow in colour. In the case of a positive reaction in the arginine-containing medium, if ammonia is found in the medium (tested for by the use of Nessler's reagent), and the organism does not possess a urease, the reaction is due to arginine dihydrolase.

Test for 3-ketolactose production (Bernaerts and De Ley, 1963)

Inoculate the organism to be tested onto a glucose yeast chalk agar slant and incubate at 25°C for 3 days. From this culture take a loopful of bacterial growth and spot-inoculate a plate of lactose yeast extract agar so that a heavy inoculum of bacteria is concentrated in a spot about 5 mm in diameter. Incubate at 28°C for 2 days. After incubation, flood with Benedict's reagent and leave at room temperature for 1 h. If 3-ketolactose has been produced, a yellow ring of cuprous oxide extending up to 2–3 cm in diameter will form around the bacterial inoculum.

Test for ability to use inorganic nitrogen as sole nitrogen source (Holding, 1960)

Lightly inoculate the organism, preferably with a straight wire, into Holding's inorganic nitrogen medium and incubate at 25°C for up to 5 days, examining daily. An organism capable of utilizing inorganic nitrogen as its sole nitrogen source will grow and produce visible turbidity. If only slight turbidity is obtained, reinoculate into another tube of the same medium to ensure that the growth was not due to a carry-over of nutrients from the original broth culture.

Detection of ability to produce a diffusible pigment that fluoresces in ultraviolet light

The production of a UV-fluorescent pigment, which is a characteristic of many *Pseudomonas* spp., is very dependent on the composition of the growth medium. The most satisfactory medium is medium 'B' of King *et al.*, (1954). After incubation for 1–5 days at 20–25°C, the medium is examined in a darkened room under a long wavelength UV lamp for fluorescence, which may be blue or green.

Test for sensitivity to penicillin

A penicillin sensitivity test disc or sensitivity test tablet is placed on a plate seeded with the test organism. Sensitivity is shown by a zone of clearing around the tablet after incubation.

Phenylalanine deaminase test (see Section 11.1.11)

(1) Inoculate a tube of phenylalanine malonate broth (Shaw and Clarke, 1955) with the organisms to be tested. Incubate for 24 h at 30° or 37°C. After incubation and examination for malonate utilization (see p. 330), add a few drops of 0.1 N hydrochloric acid and a few drops of 0.5 M ferric chloride solution. A positive reaction – due to phenylpyruvic acid produced by the action of phenylalanine deaminase – is indicated by a green colour developing in the slope and in the liquid.

(2) Alternatively (Clarke and Steel, 1966), grow the culture as a 'lawn' on a nutrient agar plate by spreading 0.2 ml of a broth culture over the surface of the plate and incubating for 24 h at 37°C (or 30°C). Place a phenylalanine disc on the surface of the

'lawn' culture and reincubate at 37°C (or 30°C) for 2 h. Test for production of phenylpyruvic acid by adding one drop of ferric chloride (10% in dilute hydrochloric acid) to the disc. A deep green-blue colour developing in about 1 min constitutes a positive reaction.

Motility test

Semi-solid media can be examined for spreading turbidity, and hanging drop preparations from broth cultures or moist agar slope cultures examined microscopically (see Section 3.3).

Indole production

See Section 11.1.4.

Citrate utilization on Simmon's citrate agar

See Section 11.2.7.

Methyl red test

See Section 11.2.5.

Voges–Proskauer test

See Section 11.2.6.

Fermentation of carbohydrates (see Section 11.2.2)

Nutrient broth has been recommended as the basal medium for fermentation tests (Report, 1958). The carbohydrate should be added aseptically as a filter-sterilized solution to the sterile basal medium.

Liquefaction of gelatin

Use ferrous chloride gelatin (see Section 11.1.7(b)), stab-inoculated, and incubated at 20°C for up to 30 days. This can be used at the same time to test for the production of hydrogen sulphide.

Hydrogen sulphide production

Use ferrous chloride gelatin (see Section 11.1.7(b)). Incubate at 20–25°C for 7 days, examining daily.

Urease test

Use Christensen's urea agar (see Section 11.1.8).

β-Galactosidase production

(1) Heavily inoculate a tube of ONPG peptone water (Lowe, 1962). Incubate at 37°C. β-Galactosidase activity is indicated by the development of a yellow colour, due to release of orthonitrophenol. The reaction will usually be detectable within 3 h, and the incubation time should not exceed 18 h.

(2) Alternatively (Clarke and Steel, 1966), grow the culture as a 'lawn' culture on a nutrient agar plate by spreading 0.2 ml of a broth culture over the surface of the plate and incubating for 24 h at 37°C (or 30°C). Place a lactose disc on the surface of the plate culture and incubate at 37°C (or 30°C) for 2 h to induce β-galactosidase synthesis. Then apply an ONPG disc to the culture so that it overlaps the lactose disc. In the case of a positive reaction, a yellow colour will develop within about 15 min. This test can be performed on the same plate and at the same time as the phenylalanine test.

Gluconate utilization

Inoculate a tube of gluconate broth, and incubate for 48 h at 37°C. After incubation, add 1.0 ml of Benedict's reagent, and place the tubes in boiling water for 10 min. A positive test is shown by the development of a yellow-brown precipitate.

Malonate utilization

Inoculate a tube of malonate broth or phenylalanine malonate broth and incubate for 24 h at 37°C. Utilization of malonate is indicated by the development of an alkaline reaction.

38.2 GRAM-NEGATIVE BACTERIA THAT GROW ON GLUCOSE YEAST EXTRACT AGAR, WORT AGAR, MALT EXTRACT AGAR OR SIMILAR MEDIA

(see also Asai, 1968; Carr, 1968; Krieg *et al.*, 1984)

Bacteria showing these cultural characteristics which have been isolated from plant products, especially those that are acidic or alcoholic, are likely to be acetic acid bacteria of the genera *Gluconobacter* (i.e. *Acetomonas*) and *Acetobacter*. The two genera can be differentiated from one another and from *Pseudomonas* as follows (see Table 38.11). Tests should be carried out at 25–30°C.

TABLE 38.11 Differentiation of *Gluconobacter* and *Acetobacter*

	Genus		
	Gluconobacter	*Acetobacter*	*Pseudomonas*
Growth at pH 4.5	+	+	–
Ethanol oxidized to CO_2	–	+	V
Lactate oxidized to CO_2	–	+	+
Flagellation	Polar or none	Peritrichous or none	Polar

Growth at pH 4.5

Ability to grow at pH 4.5 is readily checked in the solid media mentioned above, suitably acidified with sterile lactic acid or citric acid immediately before pouring the plates.

Oxidation of ethanol to carbon dioxide

This is tested by the use of bromcresol green ethanol yeast extract agar slopes (the ethanol must be added to the molten medium immediately before setting in the sloped position (see p. 399). *Gluconobacter* and *Acetobacter* produce acid from the ethanol, effecting a colour change from bluish-green to yellow, but only *Acetobacter* continues the oxidation to give a reversion in colour back to bluish-green. Cultures should be examined daily to check for pH reversion.

Oxidation of lactate to carbon dioxide

This test can be performed using calcium lactate–yeast extract agar plates. Spot-inoculate or inoculate by a single streak. *Acetobacter* grows well on this medium and produces a precipitate of calcium carbonate in the medium, whereas *Gluconobacter* grows only poorly with no white halo in the medium.

38.3 BACTERIA REQUIRING AT LEAST 12% SODIUM CHLORIDE FOR GROWTH

Obligate extreme halophiles of the genera *Halobacterium* and *Halococcus* are found in salt lakes, solar salt and foods such as fish that have been preserved with solar salt. They are strict aerobes, and characteristically produce red, pink or orange intracellular carotenoid pigments. Such microorganisms could be isolated on media containing 15% (w/v) sodium chloride used for halophilic counts on foodstuffs. In cultural studies they are usually grown on complex media containing 20% (w/v) sodium chloride. Alternatively, the chemically defined medium of Onishi *et al*. (1965) may be used.

Halobacterium is rod shaped, and may be motile by polar flagella, or non-motile. *Halococcus* is coccal and non-motile.

39

A Scheme for the Identification of Gram-positive Bacteria

Primary separation is based on microscopic examination for morphological and staining characteristics, supplemented by observation of cultural characters and by simple biochemical tests. The groups listed below are for identification purposes; they do not necessarily indicate taxonomic similarity.

1 Persistent mycelium formed of branching non-fragmenting hyphae. Aerial hyphae bear chains of exospores (conidiospores). Forms hard colonies that are partially embedded in the medium. These develop a powdery surface when the conidiospores are produced. Rarely the aerial hyphae are less well developed or absent and the spores are produced singly. .... **1** *Streptomyces* group
Not as above *2*

2 Acid-fast to 10–20% sulphuric acid or to acid–alcohol... **2** *Mycobacterium* group
Not acid-fast *3*

3 Endospores produced *4*
Endospores not produced *5*

4 Catalase positive, aerobic or facultatively anaerobic .... **3** *Bacillus*
Catalase negative, anaerobic or anaerobic–aerotolerant .... **4** *Clostridium*
Catalase negative, facultatively anaerobic .... **5** *Amphibacillus*

5 Catalase positive *6*
Catalase negative *11*

6 Cells spherical, occurring in irregular masses, or in packets of four cells or multiples of four .... **6** *Staphylococcus–Micrococcus* group
Cells rod shaped or coccobacillary or irregular rods 7
Branching or non-branching filaments or a mycelium formed. Fragmentation into shorter rods or coccobacilli frequently occurs in older cultures ....
7 *Nocardia* group

7 Ferments lactate with the production of carbon dioxide ...... **8** *Propionibacterium*
Does not ferment lactate with the production of carbon dioxide *8*

8 Good growth on nutrient agar, plate count agar or soil extract agar *9*
No growth (or very sparse growth) on nutrient agar, plate count agar or soil extract agar .... **9** Animal coryneform group

9 Aerobic .... **10** Saprophytic and plant coryneform group
Facultatively anaerobic *10*

10 Motile when grown at 20–25°C.. **11** *Listeria*
Non-motile when grown at 20–25°C .... **12** *Brochothrix*

11 Aerobic, microaerophilic or facultatively anaerobic. Grows poorly or not at all on nutrient agar.. *12*
Anaerobic.. **4** *Clostridium*

12 Cells spherical or ovoid; grows on nutritionally rich glucose-containing media... **13** *Streptococcus–Leuconostoc–Pediococcus* group
Organisms rod shaped *13*

13 Digests Loeffler's serum.. *Actinomyces pyogenes* (see **10** Animal coryneform group)
Does not digest Loeffler's serum. Grows well on nutritionally rich glucose-containing media .... **14** *Lactobacillus*

1 *Streptomyces* group *Streptomyces* characteristically produces a many-branched, persistent, mycelium that does not fragment as the culture ages. Aerial hyphae are formed, and these bear chains of exospores (conidiospores). The substrate mycelium gives a colony that is partially embedded in the agar and extremely hard, usually requiring more force to break up the colony than can normally be applied with a platinum wire loop.

When young, the colonies are usually spherical, dome-shaped, slightly shiny and with no particular distinguishing coloration, being colourless or buff to fawn. As the cultures age and conidiospores are formed, the colonies develop powdery surfaces which may be white, grey or other colours, depending on the species. The powdery surface layer of conidiospores is easily scraped off using a wire loop, leaving the hard substrate mycelium as before. The colonies may become wrinkled or folded. Some species produce pigments that diffuse into the surrounding medium. In some cases the colour of the pigment varies with the pH of the medium. Many species produce an odour that is characteristic of moist soil. For species differentiation, which is based largely on morphological and cultural characteristics, refer to *Bergey's Manual of Systematic Bacteriology, Vol. 4 (BMSB4)* (Williams *et al.*, 1989).

From the work of Gordon and Mihm (1957, 1962), Williams *et al.* (1968) and many others, it is apparent that there is a continuous spectrum of types between the genera *Mycobacterium*, *Nocardia* and *Streptomyces*. Thus the *Streptomyces* group as defined here may include certain nocardiae. Some strains, usually regarded as *Nocardia*, produce persistent, non-fragmenting mycelia, develop aerial hyphae, but do not produce conidiospores. Such organisms would be indistinguishable from asporogenous *Streptomyces*. In some cases, fragmentation of the aerial hyphae of *Nocardia* resembles spore production by *Streptomyces*. For further clarification of this complex group refer to the illustrations given in *BMSB2* (Sneath *et al*., 1986), *BMDB* (Holt *et al*., 1994) and by Williams and Wellington (1980).

Thermophilic species occur in several genera, including *Thermomonospora* and *Thermoactinomyces*. A description of the morphological and colonial characteristics of the thermophilic organisms was given by Cross (1968). A health hazard may arise when 'self-heating' occurs in moist grain (or hay) during storage, as some of these thermophiles (e.g. *Thermoactinomyces*) can cause the respiratory disease known as farmer's lung'. Air-sampling techniques may be used for isolation of the bacteria in such cases, as reported by Gregory *et al*. (1963).

2 *Mycobacterium* group The genus *Mycobacterium* consists of non-motile, non-sporing, non-branching rods. Although the organisms are Gram-positive, staining by Gram's method is difficult, probably due to the high lipid content of the mycobacteria. They are acid-fast, as determined by Ziehl–Neelsen's acid-fast staining technique. The lipids found in the acid-fast bacteria have been described by Minnikin and Goodfellow (1980). Some strains resist decolorization with alcohol as well as acid, but as this characteristic is rather variable, Cowan and Steel (Cowan and Steel, 1965; Cowan, 1974) recommended the use of acid–alcohol as the decolorizing agent, with no differentiation between organisms that are acid-fast and those that are acid- and alcohol-fast.

Mycobacteria can be divided into two main groups:

(a) *Parasites and pathogens of homoiothermic animals*. This group is typified by *M. tuberculosis*. These mycobacteria grow very slowly *in vitro*, cultures requiring up to 2 weeks' incubation or more to give visible growth, and require special media for their cultivation. They are incapable of growth, or grow extremely poorly, on media such as nutrient agar. Media that are satisfactory for their growth include solid media in which coagulated serum or egg is the solidifying agent. Because of their pathogenicity these organisms require special precautions in handling, and it is recommended that isolation and culture should be undertaken only in properly equipped clinical microbiological laboratories.

(b) *Saprophytes, and parasites and pathogens of poikilothermic animals*. These are less acid-fast than the members of group (a). They grow rapidly *in vitro*, with growth usually being visible after 2–5 days. They are capable of growing on media such as nutrient agar but usually grow better on media containing coagulated serum (e.g. Loeffler's serum medium) or egg. On solid media, particularly those containing glycerol, many saprophytic mycobacteria produce bright yellow, orange, pink or brick-red pigments, and the growth is dry and crumbly.

Thus acid-fast bacteria isolated in the laboratory during general microbiological examinations will exclude the members of group (a) except when techniques are employed specifically to detect them. There is no distinct division between the mycobacteria of group (b) and *Nocardia*. Some strains regarded as *Nocardia* are slightly acid-fast and some organisms usually included in *Mycobacterium* are only weakly acid-fast. In addition, certain mycobacteria produce branching filaments that fragment into shorter coccobacillary forms as the cultures age, thus corresponding to the morphological description of *Nocardia* given here. For a detailed description of the saprophytic mycobacteria, see Gordon and Mihm (1959).

3 *Bacillus* Species of the genus *Bacillus* are rods capable of producing endospores and of growing aerobically (although some are facultative), in most cases on nutrient agar, and are catalase positive. Difficulty may occasionally be experienced in inducing sporulation and, in this case, additional manganese should be added to the growth medium as recommended by Charney *et al.* (1951). These authors found a marked stimulatory effect on the sporulation of *B. subtilis* at concentrations of 0.1–10.0 μg manganese per ml in all media, which is obtained by the addition of 0.1–10.0 ml of a 0.4% solution of manganese sulphate ($MnSO_4 \cdot 4H_2O$) per litre of medium. Most *Bacillus* spp. are motile by means of peritrichous flagella with the exception of some of the larger-celled species and *B. anthracis*. Although the genus *Bacillus* is regarded as typically Gram-positive, a number of species may be Gram-variable or even Gram-negative, particularly those producing swollen sporangia and thus included in the morphological groups II and III of Smith and co-workers (1952) and of Gordon *et al.* (1973) described below.

Identification of *Bacillus* spp. can be carried out according to the methods of Smith *et al.* (1952) and Gordon and colleagues (1973). Preliminary subdivision of the genus into three morphological groups is made on the basis of the appearance of the sporangium: whether it is non-swollen (group I); swollen with an oval spore (group II); or swollen with a round spore (group III). Further identification is on the basis of cultural and biochemical tests (see Table 39.1).

BMSB2 (Sneath *et al*., 1986) listed 40 species and a further 27 *species incertae sedis*. The description and discussion of *Bacillus* and its species by Claus and Berkeley in *BMSB2* is invaluable for anybody wishing to characterize *Bacillus* isolates. The API50CHB kit can be used in conjunction with the API database for PCS (bioMérieux) (see also Logan and Berkeley, 1981). The following table, adapted from the data of Gordon *et al.* (1973) and Sneath *et al.* (1986), allows differentiation of the species that the author has found to be encountered most frequently in food microbiology.

Smith and co-workers (1952) recognized only two thermophilic *Bacillus* species, *B. coagulans* and *B. stearothermophilus*; cultures received as *B. calidolactis* from dairy sources were identified as one or other of these two species. *B. coagulans* may be regarded as a facultative thermophile because it shows growth at both 37° and 55°C. It may be distinguished from *B. stearothermophilus* by its failure to grow at 65°C.

4 *Clostridium* *Clostridium* species are rods capable of producing endospores but may be differentiated from *Bacillus* spp. as they are catalase negative and strict anaerobes apart from a few anaerobic, aerotolerant species capable of producing scant growth under aerobic conditions. The organisms are typically Gram-positive and may be either

TABLE 39.1 Identification of commonly encountered *Bacillus* species

Species	Group*	Growth at 50°C	Growth at 60°C	Growth at 65°C	Voges–Proskauer	Starch hydrolysis	Hugh and Leifson's	Acid from mannitol	Nitrate reduction	Citrate utilization
B. megatherium	I	–	–	–	–	+	A	V	V	+
B. cereus	I	–	–	–	+	+	F	–	+	V(+)
B. licheniformis	I	+	–	–	+	+	F	+	+	+
B. subtilis	I	V(+)	–	–	+	+	A	+	+	+
B. pumilis	I	V(+)	–	–	+	–	A	+	–	+
B. lentus	I	–	–	–	–	+	A	+	V	–
B. thuringiensis	I	–	–	–	V	+	F	–	+	+
B. coagulans	I, II	+	+	–	+	+	F	V(–)	V(–)	V(–)
B. polymyxa	II	–	–	–	+	+	F	+	+	–
B. macerans	II	V(+)	–	–	–	+	F	+	+	V(–)
B. circulans	II	V(–)	–	–	–	+	F/A	+	V(–)	V(–)
B. stearothermophilus	II	+	+	+	–	+	A	V(–)	V(+)	V(–)
B. alvei	II	–	–	–	+	+	F	–	–	–
B. brevis	II	V(+)	V(+)	–	–	V	A	V(+)	V(+)	V(–)
B. pantothenticus[†]	II, III	V	–	–	–	+	F	–	V(+)	–
B. sphaericus	III	–	–	–	–	–	A	–	–	V(–)
B. psychrophilus[‡]	III	–	–	–	–	V	A	+	+	–

*Morphological group I, ellipsoidal and cylindrical spores do not exceed diameter of the sporangium; II, ellipsoidal and cylindrical spores cause swelling of the sporangium; III, spherical spores cause swelling of the sporangium.

[†]*B. pantothenticus* can grow in the presence of 10% sodium chloride, and may sometimes occur in canned cured meats.

[‡]*B. psychrophilus* has a maximum growth temperature of around 30°C, and can sporulate at 0°C.

+, 85–100% of strains positive; V, variable reaction; V(+), variable, most strains positive; V(–): variable, most strains negative; A, 85–100% of strains obligately aerobic; F, 85–100% of strains facultatively anaerobic; F/A, 50–84% of strains facultatively anaerobic.

Adapted from Gordon *et al.* (1973) and Sneath *et al.* (1986).

non-motile or motile by means of peritrichous flagella. The sporangium may be either non-swollen or swollen with oval or spherical spores in a terminal position, or in a central or subterminal position to give a spindle-shaped or clostridial form. Difficulty may be experienced in inducing sporulation of some species as, for example, *Cl. perfringens*, which rarely sporulates *in vitro* unless special media (Ellner, 1956) are used.

A useful precaution with anaerobic culture, as pointed out by Willis (1962), is to set up a control plate incubated aerobically with each subculture of test organism. A comparison of aerobically and anaerobically incubated plates enables detection of contaminants, particularly *Bacillus* spp., capable of growing anaerobically.

BMSB2 (Sneath *et al.*, 1986) lists 90 species, but new species continue to be validly described. Species identification in *BMSB2* is by dichotomous key and tables, with initial differentiation being based on hydrolysis of gelatin and acid production from glucose. Subsequent biochemical tests include carbohydrate fermentation patterns, indole production, nitrate reduction, starch hydrolysis and production of lipase and lecithinase.

An alternative approach (Willis, 1962, 1965; Hobbs *et al.*, 1971) for the identification of a limited number of species is to use Willis and Hobbs's lactose egg-yolk milk agar medium with neutral red as pH indicator. This medium has the advantage that it combines in one medium a method of detecting the proteolytic and saccharolytic activities of the isolate together with its effect on egg-yolk agar. The egg-yolk medium enables both lecithinase and lipase activities to be detected. Lecithinase activity results in a zone of opacity extending beyond the colony, whereas lipase activity is indicated by a restricted zone of opacity and a pearly layer consisting of fatty acids, which give a surface sheen to the colony. The use of half-antitoxin plates as described by Willis (1962) enables lecithinase-producing organisms to be specifically identified.

Goudkov and Sharpe (1965), in an examination of milk and dairy products, found that clostridial spoilage was most likely to occur in low-acid sweet cheese of Swiss and Edam type and in processed cheese. Spoilage appeared to be mainly gas production by lactate fermenters such as *Cl. butyricum* or *Cl. tyrobutyricum*, and proteolysis by proteolytic clostridia such as *Cl. sporogenes* and Cl. *perfringens*. Psychrotrophic clostridia (and their significance in food microbiology was discussed by Beerens *et al.* (1965).

5 *Amphibacillus* Young cultures are Gram positive; older cultures tend to lose the stain-retaining ability. Members of the genus are motile; they produce oval, centrally placed endospores. They grow well and produce spores in both aerobic and anaerobic conditions on plate count agar adjusted to pH 10, at temperatures of 25–45°C. They are catalase negative and oxidase negative. Members of this genus have been found associated with decaying plant material. (See Niimura *et al.*, 1990.)

6 *Staphylococcus–Micrococcus* group Determination of the ability of an organism to bring about the anaerobic fermentation of glucose forms the basis of separation within this identification group. A standard procedure involving a modification of the Hugh and Leifson technique has been recommended for this purpose (Recommendations, 1965). The bromthymol blue indicator of the original medium is replaced by bromcresol purple and, because this detects only active acid producers, more clear-cut results are obtained. Catalase-positive Gram-positive cocci that produce acid throughout the medium (i.e.

ferment glucose under anaerobic conditions) are staphylococci; those that produce no acid or produce acid only at the surface of the medium (i.e. oxidize glucose) are micrococci.

Staphylococcus

BMSB2 (Sneath *et al.*, 1986) lists 28 species. The species *Staph. aureus*, *Staph. delphini*, *Staph. hyicus*, *Staph. intermedius* and *Staph. schleiferi* are coagulase positive (see Section 11.4.3), and also all of these except *Staph. delphini* produce a thermostable nuclease. Staphylococci can be identified using the APIStaph diagnostic kit (bioMérieux).

Saccharococcus

This is a thermophilic, catalase-positive, oxidase-positive, facultatively anaerobic, Gram-positive coccus. This genus has been reported in sugar solutions from sugar refineries; it has an optimum growth temperature of 65–68°C.

Micrococcus

Preliminary observations should include the examination, under high-power and oil-immersion objectives, of a hanging drop preparation prepared from an 18-h broth culture. This procedure was found by Pike (1962) to be more satisfactory for the detection of characteristic groupings of cocci and packet formation (i.e. cell division in two or three planes in space) than examination of stained preparations from agar cultures. It is also a useful method of ensuring that other coccal bacteria, such as streptococci, and soil-derived coryneforms of the genus *Arthrobacter*, which show a coccal stage in mature culture, are not falsely identified as micrococci (Pike, 1962).

There are nine species of *Micrococcus* recognized in *BMDB* (Holt *et al.*, 1994); they produce intracellular (non-diffusible) pigments from cream through yellow and orange to red. Most are halotolerant, capable of growing on media containing 5–7.5% sodium chloride. One species, *M. halobius*, and three other genera, *Marinococcus*, *Planococcus* and *Salinococcus*, are moderate halophiles, but can grow on media containing quite low concentrations (0.5–1%) of sodium chloride. These halophiles are found in marine habitats and may be found in seafoods and salted fish.

7 *Nocardia* group The genus includes:

(a) Strains consisting of rods that rarely or never show branching; they produce soft, butyrous colonies. These organisms, particularly when slightly acid-fast (e.g. not decolorized by 1–5% sulphuric acid), overlap with the saprophytic mycobacteria. On the other hand, non-acid-fast nocardiae of this type are difficult to distinguish from some of the corynebacteria, which may even show rudimentary filament formation in slide cultures.

(b) Strains producing branching filaments or rudimentary mycelia that fragment into rods and coccoid forms. Colonies tend to be of a pasty or crumbly texture.
(c) Strains capable of producing persistent mycelia, sometimes with a slight aerial mycelium (see, for example, Gordon and Mihm, 1962). Colonies of these types tend to be leathery and resistant to being broken up with a wire loop. These strains, together with *Streptomyces*, present a continuous spectrum of morphological forms such that a dividing line cannot, with certainty, be drawn between the two genera.

Within this framework of morphological variation the nocardiae are Gram-positive, obligately aerobic and largely saprophytic, although a few animal-pathogenic species of aerobic nocardiae have been described. The saprophytic strains of the nocardiae are frequently capable of utilizing as an energy source such compounds as paraffin and phenol.

8 *Propionibacterium* Propionibacteria are Gram-positive, non-acid-fast and non-motile rods which do not produce endospores. Under the microscope, Gram-stained smears often show the 'Chinese character' arrangement of cells characteristic of coryneform bacteria (see also group **10** below). There are two principal groups of *Propionibacterium* species based on habitat: those from cheese and dairy environments, naturally fermented plant materials, and soil; and those from human skin and the intestine.

Although there is a tendency towards anaerobiosis and growth is improved in an atmosphere of 5% carbon dioxide, both aerobically and anaerobically, cultures are catalase-positive, usually strongly so. Biochemically lactic acid (or lactate) and carbohydrates are fermented with the production of propionic acid, acetic acid and carbon dioxide. Production of carbon dioxide gas can be used under suitable conditions as an indicator of the presence of propionibacteria. It should be noted that, although some species of clostridia may also ferment lactate, they can be distinguished on the basis of morphological and cultural characters: lactate-fermenting clostridia are endospore-producing rods, usually motile and of a larger size than propionibacteria, catalase-negative and incapable of aerobic growth. The morphological appearance of propionibacteria depends very much on cultural conditions, in particular oxygen tension and pH of the medium, as shown by van Niel (1928). Under anaerobic conditions and in neutral media, the reaction of which remains so, the cells appear as very short rods with rounded ends sometimes resembling streptococci in appearance. Under aerobic conditions with adequate aeration and in a similar medium, long rod-shaped cells of irregular or even branched appearance may be formed.

9 Animal coryneform group The genus *Corynebacterium* includes animal parasites and pathogens, 20 species being described in *BMDB* (Holt *et al.*, 1994). These organisms grow poorly, if at all, on nutrient agar, plate count agar and similar media, but are capable of growing well on serum agar or egg-yolk agar.

The appearance of *Corynebacterium* under the microscope is very characteristic of this genus and a few others (see also group 10). The straight or slightly curved rods are frequently swollen at one or both ends. Stained by Gram's method or with methylene blue, the individual organisms usually stain irregularly, sometimes with the appearance of darkly staining granules. The arrangement of organisms is also very characteristic, with a snapping type of division causing V- or L-shaped formations of adjacent organisms. Groups or clusters of organisms thus tend to have the appearance of 'Chinese characters' or cuneiform writing.

Two coryneform bacteria, *Actinomyces pyogenes* (formerly *Corynebacterium pyogenes*) and *Corynebacterium bovis*, may be isolated from raw bovine milk plated on to blood agar, serum agar or egg-yolk agar. Whereas *A. pyogenes* is associated with a particular and relatively uncommon type of suppurative mastitis, *C. bovis* can be isolated from most freshly drawn milk samples, whether from udders that are normal or from those presenting a subclinical or clinical mastitis.

Actinomyces pyogenes is catalase-negative, haemolytic and proteolytic – slowly liquefying coagulated serum (Loeffler's serum slopes) and gelatin (nutrient gelatin with 5% serum should be used). It is urease negative. Acid is produced from a wide range of carbohydrates in Hiss's serum water sugars (see p. 40). There is sparse or no growth on Tween agar with no zones of precipitation.

C. bovis is catalase-positive and non-haemolytic. It liquefies neither coagulated serum nor gelatin. It is capable of hydrolysing urea when grown in Christensen's urea agar with added serum. Acid is produced from a number of substrates when *C. bovis* is grown in Hiss's serum water sugars, including glucose, fructose and glycerol. Maltose and/or galactose are attacked by some strains, but lactose, sucrose and salicin are not attacked. *C. bovis* grows luxuriantly on Tween 20 and Tween 80 agar with broad zones of precipitation (Harrigan, 1966).

C. ulcerans has been isolated from milk freshly drawn from normal bovine udders (Jayne-Williams and Skerman, 1966). It has also been isolated from cases of diphtheria-like diseases in human beings. Unlike *C. bovis*, this organism is haemolytic and proteolytic, but it differs from *A. pyogenes* in hydrolysing urea.

10 Saprophytic and plant coryneform group This group includes several different genera at present poorly defined, or defined by habitat rather than by taxonomic affinity, and it is therefore difficult to construct a satisfactory scheme for separation of these organisms (see Jones, 1975, 1978). *BMDB* (Holt *et al.*, 1994) uses chemical analysis of cell wall components and the menaquinones to differentiate the genera (see also Keddie and Cure, 1978; Minnikin *et al.*, 1978; Collins *et al.*, 1980; Keddie and Bousfield, 1980; Minnikin and Goodfellow, 1980).

The plant pathogens previously placed in the genus *Corynebacterium* (Buchanan *et al.*, 1974) are now included in the genera *Clavibacter* and *Curtobacterium.* Members of many of the genera are characterized by the production of yellow or orange non-diffusible pigments, including *Brevibacterium*, *Microbacterium*, *Caseobacter* and *Aureobacterium*, all found in cheese and other dairy environments; *Cellulomonas*, found in soil and on decaying plant material; and *Curtobacterium* and *Clavibacter*, both plant pathogens. *Pimelobacter*, a soil inhabitant, produces white or yellowish white colonies. *Arthrobacter* and *Kurthia* are non-pigmented.

The term 'coryneform' is used here to imply non-acid-fast, non-spore-forming rods, typically Gram-positive and frequently showing features regarded as typically coryneform, such as beaded staining, club shapes, and palisade- or Chinese letter-arrangements of the rods. In the exponential phase, irregular rods of variable length are the most common forms, with rudimentary branching often occurring, especially on richer media. In the stationary phase, cells are much shorter and more regular, even coccoid. The genera that show this cycle from rods to cocci in older cultures are *Arthrobacter*, *Brevibacterium*, *Caseobacter*, *Curtobacterium*, *Kurthia* and *Pimelobacter*. *Microbacterium*, *Cellulomonas*, *Aureobacterium* and *Clavibacter* do *not* show this cycle, although in old cultures of *Microbacterium* and *Cellulomonas* the rods tend to be somewhat shorter than in young cultures. Photomicrographs of typical coryneform organisms at the different morphological stages have been presented by Cure and Keddie (1973), and of *Arthrobacter and Brevibacterium linens* by Crombach (1974).

Brevibacterium, *Caseobacter* and *Clavibacter* are non-motile; organisms in the other genera frequently show motility.

Microbacterium lacticum, commonly isolated from milk, dairy products and dairy equipment, is characterized by a capacity to survive laboratory pasteurization in milk (63°C for 30 min) or a heat treatment in milk of 72°C for 15 sec, although such heat resistance appears *not* to be an invariable feature of these organisms.

Kurthia is found on poultry and meat. Descriptions and photomicrographs of strains of *K. zopfii* isolated from meat have been given by Gardner (1969).

The genus *Cellulomonas* consists of soil coryneforms with an ability to digest cellulose, as seen by disintegration of a strip of filter paper when incubated in a peptone water or nutrient broth culture.

It is important that older cultures of coryneform bacteria in the coccal phase are not confused with micrococci and, as pointed out by Pike (1962), examination of young broth cultures by the hanging drop method is helpful in this respect. An alternative procedure, and to follow morphogenesis, is to use the slide culture technique in which a sterile coverslip is placed over a surface-inoculated agar medium, either in a Petri dish as used by da Silva and Holt (1965) or in a slide culture. The development may then be followed microscopically, if possible by phase-contrast microscopy. Older cultures of *Arthrobacter* may show the presence of enlarged coccoid forms, sometimes known as cystites, but Stevenson (1963) concluded that these are of no particular significance in the life cycle of the organism and that the enlarged forms appear as a result of depleted cultural conditions. On transfer to fresh medium, rod forms are again produced.

The genus *Brevibacterium* is very broadly defined and includes an assemblage of typically short, unbranching, Gram-positive, non-acid-fast and non-sporing rods, isolated from dairy products and diverse other sources, including soil, salt water and fresh water. Most species are non-pigmented. One species, *Bbm. linens*, producing an orange non-diffusible pigment, is found on the surface of bacteria-ripened soft cheeses such as Limburger (see p. Section 28.7.8).

11 *Listeria* These are short, regular, Gram-positive rods, although in older cultures, longer filaments may develop and some cells lose the ability to retain the crystal violet–iodine complex. On a non-selective medium such as nutrient agar, surface colonies appear bluish-grey, but have a characteristic blue-green sheen when examined by the oblique light of Henry illumination (see Section 3.4.1). The genus has an optimum growth temperature between 30° and 37°C, but can grow over the range 1–45°C. Organisms are oxidase-positive; produce acid but no gas from glucose; are methyl red positive; Voges–Proskauer positive; do not produce indole; and do not hydrolyse urea. For species differentiation, see Section 21.9.

Note. Motile strains of the facultative anaerobe *Cellulomonas* may be traced through the diagnostic key to this group entry. However, they usually produce a yellow non-diffusible pigment, and are cellulolytic (see group **10**).

12 *Brochothrix* Organisms in this genus are rods which may remain attached to one another to give long chains; cocci develop in old cultures. Two species are listed in *BMDB* (Holt *et al.*, 1994): *B. thermosphacta* (previously named *Microbacterium thermosphactum*) and *B. campestris*. They are differentiated by *B. thermosphacta* being able to grow in media containing 8% sodium chloride or 0.05% potassium tellurite, whereas *B. campestris* cannot grow in either type of medium.

B. campestris is found in soil and on grass (Talon *et al.*, 1988).

B. thermosphacta is found in meat and meat products. It can grow at 1–4°C, and in reduced oxygen and increased carbon dioxide concentrations. Consequently it is often the dominant bacterium in refrigerated prepacked meats. In raw meat prepacked in gas-permeable films, the resulting slightly increased carbon dioxide concentration can permit as increase in the numbers of *B. thermosphacta*, whilst at the same time tending to inhibit *Pseudomonas* (see also Dainty *et al.*, 1983).

Note. Non-motile strains of the facultatively anaerobic *Cellulomonas* may be traced through the diagnostic key to this group entry. They usually produce a yellow non-diffusible pigment and are cellulolytic (see group **10**).

13 *Streptococcus–Lactococcus–Enterococcus–Leuconostoc–Pediococcus* group These genera are Gram-positive, coccal or ovoid forms, typically catalase-negative, characterized by their ability to ferment carbohydrates with the production of lactic acid. This may be either the sole product of the fermentation in which case the organism is described as homofermentative (as in the genera *Streptococcus*, *Lactococcus*, *Enterococcus* and *Pediococcus*). Alternatively the lactic acid may be accompanied by a variety of other fermentation products including acetic acid, ethanol and carbon dioxide, in which case the organism is described as heterofermentative (as in the genus *Leuconostoc*) (see Table 39.2).

The genera in this group may be distinguished from *Staphylococcus* and *Micrococcus* by their negative catalase reaction and failure to produce abundant growth on nutrient agar. They are non-pigmented, except for *Leuconostoc citreum*, *Enterococcus casseiflavus* and *E. mundtii*, which produce yellow non-diffusible pigments. Optimal growth is obtained only in media containing a suitable fermentable substrate (e.g. glucose in yeast glucose lemco broth).

Cell division in all the genera except *Pediococcus* takes place in one plane only, which gives rise to the arrangement of cells in pairs or in short or long chains. Cell division in the case of *Pediococcus* occurs in two planes to give tetrads.

The detection of heterofermentative activity by the ability to form carbon dioxide from glucose in Gibson's semi-solid medium enables the heterofermentative *Leuconostoc* to be

TABLE 39.2 Differentiation of catalase-negative, fermentative Gram-positive bacteria

	Enterococcus	*Lactococcus*	*Streptococcus*	*Pediococcus*	*Leuconostoc*
Growth at 10°C[1]	+	+	–	D	+
Growth at 45°C[1]	+	–	D[2]	D	–
Growth in the presence of 6.5% NaCl[1]	+	–	–[3]	D	D
Growth at pH 9.6[1]	+	–	–[4]	–	–
Growth at pH 4.4[1]	+	D	–	D	D
CO_2 from glucose	+	–	–	–	+
Tetrad formation	–	–	–	+	–

[1] These tests should be made using a medium such as yeast extract glucose lemco broth.
[2] *Streptococcus thermophilus* has an optimum growth temperature of 40–45°C, and can grow at 50°C.
[3] A few strains of *S. agalactiae*, *S. porcinus* and *S. cricetus* are positive.
[4] A few strains of *S. bovis* are positive.

differentiated from the other genera. Also *Leuconostoc* never produces ammonia from arginine in Abd-el-Malek and Gibson's arginine broth; therefore, a positive result contraindicates *Leuconostoc*.

Streptococcus, Lactococcus *and* Enterococcus

Suitable schemes for the full range of species of *Streptococcus* and *Enterococcus* are given in *BMDB* (Holt *et al.*, 1994). The APIStrep kit (bioMérieux) may be used, with additional tests such as the precipitin test to determine the Lancefield serological group (see Section 13.6). Most isolates obtained in food microbiology belong to the Lancefield serological groups listed below:

A *Strep. pyogenes*
B *Strep. agalactiae*
C *Strep. dysgalactiae*, *Strep. equi*
D *Strep. bovis*, *Strep. equinus*, *Enterococcus* species (including *Ent. faecalis*, *Ent. faecium*, *Ent. durans*)
N *Lactococcus*

The species of *Streptococcus* most likely to be encountered in food microbiology are: *Strep. thermophilus*, the non-pathogenic yoghurt starter organism (see Section 28.6); the mastitis-causing organisms *Strep. agalactiae* and *Strep. dysgalactiae*, and less commonly *Strep. uberis* (see Section 28.1.1.1); *Strep. pyogenes* may be occasionally involved in foodborne outbreaks of streptococcal infections as the result of contamination of food by an infected handler.

Five species are currently recognized in the genus *Lactococcus*. The former species *L. cremoris* is now included in *L. lactis* as a subspecies, *L. lactis* subsp. *cremoris*, distinguished from the two other subspecies, *L. lactis* subsp. *lactis* and subsp. *hordniae*, by not hydrolysing arginine. Of the three subspecies, only *L. lactis* subsp. *lactis* can grow in the presence of 4% sodium chloride (as can the species *L. garviae* and *L. plantarum*) or at 40°C (as can *L. garviae*), using yeast glucose lemco broth as the growth medium. The species *L. raffinolactis* is the only species producing acid from raffinose. The diacetyl-producing organisms of the former *Streptococcus lactis* subsp. *diacetylactis* of importance in the dairy industry are currently merely included in the subspecies *L. lactis* subsp. *lactis*. The diacetyl-producing ability is linked closely to the production of citratase (detected using semi-solid citrate milk agar in which citrate fermentation is indicated by gas production). This ability is determined by the possession of a plasmid. It has been suggested (Euzeby, 1997) that the subspecies be reinstated as *L. lactis* subsp. *diacetilactis.* The species *L. piscium* has been isolated only from salmonid fish.

The enterococci are used as indicator organisms (see Section 20.2); some species are occasionally found as part of the developing microflora in naturally fermented vegetable products.

Leuconostoc

These are catalase-negative Gram-positive cocci showing cell division in one plane only and in this they resemble the streptococci, but they differ in being heterofermentative and are thus capable of producing carbon dioxide from glucose. Other characters that are useful in the differentiation of *Leuconostoc* from certain other groups, in particular lactococci with which they may be associated in starter cultures, are their failure to produce ammonia from arginine and comparative inactivity in litmus milk. Garvie (1960), in an extensive study of the genus, found no strains giving reduction of litmus, a few capable of producing acid, but very few able to clot the milk, thus contrasting with the lactococci of starter cultures which give a vigorous clot and reduction.

Most *Leuconostoc* species are widely distributed on plant materials and in dairy products. However, *L. oenos* is found in wine, and unlike the other species it can initiate growth in media of pH 4.2–4.8 and can grow in a medium containing 10% ethanol. The yellow pigmented *L. citreum* has been isolated from human sources. *L. mesenteroides* produces a dextran on sucrose agar, giving profusely gummy colonies. *L. paramesenteroides* tends to be more salt tolerant than the others, and so may be found in salted vegetable products. *L. lactis* is mostly found in dairy products, and may be used as a starter organism in the dairy industry; it is a diacetyl producer (giving a positive Voges–Proskauer test in milk or in yeast glucose lemco broth) and can utilize citrate (determined using semi-solid citrate milk agar).

Pediococcus

These are homofermentative Gram-positive cocci with complex nutritional requirements and thus they resemble the streptococci, but differ from them in that cell division takes place in two planes (Günther, 1959), so that tetrads and coccal arrangements resembling those of micrococci are produced. Günther and White (1961) differentiated pediococci from micrococci on the basis of the failure of pediococci to grow in simple media, or to utilize ammonium salts as sole nitrogen source, and failure to reduce nitrate or liquefy gelatin. In addition pediococci are microaerophilic.

Pediococci are particularly associated with fermenting plant materials, and as contaminants of brewer's yeast. *Pediococcus halophilus* is found in soy sauce and pickling brines; *P. damnosus* is found in beer and brewery environments. Pediococci, especially *P. pentosaceus*, are also used as starter cultures in certain fermented meat products such as salami by manufacturers wishing to produce a less acid product than that obtained by using *Lactobacillus* as a starter culture.

Differentiation of *Pediococcus* from *Leuconostoc*, which may also be found in fermenting plant materials, can be made on the basis of morphology and on the failure of *Pediococcus* to produce carbon dioxide from glucose in Gibson's semi-solid medium. Like *Leuconostoc*, pediococci are comparatively inactive in litmus milk and rarely produce sufficient acid to cause clotting; in this they contrast sharply with *Lactococcus*.

For information on the differentiation of species within the genus refer to *BMDB* (Holt *et al.*, 1994).

14 *Lactobacillus* These are non-acid-fast, non-spore-forming, Gram-positive rods. With very few exceptions they are non-motile, and the rods frequently appear in pairs or chains, particularly in liquid media. The lactobacilli are catalase negative and microaerophilic, generally requiring layer plates for aerobic cultivation on solid media. Nutritional requirements are complex and optimal growth is obtained only in media containing a fermentable substrate and adequate growth factors, as, for example, the MRS broth of de Man *et al.* (1960). The genus includes both homofermentative and heterofermentative species, the optical activity of the lactic acid produced depending on the species. Lactobacilli are aciduric organisms and this property is utilized in the devising of selective media for their isolation as in acetate agar (Keddie, 1951) and Rogosa agar (Rogosa *et al.*, 1951), both having a final pH of 5.4 and containing acetate as selective agent.

Orla-Jensen (1943) subdivided the lactobacilli on the basis of homofermentation or heterofermentation and on optimum temperature into three groups as follows: *Streptobacterium* (homofermentative, low optimum temperature), *Thermobacterium* (homofermentative, high optimum temperature) and *Betabacterium* (heterofermentative). Although no longer accorded taxonomic status, they form convenient descriptive terms, the value of which has not been diminished by more recent extensive work using a variety of physiological and cultural tests (see also Sneath *et al.*, 1986).

Forty-four species are listed in *BMSB2* (Sneath *et al.*, 1986). Fermentation patterns using a large number of substrates are an important means of species differentiation (see Sneath *et al.*, 1986); API50CHL (bioMérieux) provides an easy way to determine these fermentation patterns. Although *BMSB2* has modified Orla-Jensen's three groups to exclude growth temperatures as a differentiating characteristic (because many recently described species will not fit into the Orla-Jensen categories), the modified groups given in *BMSB2* are not as readily distinguished in the smaller quality assurance laboratory. Consequently the original Orla-Jensen groups have been retained here, for a simpler identification scheme for use in non-critical work, as described below. To achieve proper

TABLE 39.3 Differentiation of lactobacilli (after Naylor and Sharpe, 1958b; Sharpe, 1962; Buchanan *et al.*, 1974)

Group	Gas from glucose	Ammonia from arginine	Growth at 15°	Growth at 45°	Fermentation[1] of Raffinose	Melibiose	Arabinose	Cellobiose	Gluconate	Lactose	Salicin	Sucrose	
'Streptobacterium' (homofermentative, low optimum temperature	–	–	+	V	–	–	–	V(+)	V(+)	V(+)	V(+)	V(+)	*L. casei*
					+	+	V	+	+	+	+	+	*L. plantarum*
'Thermobacterium' (homofermentative, high optimum temperature)	–	–	–	+	V	V	–	+	–	+	+	+	*L. acidophilus*
					–	–	–	–	–	+	–	–	*L. delbrueckii* subsp. *bulgaricus*
					–	–	–	V	–	+	+	+	*L. delbrueckii* subsp. *lactis*
					–	–	–	V	–	–	–	+	*L. delbrueckii* subsp. *delbrueckii*
					–	–	–	–	–	+	–	–	*L. helveticus*[3]
'Betabacterium'[2] (heterofermentative)	+	+	+	–	V(–)	+	+	–	+	V(–)	–	V	*L. brevis*
		+	–	+	+	+	V	–	+	+	–	+	*L. fermentum*
		–	+	+	–	–	–	–	+	–	–	+	*L. viridescens*
		+	+	–	V	–	–	–	+	V	–	V	*L. hilgardii*

[1] Determined using MRS fermentation medium (see p. 431).
[2] Includes *L. trichodes*, a wine spoilage organism, which grow in media containing 15–20% ethanol.
[3] *L. helveticus* ferments galactose; *L. delbrueckii* subsp. *bulgaricus* does not.

identification to species level, it will always be necessary to undertake a full examination of fermentation patterns.

To identify *Lactobacillus* isolates, first determine to which physiological subgroup it belongs. The ability to produce carbon dioxide gas from glucose in Gibson's semi-solid medium indicates whether fermentation is homofermentative or heterofermentative. The ability to grow at 15° and 45°C (in MRS broth) should also be determined. Ability to produce ammonia from arginine in a medium containing 2% glucose may also be a useful supplementary test because, as shown by Briggs (1953), most heterofermentative lactobacilli give a positive reaction. A suitable medium is MRS broth containing 0.3% arginine – ammonia production is detected by Nessler's reagent. The arginine test is also useful for differentiating heterofermentative lactobacilli from *Leuconostoc*, as cocco-bacillary forms of *Lactobacillus* may be difficult to distinguish morphologically from *Leuconostoc*.

Differentiation of the main groups of *Lactobacillus*, and identification of the species particularly likely to be isolated from food and associated samples, is given in Table 39.3. Fermentation tests should be carried out using MRS fermentation medium.

The source of the isolate may be helpful in determining its possible identity, as indicated in Table 39.4.

TABLE 39.4 *Lactobacillus* species commonly associated with named habitats

Habitat	Species likely to be found
Cheese and dairy environments	*L. brevis*
	L. delbrueckii subsp. *bulgaricus*
	L. casei
	L. delbrueckii subsp. *lactis*
	L. plantarum
Emmental and Gruyère cheese	*L. helveticus*
Fermented and acid plant products, baker's and brewer's yeast	*L. brevis*
	L. delbrueckii subsp. *delbrueckii*
	L. fermentum
	L. leichmannii
	L. plantarum
Cured meat products	*L. viridescens*
Wines and fruit juices	*L. brevis*
	L. fermentum
	L. hilgardii
	L. trichodes
Intestine of poultry, and also of human infants	*L. acidophilus*
	L. brevis
Human intestine	*L. casei*
	L. plantarum

40

A Scheme for the Identification of Yeasts and Moulds

In this Section, descriptions are given to assist in the identification of the yeasts and moulds most commonly isolated from foods and dairy products. For more comprehensive descriptions of yeasts and moulds, refer to the works given in the bibliography.

The methods of examination described in Part I (Section 14) are not always entirely satisfactory for purposes of identification. The preparation of wet-mounted stained slides from Petri-dish cultures almost invariably causes serious distortion or disintegration of sporing structures and disintegration of pseudomycelia or mycelia that are fragmenting into arthrospores. It is therefore more satisfactory to examine yeasts and moulds by slide culture, when less distortion and fragmentation can take place. The slide culture methods employed are somewhat different for yeasts than for moulds that form true mycelia. Certain fungi, such as *Aureobasidium*, *Geotrichum*, *Endomyces* and *Candida*, which are intermediate in their structure between yeasts and mycelial fungi, may on first isolation have the colonial characterization of either yeasts or moulds, depending on the cultural conditions such as medium, temperature and period of incubation. When subsequent examination suggests that the isolate may form a true mycelium *and* be capable of yeast-like growth, both methods of slide culture should be employed.

In addition to slide cultures, culture the fungi on a number of media in Petri-dish preparations. Since the nutritional requirements of fungi vary greatly, a variety of media should be used in an attempt to obtain sporulation. The media recommended are malt extract agar, Czapek–Dox agar, potato dextrose agar and Davis' yeast salt agar. At least two media, one with a high and one with a low concentration of carbohydrate, should be used. Cultures should be incubated at 25°C or other appropriate incubation temperature, and examined every other day for up to 2 weeks, using a stereoscopic microscope with incident light. The spores of a number of moulds commonly encountered in foods may, if inhaled, either induce an allergic response or establish a lung infection, so care should be taken when manipulating sporing cultures.

40.1 IDENTIFICATION OF YEASTS

Yeasts are identified on the basis of morphological and cultural characteristics and ascospore and ballistospore production, but biochemical tests are also of great importance. The details that follow are intended to assist in the identification to generic level of the yeasts likely to be isolated from foods and dairy products, based primarily on morphological characters. For further characterization and identification of yeast isolates, refer to the monographs by Barnett *et al.* (1990) and Kreger-van Rij (1984). Identification keys, based on physiological tests only, and on both physiological tests and microscopical examination, have been produced by Barnett *et al.* (1990). Suitable media for the carbon-source and nitrogen-source assimilation tests are available in dehydrated form from Difco Laboratories.

On agar media (e.g. corn meal agar, malt extract agar) young yeast colonies in 2–5 days at 25°C characteristically produce domed spherical colonies with entire edges, of a moist and butyrous consistency, with a shiny, semi-matt or matt smooth surface. The colour of the colony is normally white, cream, pink or salmon. On further incubation the colonies of some species tend to become wrinkled and of a dry and crumbly consistency. The colonies therefore resemble those of some bacteria, but most bacteria tend to grow sparsely on the low-pH or selective media used for counting or isolating yeasts and moulds. Subsurface colonies of yeasts in pour-plates often have a characteristic stellate appearance.

Microscopic examination of wet-mounted preparations, either unstained or stained with Gram's iodine or Loeffler's methylene blue, normally reveal unicellular spherical, ellipsoidal, ovoid or cylindrical organisms (see Fig. 40.1). These may reproduce vegetatively by budding or binary fission. The production of a pseudomycelium or mycelium is often revealed only by slide cultures. Slide cultures for the growth of yeasts should be carried out with malt extract agar or one of the potato extract agars.

40.1.1 Preparation of slide cultures for the examination of yeasts

Sterilize a Petri dish containing a filter paper disc in the base, on which is a U-shaped glass rod carrying two clean microscope slides. After sterilization, aseptically, add a few millilitres of a sterile 20% solution of glycerol to the filter paper. This will help to keep the slide culture from drying out during incubation. With a Pasteur pipette, add molten malt extract agar or potato dextrose agar to the surface of the slides to form a very thin layer of agar medium. Inoculate the slides with the yeast to be examined in a series of streaks with a wire loop. Place a sterile coverslip over part of the inoculated medium. Incubate at 25°C for up to 7 days with the lid of the Petri dish in place.

The slide cultures may be examined in the living state, preferably with the aid of a phase-contrast microscope, and the types of growth under the coverslip and exposed to air can be compared. Alternatively, the coverslip can be gently removed from the slide culture and mounted in a drop of lactophenol, lactophenol–picric acid or lactophenol–cotton blue on a slide. The edges of the coverslip can be sealed to the slide with paraffin wax, nail varnish or a proprietary brand of ringing cement.

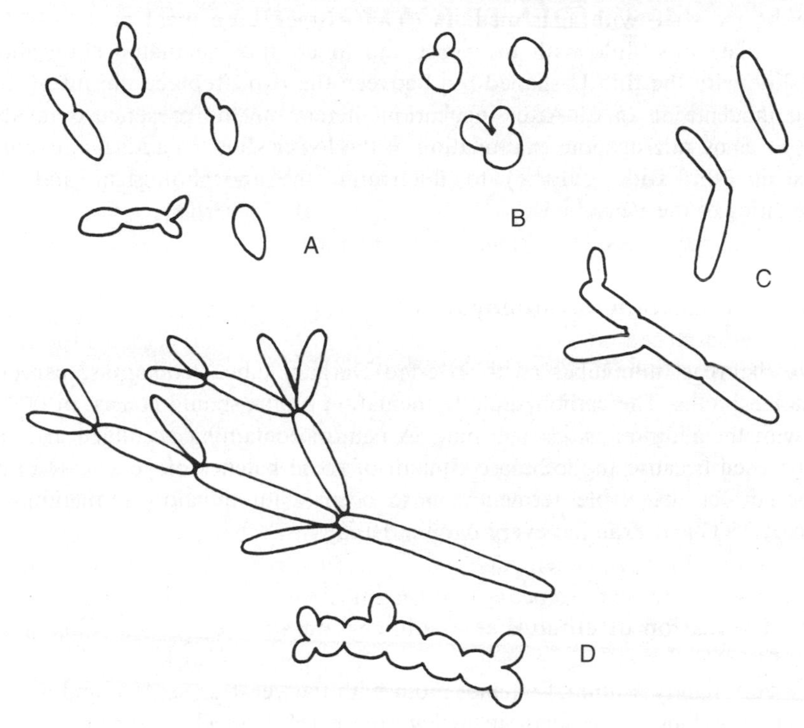

Fig. 40.1 A, Ovoid budding cells (e.g. *Kluyveromyces lactis*, *Saccharomycs cerevisiae* var. *ellipsoideus*). B, Spherical budding cells (e.g. *Saccharomyces*, *Debaryomyces*). C, Cylindrical budding cells (e.g. *Pichia*) D, Examples of pseudomycelia.

40.1.2 Morphological and physiological characters of yeasts

40.1.2.1 Ascospore production

One method of inducing ascospore production has already been described (see Section 14). Other methods include subculturing on to gypsum blocks, Gorodkowa agar or V-8 agar.

40.1.2.2 Ballistospore production

Lodder and Kreger-van Rij (1952) recommended the use of a modification of the slide culture technique to detect and examine ballistospore production by members of the Sporobolomycetaceae. In this method another very thin U-shaped glass rod is sterilized in a Petri dish together with the filter paper, U-shaped glass rod and microscope slides.

Cover only one slide with agar medium (malt extract, corn meal or potato dextrose agar). Inoculate this slide with the yeast and invert the inoculated slide above the second slide with the thin U-shaped rod between the two. Replace the lid of the Petri dish and incubate at 25°C. After incubation, determine the presence or absence of ballistospores by microscopic examination of the lower slide. In addition, examine the upper slide (the slide culture) to determine the morphological and cultural characteristics of the yeasts.

40.1.2.3 Fermentation of carbohydrates

Inoculate the yeast into tubes (with inverted Durham tubes) containing carbohydrate fermentation broths. The carbohydrate fermentation broths should consist of 0.5% yeast extract with the addition of 2% substrate. A control containing no added carbohydrate should be used because the trehalose content of some batches of yeast extract may be high enough for detectable fermentation to occur with trehalose-fermenting yeasts. Incubate at 25°C and examine every day for 10 days.

40.1.2.4 Utilization of ethanol as a sole carbon source

Inoculate very lightly a tube of ethanol broth with the yeast under test, and at the same time inoculate a tube of the medium *without* the ethanol. Incubate for up to 3 weeks at 25°C, examining at frequent intervals for growth. Because yeast extract solution is used in the medium as a source of vitamins and growth factors, the growth of the yeast being examined *must* be compared in the tubes with and without ethanol to ensure that false positives are not recorded.

40.1.2.5 Resistance to cycloheximide

Inoculate the yeast into filter-sterilized Yeast Nitrogen Base 0392 (Difco) containing 0.5% glucose and 100 μg cycloheximide per ml. Incubate at 25°C for 4 weeks, examining weekly for growth. Good growth within 1 week is taken as indicating high resistance; scant or no growth in 3 or 4 weeks is taken as indicating high sensitivity.

40.1.3 Key to the Identification of Yeasts

Ascospores produced .... Ascosporogenous yeasts
Ballistospores produced.. Family Sporobolomycetaceae
Neither ascospores nor ballistospores produced...................... Asporogenous yeasts

40.1.3.1 Ascosporogenous yeasts

These yeasts may form a pseudomycelium or a true mycelium. Vegetative reproduction is by budding, binary fission or arthrospores.

1 Vegetative reproduction by fission, with a true mycelium and arthrospores formed. No budding occurs .... **1** *Endomyces*, **2** *Schizosaccharomyces*
Vegetative reproduction by multilateral budding and sometimes fission............ *2*
Vegetative reproduction by bipolar budding. Includes............. **3** *Hanseniaspora*

2 True mycelium formed as well as budding cell .... **4** *Hyphopichia*
No true mycelium but a pseudomycelium and/or loose collections of budding cells may be formed .... *3* and **5** *Hansenula*

3 Dry dull pellicle develops rapidly in malt extract broth................................ *4*
Dry pellicle develops very slowly if at all in malt extract broth...................... *5*

4 Ellipsoidal to long cylindrical cells .... **6** *Pichia*
Round or short ovoid cells.. **7** *Debaryomyces*

5 Glucose always strongly fermented .... **8** *Saccharomyces*, *Kluyveromyces* and *Zygosaccharomyces*
Glucose not fermented, or only weakly so.............................. **7** *Debaryomyces*

1 *Endomyces* Forms a true mycelium, and is more fungal than yeast-like. Ascospores are spherical, ovoid or hat-shaped. The mycelium breaks up into arthrospores which are either cylindrical with rounded ends or ovoid (see Fig. 40.2). Vegetative reproduction is by fission.

The equivalent genus in the Fungi Imperfecti is *Geotrichum*. The species *Endomyces lactis* (also known as *Geotrichum candidum* or *Oospora lactis*) is commonly found on dairy products and is therefore called the 'dairy mould'. The colonies are white in colour and are yeast-like and butyrous, particularly when older.

2 *Schizosaccharomyces* Members of this genus are found in food products such as molasses, honey, dried fruits and fruit juices.

The cells are cylindrical, ovoid or spherical, but a mycelium that breaks up into arthrospores may be formed. Vegetative reproduction is by fission. Four or eight ascospores are formed per ascus. *Schizosaccharomyces* may have a similar appearance to *Endomyces* but in the former asci are formed after the fusion of two arthrospores, whereas in *Endomyces* asci develop in the mycelium at branches of the hyphae.

Schizosaccharomyces ferments carbohydrates. It cannot use ethanol as a sole carbon source.

3 *Hanseniaspora* The equivalent genus in the Fungi Imperfecti is *Kloeckera*. The cells are ovoid or lemon-shaped. A pseudomycelium is rarely formed. Vegetative reproduction is by budding at both poles. Ascospores (one to four per ascus) are spherical, becoming

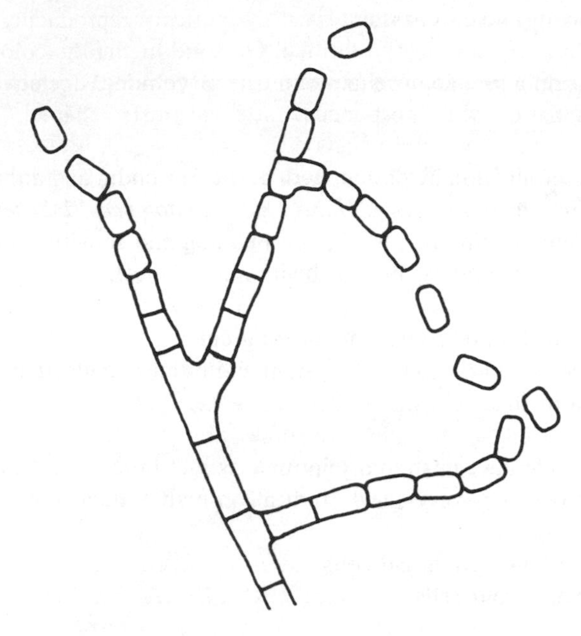

FIG. 40.2 *Endomyces.*

hat-shaped or Saturn-shaped. *Hanseniaspora* can ferment carbohydrates. It cannot use ethanol as the sole carbon source.

4 *Hyphopichia* (previously known as *Endomycopsis*) This has typical yeast-like, budding cells and in addition may form either a true mycelium that gives rise to blastospores, or a pseudomycelium, in these respects resembling *Candida*.

Ascospores may be one of a variety of shapes, depending on species, including: hat shaped, Saturn-shaped, sickle-shaped, spherical or ovoid.

A pellicle is usually formed on liquid media. *Hyphopichia* attacks sugars oxidatively and is only very weakly fermentative. It grows only slightly with ethanol as the sole carbon source. Members of the genus have been isolated from bread, fruit and vegetables.

5 *Hansenula* This genus produces on solid media white or cream, dull or shiny colonies, which are usually wrinkled.

The cells may be spherical, ovoid, ellipsoidal or cylindrical, and budding occurs. Pseudomycelia are frequently formed, particularly by species with cylindrically shaped cells. In liquid media a dry pellicle with a dull appearance is usually formed. Ascospores (one to four per ascus) are spherical, hat-shaped or Saturn-shaped.

Glucose is fermented vigorously, other carbohydrates being fermented according to species. Some species produce aromatic esters amongst their fermentation products. Ethanol can be utilized as a sole carbon source. This genus is of importance as a contaminant in the brewing industry and may cause spoilage of pickling brines.

6 *Pichia* Cells are ovoid, ellipsoidal or cylindrical, reproducing vegetatively by budding. A pseudomycelium may be formed. On solid media the colonies are white, or cream, dull and usually wrinkled. In liquid media a wrinkled, dry pellicle is formed.

Ascospores (one to four per ascus) are variously shaped, being spherical, hemispherical, hat-shaped or Saturn-shaped.

Carbohydrates are not usually fermented, or only weakly so. Many species can use ethanol as a sole carbon source. Like *Hansenula*, *Pichia* is an important contaminant of beers and pickled products.

7 *Debaryomyces* The cells are spherical or ovoid and show multilateral budding; sometimes a pseudomycelium is formed. On solid media the colonies are off-white, yellow or brown, and dull to shiny with a wrinkled appearance. A dry pellicle may be produced in liquid media.

Ascospores (usually one per ascus) are spherical or ovoid.

Carbohydrates are not fermented, or only weakly so. Some species can use ethanol as a sole carbon source. *Debaryomyces* has a high salt tolerance and can frequently be isolated from pickled and salted foods.

8 *Saccharomyces, Kluyveromyces* and *Zygosaccharomyces* The members of these genera usually grow as unicellular spherical, ovoid or longer cells with rounded ends, but a pseudomycelium may sometimes be formed. Vegetative reproduction is by budding. On solid media, colonies are usually white or cream, domed, smooth, semi-matt to shiny, up to 5 mm in diameter, and of a butyrous consistency. In liquid media, sediment is formed, sometimes with ring growth at the air–liquid–glass junction; a pellicle is not formed.

Ascospores (one to four per ascus) are usually spherical or ovoid. In *Kluyveromyces* the mature asci rupture to release the ascospores. In the case of *Saccharomyces*, rupture of asci does not occur on maturity.

Carbohydrates are fermented vigorously (see Table 40.1). Species are defined mainly on the basis of carbohydrate fermentation and carbon-source utilization (see Barnett *et al.*, 1990). Although morphological differences between species can be seen, there is usually too great a morphological variation within a species for characteristics such as cell shape to be used with any degree of reliability. *S. cerevisiae* is the 'top yeast' used in the brewing industry. *S. cerevisiae* var. *ellipsoideus* is the yeast responsible for vinous fermentations, and is supposedly distinguished from *S. cerevisiae* on the basis of its more ellipsoidal shape, compared with the spherical or short ovoid cell characteristic of *S. cerevisiae*, although there appears to be little differentiation between the two. *S. cerevisiae* is also the typical baker's yeast.

Several other species responsible for the production of fermented beverages have been described. *S. carlsbergensis*, for example (which is included in the species *S. uvarum*), is the 'bottom yeast' used in the manufacture of lager.

The strains of *Kluyveromyces marxianus* previously included in *K. (Saccharomyces) fragilis* and *K. (Saccharomyces) lactis* are found in certain fermented milks.

S. rouxii and *Zygosaccharomyces bisporus* are osmophilic or osmotolerant, and can cause spoilage of jams, honeys and similar products.

TABLE 40.1 Differentiation of common species of *Saccharomyces*, *Kluyveromyces* and *Zygosaccharomyces*

Organism	Glucose fermentation	Maltose fermentation	Galactose fermentation	Sucrose fermentation	Lactose fermentation	Fructose fermentation	Utilization of ethanol as sole carbon source	Cycloheximide resistance
S. cerevisiae	+	V	V	V	−	+	V	−
S. uvarum (S. carlsbergensis)	+	V	V	+	−	+	−	−
*Kluyveromyces marxianus**	V	V	+	V	V	+	+	+
Zygosaccharomyces rouxii	+	V	−	−	−	+	V	−
Z. bisporus	+	−	−	−	−	−	+	−

* Synonyms include *K. fragilis* and *K. lactis*.

40.1.3.2 Sporobolomycetaceae

Ballistospores are formed and discharged forcibly into the air when ripe, and these may give rise to 'mirror images' of the colonies on the lid of the Petri dish or the lower slide when these yeasts are grown in plate cultures or two-slide slide cultures. The ballistospores may be spherical, ovoid, kidney shaped, bean shaped or sickle shaped (see Fig. 40.3).

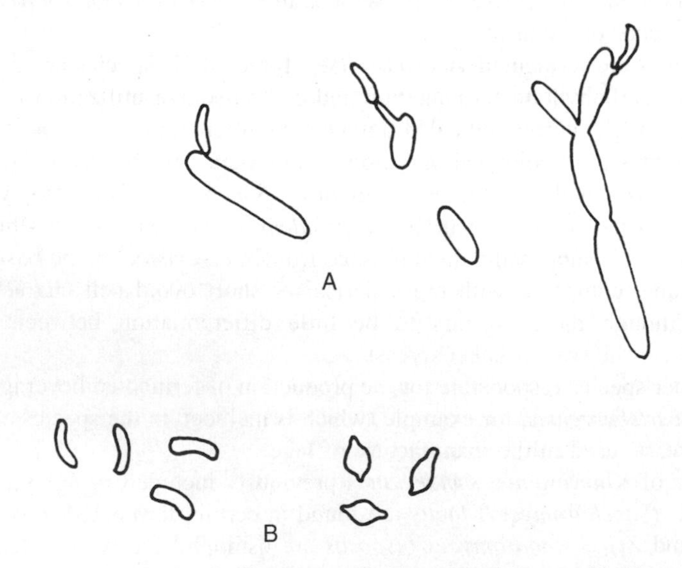

FIG. 40.3 A, Yeast cells bearing ballistospores. B, Two typical forms of ballistospore.

In liquid media growth is predominantly on the surface, although slight sediment may be formed. Neither *Sporobolomyces* nor *Bullera* can ferment carbohydrates.

Colonies red, pink or salmon-pink in colour, although sometimes very pale or even cream. A pseudomycelium or true mycelium may be formed. Ballistospores kidney or sickle shaped .... **9** *Sporobolomyces*

Colonies colourless, white, cream or pink. A pseudomycelium or true mycelium is never formed. Ballistospores spherical or ovoid.................................. **10** *Bullera*

40.1.3.3 Asporogenous yeasts

The identification of an organism to generic level may be difficult in this family owing to the relatively slight differences between genera and the extreme morphological variability that can be observed in a culture. It is recommended, therefore, that, for confirmation of the identity of a culture and for differentiation at specific level, the monographs of Kreger-van Rij (1984) and Barnett *et al.* (1990) be consulted for physiological keys.

1 Forms budding cells and pseudomycelia, and also a true mycelium that fragments into arthrospores .... **11** *Trichosporon*
No arthrospores produced, although a true mycelium may be formed in addition to budding cells and pseudomycelia... *2*

2 Pseudomycelium and/or a true mycelium ....**12** *Candida*
Predominantly unicellular, although a very rudimentary pseudomycelium may be formed *3*

3 Brightly coloured red, orange or yellow................................ **13** *Rhodotorula*
Not brightly coloured (may be white, grey, cream, yellowish or brown)........... *4*

4 Spherical or ovoid cells pointed at one end ('ogive' shaped) ... **14** *Brettanomyces*
Spherical or ovoid cells.. **15** *Torulopsis*

11 *Trichosporon* *Trichosporon* is morphologically very variable with budding cells, a pseudomycelium, or a true mycelium that fragments into arthrospores. Carbohydrates are not usually fermented.

12 *Candida* These are morphologically very variable, ranging from budding unicellular organisms to pseudomycelia or sometimes a true mycelium. Blastospores are frequently formed on the mycelia (see Fig. 40.4).

Reproduction is by budding or by fission. On solid media colonies are off-white or cream. Species are differentiated partly by morphological characteristics and partly by carbohydrate fermentation patterns.

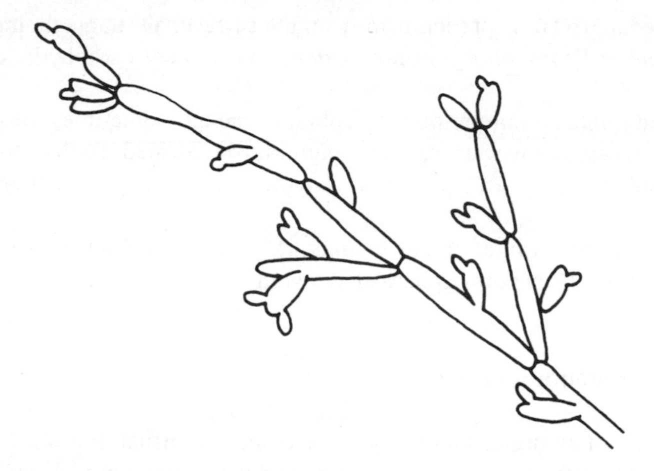

Fig. 40.4 *Candida* showing blastospore production.

C. albicans is an animal parasite or pathogen and is capsulate. Other species may also be pathogenic for animals. A number of species may be isolated from foods and can cause spoilage of fermented beverages and pickled products. *C. kefyr* (synonyms include *Torulopsis kefyr* and *Candida pseudotropicalis* var. *lactosa*) is involved in the production of kefir and may be isolated from kefir grains, and sometimes from buttermilk.

13 *Rhodotorula* This usually consists of budding cells, but occasionally a very rudimentary pseudomycelium may be formed. The colonies on solid media are brightly coloured red, pink, orange or yellow. Members of this genus do not ferment carbohydrates. They are frequently involved in the spoilage of a wide range of food products.

14 *Brettanomyces* Usually consists of budding cells that are 'ogive'-shaped (i.e. spherical or ovoid, but pointed at one end). A pseudomycelium may be formed. On solid media, colonies are white to yellowish, and may be glistening, moist and smooth, or dull and wrinkled.

Carbohydrates are fermented by some species. Acetic acid can be produced aerobically from glucose. Ethanol can be utilized. This organism can cause spoilage in beer, cider and wine.

15 *Torulopsis* This consists of spherical or ovoid, budding cells. A mycelium is not formed. A rudimentary pseudomycelium is rarely formed. It is rarely capsulate. On solid media, colonies are off-white, cream, yellowish or brownish, smooth, and may be shiny or dull.

Carbohydrates may be fermented. Some species utilize ethanol as a sole carbon source. Some are osmophilic or halophilic and can cause spoilage in pickled products and foods with a high sugar concentration.

40.2 IDENTIFICATION OF FILAMENTOUS MICROFUNGI

Moulds are differentiated on the basis of morphological and cultural characteristics so it is necessary to employ methods of culture and examination that, as far as possible, avoid distortion of sporing structures. In addition, the formulation of the medium on which a mould is grown has a profound influence on the colonial appearance and on the development of aerial sporing structures, chlamydospores, sclerotia, etc. It is advisable, therefore, during the identification of a mould, to grow it on a wide variety of media. As mentioned earlier, these may include malt extract agar, Czapek–Dox agar, potato dextrose agar, potato carrot dextrose agar and Davis's yeast salt agar.

Colonial and cultural characteristics are determined on plate cultures by centrally inoculating a poured plate with spores (obtained with a wire loop from the surface of a colony) or with a small piece of agar containing substrate mycelium from a colony of the mould.

Morphological characteristics sometimes can be determined by wet preparations from plate cultures (see Section 14). Very often, however, serious distortion of sporing structures results from such preparations, with the majority of spores being detached. If this prevents identification, slide cultures should be prepared. There are two main methods of slide culture that can be used. In addition, moulds that have fragmenting mycelia can be examined using the method of slide culture already recommended for yeasts. The methods described below do not include a method of fixing the fungal growth to the slides or coverslips as such fixing usually leads to distortion. If fixation is found necessary, brief heat fixation may be tried. Alternatively, fixation may be accomplished using Bouin's fixative for 1 h before removing the agar from the slide or coverslip.

40.2.1 Slide culture methods

40.2.1.1 Agar block slide cultures

Pour 10 ml of medium into a Petri dish and allow it to set. Sterilize a Petri dish containing a filter paper disc in the base, on which is a U-shaped glass rod carrying two clean microscope slides. After sterilization, aseptically add a few millilitres of a sterile 20% solution of glycerol to the filter paper, to keep the slide culture from drying up during incubation. From the solidified medium, aseptically cut blocks of 1 cm^2 and mount them on the microscope slides. Inoculate the four cut surfaces of each block with the mould to be examined and place a sterile coverslip over the block. Incubate the slide cultures in the Petri dishes with the lids in place. The slide cultures may be removed from time to time and examined microscopically under the low-power or high-power dry objectives, the coverslips helping to prevent contamination at these times. When a satisfactory development of the mould is observed, wet preparations can be made, as the mould tends to adhere to coverslip and slide. Carefully remove the coverslips from the agar blocks and very carefully mount the coverslips in lactophenol, lactophenol–cotton blue or lactophenol–picric acid on fresh slides. In addition, the agar blocks can be removed from the original slides and these slides also examined with the staining mountant and fresh coverslips.

40.2.1.2 Johnson's slide culture method (Johnson, 1946)

Sterilize a Petri dish containing filter paper, glass rod and slides, and add sterile 20% glycerol as already described. Place aseptically two narrow parallel strips of sterile Vaspar about 2 cm apart across the width of a slide. Centrally between the two strips, place a fungal inoculum on the slide and then cover with a coverslip placed across the Vaspar strips (see Fig. 40.5). Using a sterile Pasteur pipette, run sterile melted agar medium (cooled to 55°C) between the slide and the coverslip, until the advancing front of medium just reaches the inoculum (Fig. 40.5(b)). This thus provides the mould with a growing edge. Lastly, seal the edge of the coverslip over the agar medium, to prevent the medium from drying out (Fig. 40.5(c)). The fourth side of the coverslip is left open.

Incubate the slide culture in the Petri dish with the dish cover replaced.

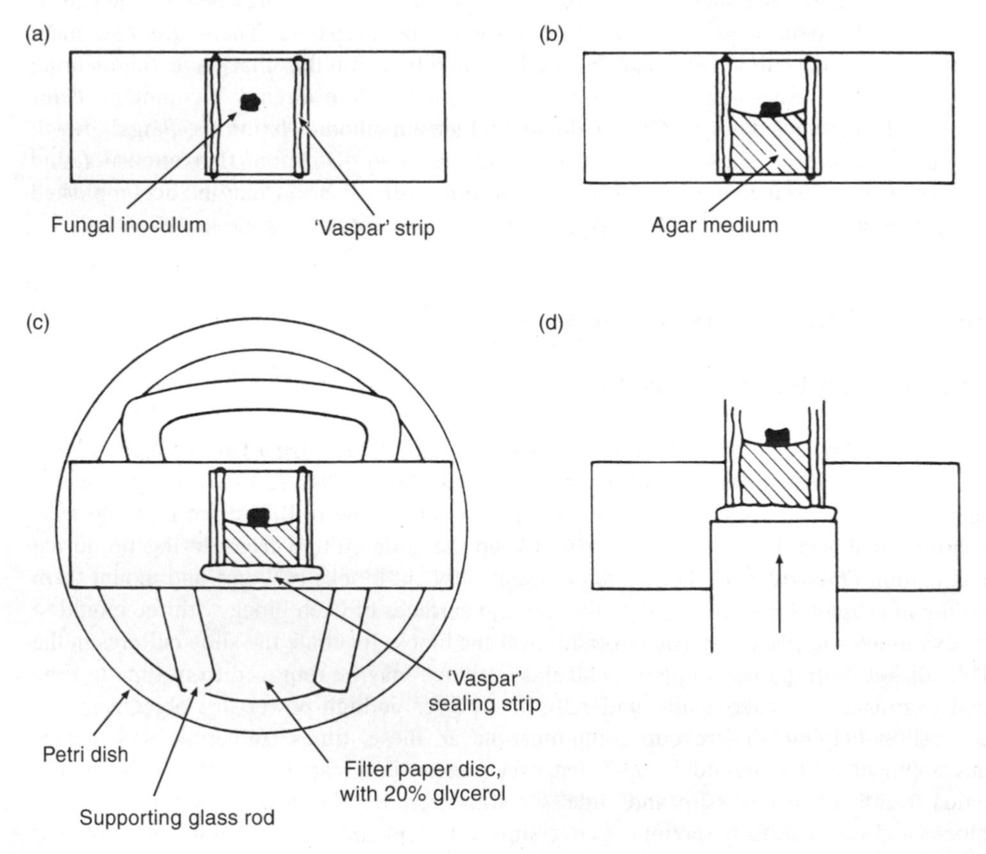

FIG. 40.5

The culture can be examined microscopically from time to time and, when required, a wet, stained preparation of the coverslip can be examined. To remove the coverslip, cut off the Vaspar seal at the edge of the slide and carefully push the coverslip off the slide using another slide, (Fig. 40.5(d)). Then cut off the Vaspar strips at each side. Carefully cut away the agar, leaving the fungal growth on the coverslip. Mount the coverslip in one of the lactophenol mounting media.

40.2.2 Key to the identification of moulds

The moulds described are those most commonly isolated from foods, dairy products, etc. and identification is primarily on the basis of the morphology of sporing structures. If difficulty is found in an identification, the works by Pitt and Hocking (1997) and Samson and van Reenen-Hoekstra (1988) provide keys, photomicrographs and descriptions of a wider range of the microfungi found in foods.

1 Coenocytic mycelium (septa absent or only present in, for example, chlamydospore formation)........ *2*
Mycelium possessing many septa........ *4*

2 Sporangiophores each bear many cylindrical sporangia, each containing a chain of spores........ **1** *Syncephalastrum*
Sporangiophores each bear terminally a single large globose sporangium containing many spores........ *3*

3 Stoloniferous type of spread, with sporangiophores developing from the nodes at which thick tufts of rhizoids develop........ **2** *Rhizopus*
Long sporangiophores carry dichotomously branched short sporangiophores bearing small sporangioles (containing few spores), usually in addition to a large terminal sporangium containing numerous spores........ **3** *Thamnidium*
Sporangiophores each bear terminally a single large globose sporangium containing many spores. Stoloniferous growth does not occur........ **4** *Mucor*

4 Vegetative hyphae clear and transparent, colourless or brightly coloured........ *5*
Vegetative hyphae darkly coloured, not clear and transparent........ *12*

5 Asexually produced spores are unicellular........ *6*
Asexually produced spores of two or more cells........ *11*

6 Spores borne singly on extremely short projections from near the ends of hyphae........ **5** *Sporotrichum*
Spores borne in clusters or chains........ *7*

7 Conidia formed in grape-like clusters. (Sclerotia, but not conidia, are formed on media of high C : N ratio, colonies being white and fluffy with black sclerotia at or below the surface of the medium. On media with a low C : N ratio, colonies are greenish-grey with conidia.) .... **6** *Botrytis*
Spores formed in branched chains .... **7** *Monilia*
Spores formed in unbranched chains *8*

8 Conidiophores non-septate at first, but then become septate and produce arthrospores .... **8** *Wallemia*
Non-septate conidiophores arising from thick-walled foot cells. Conidia borne by phialides arising from a terminal swelling on the conidiophore .. **9** *Aspergillus*
Condiophores septate, no specialized foot cells.. *9*
Conidiophores septate, with foot cells .... **10** *Stachybotrys*

9 Spores truncated, each with a thickened basal ring .... **11** *Scopulariopsis*
Spores produced on phialides, and not possessing thickened basal rings. Phialides in brush-like clusters *10*
Spores produced on three to seven swollen phialides at apex of conidiophore. Dark spores, usually becoming enveloped in slime.. **10** *Stachybotrys*

10 Clusters of eight-spored asci are formed, without an outer retaining wall (peridium).. **12** *Byssochlamys*
Ascospores, if formed, contained within perithecia .... **13** *Penicillium*

11 Conidia two-celled, pear-shaped .... **14** *Trichothecium*
Conidia many-celled, spindle- or sickle-shaped.. **15** *Fusarium*

12 Blastospores formed on any part of the mycelium; the colonies are at first slimy, and then become dark greenish-black and leathery .... **16** *Aureobasidium*
Spores of one or two cells.. **17** *Cladosporium*
Spores of more than two cells.. *13*

13 Spores with cross walls; spores usually bent or curved, with one or more of the middle cells thickened .... **18** *Curvularia*
Spores with cross walls and longitudinal septa; spores usually pear-shaped........ **19** *Alternaria*

1 *Syncephalastrum* Possesses a non-septate mycelium of hyphae of large diameter. The sporangiophores, which are usually branched but with the branches not cut off by septa, have a terminal swelling. Each terminal swelling bears a number of cylindrical sporangia, each of which contains a chain of spores (see Fig. 40.6).

The colonies are at first white, but later become grey to black as a result of the production of sporangia, which are black in colour.

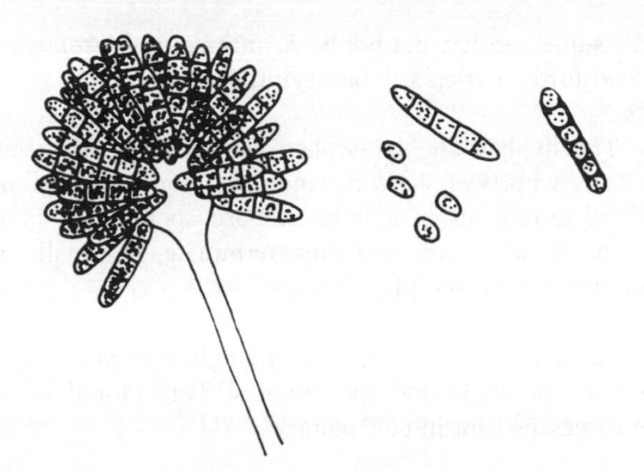

FIG. 40.6 *Syncephalastrum*.

2 *Rhizopus* *Rhizopus* has a mycelium of non-septate hyphae of large diameter. Growth on plate cultures is extremely rapid, by the development of stolons. Where a stolon touches the medium, a tuft of short, rather thick, rhizoids grows into the medium, and from this 'rooted' node sporangiophores develop. Each sporangiophore bears a terminal black spherical sporangium (see Fig. 40.7).

The mycelium is thick and similar to cottonwool, and when sporangia are present these are visible to the naked eye as black pin-heads. A Petri dish frequently may be rapidly filled with the aerial mycelium. The commonest species is *R. nigricans*. Chlamydospores

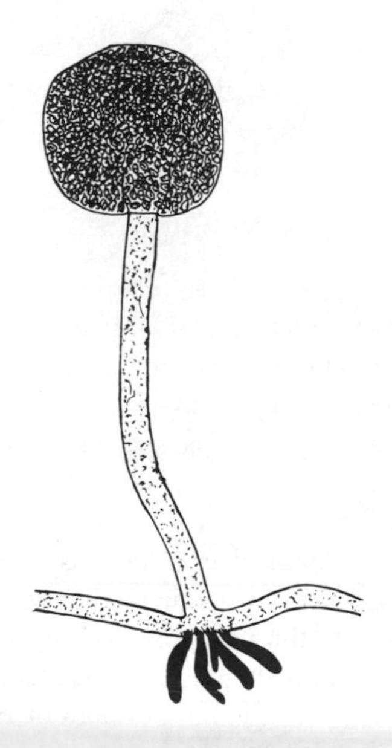

FIG. 40.7 *Rhizopus*.

may be formed by some species, but not by *R. nigricans*. *Rhizopus* is a fairly common contaminant of foodstuffs, particularly bakery products.

3 *Thamnidium* The hyphae are non-septate and of large diameter. The long main sporangiophores carry whorls of short, dichotomously-branched, sporangiophores. Each long sporangiophore terminates in a large globose sporangium containing numerous spores, whereas the short sporangiophores terminate in small sporangioles, each containing one to four spores (see Fig. 40.8).

On solid culture media, under favourable conditions, the mycelium formed is loose and cotton-wool-like, up to 1 cm or more in height, light grey at first, darkening as a result of the production of sporangia and sporangioles. This mould is occasionally found growing on meat carcasses kept in cold storage.

FIG. 40.8 *Thamnidium*.

4 *Mucor* The hyphae are non-septate and of large diameter, the younger the hyphae the smaller the diameter. The sporangiophores are erect and may be unbranched or branched, each sporangiophore or branch bearing a single globose sporangium containing a large number of spherical or ellipsoidal spores. The sporangiophores never arise from nodes on stolons, and rhizoids are absent from the mycelium. Chlamydospores may be formed by some species (see Fig. 40.9).

Usually growth on a solid medium gives a loose cottonwool-like aerial mycelium, at first white or grey in colour, later becoming darker as sporangia are produced. The sporangia are often just visible to the naked eye as darkly coloured or black pin-heads scattered over the aerial mycelium. It should be noted that a few species do not produce a white or grey mycelium at first, being yellow, orange or bluish in colour. In addition, a few species produce yellow or orange sporangia.

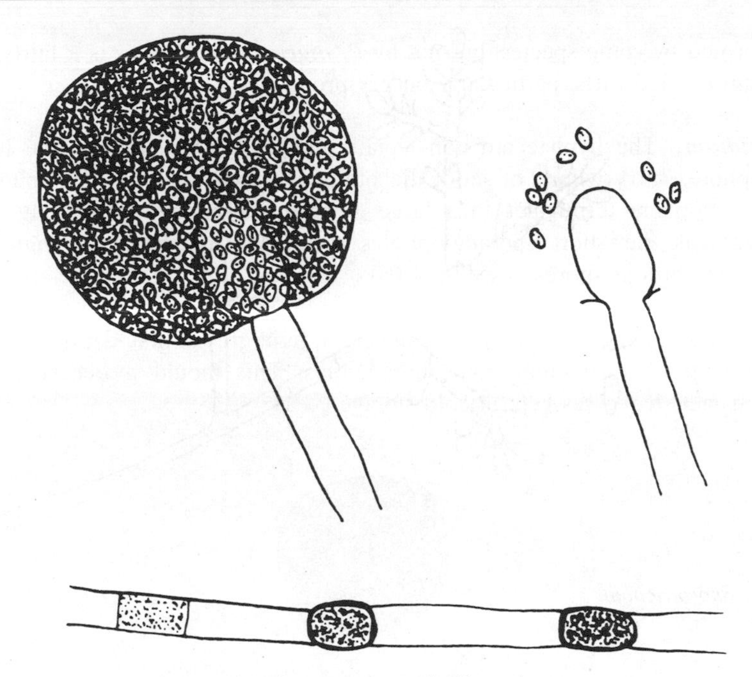

FIG. 40.9 *Mucor*: entire and dehisced sporangia, and chlamydospores.

Mucor is a very common soil-dwelling organism and therefore may be found contaminating a variety of food products.

5 *Sporotrichum (Aleurisma)* Proper conidiophores are not formed; conidia are borne singly on very short projections which arise toward the ends of the hyphae (see Fig. 40.10). The conidia may be spherical or ovoid, colourless or brightly coloured and transparent.

Colonies may be white, grey, yellow, green, pink or red, depending on the species. As the conidia mature, the colonies change from being velvety to being powdery. The mycelium is fairly closely adpressed to the surface of the medium.

The genus includes soil-dwelling saprophytes and also dermatophytes. *Sporotrichum* has been found as a contaminant on the surface of meat carcasses being kept in cold stores.

6 *Botrytis* Conidiophores are irregularly branched, and the apical cells are frequently enlarged. Conidia are borne singly on short sterigmata, but the positions and numbers of the sterigmata cause the conidia to be arranged in grape-like clusters (see Fig. 40.11).

Botrytis cinerea is the species most commonly encountered. Colonies are frequently white and fluffy at first. On media with a low C : N ratio, conidia are formed. The conidia are usually grey or greenish-grey in mass, hyaline and one-celled, and the colonies consequently change from being white and fluffy to being greenish-grey and dusty. Media with a high C : N ratio, on the other hand, suppress the production of conidia, but encourage the development of sclerotia. Colonies on such media tend to remain white and

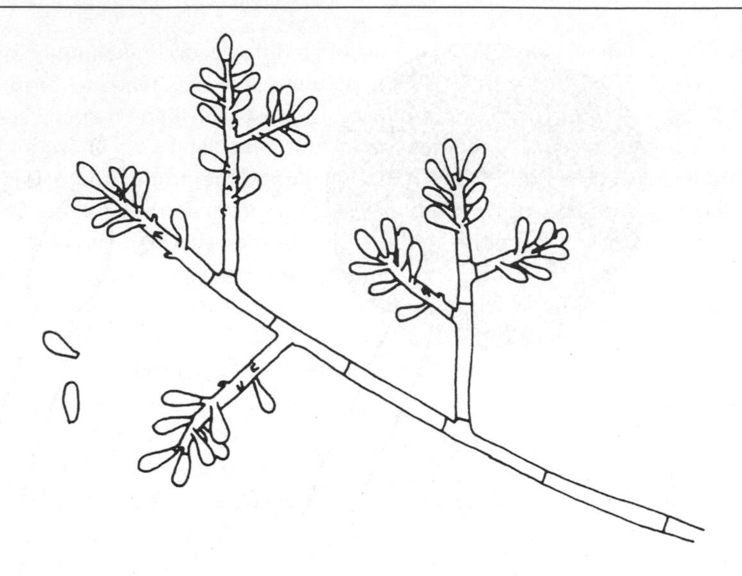

FIG. 40.10 *Sporotrichum.*

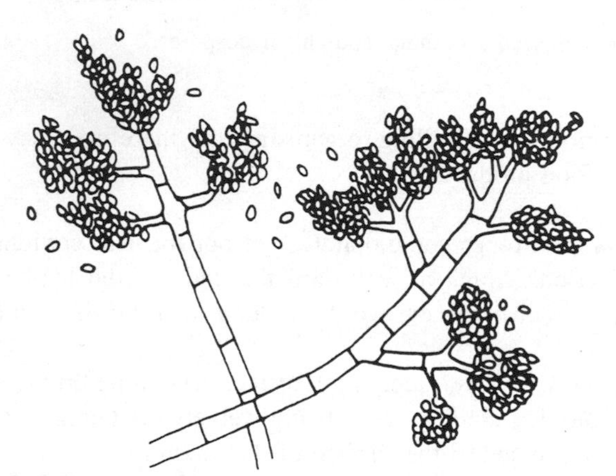

FIG. 40.11 *Botrytis.*

fluffy, with numerous flat greenish-black sclerotia up to about 3 mm long being formed at or below the surface of the medium. Media with high and low C : N ratios can be made by using malt extract agar with modifications to the concentrations of malt extract and peptone. Reduction in the amount of peptone provides a medium with a high C : N ratio, whereas reducing the amount of malt extract and increasing the amount of peptone produces a medium with a low C : N ratio.

Botrytis cinerea is saprophytic or plant-parasitic. It causes 'grey mould' of many plants, and a similar spoilage of fruit and vegetable products. As a saprophyte it is common in soils. Therefore, it is frequently the cause of mould spoilage of strawberries.

7 *Monilia* Simple or branched chains of conidia occur, branching being due to budding of the conidia (see Fig. 40.12). The mycelium as it ages tends to form arthrospores.

Monilia sitophila is the commonest and most well known species. It grows best on complex organic media. Colonies are at first white and very loose and fluffy, but they rapidly become pale pink to red in colour due to the production of masses of coloured conidia. The conidia can be produced in great profusion – covering the lid of the Petri dish and even being found on the outside between the dish and its cover. *M. sitophila* can therefore easily become a contaminant in microbiological laboratories and incubators, and its nuisance value is increased by its ability to grow well on nutrient agar and similar bacteriological media and at temperatures up to and including 37°C. It is most frequently associated with spoilage of bakery products – particularly wrapped, sliced bread, although inhibited by the inclusion of sorbate – causing a pink fluffy growth which is extremely characteristic and easy to recognize.

It should be noted that the conidia are very easily detached and the formation and structure of the chains of conidia, in particular the characteristic branching due to conidial budding, can best be seen in slide culture.

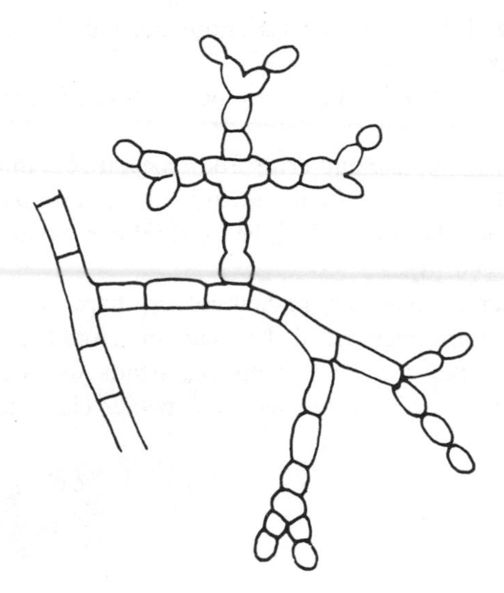

Fig. 40.12 *Monilia.*

8 *Wallemia* Conidiophores are usually non-pigmented and produced erect from the substrate mycelium. The conidiophore terminates in a hypha which is non-septate at first, but develops septa, and produces endospores (or arthrospores) which are usually pigmented (see Fig. 40.13).

Wallemia sebi (previously known as *Sporendonema ichthyophaga*) may be found in bakery goods and in foods with high sugar or salt content. It produces brown colonies.

9 *Aspergillus* There are a large number of species in this genus and they are amongst the commonest isolates from soils, spoiled foods, etc. Identification of an isolate to the

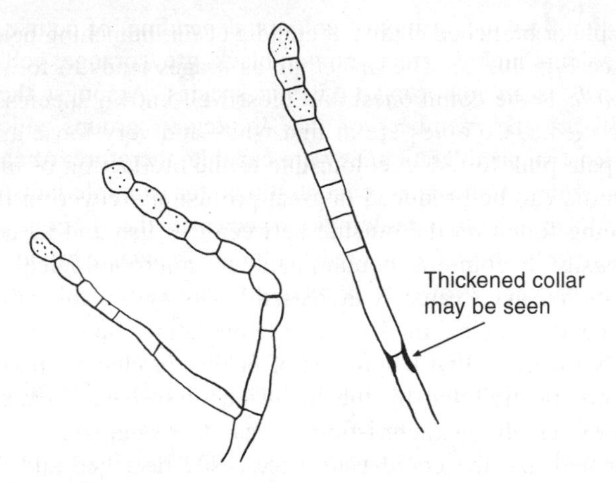

Fig. 40.13 *Wallemia.*

specific level is made difficult by the variation that can be found in a single strain according to the cultural conditions.

If identification of an isolate is required to specific level, Pitt and Hocking (1997) may be consulted.

Conidiophores are non-septate and arise from specialized, thick-walled, hyphal cells (known as foot-cells). Each conidiophore ends in a terminal enlarged ellipsoidal, hemispherical or spherical swelling which bears phialides either at the apex or over the entire surface (see Fig. 40.14).

Conidia, which are unicellular, vary in their colour, shape and wall marking, according to species. The conidia form unbranched chains arising from the tips of the phialides, the chains being arranged into radiate or mop-like heads or into columnar heads (the structure of the head once again being used as a species characteristic).

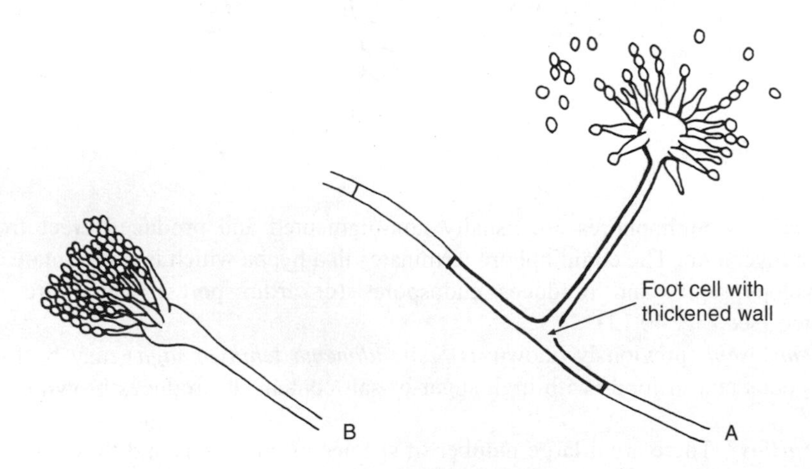

Fig. 40.14 *Aspergillus.* A, Radiate head. B, Columnar head.

Colonies may be of a wide variety of colours, depending, of course, on the colour of the conidia. The colours include green, brown, black, grey, orange, yellow and off-white.

Aspergillus flavus is an important toxigenic species. Amongst the more important spoilage aspergilli are the members of the *A. glaucus* group, which are frequently osmophilic or at least sugar tolerant. They are capable, therefore, of causing spoilage of food products containing high sugar concentrations, for example jams, and such usually indestructible products as 'plum puddings'. Species in the *A. glaucus* group may be recognized by their greyish-green or green colonies in which bright yellow or orange cleistothecia develop, when grown on a medium such as Czapek–Dox agar with 20% sucrose or glucose.

A. fumigatus is a species that causes a respiratory disease, aspergillosis, in poultry, horses and other farm and domestic animals, and in humans. The disease in humans is usually occupational, being known as farmer's lung, tea-taster's cough, etc. The colonies of this species are dark green and usually velvety. The phialides are produced at the apex of the vesicle, thus giving rise to a columnar head of conidia. The conidia are spherical with rough surfaces.

The *A. niger* group is very common in a variety of habitats. As the name suggests, the heads are black or very darkly coloured, resulting in colonies that are black or dark brown. The spherical vesicle bears phialides over its entire surface and thus gives rise to a 'mop-like' head of conidia.

10 *Stachybotrys* Although the vegetative mycelium has transparent hyphae, the conidiophores become dark as they mature and the spores are usually dark in colour. The phialides which bear the conidia are characteristically swollen, and borne as an apical cluster of only three to seven on each conidiophore. The spores may be spherical, ovoid or cylindrical (Fig. 40.15), and frequently become enveloped in slime.

Some species produce spores that are not dark in colour, but cause the colonies to appear pink or salmon. The common species of *Stachybotrys* are cellulolytic.

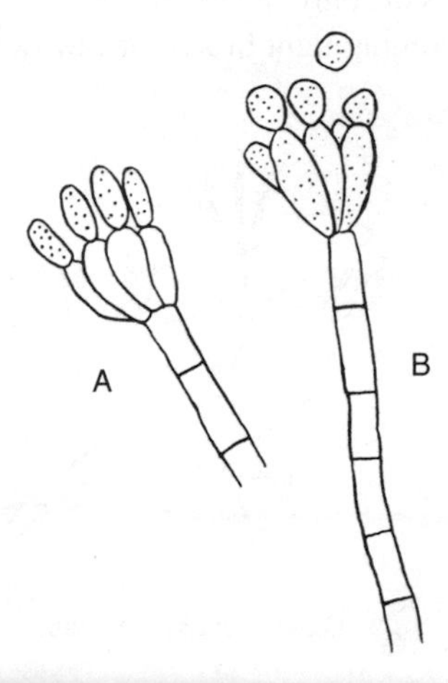

FIG. 40.15 *Stachybotrys*. A, Cylindrical spores. B, spherical spores.

11 *Scopulariopsis* The conidia are most characteristic, being truncated spheres, with a thickened basal ring around the truncation. The conidial wall is frequently roughened. The conidia tend to remain together in chains. Conidiophores bear a slight resemblance to those of *Penicillium* in the arrangement of the conidium-bearing structures (not true phialides), but single conidium-bearing 'phialides' may occur scattered over the aerial hyphae (Fig. 40.16).

The colonies may be coloured cream, yellow, brown, or chocolate-brown but they are never green. *Scopulariopsis* grows well on media (and substrates) containing high concentrations of protein, and may thus cause spoilage of foods such as meat.

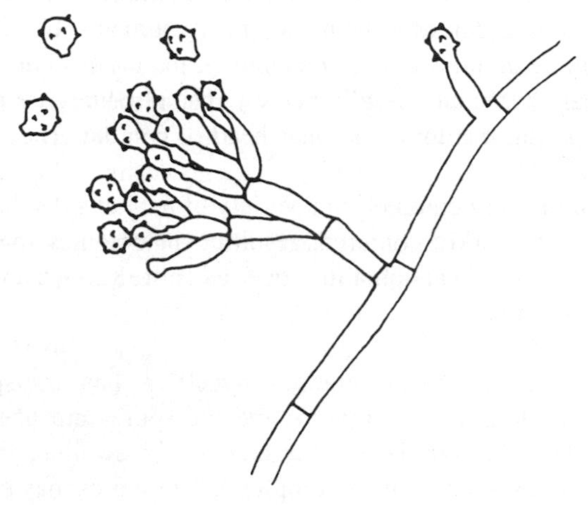

FIG. 40.16 *Scopulariopsis*.

12 *Byssochlamys* The conidiophore is branching and septate, with the ovoid conidia being borne on small brush-like clusters of phialides (Fig. 40.17). Eight-spored asci are produced. *Byssochlamys fulva* is the only species of interest. The colonies on solid media are loose and cottonwool-like, becoming light brown or tawny when the conidia are produced.

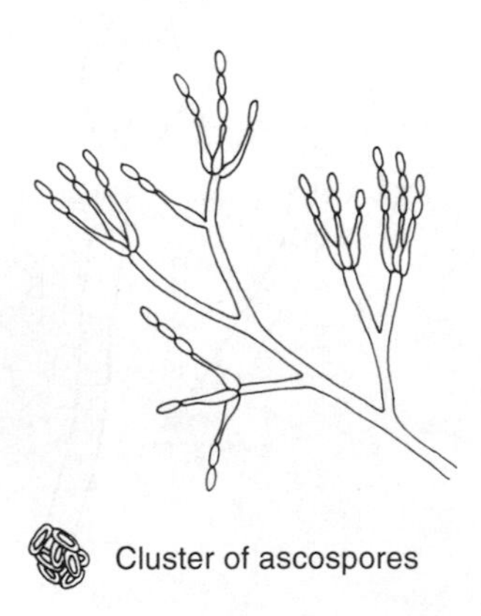

FIG. 40.17 *Byssochlamys*.

The ascospores are more heat resistant than any other fungal spores and this characteristic, together with the ability of *B. fulva* to grow in low oxygen concentrations, has led to the mould being the cause of spoilage of canned fruit products.

13 *Penicillium* As in the case of the genus *Aspergillus* there is a very large number of species, and for species differentiation Pitt and Hocking (1997) may be consulted.

The conidiophore is branched to form a brush-like conidial head. According to the type of branching, the genus is divided into the Monoverticillata, in which each head consists of a single whorl of phialides; the Asymmetrica in which there is more than one branch in the conidiophore, branching being more or less asymmetrical; and the Biverticillata-Symmetrica in which the head consists of a compact whorl of metulae each bearing a number of phialides, the whole head being symmetrical about the conidiophore (Fig. 40.18).

With the production of conidia, colonies usually become green, grey-green, blue-green or yellow-green, although P. *caseicolum* remains white. Some species produce sclerotia or perithecia and this may cause colonies to develop areas of a different colour, for example pink, yellow or orange. Some species produce columnar chains of conidia that are readily visible under low-power magnification.

P. camemberti and *P. caseicolum* are of importance in the production of Camembert, Brie and similar cheeses. Colonies of P. *caseicolum* remain white, whereas those of *P. camemberti* gradually become pale grey or grey-green. *P. roqueforti* is used in the production of Roquefort, Gorgonzola and similar blue-veined cheeses.

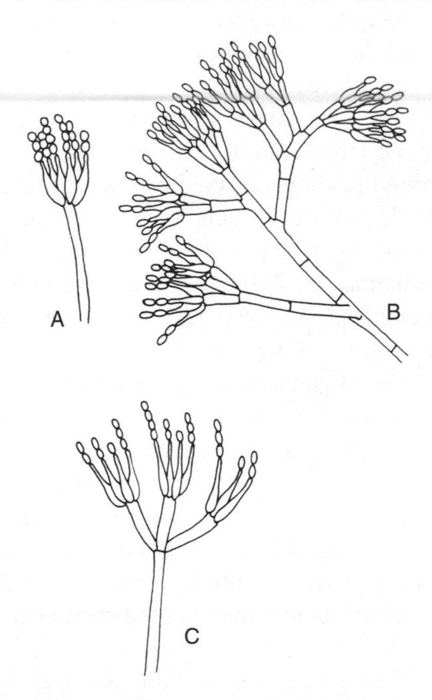

FIG. 40.18 *Penicillium*. Head characteristics of: A, the Monoverticillata; B, the Asymmetrica; and C, the Biverticillata-Symmetrica.

14 *Trichothecium* The long, slender septate conidiophores produce two-celled ovoid or pear-shaped conidia at the tip, the conidia tending to remain together to form a small cluster or short chain (Fig. 40.19). When chains are formed the conidia are not arranged end to end.

Colonies may be white or pale pink. *T. roseum*, the single species, may cause spoilage of certain fruit and vegetable products.

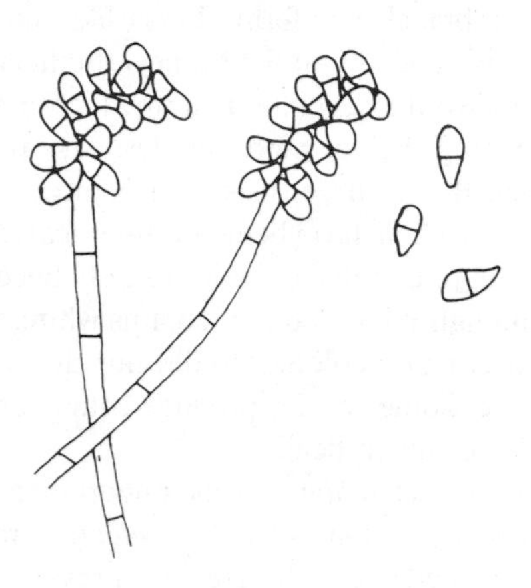

FIG. 40.19 *Trichothecium.*

15 *Fusarium* The macroconidia produced by members of this genus are very characteristic, being multicellular and spindle- or sickle-shaped. However, in addition to macroconidia, one-celled ovoid microconidia may be formed. In the absence of macroconidia identification becomes rather difficult. The formation of conidia and the colonial characteristics depend largely on the medium used, but no general recommendations can be made concerning the choice of medium as the correct cultural conditions for the formation of macroconidia vary according to the species. However, potato dextrose agar and potato carrot dextrose agar may be tried with and without glucose.

Colonies on suitable media may be fluffy and spreading, and often coloured rose-pink, salmon-pink, brown, purple or yellow. A typical appearance is for the central part of the colony to be fairly flat and pink in colour with fluffier, white, peripheral parts. On media containing high concentrations of sugar, cultures tend to become slimy.

Chlamydospores are frequently formed in the mycelium and even in the conidia (see Fig. 40.20). For species differentiation, see Pitt and Hocking (1997).

16 *Aureobasidium* The mature mycelium is dark coloured – although hyaline when younger – and very noticeably septate. Single blastospores are borne laterally on the hyphae. The blastospores may produce buds (see Fig. 40.21). The mature hyphae may show variation with some sections being thick-walled and dark, other sections being thin-walled and hyaline.

The colonies are most characteristic, being slimy and lightly coloured or dirty white when young, but rapidly becoming darker. The mature colony is a very dark greenish-black, leathery and shiny, adpressed to the medium. The single species, *A. pullulans*, is a

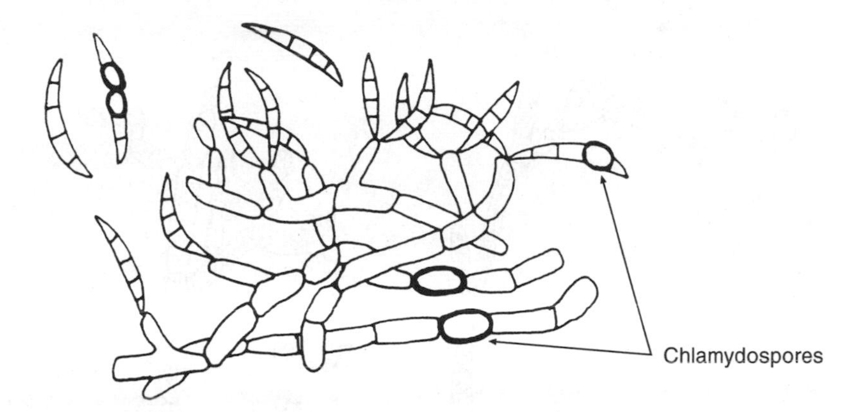

Fig. 40.20 *Fusarium.*

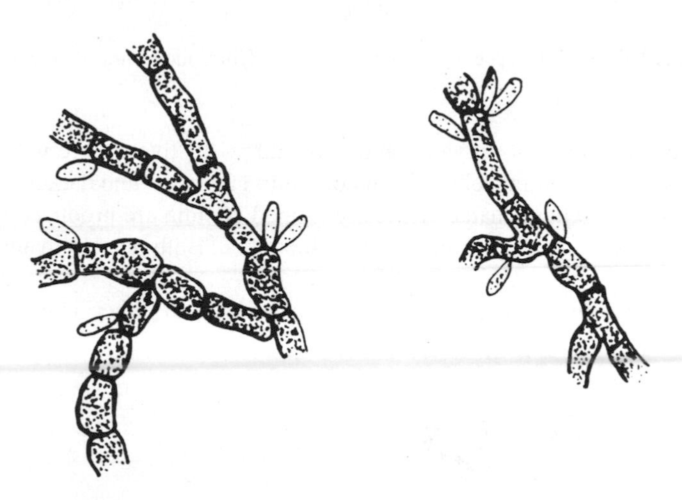

Fig. 40.21 *Aureobasidium.*

fairly common isolate from soils and may be a contaminant on certain types of food product.

17 *Cladosporium* The conidia are one- or two-celled and produced in branched chains on the conidiophores. The branching of the conidial chains is due to the ability of the conidia to reproduce by budding (see Fig. 40.22). Mycelium and conidia are dark in colour.

Colonies are darkly coloured green, olive-green, brown or black, and usually of a fairly thick, velvety nature.

This genus includes the forms producing one-celled conidia, previously described as *Hormodendrum*.

Members of *Cladosporium* are common in soil and as contaminants causing spoilage of a wide variety of foods, including meat kept in cold storage.

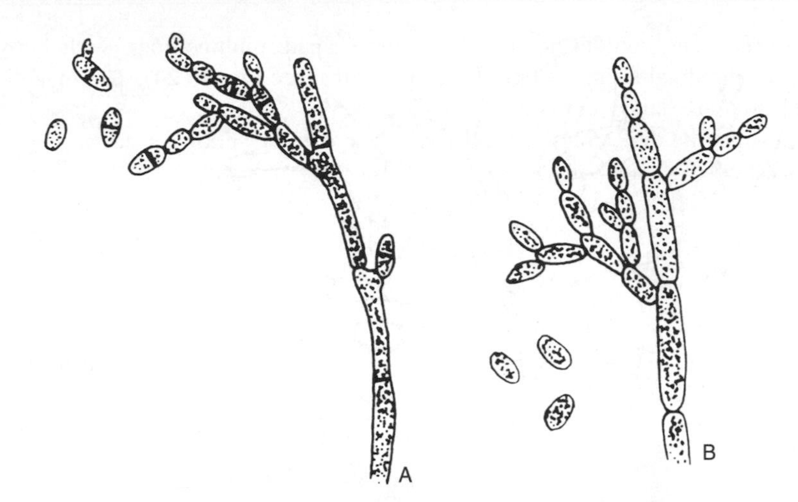

FIG. 40.22 A, 'Cladosporium' type of *Cladosporium*. B, 'Hormodendrum' type of *Cladosporium*.

18 *Curvularia* The bent or curved conidia are three- to five-celled, with cross septa only. One or two of the central cells in each conidium are larger and darker than the other cells (see Fig. 40.23). These characteristically shaped conidia are produced in clusters at the tips of septate conidiophores, which may be branched. Both mycelium and conidia are darkly coloured.

Colonies may be olive-green, brown or brownish-black, of either a velvety or a fluffy texture.

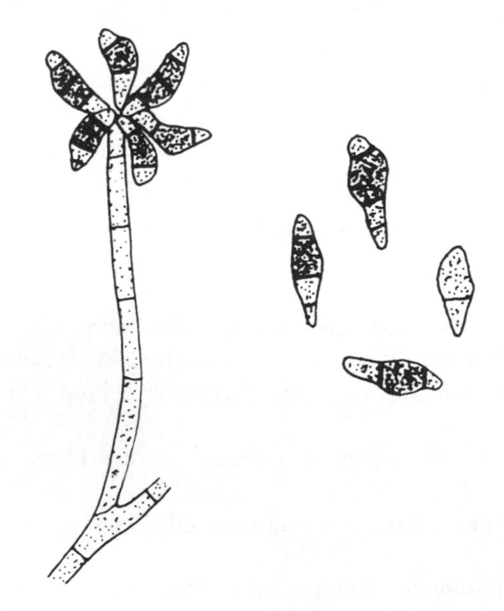

FIG. 40.23 *Curvularia*.

19 *Alternaria* The conidia are pear- or club-shaped, multicellular, with both cross-walls and longitudinal septa, and produced in chains (see Fig. 40.24). The mycelium and mature conidia are darkly coloured.

Colonies may be dark green, greenish-black, brown or yellowish-brown.

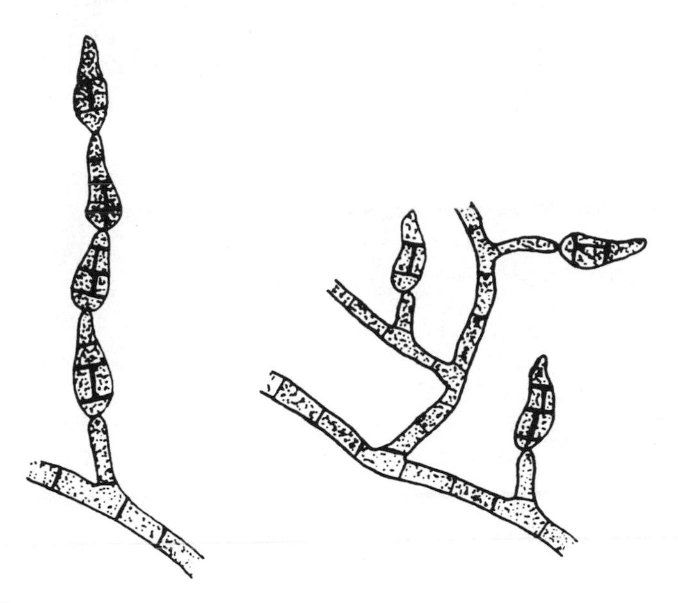

FIG. 40.24 *Alternaria.*

APPENDIX 1

Recipes for Stains, Reagents and Media

CHEMICAL HAZARDS

In the preparation of stains, reagents and media it is inevitable that many dangerous chemicals will be used. Correct handling procedures will minimize the hazards, whilst providing more pleasant working conditions. For example, it should be a strictly observed rule that balances and work benches are kept scrupulously clean. Remember that a worker will know the identity of the chemicals that he or she is using – and should know the hazards presented by them – but if the apparatus and work benches become contaminated, the next person to use them may not be as fortunate. Refer to Section 2.1 for general safety management procedures.

The following list indicates some of the hazardous chemicals that are used in the subsequent recipes for stains, reagents and media. There are four distinct types of hazardous chemical:

1. A number of these substances are primarily a fire hazard (e.g. ethanol, potassium nitrate).
2. Some are corrosive, cause burns and are dangerous if ingested because of the severe local damage that may be caused to the mouth, oesophagus and stomach (e.g. strong acids and alkalis).
3. Many are toxic, and note should be made of the possibilities and the insidious consequences of chronic toxicity effects. For example, the worker frequently using the earlier version of Newman's stain containing tetrachloroethane became accustomed to the odour of low concentrations of tetrachloroethane in the laboratory atmosphere, and sometimes failed to use a fume cupboard, overlooking the long-term harm that would ensue.

   Great care should be taken when handling chemicals that may be absorbed through the skin (e.g. aniline, xylene, mercury salts).
4. A large number of chemical compounds are now suspect as possible carcinogens. It is a wise precaution to regard, for example, all dyes (e.g. carbol fuchsin) and indicators with suspicion, and to use forceps for handling microscope slides to avoid staining the hands. The incorporation of these compounds into media can usually be achieved using standard stock solutions so that the weighing of dry ingredients is minimized (this practice also facilitates the incorporation of the small quantities of such compounds usually required in media). Asbestos heat-resistant Bunsen burner mats are still available in a number of countries, but the alternative, safe, ceramic heat-protecting mats should be used.

Liquids should *not* be pipetted by mouth, and safety pipette fillers should always be used. The accidental ingestion of poisons is unlikely to occur if this precaution is always observed. In the unlikely event of such ingestion, vomiting should generally be induced in conscious victims *unless* the poison is corrosive or is a hydrocarbon (but the recommendations of national first-aid organizations should be observed). Phenol is an exception in that it is both corrosive and toxic, and vomiting should be induced in this case. Vomiting can be induced by drinking a tumbler (200 ml of tepid water containing two tablespoons (30 g) of sodium chloride.

When corrosive poisons have been ingested, the victim should drink a demulcent, such as milk, and large quantities of water.

The best treatment for chemical burns (or thermal burns) to the skin is irrigation by large quantities of water.

In all cases seek medical assistance immediately.

Compound	*Principal nature of hazard*
Acetic acid, glacial	Corrosive; causes burns; irritant vapour; flammable
Acetone	Toxic; harmful vapour; flammable
Ammonia (0.880)	Corrosive; causes burns; harmful vapour. Store in a cool place
Amyl alcohol (pentanol)	Irritant vapour; toxic; flammable
Aniline	Chronic toxicity; can be absorbed through the skin; toxic vapour
Chloral hydrate	Harmful vapour; dangerous by ingestion
Chromic acid	Corrosive, causes skin ulcers; powerful oxidizing agent
Copper salts	Avoid inhalation of dust, or ingestion; acute toxicity
Cycloheximide	Toxic; skin irritant
Ethanol	Highly flammable; intoxicant
Formaldehyde (formalin)	Irritant vapour; flammable; chronic toxicity. Avoid contact with chlorine-containing compounds as a volatile carcinogen may be formed
Hydrochloric acid, concentrated	Corrosive; causes burns; irritant gas
Hydrogen peroxide	Corrosive; causes burns; irritant
Iodine	Corrosive; causes burns; harmful vapour
Lactic acid	Irritant; causes burns
Lead acetate	Acute and chronic toxicity; avoid breathing dust
Mercury and mercury salts	Chronic toxicity; can be absorbed through the skin; harmful vapour (from elemental mercury)
Methanol	Chronic toxicity; harmful vapour; flammable
o-nitrophenyl-β-D-galactopyranoside	Chronic toxicity; may be absorbed through the skin
Orthophosphoric acid	Corrosive; causes burns
Oxalates	Toxic by ingestion
Phenol	Corrosive; causes burns; acute and chronic toxicity; can be absorbed through the skin; harmful vapour
Picric acid	Chronic toxicity; toxic on skin contact or by ingestion; explosive when dry
Potassium dichromate	Chronic toxicity; causes skin ulcers; harmful dust; oxidizing agent – fire hazard
Potassium hydroxide	Corrosive; causes burns

Compound	*Principal nature of hazard*
Potassium nitrate	Oxidizing agent – fire hazard
Potassium tellurite	Chronic toxicity; harmful dust
Silver nitrate	Corrosive; causes burns; toxic
Sodium azide	Toxic by skin absorption or ingestion; explosive
Sodium hydroxide	Corrosive; causes burns
Sodium hypochlorite	Corrosive; causes burns
Sodium hydrogen selenite	Chronic toxicity; possibly teratogenic (Robertson, 1970); can be absorbed through the skin
Sulphuric acid, concentrated	Corrosive; causes burns
Tetrachloroethane	Extremely harmful vapour; chronic toxicity
Thallium acetate	Chronic toxicity; can be absorbed through the skin
Trichloroethane	Harmful vapour
Xylene	Chronic toxicity; harmful vapour; can be absorbed through the skin; flammable

Note. Many dehydrated media exist as very fine powders. Care should be taken in dispensing these that the dust is not inhaled, particularly if the medium is known to contain a toxic chemical.

The personnel trained in first aid should attempt to acquaint themselves with the recommended first-aid procedures for all toxic substances in common use in the laboratory. *Hazards in the Chemical Laboratory* (S. G. Luxon, 1992) or equivalent text should be displayed in a prominent position in every laboratory.

In the recipes that follow, [H] indicates a hazardous chemical from the above list. Other constituents may also be hazardous – particularly some of the antibiotics and other inhibitors used in selective media.

STAINS

Acid–Alcohol

(Decolorizing agent in the acid-fast staining technique)

Concentrated hydrochloric acid[H]	3 ml
Ethanol (95%)[H]	97 ml

(Cowan, 1974)

Bartholomew and Mittwer's Spore Stain

Stain

Saturated aqueous solution of malachite green

Counterstain

Safranin	0.25 g
Distilled water	100 ml

(Bartholomew and Mittwer, 1950)

Carbol Fuchsin, Ziehl-Neelsen

(For acid-fast staining)

Basic fuchsin[H]	1.0 g
Ethanol (95%)[H]	10 ml
Phenol[H] (5% aqueous solution)	100 ml

Crystal Violet Stain

(For Gram's staining method and/or simple staining)

Crystal violet (gentian violet)	0.5 g
Distilled water	100 ml

Fontana's Flagella Stain (Modified)

Mordant

Tannic acid (10% aqueous solution)	10 ml
Aluminium potassium sulphate (potash alum), saturated aqueous solution	5 ml
Aniline water (saturated aqueous solution of aniline[H])	1 ml
Ferric chloride (5% aqueous solution)	1 ml

Mix the tannic acid and potash alum solutions, then add the aniline water, taking care not to add any drops of undissolved aniline. At this stage a precipitate forms. Shake the tube until the precipitate redissolves, and then add 1 ml of the ferric chloride solution. Allow the mordant to stand for at least 10 min before use. Use the same day.

Ammoniacal silver nitrate solution

Silver nitrate[H]	5.0 g
Distilled water	100 ml

To 90 ml of the silver nitrate solution slowly add strong ammonia solution[H] (specific gravity 0.880) until the precipitate that forms *just* redissolves. Add some of the remaining silver nitrate solution drop by drop, shaking between the addition of each drop, until the solution remains slightly turbid after shaking. Store the solution in dark bottles, and in a cool place protected from light. Do not store for too long (up to 3 weeks, if refrigerated), do not expose to sunlight during storage, and do not store at eye level – *this solution can become spontaneously explosive*. It is advisable to overwrap glass bottles with plastic film.

(Rhodes, 1958)

Gram's Iodine Solution

Iodine[H]	1.0 g
Potassium iodide	2.0 g
Distilled water	300 ml

Lactophenol

(For wet microscopic preparations of moulds)

Lactic acid[H]	100 ml
Phenol[H]	100 g
Glycerol	200 ml
Water	100 ml

Dissolve the phenol in the water without heat, then add the lactic acid and the glycerol.

Lactophenol–Cotton Blue

(For wet-mounting and staining of moulds)

Cotton blue solution

Saturated solution of cotton blue[H] (soluble aniline blue)	5 ml
Glycerol	10 ml
Water	85 ml

Mix equal parts of lactophenol and cotton blue solution.

Lactophenol–Picric Acid

(For wet-mounting and staining of moulds)

Lactic acid[H]	100 ml
Phenol[H]	100 g
Glycerol	200 ml
Picric acid[H], saturated aqueous solution	100 ml

Dissolve the phenol in the picric acid solution without heat, then add the lactic acid and the glycerol.

Leifson's Flagella Stain

Solution A

Basic fuchsin[H]	1.2 g
Ethanol (95%)[H]	100 ml

Dissolve with frequent shaking. Store in a tightly stoppered bottle to prevent evaporation.

Solution B

Tannic acid	3.0 g
Distilled water	100 ml

If it is intended to keep this for some time as a stock solution, add phenol[H] to a concentration of 0.2% to prevent microbial growth.

Solution C

Sodium chloride	1.5 g
Distilled water	100 ml

These solutions are stable at room temperature. To prepare the working solution, mix equal quantities of solutions A, B and C, and store in a tightly stoppered bottle in the refrigerator (stable for several weeks) or deep freezer (stable for months). If stored deep frozen, the stain should be well shaken after thawing because the ethanol tends to separate from the water.

Counterstain

Methylene blue	1.0 g
Distilled water	100 ml

(Leifson, 1951, 1958)

Loeffler's Methylene Blue

(For simple staining)

Potassium hydroxide[H] solution (1%)	1.0 ml
Methylene blue, saturated solution in 95% ethanol	30 ml
Distilled water	100 ml

Malachite Green

(For staining membrane filters)

Malachite green	0.01 g
Distilled water	100 ml

(Fifield and Hoff, 1957)

Newman's Stain, Modified

(For staining bacteria and bovine cells in milk)

Solution A

Methylene blue	0.6 g
Ethanol (95%)[H]	54 ml
1,1,1-Trichloroethane[H]	40 ml
Glacial acetic acid[H]	6 ml

Add the ethanol to the trichloroethane and heat in a water-bath in a fume cupboard, to a temperature not exceeding 70°C. Add the mixture to the methylene blue, and shake until the dye has completely dissolved. Cool, add the acetic acid very slowly and filter.

Solution B

Basic fuchsin[H]	0.25 g
Ethanol[H] (70%)	70 ml

Dissolve the dye in the ethanol and filter.

Mix the two solutions thoroughly and store in an air-tight glass stoppered bottle. This mixture will keep indefinitely if not allowed to evaporate.

The original recipe employed tetrachlorethane, but, because of the extreme toxicity of this solvent, it is recommended that the less toxic trichloroethane is used. Nevertheless, because of the toxic effects of trichloroethane, this staining solution should be prepared, stored and used in a fume cupboard.

(modified from Charlett, 1954)

Safranin Solution

(Counterstain for Gram's staining method)

Safranin	0.25 g
Distilled water	100 ml

REAGENTS

Acidic Ferric Chloride Solution

(For the phenylalanine deaminase test)

Ferric chloride	12.0 g
Concentrated hydrochloric acid[H]	2.5 ml
Distilled water	100 ml

(MacFaddin, 1980)

Andrade's Indicator

(Indicator for acid production from carbohydrates)

Acid fuchsin[H]	0.5 g
Sodium hydroxide[H], N solution	16 ml
Distilled water	100 ml

Dissolve the acid fuchsin in the distilled water and add the sodium hydroxide solution. Allow to stand overnight at room temperature. If the fuchsin has not decolorized to a straw colour, add a further 1–2 ml of sodium hydroxide solution.

(Silverton and Anderson, 1961)

Benedict's Reagent

(For detecting reducing sugars, and used in the test for 3-ketolactose)

Copper sulphate, hydrated[H]	17.3 g
Sodium carbonate, anhydrous	100 g
Sodium citrate, hydrated	173 g
Distilled water	to 1 litre

Dissolve the sodium carbonate and sodium citrate in 600 ml of distilled water, filter and make up to 850 ml. Dissolve the copper sulphate in 100 ml of distilled water and make up to 150 ml. Add the copper sulphate solution to the carbonate and citrate solution very slowly and with constant stirring.

(Benedict, 1908–1909)

Bouin's Fixative

(For fixing slide cultures)

Picric acid[H] (saturated aqueous solution)	75 ml
Formaldehyde[H] (40% solution)	25 ml
Glacial acetic acid[H]	5 ml

Chromic Acid Cleaning Solution

	Recipe 1	*Recipe 2*
Potassium dichromate[H]	63.0 g	100 g
Distilled water	35 ml	750 ml
Concentrated sulphuric acid[H]	960 ml	250 ml

Add the potassium dichromate to the water in a 2-litre flask. Slowly and carefully add the acid. Chromic acid cleaning solution can be used repeatedly until it begins to turn green, when it should be discarded. Great care should be taken when preparing and handling this cleaning solution as it is extremely corrosive. *Use protective clothing and face protection.*

Griess–Ilosvay's Reagent (Modified)

(To test for the presence of nitrite)

Reagent 1

Sulphanilic acid	1.0 g
Acetic acid[H], 5N	100 ml

Reagent 2

α-Naphthol[H]	1.0 g
Ethanol[H] (95%)	100 ml

(McLean and Henderson, 1966)

Because of the potential hazard from this reagent, it is preferable to use nitrite detection test papers (e.g. as produced by Merck), if these can be obtained.

Kovacs's Indole Reagent

(To test for indole production in tryptone water)

Pentanol (amyl or isoamyl alcohol)[H]	150 ml
p-Dimethylaminobenzaldehyde	10.0 g
Concentrated hydrochloric acid[H]	50 ml

Dissolve the aldehyde in the alcohol, then slowly add the acid. Store in the refrigerator. Note that pentanol is volatile and its vapour is extremely irritant, so this should be used in a fume cupboard.

(Report, 1958)

Kovacs's Oxidase Test Reagent

(To test for the presence of cytochrome oxidase in a bacterial culture)

Tetramethyl-*p*-phenylenediamine hydrochloride[H]	0.1 g
Ascorbic acid	0.01 g
Distilled water	10.0 ml

The reagent may be kept in a *dark* bottle in a refrigerator, but must be discarded when auto-oxidation has caused it to become purple.

Because of the toxic nature of this reagent, it is recommended that commercially available dipsticks or slides be used wherever possible (see Section 11.4.2).

(Kovacs, 1956; Steel 1962)

Lactose Discs

(Inducer discs for ONPG test)

Prepare 1-cm diameter discs from filter paper. Dissolve 1.5 g lactose in 10 ml of distilled water. Dip each disc once, using forceps, and drain by touching the disc to the side of the bottle. Dry at 37°C for 3 h. Dip a second time and again dry at 37°C.

(Clarke and Steel, 1966)

Methyl Red Solution

(For the methyl red test)

Methyl red	0.1 g
Ethanol[H] (95%)	300 ml
Distilled water	to 500 ml

Dissolve the methyl red in the ethanol and make up to 500 ml with distilled water.

Nessler's Reagent

(To test for the production of ammonia)

Potassium iodide	7.0 g
Mercuric iodide[H]	10.0 g
Potassium hydroxide[H]	10.0 g
Distilled water	to 100 ml

Dissolve the potassium iodide and mercuric iodide in 40 ml of distilled water. Dissolve the potassium hydroxide in 50 ml of distilled water, and allow to cool. Mix the two solutions and add distilled water to 100 ml. Allow the precipitate to settle, decant the clear supernatant liquid into a reagent bottle, and discard the precipitate.

ONPG Discs

(To test for β-galactosidase activity)

Prepare 1-cm diameter discs from filter paper. Dissolve 20 mg of *o*-nitrophenyl-β-D-galactoside[H] in 10 ml of distilled water. Using plastic forceps dip each disc once, and drain by touching the disc to the side of the bottle. Dry at 37°C for 3 h.

(Clarke and Steel, 1966)

Phenylalanine Discs

(To test for phenylalanine deaminase activity)

Prepare 1-cm diameter discs from filter paper. Dissolve 0.1 g of DL-phenylalanine in 10 ml of distilled water. Dip each disc once, drain by touching the disc to the side of the bottle. Dry at 37°C for 3 h.

(Clarke and Steel, 1966)

Potassium Hydroxide Saline Solution

(For pretreatment of foods in the isolation of *Yersinia*)

Potassium hydroxide[H]	2.5 g
Sodium chloride	8.5 g
Distilled water	1000 ml

Dissolve the ingredients in the water, dispense into screw-capped bottles and sterilize by autoclaving at 121°C for 15 min.

(Aulisio *et al.*, 1980; ISO 10273:1994)

Turbidity Standards for Nephelometry

Tube no.	1% Barium chloride	!% Sulphuric acid
1	0.1	9.9
2	0.2	9.8
3	0.3	9.7
4	0.4	9.6
5	0.5	9.5
6	0.6	9.4
7	0.7	9.3

Using a set of optically clean tubes of standard diameter and standard wall thickness (nephelometer tubes), add the above amounts of reagents. Stopper with black Astell seals or similar. Keep refrigerated. Shake well before use.

(Balows *et al.*, 1991)

Voges–Proskauer Test Reagents

Barritt's modification (Barritt, 1936)

(a) Potassium hydroxide[H] (16% solution)
(b) α-Naphthol[H], 6% solution in 95% ethanol[H]

MEDIA

The methods for the preparation of the media used in this book are given below. In the case of salts with more than one hydrated form, the hydrate to be used is specified in the recipe where the concentration of the ingredient is likely to be critical. With some media it is much more convenient to use the dehydrated version (e.g. in the case of violet red bile agar) or the ready-prepared version (e.g. as in the case of Dorset's egg medium) which may be commercially available, unless large quantities are required regularly. In addition, when attempting to follow standard methods, it is usually easier to use dehydrated media where these are obtainable (although quality assurance checks on the made-up media should still be performed; see Section 2.5). The recipes for some media mentioned in Parts II and III, for example lysine agar (Section 31) and the selective media for *Campylobacter* (Section 21.8), are not listed here because the complexity of the media makes it inappropriate for a microbiological quality assurance laboratory to attempt preparation of these media from basic constituents.

Most of the media described here are available in either dehydrated or ready-prepared form from many manufacturers, including Oxoid, Difco Laboratories, Baltimore Biological Laboratories (BBL), Merck and bioMérieux. Some media are available as a dehydrated product in a basal form that requires the addition of substrate or supplement. These supplements are usually also available from the media manufacturer. However, if a standard procedure is being followed (e.g. an ISO method), the formulation should be checked for conformance; this is especially important in the case of complex, multiple-antibiotic supplements. Dehydrated media are frequently provided in powder rather than granular form. If preparing such a dehydrated medium of a formulation that contains a toxic constituent (particularly if there is a hazard of chronic toxic effects), care should be taken to avoid inhaling or ingesting dust that may contaminate the air. At least one manufacturer (Merck) has attempted to produce in granular form most of its selective media containing toxic components. However, it is important to remember that there will still be some dust from a granular product so appropriate precautions (e.g. wearing a face mask) should still be taken when dispensing these media. It should be noted that many ingredients of selective media, including many antibiotics, are toxic or irritant.

When preparing media, care should be taken that each ingredient or the dehydrated medium is adequately mixed into the water. If mixing is inadequate, stratification of dense concentrated solutions will occur and on heating there will be deleterious changes caused by caramelization, hydrolysis, etc. When using dehydrated media it is good practice to allow the powder to soak in the water for 15 min with frequent mixing by agitation, before heating. Agar media being dissolved by the application of heat to the base of the container should be agitated to prevent the agar granules from settling out and being burned. The manuals provided by the manufacturers of dehydrated media give detailed recommendations on the procedures for reconstitution and sterilization, and these should be followed closely.

The recipes for some media recommend heat treatments insufficient to ensure sterility, often because the medium is susceptible to heat degradation. In such cases it is often necessary to incubate the media for 2 days at 37°C and for 2 days at 30°C to detect any residual contaminants; if the media are to be used for the growth of psychrotrophs or thermophiles, preincubation at the appropriate temperature should also be carried out. Alternatively uninoculated controls should be incubated concurrently with the tests. Such procedures are an advisable precaution with *all* media.

In most recipes the type of closure to be used with test-tubes has not been specified. Non-absorbent cotton-wool, autoclavable polypropylene, alloy or stainless steel caps may be used according to personal preference. In some situations cotton-wool may be contraindicated; for example, there may be inhibitory effects on fastidious organisms because of the adventitious introduction of fatty acids (Meynell and Meynell, 1970), or turbidity measurements made to assess growth may be disturbed by the introduction of cotton-wool fibres into the medium.

Rubber bung closures (especially of the vented Astell seal design) are suitable for tests on many organisms, but the obstruction to gas exchange with the external atmosphere may affect the growth and biochemical test reactions, especially of obligate aerobes. Such closures do have the advantage of extending the shelf-life of media by preventing evaporation (as does the use of screw-capped bottles), but this in itself may disguise the unsuitability of media subjected to prolonged storage because of chemical changes.

Acetate Agar

(A selective medium for the isolation or enumeration of lactobacilli)

1. Basal medium

Peptone	10.0 g
Meat extract (e.g. Lab-Lemco from Oxoid)	8.0 g
Yeast extract (Difco)	5.0 g
Agar (Oxoid No. 3)	15.0 g

Triammonium citrate	2.0 g
Salts solution[§]	5.0 ml
D-Glucose	10.0 g
Tween 80	0.5 ml
Distilled water	1 litre

pH 5.4

§ *Salts solution*

Magnesium sulphate, hydrated ($MgSO_4 \cdot 7H_2O$)	8.0 g
Manganese sulphate, hydrated ($MnSO_4 \cdot 4H_2O$)	2.0 g
Distilled water	100 ml

Dissolve the peptone, meat extract, yeast extract and agar in 1 litre of distilled water by autoclaving at 121°C for 15 min. Then add the citrate and salts solution and mix well. Adjust the pH to 5.4 and filter if necessary. Add the glucose and Tween 80 and mix thoroughly. Dispense in 90-ml amounts in screw-capped bottles. Sterilize by autoclaving at 121°C for 15 min.

2. Preparation of 2 M acetic acid–sodium acetate buffer at pH 5.4

Dissolve 23.3 g of sodium acetate (hydrated) and 1.7 g of glacial acetic acid[H] in distilled water and make up to 100 ml. Check the pH by diluting a portion 1 : 10 in distilled water before using the pH meter (dilution is necessary because the high concentration of sodium ions may affect the glass electrode). Distribute in screw-capped containers in 10-ml amounts, and sterilize by autoclaving at 115°C for 20 min.

3. Preparation of the medium for use

The medium may be used with *or* without the acetic acid–sodium acetate buffer (see the appropriate section in Parts II and III). If the medium is to be used with buffer added, add aseptically 10 ml of the buffer solution to 90 ml of the medium, which has been melted and cooled to 50°C. Mix gently by careful inversion to avoid frothing, and pour double-layer plates in the usual way. The final pH of the complete medium should be 5.4 ± 0.05.

(Keddie, 1951 and personal communication)

Aesculin Hydrolysis Medium

(For identification of *Aeromonas*)

Peptone	10 g
Sodium citrate	1.0 g
Aesculin	1.0 g
Iron (III) citrate	0.05 g
Distilled water	1 litre

pH 7.0

Dissolve the ingredients in the water, distribute in test-tubes, and sterilize at 121°C for 15 min.

(Sneath, 1966)

Alkaline Saline Peptone Water

(For isolation of *Vibrio*)

Peptone	10.0 g
Sodium chloride	30.0 g
Distilled water	to 1 litre

Adjust to pH 8.5 with sodium hydroxide solution. Distribute as required, and sterilize at 121°C for 15 min.

Arginine Broth

(A medium to test for the production of ammonia from arginine by streptococci)

Tryptone	5.0 g
Yeast extract	2.5 g
D-Glucose	0.5 g
Dipotassium hydrogen phosphate	2.0 g
L-Arginine monohydrochloride	3.0 g
Distilled water	1 litre

pH 7.0

Dissolve the ingredients in the water and sterilize at 121°C for 15 min.

(Abd-el-Malek and Gibson, 1948)

Arginine MRS Broth

(A medium to test for the production of ammonia from arginine by lactobacilli)

MRS broth plus 0.3% (w/v) L-arginine monohydrochloride. Add the arginine during the preparation of the MRS broth before sterilization.

(Sharpe, 1962)

Arginine Tetrazolium Agar

(A medium to differentiate *Lactococcus lactis* subsp. *cremoris* from the other subspecies of *Lactococcus lactis*)

Tryptone	5.0 g
Yeast extract	5.0 g
L-Arginine monohydrochloride	2.0 g
Dipotassium hydrogen phosphate	2.0 g
Glucose	0.5 g
Agar	15.0 g
Distilled water	to 1 litre

pH 6.0

Dissolve the ingredients in the water by steaming. Mix, adjust the pH to 6.0, distribute in 100-ml amounts in screw-capped bottles, and sterilize by autoclaving at 121°C for 20 min.

For use melt 100 ml of medium, and cool to 50°C. Add 1 ml of a 0.5% filter-sterilized solution of triphenyl tetrazolium chloride[H], mix well and pour plates. Dry plates before use, and spread 0.1-ml inocula.

(Turner *et al.*, 1963)

Baird-Parker's Medium (Egg-yolk Tellurite Glycine Agar)

(A selective and diagnostic medium for isolation and enumeration of *Staphylococcus aureus*)

Tryptone (Difco)	10.0 g
Lab-Lemco meat extract (Oxoid)	5.0 g
Yeast extract (Difco)	1.0 g
Lithium chloride, hydrated	5.0 g
Agar (Difco)	20.0 g
Sulphadimidine sodium salt (0.2% solution)	25.0 ml
Distilled water	975 ml

pH 6.8–7.0

Dissolve the ingredients by steaming. Adjust to pH 6.8. Dispense without filtration in 90-ml amounts in screw-capped bottles and sterilize by autoclaving a 121°C for 15 min.

Before pouring the plates, to each 90 ml of basal medium, melted and cooled to 50°C, add aseptically the following solutions (all previously sterilized by filtration) in the amounts given:

(a) 20% (w/v) solution of glycine	6.5 ml
(b) 1% (w/v) solution of potassium tellurite[H] (e.g. from Sigma or Merck)	1.1 ml
(c) Egg-yolk emulsion	5.4 ml

Mix well and pour into Petri dishes in 12-ml amounts. These poured plates may be stored in plastic bags at 4°C for up to 1 month. Before use 0.5 ml of a 20% (w/v) solution of sodium pyruvate should be spread aseptically over the surface of each plate. The plates are then dried (with the medium *surface* uppermost) at 50°C for 1 h, before inoculation.

Note.

1. The 0.2% solution of sulphadimidine sodium salt is prepared by dissolving 0.5 g of pure sulphadimidine (sulphamezathine) in 25 ml of 0.1 N sodium hydroxide solution and making up to 250 ml with distilled water.
2. The source of potassium tellurite may affect the selectivity and sensitivity of this medium. It is possible to obtain standardized sterile potassium tellurite solution or egg-yolk tellurite mixture from manufacturers of dehydrated media. However, if the potassium tellurite solution is prepared in the laboratory, each bottle of potassium tellurite from any batch not previously tested should be validated by checking the suitability of the medium against known stock cultures of *Staphylococcus aureus*.
3. The method of preparation of the egg-yolk emulsion is given on p. 412. It must be stored in a refrigerator.
4. The sodium pyruvate solution must be stored in a refrigerator and used within 1 month.
5. As indicated, this medium is available in dehydrated form, as a basal medium, to which must be added egg-yolk, tellurite and the sulphadimidine salt. A ready-prepared egg-yolk tellurite emulsion is available from, for example, Oxoid or Difco, and can be used in conjunction with these dehydrated media.

(Baird-Parker, 1962, 1969; Smith and Baird-Parker, 1964; Holbrook *et al.*, 1969)

Barnes' Thallium Acetate Tetrazolium Glucose Agar

(For the detection and enumeration of faecal streptococci)

Basal medium

Peptone (Difco)	10.0 g
Yeast extract	10.0 g
Agar	15.0 g
Distilled water	1 litre

pH 6.0

Dissolve the ingredients in the water by steaming, cool to 50–60°C and adjust to pH 6.0. Distribute in 95-ml amounts in screw-capped bottles, and autoclave at 121°C for 20 min.

Preparation of complete medium

To 95 ml of molten medium, cooled to 45–50°C, add the following:

1. 2 ml of a 5% aqueous solution of thallium acetate[H] (previously sterilized by autoclaving at 115°C for 15 min)
2. 1 ml of a 1% aqueous solution of 2,3,5-triphenyl tetrazolium chloride[H] (previously sterilized by filtration or by steaming for 30 min)
3. 5 ml of a 20% solution of glucose (previously sterilized by filtration or by steaming for 30 min on three successive days)

Mix the medium well after each addition and finally pour the plates as required.

(Barnes, 1956)

Blood Agar

Sterile defibrinated blood	5.0 ml
Nutrient agar (modified)	100 ml

The nutrient agar base should contain an extra 3.5 g of sodium chloride per litre of medium.

Blood agar bases are also available commercially as dehydrated media. A suitable basal medium for fastidious organisms is Columbia agar.

Horse blood is most suitable for streptococci, but for studying the growth and haemolytic reactions of staphylococci, sheep, rabbit or ox blood may be required.

Liquefy the basal medium, cool to 45–50°C and add aseptically the sterile blood. Mix well and pour the plates. Layer plates may be prepared by pouring into each plate 5 ml of blood agar on top of a thin (5–10 ml) layer of the basal medium previously poured and allowed to solidify.

Brilliant Green Lactose Bile Broth

(Selective medium for isolating and counting coliform organisms and *Escherichia coli*)

Peptone	10.0 g
Lactose	10.0 g
Ox bile (Oxoid L50 or equivalent)[§]	20.0 g
Brilliant green (1% aqueous solution)	13.3 ml
Distilled water	to 1 litre

pH 7.4

[§] Different proprietary preparations may not have equivalent activity at the same concentration.

Dissolve the peptone and lactose in 500 ml of distilled water. Dissolve the ox bile in 200 ml of distilled water. Mix the two solutions, add distilled water to 950 ml, and adjust to pH 7.4. Add 13.3 ml of a 1% aqueous solution of brilliant green and then make up the volume to 1 litre with distilled water. Dispense into test-tubes containing inverted Durham tubes, plug and sterilize by autoclaving at 121°C for 15 min.

(Mackenzie *et al.*, 1948)

Brilliant Green Phenol Red Agar

(For the detection and isolation of *Salmonella*, other than *S. typhi*)

Peptone	10.0 g
Yeast extract	3.0 g
Sodium chloride	5.0 g
Lactose	10.0 g
Sucrose	10.0 g
Phenol red (0.2% aqueous solution)	40.0 ml
Brilliant green (1.0% aqueous solution)	12.5 ml
Agar	15.0 g
Distilled water	948 ml

pH 6.9

Dissolve the solid ingredients in the water by steaming, then add the solutions. Dispense in 100-ml amounts and autoclave at 121°C for 15 min. *Avoid over-heating*. The final pH should be approximately 6.9, and the medium will be orange in colour.

To use, pour thick plates of 15 ml and dry the surface of the plates before inoculation. It should be noted that, if bottles of medium have not been required after melting, a subsequent re-melting is likely to lead to a decrease in selectivity.

(Thatcher and Clark, 1968)

Bromcresol Green Ethanol Yeast Extract Agar

(For the differentiation of *Gluconobacter* (*Acetomonas*) and *Acetobacter*)

Basal medium

Yeast extract (Difco)	30.0 g
Bromcresol green (2.2% aqueous solution)	1.0 ml
Agar	20.0 g
Distilled water	1 litre

Dissolve the yeast extract and agar in the water by steaming, and add the bromcresol green solution. Mix well and dispense in 6.5-ml amounts in 30-ml (1 oz) screw-capped bottles. Autoclave at 121°C for 15 min.

Preparation of complete medium

Melt the bottles of medium, cool to 45–50°C and add aseptically to each bottle 0.3 ml of 50% ethanol (previously sterilized by filtration). Mix and allow to set in a sloped position.

(Carr, 1968)

Buffered Peptone Water

(For non-selective resuscitation)

Peptone	10.0 g
Sodium chloride	5.0 g
Disodium hydrogen phosphate	3.5 g
Potassium dihydrogen phosphate	1.5 g
Distilled water	to 1 litre

pH 7.2 ± 0.2

Dissolve the ingredients in the water, distribute as required, and sterilize by autoclaving at 121°C for 15 min.

Butter-fat Agar

Butter-fat	5 ml
Yeast extract agar or nutrient agar	100 ml

Before sterilization of the basal medium, adjust the reaction to pH 7.8. Prepare the butter-fat by warming fresh, unsalted butter in a beaker in a 50°C water-bath until liquid. Separate the butter-fat from the curd by careful pipetting, by filtration, or with a separating funnel. Dispense the butter into screw-capped bottles and sterilize by autoclaving at 121°C for 20 min.

To 100 ml of molten, sterile basal medium at 45°C add aseptically 5 ml of molten sterile butter-fat. Emulsify the medium by shaking vigorously. Pour plates containing 15-ml amounts.

(Berry, 1933)

Calcium Lactate Yeast Extract Agar

(For the differentiation of *Gluconobacter* from *Acetobacter*)

Yeast extract (Difco)	20.0 g
Calcium lactate	20.0 g
Agar	15.0 g
Distilled water	1 litre

Mix the ingredients and dissolve by steaming. Mix and distribute as required. Sterilize by autoclaving at 121°C for 15 min.

(Carr, 1968)

Calgon–Ringer's Solution

(For the suspension of alginate swabs)

Prepare as Ringer's solution, quarter-strength (see p. 454) and add sodium hexametaphosphate ('Calgon') at 10 g per litre. Dispense in 10-ml amounts in screw-capped bottles, and sterilize by autoclaving at 121°C for 15 min.

(Higgins, 1950)

Carbohydrate Fermentation Broths

Basal medium 1 (general purpose medium)

Tryptone	10.0 g
Sodium chloride	5.0 g
Bromcresol purple (1% solution)[§]	2.5 ml
Distilled water	1 litre

[§] Andrade's reagent to a final concentration of 1%, or phenol red to a final concentration of 0.01% may be used in place of the bromcresol purple if desired.

Add the ingredients (except the indicator) to the water and dissolve by steaming. Adjust to pH 7–2 and then add the indicator. Distribute in 3- or 5-ml amounts in screw-capped bottles or test-tubes, each bottle or tube being provided with an inverted Durham tube. Sterilize at 121°C for 20 min.

Basal medium 2

Recommended in Report (1958) for tests on the Enterobacteriaceae:

Nutrient broth: pH indicator, distribution, etc., as described above.

Basal medium 3

Recommended as the basal medium for tests on yeasts:

0.5% solution of yeast extract powder: pH indicator, distribution, etc., as described above.

Basal medium 4 (Hiss's serum water)

Recommended for tests on fastidious animal parasites and pathogens (e.g. *Corynebacterium*):

Mix one part of sterile bovine serum with three parts of sterile distilled water, adjust to pH 7.6; to each 400 ml of medium add 20 ml of a 0.2% phenol red solution. Dispense in 2.5-ml amounts in screw-capped 7-ml (bijou) bottles and sterilize by steaming for 20 min on three successive days.

Phenol red solution is prepared by dissolving 1.0 g of phenol red in 28.4 ml of 0.1 N sodium hydroxide with gentle heating. Distilled water should then be added to approximately 400 ml, 28.4 ml of 0.1 N hydrochloric acid added, and the solution made up to 500 ml with distilled water and filtered before use.

Basal medium 5

Recommended for fermentation tests on lactobacilli. See MRS Fermentation Medium.

Substrate solutions

Prepare the substrates as 10% solutions in distilled water (salicin has low solubility and should be prepared as a 5% solution), and sterilize by filtration.

Preparation of complete fermentation broth

To each 2.5-, 3- or 5-ml amount of sterile basal medium, add aseptically 0.25, 0.3 or 0.5 ml respectively of the desired substrate solution in the case of base media 1, 2 and 4. In the case of fermentation broths for tests on yeasts using medium 3, and for tests on lactobacilli using MRS fermentation medium, the substrate should be added to a final concentration of 2%. Salicin solution should be added in double the above quantities to achieve an adequate final concentration of substrate.

Cefixime–Cefsulodin–Vancomycin Buffered Peptone Water

(For selective enrichment of *Escherichia coli* O157)

To sterile buffered peptone water (see p. 399), immediately before use add aseptically sterile cefixime–cefsulodin–vancomycin supplement, equivalent to 1 ml of supplement per litre of buffered peptone water (BPW).

Preparation of cefixime–cefsulodin–vancomycin supplement

Vancomycin hydrochloride	0.8 g
Cefsulodin	1.0 g
Cefixime solution (50 mg in 10 ml ethanol)	1.0 ml
Distilled water	100 ml

Dissolve the solid ingredients in the water, and add the cefixime solution. Sterilize by membrane filtration. Distribute aseptically in 1-ml amounts in sterile bijou bottles and keep frozen (preferably at –80°C if a low temperature freezer is available) until required.

(after DoE/SCA, 1996)

Cefixime–Tellurite Sorbitol MacConkey's Agar (CT-SMAC)

(For the isolation and differentiation of *Escherichia coli* O157)

To 100 ml of Sorbitol MacConkey's agar (see p. 461), melted and cooled to 50–55°C, add 1 ml of the cefixime–potassium tellurite solution, to give final concentrations of 0.05 mg cefixime per litre, and 2.5 mg potassium tellurite per litre (Roberts *et al.*, 1995). Mix well and pour into Petri dishes. (Plates may be stored refrigerated in plastic bags for up to 1 week. They should be restored to ambient temperature before inoculation.)

Cefixime–potassium tellurite solution

Potassium tellurite	0.25 g
Cefixime solution (50 mg in 10 ml ethanol)	1.0 ml
Distilled water	100 ml

Dissolve the potassium tellurite in the water, and add the cefixime solution. Sterilize by membrane filtration. Distribute aseptically in 1-ml amounts in sterile bijou bottles and keep frozen (preferably at –80°C if a low temperature freezer is available) until required.

Alternatively, commercially available sterile cefixime and potassium supplements may be used, and added to give the correct final concentrations.

(Zadik *et al.,* 1993; Roberts *et al.,*, 1995)

Cefsulodin Irgasan Novobiocin Agar (CIN)

(For the isolation of *Yersinia enterocolitica*)

Special peptone (Oxoid)	20.0 g
Yeast extract (Difco)	2.0 g
Mannitol	20.0 g
Sodium pyruvate	2.0 g
Sodium chloride	1.0 g
Sodium desoxycholate	0.5 g
Magnesium sulphate, hydrated ($MgSO_4 \cdot 7H_2O$), 0.1% stock solution	10 ml
Agar	12.0 g
Distilled water	950 ml

Boil to dissolve the ingredients. Cool to 55–60°C. Add the following supplements which have been previously sterilized by filtration. (The stock solutions of antibiotics may be prepared and filter-sterilized before use, and stored at –70 to –80°C until required.)

1. 10 ml of irgasan DP300 (2,4,4′-trichloro–2′-hydroxydiphenyl ether) (Ciba-Geigy) solution (0.04% in 95% ethanol)
2. 10 ml of cefsulodin (Takeda) solution (150 mg per 100 ml water)

3. 10 ml of novobiocin (Upjohn) solution (25 mg per 100 ml water)
4. 10 ml neutral red solution (0.3% in water)
5. 10 ml crystal violet solution (10 mg per 100 ml water)

Mix, and pour plates. The pH of the final medium should be 7.4 ± 0.2.

(Schiemann, 1979)

Cephaloridine–Fucidin–Cetrimide Agar (CFC)

(A selective medium for the isolation of *Pseudomonas* from foods)

To 100 ml of sterile molten Heart Infusion Agar, cooled to 50–55°C, add the following solutions (all are 1% aqueous solutions, previously sterilized by membrane filtration):

Cephaloridine (e.g. Ceporin, Glaxo)	0.5 ml
Sodium fucidate (e.g. Fucidin, Leo Laboratories)	0.1 ml
Cetyltrimethyl ammonium bromide (Cetrimide) (e.g. Sigma)	0.1 ml

Mix well and pour into Petri dishes. Prepared plates can be stored in plastic bags in the refrigerator or at ambient temperature for up to 1 month without loss of selectivity (but ensure that the plates do not become dehydrated on storage).

(Mead, 1985)

Cheese Agar

(For the isolation of *Brevibacterium linens*)

Ripened cheese	100.0 g
Potassium citrate	10.0 g
Peptone	10.0 g
Sodium chloride	50.0 g
Sodium oxalate[H]	2.0 g
Agar	15.0 g
Distilled water	1 litre

pH 7.4

Dissolve the potassium citrate in 300 ml of distilled water and add the cheese. Warm the suspension to 50°C, then transfer to a funnel with rubber tubing and spring clip and allow to stand for 30 min to allow separation of the fat. Run off the aqueous portion and add to the remainder of the ingredients dissolved in 700 ml of distilled water. Adjust the pH to 7.4 and distribute as required. Sterilize at 121°C for 25 min. When using the medium it must be thoroughly agitated to distribute suspended cheese solids.

(Albert *et al.*, 1944)

Christensen's Urea Agar

(Especially for differentiation within the Enterobacteriaceae)

Peptone	1.0 g
Sodium chloride	5.0 g
Potassium dihydrogen phosphate	2.0 g
D-Glucose	1.0 g
Phenol red (0.2% aqueous solution)	6.0 ml
Agar	20.0 g
Distilled water	1 litre

pH 7.0

Dissolve the ingredients in the water by steaming. Distribute in bottles or test-tubes in sufficient quantities to enable slants to be prepared, and sterilize by autoclaving at 121°C for 15 min. Cool to 50°C and to each tube add aseptically sufficient sterile 20% urea solution (previously sterilized by filtration) to give a final concentration of 2%. Allow the medium to set in the sloped position.

(Christensen, 1946)

Columbia Agar

(A basal medium suitable for preparation of blood agar)

Proteose peptone	23.0 g
Starch	1.0 g
Sodium chloride	5.0 g
Agar	15.0 g
Distilled water	1 litre

Add the ingredients to the water, mix, allow to soak for 15 min, with occasional mixing. Dissolve by steaming or boiling. Dispense in 100-ml amounts in screw-capped bottles, and sterilize at 121°C for 15 min.

The above recipe is a simplified version; most manufacturers of dehydrated media produce versions of Columbia agar containing a mixture of special peptones to provide a more nutritious environment for fastidious pathogens.

CPS Medium

(For the isolation and enumeration of aquatic bacteria)

Soluble casein (Merck)	0.5 g
Peptone (Difco)	0.5 g

Soluble starch	0.5 g
Dipotassium hydrogen phosphate	0.2 g
Magnesium sulphate, hydrated ($MgSO_4 \cdot 7H_2O$)	0.05 g
Ferric chloride (0.01% solution)	4 drops
Glycerol	1 ml
Agar (Difco)	15.0 g
Distilled water	1 litre

pH 6.9–7.0

Dissolve the casein, peptone, starch and agar in the water by steaming. Add the remaining ingredients to the hot medium, mix well and filter. Distribute as required and sterilize by autoclaving at 121°C for 20 min.

(Collins *et al.*, 1973)

Crossley's Milk Peptone Medium

(For sterility testing of canned evaporated milk)

Skim milk powder	100 g
Peptone	10 g
Bromcresol purple (1% aqueous solution)	10 ml
Distilled water	1 litre

Mix the milk powder to a smooth paste with a little of the water and gradually stir in the rest of the water. Dissolve the peptone and mix the bromcresol purple into the medium. Distribute in 10-ml amounts in test-tubes or as required. Sterilize by autoclaving at 121°C for 5 min, followed by steaming for 30 min on each of two successive days. Incubate at 37°C for 2 days and at 30°C for 2 days before use to check sterility.

(Crossley, 1941)

Crystal Violet Agar

(A medium for the isolation of Gram-negative bacteria)

Yeast extract agar, plate count agar or nutrient agar	100 ml
Crystal violet (0.05% aqueous solution)	0.4 ml

Add aseptically the crystal violet solution (previously sterilized by filtration) to the sterile molten medium immediately before pouring the plates. This gives a final concentration of crystal violet of 2 p.p.m.

(Holding, 1960)

Cystine Broth

(For detecting the ability to produce hydrogen sulphide)

Add cystine to a final concentration of 0.01% to a basal medium of peptone water or nutrient broth. Distribute in 10-ml amounts in test-tubes, plug and sterilize at 121°C for 15 min.

Czapek–Dox Agar

(For the culture of yeasts and moulds)

Sodium nitrate	2.0 g
Potassium chloride	0.5 g
Magnesium sulphate, hydrated ($MgSO_4 \cdot 7H_2O$)	0.5 g
Dipotassium hydrogen phosphate	1.0 g
Ferrous sulphate, hydrated ($FeSO_4 \cdot 7H_2O$)	0.01 g
Sucrose	30.0 g
Agar	15.0 g
Distilled water	to 1 litre

Dissolve ingredients in water by steaming. Dispense as required, and sterilize at 115°C for 20 min.

Davis's Yeast Salt Agar

(For the isolation and enumeration of yeasts and moulds)

Ammonium nitrate	1.0 g
Ammonium sulphate	1.0 g
Sodium monohydrogen phosphate, anhydrous	4.0 g
Potassium dihydrogen phosphate	2.0 g
Sodium chloride	1.0 g
D-Glucose	10.0 g
Yeast extract powder	1.0 g
Agar	20.0 g
Distilled water	1 litre

pH 6.6

Dissolve all the ingredients by steaming. Check the reaction and adjust if necessary to pH 6.6. Distribute in 100-ml amounts in screw-capped bottles and sterilize by autoclaving at 115°C for 20 min.

The pH can be adjusted to 3.5 if required, by the addition aseptically of 5.7 ml of sterile 10% citric acid solution (previously sterilized by filtration) to each 100 ml of molten sterile medium at 50°C immediately before pouring the plates.

As an alternative to acidification for the inhibition of bacteria, antibiotics may be added as sterile solutions to the molten medium immediately before pouring the plates (see p. 138).

(Davis, 1958)

Davis's Yeast Salt Broth

This should be prepared in a similar way to the agar medium to the same recipe, or as a double-strength medium as required, except that the agar is omitted from the recipe. Distribute the single-strength medium in test-tubes, and the double-strength medium in suitable containers, insert inverted Durham tubes, and sterilize by autoclaving at 115°C for 20 min.

(Davis, 1970)

Dichloran–Rose Bengal–Chloramphenicol Agar (DRBCA)

(For the detection and isolation of fungi in foods)

D-Glucose	10.0 g
Mycological peptone	5.0 g
Potassium dihydrogen phosphate	1.0 g
Magnesium sulphate, hydrated ($MgSO_4 \cdot 7H_2O$)	0.5 g
Rose bengal, 5% (w/v) aqueous solution	0.5 ml
Dichloran, 0.2% (w/v) ethanolic solution	1.0 ml
Chloramphenicol, 1% (w/v) ethanolic solution	10.0 ml
Agar	15.0 g
Distilled water	1 litre

Steam to dissolve. Dispense and sterilize by autoclaving at 121°C for 15 min. Dispense in 15-ml amounts into sterile Petri dishes.

Use immediately, or store refrigerated in the dark. This medium, when made, *must* be kept in the dark, because under the influence of light Rose Bengal may generate cytotoxic products which will result in the medium becoming too inhibitory (Banks *et al.*, 1985)

(Pitt and Hocking, 1985)

Dichloran 18% Glycerol Agar (DG18)

(For the enumeration and isolation of xerophilic fungi in foods)

D-Glucose	10.0 g
Mycological peptone	5.0 g
Potassium dihydrogen phosphate	1.0 g

Magnesium sulphate, hydrated ($MgSO_4 \cdot 7H_2O$)	0.5 g
Dichloran, 0.2% (w/v) ethanolic solution	1.0 ml
Chloramphenicol, 1% (w/v) ethanolic solution	10.0 ml
Agar	15.0 g
Distilled water	to 1 litre
Glycerol	220 ml

Suspend the ingredients except the dichloran, chloramphenicol and glycerol in about 800 ml of water. Heat to dissolve. Add the ethanolic solutions of dichloran and chloramphenicol to the mixture, and make up to 1 litre with water (at 55°C). Add the glycerol (preheated to 55°C), mix thoroughly, and sterilize by autoclaving at 121°C for 15 min. Cool to 50°C, mix well, and dispense into sterile Petri dishes. Allow to set, and dry.

Use immediately, or can be stored for up to 7 days in the refrigerator.

(Hocking and Pitt, 1980)

Differential Reinforced Clostridial Medium

(For the detection and enumeration of sulphite-reducing clostridia by a multiple tube technique)

Basal medium

Peptone	10.0 g
Lab-Lemco meat extract (Oxoid)	10.0 g
Sodium acetate, hydrated	5.0 g
Yeast extract (Difco)	1.5 g
Soluble starch	1.0 g
D-Glucose	1.0 g
L-Cysteine monohydrochloride	0.5 g
Distilled water	1 litre

pH 7.1–7.2

Dissolve the peptone, meat extract, sodium acetate and yeast extract in 800 ml of distilled water. Mix the soluble starch to a paste with a little of the remaining 200 ml of water, boil the rest of the 200 ml of water and stir it into the paste. Mix the two solutions and steam for 30 min to complete the solution of the ingredients. After the steaming add the glucose and cysteine, mix and adjust the pH to 7.1–7.2. Filter through hot paper pulp (see p. 445), dispense in 25-ml amounts in 30-ml (1 oz) screw-capped (McCartney) bottles. Sterilize by autoclaving at 121°C for 15 min.

Sodium sulphite solution

Prepare a 4% (w/v) solution of sodium sulphite (anhydrous) and sterilize by filtration. This solution may be stored in fully filled screw-capped bottles in the refrigerator for up to 14 days.

Ferric citrate solution

Prepare a 7% (w/v) solution of ferric citrate (scales), heating briefly to dissolve. Sterilize by filtration. It may be stored in fully filled screw-capped bottles in the refrigerator for up to 14 days.

Preparation of complete medium

When the DRCM is required, steam and cool the basal medium, and then add aseptically to each bottle of medium 0.5 ml of a freshly prepared mixture of equal volumes of sodium sulphite solution and ferric citrate solution.

The medium may be made more selective for sulphite-reducing *Clostridium* spp. by also adding aseptically to each bottle 0.9 ml of a solution of 10 000 units of polymyxin B sulphate in 5 ml of sterile distilled water.

(Gibbs and Freame, 1965; Freame and Fitzpatrick, 1971)

EC Broth

(For the detection of coliform bacteria and/or *Escherichia coli*)

Tryptose (or trypticase)	20.0 g
Lactose	5.0 g
Bile salts (No. 3: Oxoid L56 or equivalent)§	1.5 g
Dipotassium hydrogen phosphate	4.0 g
Potassium dihydrogen phosphate	1.5 g
Sodium chloride	5.0 g
Distilled water	1 litre

pH 6.8–6.9

§ Different proprietary preparations may not have equivalent activity at the same concentration.

Dissolve the ingredients in the water with minimal steaming, distribute in 5-ml amounts in 150 × 16-mm test-tubes with inverted Durham tubes. Autoclave at 121°C for 15 min.

(Hajna and Perry, 1943)

Edwards's Aesculin Crystal Violet Blood Agar

(For the isolation of animal parasitic streptococci)

Nutrient agar (pH 7.4)	100 ml
Crystal violet, 0.05% solution (sterile)	0.4 ml
Defibrinated ox blood (sterile)	5.0 ml
Aesculin	0.1 g

Add the aesculin to the molten nutrient agar and sterilize by steaming for 30 min on each of three successive days.

To 100 ml of molten sterile aesculin nutrient agar cooled to 50°C, add with aseptic precautions 0.4 ml of sterile 0.05% crystal violet solution (previously sterilized by autoclaving or filtration) and mix well. Next add aseptically 5 ml of sterile defibrinated ox blood, mix thoroughly and pour into Petri dishes.

(Edwards, 1933)

Egg-yolk Agar

(For the detection of lecithinase production)

Egg-yolk emulsion	10 ml
Sodium chloride	1.0 g
Yeast extract agar or nutrient agar	100 ml

To the basal medium add the extra salt and sterilize by autoclaving at 121°C for 20 min. Cool to 45°C and add aseptically the sterile egg-yolk emulsion, mix well and pour plates.

The egg-yolk emulsion (see p. 412) may be prepared after the method of Billing and Luckhurst (1957). Alternatively, egg-yolk emulsions are available commercially from most suppliers of dehydrated media and media supplements.

Egg-yolk Broth

(For the detection of lecithinase production)

Nutrient broth	100 ml
Sodium chloride	1.0 g
Egg-yolk emulsion, sterile (see p. 412)	10 ml

Add the sodium chloride to the nutrient broth, distribute in 5-ml amounts in 150 × 16-mm test-tubes, and sterilize at 121°C for 15 min. When cold, add with aseptic precautions 0.5 ml of sterile egg-yolk emulsion to each tube of medium.

The egg-yolk emulsion may be prepared after the method of Billing and Luckhurst (1957; see below), or obtained as a commercial preparation from most suppliers of media.

Egg-yolk Emulsion

Separate the yolks from the whites of the eggs by pipette and add four parts of distilled water to one part of egg yolk. Mix thoroughly and heat in a waterbath at 45°C for 2 h. Centrifuge to remove the precipitate (alternatively, stand the mixture overnight in the refrigerator). Decant the supernatant liquid. Sterilize by filtration through stacked graded membrane filters (see p. 21).

(Billing and Luckhurst, 1957)

Egg-yolk-free Tryptose Sulphite–Iron Citrate–Cycloserine Agar (SICA)

(For enumeration of mesophilic sulphite-reducing clostridia)

Basal medium

Tryptose	15.0 g
Soya peptone	5.0 g
Meat extract	5.0 g
Yeast extract	5.0 g
Glucose	2.0 g
Agar	15.0 g
Distilled water	900 ml

Dissolve the ingredients in the water by steaming or boiling. Mix well, and dispense in 90-ml amounts in screw-capped bottles. Sterilize at 121°C for 15 min.

Stock solution of disodium disulphite and iron (III) citrate

Disodium disulphite, anhydrous	1.0 g
Iron (III) ammonium citrate	1.0 g
Distilled water	100 ml

Mix and dissolve, sterilize by membrane filtration. Store refrigerated for up to 4 weeks.

Stock solution of D-cycloserine

D-Cycloserine (crystalline)	0.4 g
Distilled water	10 ml

Dissolve and mix, and sterilize by membrane filtration. Store in the refrigerator for up to 4 weeks.

Preparation of complete medium

To each 90 ml of medium, melted and cooled to 50°C, add aseptically 10 ml of sterile sodium disulphite–iron (III) ammonium citrate solution, and 1.0 ml of sterile D-cycloserine solution, mix well and pour into the prepared Petri dishes.

(after ISO 7937:1997)

Enterobacteriaceae Enrichment (EE) Broth

Peptone	10.0 g
Glucose	5.0 g
Sodium sulphate, hydrated ($Na_2SO_4 \cdot 2H_2O$)	8.0 g
Potassium dihydrogen phosphate	2.0 g
Dehydrated ox bile[§]	20.0 g
Brilliant green (0.135% solution)	10 ml
Distilled water	1 litre

pH 7.2 ± 0.2

[§] Different proprietary preparations may not have equivalent activity at the same concentration.

Dissolve the ingredients in the water, mix well, and distribute as required in flasks or tubes. Heat at 100°C for 30 min.

(Mossel *et al.*, 1963)

Ethanol Broth

(For the determination of the utilization of ethanol by yeasts)

Ammonium sulphate	1.0 g
Potassium dihydrogen phosphate	1.0 g
Magnesium sulphate, hydrated ($MgSO_4 \cdot 7H_2O$)	0.5 g
Distilled water	1 litre

Dissolve the salts in the water, distribute in 5-ml amounts in test-tubes and sterilize at 121°C for 15 min. When cool, add aseptically to each tube 0.3 ml of sterile 50% ethanol and *one drop* of a sterile 5% solution of yeast extract. The ethanol and yeast extract solution are sterilized beforehand by filtration.

Growth of the yeast being examined should be compared with growth in the same medium with the ethanol omitted.

(Lodder and Kreger-van Rij, 1952; Lodder, 1970)

Ferrous Chloride Gelatin

(For the simultaneous determination of hydrogen sulphide production and liquefaction of gelatin)

Ferrous chloride (10% solution)	0.5 ml
Nutrient gelatin, sterile	100 ml

Heat the sterile nutrient gelatin to 100°C in a steamer, and add the *freshly prepared* ferrous chloride solution. Immediately dispense aseptically in sterile 130 × 12-mm test-tubes to give an 8-cm depth of medium. Seal with sterile air-tight stoppers (e.g. rubber bungs or Astell roll-tube closures, sterilized by immersion in boiling water for 20 min), and cool the tubes rapidly by partial immersion in cold water.

(Report, 1958)

Fortified Nutrient Agar

(For spore production by *Bacillus stearothermophilus*)

Nutrient agar powder (Difco or equivalent)	15 g
Plain agar powder (Difco or equivalent)	5 g
D-Glucose	0.5 g
Manganese sulphate, hydrated ($MnSO_4 \cdot 4H_2O$)	30 mg
Distilled water	1 litre

Dissolve the ingredients in the water by steaming. Dispense in Roux bottles or in 500-ml screw-capped medical flat bottles in sufficient amount to provide an approximately 2-cm layer when the bottles are laid on their sides. Sterilize by autoclaving at 121°C for 25 min, and place the bottles on their sides so that the medium sets to provide the maximum surface area.

(Finley and Fields, 1962)

Fraser Broth and Half Fraser Broth

(For enrichment of *Listeria monocytogenes*)

Meat peptone	5.0 g
Tryptone	5.0 g
Beef extract	5.0 g
Yeast extract	5.0 g
Sodium chloride	20.0 g
Disodium hydrogen phosphate, hydrated ($Na_2HPO_4 \cdot 2H_2O$)	12.0 g
Potassium dihydrogen phosphate	1.35 g
Aesculin	1.0 g
Distilled water	1 litre

pH 7.2 ± 0.2

Dissolve the ingredients in the water, by steaming or boiling. Adjust the pH so that after sterilization it will be 7.2 ± 0.2 at 25°C (this needs to be determined by experience for the brands of constituent used).

Dispense in appropriate volumes (e.g. 100-ml amounts for the primary enrichment Half Fraser broth, 10-ml amounts in test-tubes for Fraser broth). Sterilize by autoclaving at 121°C for 15 min.

Lithium chloride solution

Lithium chloride	3.0 g
Distilled water	10 ml

Add the lithium chloride to the water with careful stirring (this is a strongly exothermic reaction so take all precautions, including eye protection and wearing rubber gloves). Sterilize by membrane filtration.

Acriflavine solution

Acriflavine hydrochloride	0.25 g
Distilled water	100 ml

Dissolve the acriflavine hydrochloride in the water and sterilize by membrane filtration.

Nalidixic acid solution

Nalidixic acid sodium salt	0.1 g
Sodium hydroxide solution (0.05 mol l^{-1})	10 ml

Dissolve and mix, and sterilize by membrane filtration.

Ammonium iron (III) citrate solution

Ammonium iron (III) citrate	5.0 g
Distilled water	100 ml

Dissolve and mix, and sterilize by membrane filtration. The ammonium iron (III) citrate used should be the more pure green form (around 18% iron content).

Preparation of Half Fraser Broth

To each 100 ml of basal medium immediately before use add with aseptic precautions:

Lithium chloride solution	1.0 ml
Nalidixic acid solution	0.1 ml
Acriflavine solution	0.5 ml
Ammonium iron (III) citrate solution	1.0 ml

Preparation of Fraser Broth

To each 10 ml of basal medium immediately before use add with aseptic precautions:

Lithium chloride solution	0.1 ml
Nalidixic acid solution	0.02 ml
Acriflavine solution	0.1 ml
Ammonium iron (III) citrate solution	0.1 ml

To achieve the appropriate level of precision in the amounts of supplements added, use positive displacement pipetting devices with disposable plastic tips (e.g. Eppendorf or Gilson).

(ISO 11290–1:1996)

Frazier's Gelatin Agar

(For the determination of gelatinase production (proteolysis))

Gelatin	0.4 g
Yeast extract agar or nutrient agar	100 ml

Add the gelatin to the molten agar medium and steam with occasional mixing until completely dissolved. Sterilize by autoclaving at 115°C for 20 min.

(Smith *et al.*, 1952)

Gardner's Streptomycin Thallous Acetate Cycloheximide Agar

(For isolation and enumeration of *Brochothrix thermosphacta*)

Basal medium

Peptone	20.0 g
Yeast extract	2.0 g
Glycerol	15.0 g
Disodium hydrogen phosphate	1.0 g
Magnesium sulphate, hydrated ($MgSO_4 \cdot 7H_2O$)	1.0 g
Agar	13.0 g
Distilled water	1 litre

pH 7.0 ± 0.2

Dissolve the ingredients in the water by steaming or boiling. Distribute in 100-ml amounts in screw-capped bottles and sterilize at 121°C for 15 min.

Supplements solution

Streptomycin sulphate	0.5 g
Thallium (I) acetate[H]	0.05 g
Cycloheximide[H]	0.05 g
Distilled water	100 ml

Dissolve the ingredients in the water, and sterilize by membrane filtration.

Preparation of complete medium

To 100 ml of medium, melted and cooled to 50°C, add aseptically 1 ml of the sterile supplements solution, mix well and pour plates as required.

(Gardner, 1966)

Gibson's Semi-solid Tomato Juice Medium

(For the detection of carbon dioxide production from glucose by lactic-acid bacteria)

Yeast extract	2.5 g
D-Glucose	50.0 g
Tomato juice, pH 6.5[§]	100 ml

Reconstituted skim milk	800 ml
Nutrient agar	200 ml

§ or manganese sulphate ($MnSO_4 \cdot 4H_2O$), 0.4% solution 10 ml

Mix the tomato juice or manganese sulphate solution with the reconstituted skim milk, add the yeast extract and glucose and heat in the steamer. While still hot, add the molten nutrient agar and mix well. Check the pH and adjust if necessary to pH 6.5. Distribute in test-tubes to a depth of 5–6 cm, and sterilize by steaming for 30 min on each of three successive days.

(Gibson and Abd-el-Malek, 1945; Stamer *et al.*, 1964)

Giolitti and Cantoni's Tellurite Mannitol Glycine Broth

(For the selective enrichment of *Staphylococcus aureus*)

Tryptone	10.0 g
Meat extract	5.0 g
Yeast extract	5.0 g
Lithium chloride	5.0 g
Mannitol	20.0 g
Sodium chloride	5.0 g
Glycine	1.2 g
Sodium pyruvate	3.0 g
Distilled water	1 litre

pH 6.9 ± 0.2

Dissolve the ingredients in the water with heating, mix well. Cool to 25°C and adjust the pH. Distribute as required, and sterilize at 115°C for 20 min.

The sterile basal medium may be stored refrigerated for up to 15 days.

Potassium tellurite solution

Potassium tellurite[H]	1.0 or 10.0 g*
Distilled water	to 100 ml

Dissolve, mix and sterilize by membrane filtration. The quality of potassium tellurite is critical. Potassium tellurite from Merck has been found satisfactory; its inhibitory action for known strains of *Staph. aureus* should be determined, especially if other sources for the tellurite need to be used.

* If the enrichment is to be achieved from adding a 10^{-1} dilution, then the lower concentration of potassium tellurite is used; if the enrichment broth is to be inoculated

with undiluted food (up to 1 g of food per 20 ml of broth or 5 g food per 100 ml of broth) then the higher concentration of potassium tellurite should be used to suppress the higher numbers of contaminants that will find themselves in a medium more substantially modified by the added food.

Preparation of complete medium

To the sterile basal medium add with aseptic precautions an appropriate volume to an equivalent of 0.5 ml of sterile potassium tellurite solution per 100 ml of basal medium (e.g. 0.1 ml per 20 ml). The choice of concentration of potassium tellurite solution is determined by the type and amount of inoculum (see above). Use the complete medium on the day of preparation.

(Giolitti and Cantoni, 1966)

Gluconate Broth

(For the determination of gluconate utilization, especially by members of the Enterobacteriaceae)

Peptone	1.5 g
Yeast extract	1.0 g
Dipotassium hydrogen phosphate	1.0 g
Potassium gluconate	40.0 g
Distilled water	1 litre

pH 7.0

Dissolve the ingredients in the water, distribute in 5-ml amounts in test-tubes, and sterilize at 115°C for 10 min.

(Shaw and Clarke, 1955)

Glucose Azide Broth

(For the enumeration of faecal streptococci by the multiple tube technique)

Peptone (Difco)	10.0 g
Sodium chloride	5.0 g
Dipotassium hydrogen phosphate	5.0 g
Potassium dihydrogen phosphate	2.0 g
D-Glucose	5.0 g
Yeast extract	3.0 g
Sodium azide[H]	0.25 g

Bromcresol purple (1.0% solution)	3 ml
Distilled water	1 litre

pH 6.6–6.8

Dissolve the ingredients in the water. Distribute in 5-ml amounts in 150 × 16-mm test-tubes. Sterilize by autoclaving at 121°C for 15 min. For inocula of large amounts of sample or diluent (5 ml or more), a double-strength medium should be prepared and distributed in amounts equal in volume to the inocula to be used. Double-strength medium is prepared in a similar manner to that described above, the ingredients being dissolved in half the quantity of distilled water.

Note. The effectiveness of this medium should be checked from time to time using stock cultures at low inoculum levels; it has been found that variations in the nutrients composition (particularly the peptone) from batch to batch may affect selectivity and/or sensitivity.

(Hannay and Norton, 1947)

Glucose Phosphate Broth

(Used for the methyl red test and Voges–Proskauer test, particularly for the differentiation of the Enterobacteriaceae)

Peptone	5.0 g
D-Glucose	5.0 g
Dipotassium hydrogen phosphate	5.0 g
Distilled water	1 litre

pH 7.5

Dissolve the ingredients in the water. Adjust to pH 7.5 and distribute in 5-ml amounts in 150 × 16-mm test-tubes. Sterilize by autoclaving at 115°C for 20 min.

Glucose Tryptone Agar (Dextrose Tryptone Agar)

(For the detection and enumeration of ‘flat-sour’ spoilage organisms)

Tryptone	10.0 g
D-Glucose	5.0 g
Bromcresol purple (1% solution)	4.0 ml
Agar	15.0 g
Distilled water	1 litre

pH 7.0

Add the ingredients to the water and heat in a steamer until dissolved. Adjust to pH 7.0, distribute as required and sterilize at 121°C for 15 min.

(Hersom and Hulland, 1969)

Glucose Tryptone Broth (Dextrose Tryptone Broth)

(For the detection and enumeration of 'flat-sour' spoilage organisms)

Prepare in the same way as glucose tryptone agar, except that the agar is omitted. Dispense in 5-ml or 10-ml amounts in test-tubes or other containers, and sterilize at 121°C for 15 min. For the detection of low concentrations of organisms in foods, double-strength medium may be prepared, and distributed in amounts equal in volume to the inocula to be used.

(Hersom and Hulland, 1969)

Glucose Yeast Chalk Agar

(Used in the 3-ketolactose test)

Yeast extract	10.0 g
D-Glucose	20.0 g
Calcium carbonate	20.0 g
Agar	15.0 g
Distilled water	1 litre

Add the ingredients to the water and dissolve by steaming. Distribute in 7-ml amounts in 150 × 16-mm test-tubes, with frequent mixing to retain the calcium carbonate in suspension. Sterilize at 115°C for 20 min, and set in the slanted position.

(Bernaerts and De Ley, 1963)

Gorodkowa Agar (Modified)

(For the sporulation of yeasts)

Peptone	10.0 g
D-Glucose	1.0 g
Sodium chloride	5.0 g
Agar	20.0 g
Distilled water	1 litre

Dissolve the ingredients in the water by steaming, distribute as required and sterilize at 121°C for 15 min; set as slants.

(Lodder and Kreger-van Rij, 1952; Lodder, 1970)

Gypsum Blocks

(For the promotion of sporulation by yeasts)

Prepare 3 × 3 × 1-cm plaster of Paris blocks from a mixture of eight parts calcium sulphate hemihydrate and three parts water. After setting place the blocks in sterile Petri dishes and sterilize in a hot air oven at 120°C for 2 h.

(Lodder and Kreger-van Rij, 1952; Lodder, 1970)

Hajna's GN Broth

(For the selective enrichment of enterobacteria)

Tryptose	20.0 g
D-Glucose	1.0 g
Mannitol	2.0 g
Sodium citrate	5.0 g
Sodium desoxycholate	0.5 g
Dipotassium hydrogen phosphate	4.0 g
Potassium dihydrogen phosphate	1.5 g
Sodium chloride	5.0 g
Distilled water	1 litre

Dissolve the ingredients in the water with minimum heating. Adjust the pH of the medium to 7.0 if necessary. Distribute as required and sterilize at 115°C for 15 min. It is important to avoid excessive heating of this medium.

(Croft and Miller, 1956)

Hayes' Medium

(For the demonstration of gliding motility)

Lab-Lemco beef extract (Oxoid)	1.0 g
Peptone	2.5 g
Sodium chloride	5.0 g
Agar	20.0 g
Distilled water	1 litre

pH 7.2

Dissolve the ingredients in the water by steaming, distribute and sterilize at 121°C for 20 min.

For the examination of marine organisms the sodium chloride and distilled water should be replaced by aged sea water (750 ml) and distilled water (250 ml).

(Hayes, 1963)

Hektoen Enteric Agar

(For the selective isolation of *Salmonella* and *Shigella*)

Proteose peptone	12.0 g
Meat extract	3.0 g
Lactose	12.0 g
Sucrose	12.0 g
Salicin	2.0 g
Sodium chloride	5.0 g
Bile salts§	9.0 g
Sodium thiosulphate	5.0 g
Ammonium iron (III) citrate*	1.5 g
Acid fuchsin	0.1 g
Bromthymol blue (0.4 % aqueous solution)	16 ml
Andrades' indicator (see p. 386)	20 ml
Agar	14.0 g
Distilled water	1 litre

pH 7.5 ± 0.2

§ Different proprietary preparations may not have equivalent activity at the same concentration.
* The ammonium iron(III) citrate used should be the more pure green form (around 18% iron content).

Mix the ingredients in the water. Allow to soak for 15 min, with occasional mixing. Heat to boiling for a few seconds until the agar is dissolved. Do not autoclave. Cool to 55°C and pour into Petri dishes.

Citrobacter and *Proteus* may be further inhibited by incorporating novobiocin in the medium just before pouring the plates (Hoben *et al.*, 1973). To each 1 litre of medium add 1 ml of a sterile novobiocin solution (150 mg novobiocin dissolved in 10 ml of distilled water, sterilized by membrane filtration).

(after King and Metzger, 1968)

Holding's Inorganic Nitrogen Medium

(For the differentiation of certain Gram-negative bacteria)

D-Glucose	5.0 g
Sodium citrate, hydrated	1.0 g
Sodium acetate, hydrated	1.0 g
Sodium succinate, hydrated	1.0 g
Calcium gluconate, hydrated	1.0 g
Ammonium dihydrogen phosphate	1.0 g
Dipotassium hydrogen phosphate	0.08 g
Potassium dihydrogen phosphate	0.02 g

Potassium nitrate[H]	1.0 g
Distilled water	1 litre

Dissolve the ingredients in the water, heat to 107°C and, as soon as the desired pressure of 34.5 kN m^{-2} (5 lb $inch^{-2}$) is reached, stop heating. Filter, distribute in 5-ml amounts in 150 × 16-mm test-tubes, and sterilize at 115°C for 20 min.

(Holding, 1960)

Hugh and Leifson's Medium

(For differentiating oxidative and fermentative metabolism of carbohydrates, by Gram-negative bacteria)

Recipe 1 (Hugh and Leifson, 1953)

Peptone	2.0 g
Sodium chloride	5.0 g
Dipotassium hydrogen phosphate	0.3 g
Bromthymol blue (1% aqueous solution)	3.0 ml
Agar	3.0 g
Distilled water	1 litre

pH 7.1 ± 0.2

Recipe 2 (Scholefield, 1964)

Tryptone	1.0 g
Yeast extract	1.0 g
Sodium chloride	5.0 g
Dipotassium hydrogen phosphate	0.3 g
Bromthymol blue (1% aqueous solution)	3.0 ml
Acid fuchsin (1% solution)	1.5 ml
Agar (Oxoid No. 3)	4.5 g
Distilled water	1 litre

pH 7.0 ± 0.2

Add the ingredients, except indicators, to the water and dissolve by steaming. Add the indicator(s) and mix well. Adjust the pH to 6.8 with thorough mixing, using a pH meter. (After autoclaving, the final pH of the medium should be 7.0–7.1.) Dispense in 10-ml amounts in 150 × 16-mm test-tubes, and close with Astell seals (Astell–Hearson), or other suitable tube closures. Leave the closures loose, and sterilize by autoclaving at 121°C for 15 min. To each tube of molten medium add aseptically 1 ml of a sterile 10%

solution of the desired substrate solution, mix well (without aeration) and allow to set. The carbohydrate normally employed in this medium is glucose, and it may be incorporated into the medium at the time of preparation if differential studies of reactions with a range of substrates are not being undertaken.

The second recipe has been found to allow the use of a single tube only, provided the tubes of media are stood in a boiling water-bath for 10 min and then cooled rapidly immediately before inoculation; in addition, a very clear colour change from blue-green to orange-red results from the production of acid.

Hugh and Liefson's Medium, Modified

(For differentiating oxidative and fermentative metabolism of glucose and mannitol by *Staphylococcus* and *Micrococcus*)

Tryptone	10.0 g
Yeast extract (Difco)	1.0 g
D-Glucose or mannitol	10.0 g
Bromcresol purple (1% solution)	4 ml
Agar	2.0 g
Distilled water	1 litre

pH 7.0 ± 0.2

Dissolve the ingredients in the water by steaming. Adjust to pH 7.0, and dispense in 150×16-mm test-tubes, filling them two-thirds full. Autoclave at 115°C for 20 min.

Immediately before use, steam the medium for 10–15 min to expel dissolved oxygen, and solidify by placing the tubes in cold or iced water.

(Recommendations, 1965)

Irgasan–Ticarcillin Potassium Chlorate Broth (ITPC)

(For selective enrichment of *Yersinia enterocolitica*)

Basal medium

Tryptone	10.0 g
Yeast extract	1.0 g
Magnesium chloride, hydrated ($MgCl_2 \cdot 6H_2O$)	60.0 g
Sodium chloride	5.0 g
Malachite green (0.2% aqueous solution)	5 ml
Potassium chlorate	1.0 g
Distilled water	1 litre

pH 6.9 ± 0.2

Dissolve the ingredients in the water, distribute in 100-ml amounts in flasks that minimize the surface area. Sterilize at 121°C for 15 min.

Irgasan solution

Irgasan (Ciba-Geigy; 2,3,4′-trichloro–2′-hydroxydiphenyl ether)	10 mg
Ethanol (95%)	10 ml

Dissolve the Irgasan in the ethanol and sterilize by positive-pressure membrane filtration (e.g. by using a Millipore Swinnex attachment for hypodermic syringe). This solution can be stored frozen.

Ticarcillin solution

Ticarcillin, disodium salt (Sigma)	10 mg
Distilled water	10 ml

Dissolve the Ticarcillin in the water, and sterilize by membrane filtration.

Preparation of complete medium

To each 100 ml of basal medium, heated to drive off dissolved oxygen and then cooled, add, with aseptic precautions, 0.1 ml of Irgasan solution and 0.1 ml of Ticarcillin solution. Use the same day, whilst the medium is still oxygen deficient.

(Wauters *et al.*, 1988; ISO 10273:1994)

Kanamycin Aesculin Azide Agar

(For the isolation of group D streptococci)

Tryptone	20.0 g
Yeast extract	5.0 g
Sodium chloride	5.0 g
Sodium citrate	1.0 g
Ammonium iron (III) citrate*	0.5 g
Kanamycin sulphate	0.02 g
Aesculin	1.0 g
Sodium azide[H]	0.15 g
Agar	15.0 g
Distilled water	1 litre

pH 7.1 ± 0.2

* Ammonium iron (III) citrate used should be the more pure green form (around 18% iron content).

Mix the ingredients in the water, and dissolve by steaming or boiling. Distribute into screw-capped bottles and sterilize at 121°C for 15 min.

(Mossel *et al.*, 1978)

King, Ward and Raney's Medium

(For demonstration of fluorescin production by *Pseudomonas*)

Proteose peptone No. 3 (Difco)	20.0 g
Dipotassium hydrogen phosphate	1.5 g
Magnesium sulphate, hydrated ($MgSO_4 \cdot 7H_2O$)	1.5 g
Glycerol	10.0 g
Agar	15.0 g
Distilled water	1 litre

pH 7.2 ± 0.2

Dissolve the ingredients in the water by steaming. Dispense as required (the medium may be used as poured plates or as slants in test-tubes) and sterilize by autoclaving at 121°C for 15 min.

(King *et al.*, 1954)

Kligler's Iron Agar

(A medium used in identification of members of the Enterobacteriaceae)

Peptone	20.0 g
Yeast extract	3.0 g
Lactose	10.0 g
Glucose	1.0 g
Sodium chloride	5.0 g
Ammonium iron (III) citrate*	0.5 g
Sodium thiosulphate	0.5 g
Phenol red	0.05 g
Agar	15.0 g
Distilled water	1 litre

pH 7.4 ± 0.2

* Ammonium iron (III) citrate used should be the more pure green form (around 18% iron content).

Mix the ingredients in the water, dissolve by steaming or boiling. Distribute into test-tubes in sufficient volume that the medium can be set as slants with deep (at least 3 cm) butts. Autoclave at 121°C for 15 min.

(MacFaddin, 1985)

Knisely's Chloral Hydrate Agar

(For the identification of *Bacillus cereus*)

Distribute Heart Infusion Agar (Difco) in 100-ml amounts and sterilize as recommended by the manufacturer.

Chloral hydrate solution

Prepare a 10% solution of chloral hydrate[H], and sterilize by filtration.

Preparation of complete medium

Add 2.5 ml of chloral hydrate solution (using a safety pipette filler or a positive displacement pipette) to each 100 ml of molten medium at 45–50°C immediately before pouring the plates.

(Knisely, 1965)

Lactose Yeast Extract Agar

(Used in the 3-ketolactose test)

Lactose	10.0 g
Yeast extract	1.0 g
Agar	20.0 g
Distilled water	1 litre

Dissolve the ingredients in the water by steaming. Distribute as required and sterilize at 115°C for 20 min.

(Bernaerts and De Ley, 1963)

Lauryl Sulphate Tryptose Broth

(For detection and enumeration of coliform organisms)

Tryptose	20.0 g
Dipotassium hydrogen phosphate	2.75 g
Potassium dihydrogen phosphate	2.75 g
Sodium chloride	5.0 g
Lactose	5.0 g
Sodium dodecyl sulphate (sodium lauryl sulphate)	0.1 g
Distilled water	1 litre

Dissolve the ingredients in the water by heating in a steamer. Distribute into tubes or bottles each containing an inverted Durham tube and sterilize in a steamer at 100°C for 30 min on each of three successive days, or by autoclaving at 121°C for 15 min.

(Mallman and Darby, 1941)

Liquid Paraffin

(Sterile, for covering cultures)

Dispense liquid paraffin in Erlenmeyer flasks in shallow layers and sterilize in the hot air oven at 160°C for 2 h.

Litmus Milk

Add sufficient litmus solution to reconstituted skim milk to give a pale mauve colour (10 ml of 4% litmus solution per litre of milk). Dispense as required (in relatively small volumes) and sterilize at 121°C for 5 min, followed by steaming for 30 min on each of the two following days. Autoclaving on the first day has been found to reduce substantially the number of spoiled tubes or bottles resulting from residual spores. Always preincubate before use, as described on p. 83.

Loeffler's Serum

(For the culture of fastidious animal parasites)

Nutrient broth plus 1% D-glucose	250 ml
Sterile serum	750 ml

Sterilize the nutrient broth with 1% added glucose by steaming at 100°C for 30 min on each of three successive days, or by autoclaving at 115°C for 20 min. Mix aseptically with the sterile serum in a sterile flask and distribute aseptically into sterile 7-ml (¼-oz or bijou) screw-capped bottles. Place the bottles (with caps screwed on tightly) in a sloping position in an oven or inspissator and heat slowly to 85°C to coagulate the serum. The medium may be sterilized by heating at 85°C for 20 min on each of three successive days.

Note. An alternative method of inspissation is to heat the medium to 80°C and maintain this temperature for 2h.

L-S Differential Medium

(For the differentiation of thermophilic yoghurt starter bacteria)

Casein peptone	10.0 g
Soya peptone	5.0 g
Meat extract	5.0 g
Yeast extract	5.0 g
Glucose	20.0 g
Sodium chloride	5.0 g
L-Cysteine hydrochloride 1-hydrate	0.3 g
Agar	13.0 g
Distilled water	890 ml

Mix and boil to dissolve the ingredients. Sterilize at 121°C for 20 min. Cool the medium to 50°C, and add 100 ml of a sterile 10% milk powder solution (sterilized at 121°C for 5 min) warmed to 50°C, and 10 ml of a sterile 2% solution of 2,3,5-triphenyltetrazolium chloride (sterilized by membrane filtration), warmed to 50°C.

(Eloy and Lacrosse, 1976; Corry *et al.*, 1995a)

M-Endo Broth

(For the isolation of coliforms by membrane filtration)

Tryptose	10.0 g
Thiopeptone	5.0 g
Trypticase or Casitone	5.0 g
Yeast extract	1.5 g
Lactose	12.5 g
Sodium chloride	5.0 g
Dipotassium hydrogen phosphate	4.375 g
Potassium dihydrogen phosphate	1.375 g
Sodium dodecyl sulphate (sodium lauryl sulphate)	0.05 g
Sodium desoxycholate	0.1 g
Sodium sulphite	2.1 g
Basic fuchsin	1.05 g
Distilled water	1 litre

pH 7.2 ± 0.2

Dissolve constituents. Heat to boiling. Do not autoclave. The medium may be stored refrigerated for up to 4 days. Dispense on to sterile absorbent membrane filtration pads in MF containers or Petri dishes.

(Membrane Endo Broth or Membrane Endo Agar is available in dehydrated form from a number of manufacturers.)

After 24 h at 35° or 37°C, coliforms produce red colonies with a golden-metallic sheen.

(American Public Health Association, 1995)

MRS Agar

(For the culture of lactobacilli)

MRS broth plus 1.5% agar

Dissolve ingredients for MRS broth (see below) with the exception of glucose by heating in the steamer, and adjust pH to 6.2–6.6. Add the agar and dissolve at 121°C for 5 min. Dissolve the glucose in the molten agar medium and distribute as required. Sterilize at 121°C for 15 min.

(de Man, Rogosa and Sharpe, 1960)

MRS Broth

(For the culture of lactobacilli)

Peptone	10.0 g
Meat extract	10.0 g
Yeast extract	5.0 g
Glucose	20.0 g
Tween 80 (polyoxyethylene(20) sorbitan monoleate)	1.0 g
Dipotassium hydrogen phosphate	2.0 g
Sodium acetate	5.0 g
Triammonium citrate	2.0 g
Magnesium sulphate, hydrated ($MgSO_4 \cdot 7H_2O$)	0.2 g
Manganese sulphate, hydrated ($MnSO_4 \cdot 4H_2O$)	0.05 g
Distilled water	1 litre

Dissolve the ingredients in the distilled water by steaming. Adjust the pH to 6.2–6.6. Distribute as required and sterilize at 121°C for 15 min. The final pH after sterilization should be 6.0–6.5.

(de Man, Rogosa and Sharpe, 1960)

MRS Fermentation Medium

(For fermentation studies of lactobacilli)

Prepare MRS broth as usual, but omit the glucose and meat extract. Adjust the pH to 6.2–6.5. Add 0.004% chlorphenol red as indicator. Distribute in test-tubes or as required.

Prepare 10% solutions of the test substrates and sterilize by filtration. Add aseptically the required substrate to give a final concentration of 2%.

(de Man *et al.*, 1960)

MacConkey's Agar

(For isolation and growth of enterobacteria, especially the coliform bacteria)

Peptone	20.0 g
Bile salts[§]	5.0 g
Sodium chloride	5.0 g
Lactose	10.0 g
Neutral red (1% aqueous solution)	7.0 ml
Agar	15.0 g
Distilled water	1 litre

[§] Different proprietary preparations may not have equivalent activity at the same concentration.

Dissolve the peptone, bile salts and sodium chloride in the water by steaming. Cool and adjust to pH 7.4. Add the agar and dissolve by autoclaving. Filter through hot paper pulp (see p. 445). Adjust the pH to 7.4. Add the lactose and neutral red and steam until dissolved. Mix well and distribute in bottles or test-tubes as required. Sterilize at 115°C for 15 min.

(Report, 1969)

Brilliant green MacConkey's agar

Add 3.3 ml of 1% brilliant green solution at the same time as the lactose and neutral red (Harvey and Price, 1974).

MacConkey's Broth

(For the detection and enumeration of lactose-fermenting enterobacteria by the multiple tube technique)

Peptone	20.0 g
Bile salts[§]	5.0 g
Sodium chloride	5.0 g
Lactose	10.0 g
Bromcresol purple (1% solution)	1 ml
Distilled water	1 litre

pH 7.4

[§] Different proprietary preparations may not have equivalent activity at the same concentration.

Add to the water the peptone, bile salts and sodium chloride and heat in a steamer for 1–2 h. Add the lactose and dissolve by heating for a further 15 min. Cool and filter. Adjust to pH 7.4. Add the indicator. Mix well. Distribute in 5-ml amounts in 150 × 16-mm test-tubes provided with inverted Durham tubes. Sterilize by autoclaving 115°C for 15 min.

Double-strength medium can be prepared in a similar manner, by dissolving the ingredients in half the quantity of distilled water. Distribute in amounts equal in volume to the inocula to be added.

(Report, 1969)

MacConkey's Broth (for Membrane Filtration)

(For the detection and enumeration of lactose-fermenting enterobacteria by membrane filtration)

Peptone	10.0 g
Bile salts[§]	4.0 g
Sodium chloride	5.0 g
Lactose	30.0 g
Bromcresol purple (1% solution)	12 ml
Distilled water	1 litre

pH 7.4

[§] Different proprietary preparations may not have equivalent activity at the same concentration.

Prepare in a manner similar to standard MacConkey's broth, but adjust the pH to 7.4, and distribute in suitable storage containers (e.g. screw-capped bottles) before sterilizing.

(Taylor *et al.*, 1955)

Malonate Broth

(For the determination of malonate utilization by the Enterobacteriaceae)

Yeast extract	1.0 g
Ammonium sulphate	2.0 g
Dipotassium hydrogen phosphate	0.6 g
Potassium dihydrogen phosphate	0.4 g
Sodium chloride	2.0 g
Sodium malonate	3.0 g
Bromthymol blue (1% solution)	2.5 ml
Distilled water	1 litre

Dissolve the ingredients in the water, distribute in 5-ml amounts in 150 × 16-mm test-tubes and sterilize at 121°C for 15 min.

(Report, 1958)

Malt Extract Agar

(For the culture of yeasts and moulds)

Malt extract	30.0 g
Mycological peptone	5.0 g
Agar	15.0 g
Distilled water	1 litre

pH 5.4

Dissolve the ingredients in the water by steaming. Distribute as required and sterilize by autoclaving at 121°C for 15 min.

To inhibit the growth of bacteria, antibiotics may be added as sterile solutions to the molten medium immediately before pouring the plates (see Section 14), or the medium may be acidified to pH 3.5. Acidification may be achieved by adding aseptically sterile 10% lactic acid (or citric acid) solution to the molten medium immediately before pouring the plates. The exact amount of acid to be added will depend on the make or even the batch of the constituents used.

Sucrose (20%) may be added to this medium to make it suitable for osmophilic counts.

Maltose Azide Broth

(For detection and enumeration of faecal streptococci in foods, etc., using the multiple tube technique)

Proteose peptone No. 3 (Difco)	10.0 g
Yeast extract	10.0 g
Sodium chloride	5.0 g
Sodium glycerophosphate, hydrated	10.0 g
Maltose	20.0 g
Lactose	1.0 g
Sodium azide[H]	0.4 g
Sodium carbonate (AR grade)	0.636 g
Bromcresol purple (1% aqueous solution)	1.5 ml
Distilled water	1 litre

pH 7.2

Dissolve the ingredients in the water by steaming. Adjust to pH 7.2. Distribute in 10-ml amounts in 150 × 16-mm test-tubes and sterilize at 121°C for 10 min.

For large volumes of inocula (5 ml and over) use the appropriate amounts of 1½-strength broth (e.g. for 10-ml inocula use 20-ml amounts of 1½-strength broth).

(Kenner *et al.*, 1961)

Maltose Azide Tetrazolium Agar

(For detection and enumeration of faecal streptococci)

This has the same basic recipe as maltose azide broth with the addition of 2% agar. Distribute in 100-ml amounts and sterilize at 121°C for 15 min. Immediately before pouring, add aseptically to each 100 ml of molten agar medium 1 ml of a sterile 1% (w/v) solution of triphenyltetrazolium chloride[H] (previously sterilized by filtration). The TTC solution should be stored in a dark bottle in the refrigerator, and boiled for 5 min each time immediately before use.

(Kenner *et al.*, 1961)

Mannitol Egg-yolk Phenol Red Polymyxin Agar

(For the detection and differentiation of *Bacillus cereus*)

Basal medium

Peptone	10.0 g
Meat extract	1.0 g
D-Mannitol	10.0 g
Sodium chloride	10.0 g
Phenol red (0.2% solution)	12.5 ml
Agar	15.0 g
Distilled water	887.5 ml

pH 7.1

Dissolve all the ingredients in the water by steaming. Distribute in 90-ml amounts in screw-capped bottles, and sterilize at 121°C for 15 min.

Polymyxin B sulphate solution

Dissolve 50 mg of polymyxin B sulphate in 50 ml of distilled water. Sterilize by membrane filtration.

Preparation of complete medium

To 90 ml of molten medium, cooled to 45–50°C, add with aseptic precautions 10 ml of egg-yolk emulsion and 1 ml of sterile polymyxin B sulphate solution. The final concentration of antibiotic in the medium is thus 10 μg per ml of medium. Mix well, and pour plates with about 15 ml medium in each plate. Dry for 1 h at 45°C before use in order that the medium will absorb the inoculum liquid during surface colony count procedures.

(Mossel *et al.*, 1967)

Milk Agar (10% Milk)

(For detecting proteolytic activity)

Reconstituted skim milk	10 ml
Yeast extract agar or nutrient agar	100 ml

Add the milk to the molten agar medium, mix well, dispense as required and sterilize by autoclaving at 115°C for 20 min. Alternatively, add aseptically 1 ml of sterile skim milk (sterilized as for litmus milk; see p. 429) to 10 ml of sterile molten medium, mix well, and pour into a Petri dish.

Milk Agar (30% Milk)

(For the detection of caseolytic (proteolytic) activity)

Mix aseptically 10 ml of a hot sterile 2.5% solution of agar with 5 ml of hot sterile reconstituted skim milk and pour into a Petri dish. This medium may be overlaid as a thin layer on a layer of 10 ml of plain agar previously poured and allowed to solidify.

(Smith *et al.*, 1952)

Milk Salt Agar

(For the selective isolation of *Staphylococcus aureus* from Giolitti and Cantoni's enrichment medium cultures)

Peptone	5.0 g
Meat extract	3.0 g
Sodium chloride	65.0 g
Agar	15.0 g
Distilled water	1 litre

Soak the ingredients in the water; dissolve by steaming or boiling. Adjust the pH so that after autoclaving the pH will be 7.4 ± 0.1 (this needs to be determined by experiment), distribute in 100-ml amounts in screw-capped bottles, and sterilize at 121°C for 15 min.

To prepare the complete medium, melt a bottle of 100 ml basal medium, cool to 55°C, and add 10 ml of sterile skim milk (sterilized as for litmus milk; see p. 429), warmed to 50°C. Mix well and pour into Petri dishes.

(ICMSF, 1978)

Minimal Nutrients Recovery Medium

(For the resuscitation of injured bacteria from foods prior to selective isolation)

Disodium hydrogen phosphate	7.0 g
Potassium dihydrogen phosphate	3.0 g
Sodium chloride	0.5 g
Ammonium chloride	1.0 g
Magnesium sulphate, hydrated ($MgSO_4 \cdot 7H_2$)	0.25 g
Glucose	2.0 g
Distilled water	1 litre

Dissolve the ingredients in the water, dispense as required, and sterilize by autoclaving at 121°C for 15 min.

(Gomez *et al.*, 1973; Wilson and Davies, 1976)

Modified Alkaline Peptone Water

(For the isolation of *Vibrio*)

Peptone	10.0 g
Sodium chloride	10.0 g
Magnesium chloride hexahydrate	4.0 g
Potassium chloride	4.0 g
Distilled water	1 litre

pH 8.6 ± 0.2

Dissolve the ingredients in the water, mix, and adjust the pH to 8.6 at 25°C. Distribute as required and sterilize at 121°C for 15 min.

(Roberts *et al.*, 1995)

Modified Cellobiose Polymyxin B Colistin (MCPC) Agar

(A selective and differential medium for *Vibrio cholerae* and *Vibrio vulnificus*)

Basal medium

Peptone	10.0 g
Meat extract	5.0 g
Sodium chloride	20.0 g
Bromthymol blue	40 mg

Cresol red	40 mg
Agar	15.0 g
Distilled water	900 ml

pH 7.6

Dissolve the ingredients in the water, and adjust to give a pH of 7.6 after autoclaving. Distribute in 100-ml amounts in screw-capped bottles and sterilize by autoclaving at 121°C for 15 min.

Supplements solution

Cellobiose	15.0 g
Polymyxin B sulphate	100 000 units
Colistin methanesulphonate sodium salt (Sigma) (polymyxin E)	1.36 million units
Distilled water	100 ml

Dissolve the ingredients in the water and sterilize by membrane filtration.

Preparation of complete medium

To 100 ml of molten basal medium, cooled to 55°C, add with aseptic precautions 10 ml of the sterile supplements solution prewarmed to 50°C, mix and pour into Petri dishes.

(Massad and Oliver, 1987)

Modified *Salmonella–Shigella* Agar

(For the isolation of *Yersinia enterocolitica*)

Meat extract	5.0 g
Meat peptone	5.0 g
Yeast extract	5.0 g
Lactose	10.0 g
Bile salts[§]	8.5 g
Sodium desoxycholate	10.0 g
Calcium chloride	1.0 g
Sodium citrate, hydrated	10.0 g
Sodium thiosulphate, pentahydrate	8.5 g
Iron (III) citrate	1.0 g
Brilliant green	0.0003 g
Neutral red	0.025 g
Agar	13.5 g
Distilled water	1 litre

pH 7.4

[§] Different proprietary preparations may not have equivalent activity at the same concentration.

Dissolve all the ingredients in the water by boiling. Adjust the pH so that the pH of the set medium is 7.4 at 25°C. Pour 20 ml amounts into sterile Petri dishes.

Note. This medium must not be autoclaved.

(ISO 10273:1994)

Møller's Decarboxylase Medium

(For the determination of decarboxylase activity, especially by members of the Enterobacteriaceae)

Peptone	5.0 g
Beef extract	5.0 g
D-Glucose	0.5 g
Pyridoxal	5 mg
Bromthymol blue (1% aqueous solution)	1.0 ml
Cresol red (0.2% aqueous solution)	2.5 ml
Distilled water	1 litre

pH 6.0

Dissolve the solid ingredients in the water by steaming. Adjust to pH 6.0 and add the indicator solutions. Prepare the complete media as follows.

For the control medium (no added amino acid) distribute in 3-ml amounts in 100 × 12-mm test-tubes, add a 5-mm depth of liquid paraffin and sterilize by autoclaving at 115°C for 10 min.

To prepare the arginine, lysine or ornithine medium, add the appropriate amino acid as the hydrochloride to a final concentration of 1% w/v (if the L-amino acid is used) or 2% w/v (if the DL-amino acid is used). Readjust to pH 6.0, distribute in 3-ml amounts in 100 × 12-mm test-tubes. Add a 5-mm depth of liquid paraffin and sterilize at 115°C for 10 min.

(Møller, 1955)

Neomycin Blood Agar

(For the selective isolation of *Clostridium perfringens*)

To a bottle of molten blood agar (see p. 397) add, immediately before pouring the plates, 1 ml of sterile neomycin sulphate solution (100 mg neomycin sulphate dissolved in 10 ml of distilled water, sterilized by membrane filtration).

(Roberts *et al.*, 1995)

Neutral Red Chalk Lactose Agar

(For the detection of lactic streptococci in milk and milk products)

Peptone	3.0 g
Meat extract	3.0 g
Yeast extract	3.0 g
Agar	15.0 g
Lactose	10.0 g
Calcium carbonate, precipitated	15.0 g
Neutral red (1% aqueous solution)	5 ml[§]
Distilled water	1 litre

[§] Or 2.5 ml of a 1% aqueous solution of bromcresol purple for bromcresol purple chalk lactose agar.

Dissolve the peptone, meat extract, yeast extract and agar in the distilled water by steaming. Allow to cool and adjust if necessary to pH 6.8. Filter if necessary through hot paper pulp (see p. 445). Add the lactose, calcium carbonate and indicator, and mix well to dissolve the lactose. Dispense in 100-ml or 10-ml amounts in screw-capped bottles, keeping the chalk dispersed. Sterilize by autoclaving at 121°C for 20 min.

When using the medium, the chalk must be resuspended before pouring the plates.

(Chalmers, 1962)

Nitrate Peptone Media

(For detecting nitrate-reducing ability of bacteria)

1. Nitrate peptone water

Potassium nitrate[H] (AR grade)	0.2 g
Peptone water	1 litre

Dissolve the potassium nitrate in the peptone water, distribute in 5-ml amounts in 150 × 16-mm test-tubes, each provided with an inverted Durham tube. Sterilize by autoclaving at 121°C for 15 min.

2. Semi-solid nitrate peptone medium

For the determination of nitrate reduction by anaerobes (e.g. as in the identification of *Clostridium perfringens*), 0.3% agar should also be incorporated into the medium.

Nutrient Agar

(A general purpose culture medium for bacteria)

Nutrient broth	1 litre
Agar	15.0 g

pH 7.2 ± 0.2

Dissolve the agar in the nutrient broth by autoclaving at 121°C for 20 min. Adjust the pH to 7.2. Filter through paper pulp (see p. 445). Distribute as required and sterilize at 121°C for 20 min.

Nutrient Broth

(A general purpose culture medium for bacteria)

Peptone	10.0 g
Lab-Lemco meat extract (Oxoid)	10.0 g
Sodium chloride	5.0 g
Distilled water	1 litre

pH 7.2 ± 0.2

Dissolve the ingredients in the water by steaming. Cool, adjust to pH 7.6, and autoclave at 121°C for 15 min. Filter and adjust to pH 7.2. Distribute as required, and sterilize at 121°C for 20 min.

Nutrient Gelatin

(For detection of proteolytic activity)

Gelatin	150.0 g
Nutrient broth	1 litre

pH 7.2 ± 0.2

Add the gelatin to the nutrient broth and steam until dissolved. Adjust to pH 7.2. Distribute in 100 × 12-mm test-tubes, and sterilize by autoclaving at 115°C for 20 min.

Olive Oil Agar

(Used for detection of lipolytic activity)

Olive oil	5 ml
Yeast extract agar or nutrient agar	100 ml

Before sterilization of the basal medium, adjust the reaction to pH 7.8.

Dispense the olive oil in screw-capped bottles and sterilize by autoclaving at 115°C for 20 min.

To 100 ml of sterile basal medium, melted and cooled to 45°C, add aseptically 5 ml of sterile olive oil. Emulsify the medium by shaking vigorously, and pour plates containing about 15-ml amounts.

(Berry, 1933; Jones and Richards, 1952)

ONPG Peptone Water

(For the detection of β-galactosidase activity)

ONPG Solution

o-Nitrophenyl-β-D-galactopyranoside[H]	0.6 g
0.01 M sodium phosphate buffer, pH 7.5	100 ml

Dissolve at room temperature and sterilize by filtration.

Preparation of medium

Add aseptically one part of ONPG solution to three parts of sterile peptone water (prepared to pH 7.5). Distribute aseptically in 2-ml amounts into sterile 100 × 12-mm test-tubes. (Alternatively the peptone water, pH 7.5, can be prepared and distributed and sterilized in 1.5-ml amounts in the test-tubes, and the ONPG solution added as 0.5-ml amounts.)

Check for sterility of medium by incubating at 37°C for 24 h.

(Lowe, 1962)

Orange Serum Agar

(For isolation and enumeration of yeasts and moulds in foods)

This medium is based on orange serum and, because such an extract prepared in small amounts in the laboratory will show considerable batch-to-batch variation, it is recommended that the commercially available dehydrated media are used, particularly if the medium is being used for enumeration of microfungi in foods.

(See Murdock *et al.*, 1952; Hayes and Reister, 1952)

Osmophilic Agar

(For the growth of osmophilic and osmotolerant yeasts and moulds)

This medium is prepared by dissolving a dehydrated Wort Agar (available from most manufacturers of deydrated media) in a 45° Brix syrup containing 35 g of sucrose and 10 g of glucose in 100 ml of solution (for preparation of the medium from basic ingredients, see Wort Agar). Sterilize the medium by autoclaving at 108°C for 20 min, cool to 45–50°C and pour plates as required. It is advisable to avoid unnecessary heating of this medium, so it is preferable to make the medium as required.

(Scarr, 1959; Beech and Davenport, 1969)

Oxford Agar

(For the selective isolation of *Listeria monocytogenes*)

Basal medium

Proteose peptone	23.0 g
Starch	1.0 g
Sodium chloride	5.0 g
Aesculin	1.0 g
Ammonium iron (III) citrate*	0.5 g
Lithium chloride	15.0 g
Agar	15.0 g
Distilled water	1 litre

pH 7.0 ± 0.2

* The ammonium iron (III) citrate used should be the more pure green form (around 18% iron content).

Dissolve the ingredients in the water, with steaming or boiling. Dispense in 100 ml amounts in screw-capped bottles and sterilize at 121°C for 15 min.

Supplements solution

Cycloheximide[H]	400 mg
Colistin methanesulphonate sodium salt (Sigma) (polymyxin E)	20 mg
Acriflavine hydrochloride	5 mg
Cefotetan (ICI)	2 mg
Phosphomycin disodium salt (Sigma)	10 mg
Ethanol (95%)	5.0 ml
Distilled water	5.0 ml

Dissolve the components in the ethanol–water mixture, and sterilize by membrane filtration.

Preparation of complete medium

To a bottle of 100 ml of basal medium, melted and cooled to 47–50°C, add with aseptic precautions 1 ml of the sterile supplements solution, mix well and pour plates. The prepared plates may be kept in plastic bags (to minimize dehydration), refrigerated and in the dark for up to 2 weeks.

(Curtis *et al.*, 1989)

Packer's Crystal Violet Azide Blood Agar

(For the detection and enumeration of faecal streptococci in foods)

Basal medium

Tryptose	15.0 g
Meat extract	3.0 g
Sodium chloride	5.0 g
Agar	15.0 g
Distilled water	1 litre

pH 6.9 ± 0.2

Dissolve the ingredients in the water by steaming. Distribute in 100-ml amounts in screw-capped bottles, and sterilize by autoclaving at 121°C for 15 min.

Preparation of the complete medium

To 100 ml of molten basal medium cooled to 45–50°C add aseptically:

(a) 0.4 ml of a 0.05% aqueous solution of crystal violet (previously sterilized by membrane filtration or at 121°C for 15 min);
(b) 1.0 ml of a 5% aqueous solution of sodium azide[H] (previously sterilized by filtration);
(c) 5.0 ml of fresh defibrinated sheep blood.

Mix well and pour plates.

Note. Azide Blood Agar Base is available as a commercially dehydrated medium; this can be made into a similar medium to the above by the addition of crystal violet and blood as described above, but the sodium azide concentration is 0.02%, compared with 0.05% in Packer's medium.

(Packer, 1943; Mossel *et al.*, 1957; Thatcher and Clark, 1968)

Paper Pulp for Filtration of Media

Soak two large (46 × 57-cm) sheets of filter paper (e.g. Whatman No. 1) in water and mash to a pulp. Bring to the boil in a large beaker and pour into a Buchner-type filter funnel while applying suction. Replace the receiving flask by a clean flask and immediately filter the agar medium.

Peptone Sorbitol Bile Salts Broth

(For the selective enrichment of *Yersinia enterocolitica*)

Peptone	5.0 g
Sorbitol	10.0 g
Sodium chloride	5.0 g
Disodium hydrogen phosphate	8.23 g
Sodium dihydrogen phosphate, hydrated ($NaH_2PO_4 \cdot H_2O$)	1.2 g
Bile salts§	1.5 g
Distilled water	1 litre

pH 7.6

§ Different proprietary preparations may not have equivalent activity at the same concentration.

Dissolve the ingredients in the water. Dispense in the required volumes, and sterilize at 121°C for 15 min.

(ISO 10273:1994)

Peptone Water

(Suitable for the indole test)

Tryptone or tryptose	10.0 g
Sodium chloride	5.0 g
Distilled water	1 litre

pH 7.1 ± 0.2

Dissolve the peptone and sodium chloride in the water by steaming. Adjust to pH 7.2, and dispense in 5-ml amounts in 150 × 16-mm test-tubes and sterilize by autoclaving at 121°C for 15 min.

Peptone Water Diluent

(For enumeration and isolation techniques)

Peptone	1.0 g
Distilled water	1 litre

pH 7.1 ± 0.2

Dissolve the peptone in the water, adjust to pH 7.0, dispense as required and sterilize by autoclaving at 121°C for 20 min.

(Straka and Stokes, 1957)

Phenolphthalein Phosphate Agar

(For the detection of phosphatase production, particularly by staphylococci)

Phenolphthalein phosphate (1% solution)	1 ml
Plate count agar, yeast extract agar or nutrient agar	100 ml

To the sterile molten basal medium cooled to 45°C, add aseptically 1 ml of a sterile 1% solution of phenolphthalein phosphate (previously sterilized by filtration), mix, and pour plates as required.

Phenolphthalein phosphate agar with polymyxin (Gilbert *et al.*, 1969)

The above medium can be made suitable for the detection of phosphatase-producing staphylococci in foods, by the addition, immediately before pouring the plates, of 12 500 units of polymyxin B sulphate per 100 ml of medium. (Prepare a stock solution of polymyxin B sulphate solution by dissolving 125 000 units in 10 ml of distilled water and sterilize by membrane filtration.)

Phenylalanine Agar

(For the phenylalanine deaminase test)

Yeast extract	3.0 g
Dipotassium phosphate (K_2HPO_4)	1.0 g
Sodium chloride	5.0 g
DL-Phenylalanine	2.0 g
Agar	12.0 g

pH 7.3 ± 0.2

Dissolve the ingredients in the water by steaming or boiling. Dispense in 4-ml amounts in test-tubes. Sterilize at 121°C for 15 min, and allow to set as long slants.

(MacFaddin, 1980)

Phenylalanine Malonate Medium

(Combined medium for the determination of phenylalanine deamination and malonate utilization)

Ammonium sulphate	2.0 g
Dipotassium hydrogen phosphate	0.6 g
Potassium dihydrogen phosphate	0.4 g
Sodium chloride	2.0 g
Sodium malonate	3.0 g
DL-Phenylalanine	2.0 g
Yeast extract	1.0 g
Bromthymol blue (1% solution)	2.5 ml
Distilled water	1 litre

Dissolve the ingredients in the water by steaming. Distribute in 5-ml amounts in 150 × 16-mm test-tubes, and sterilize at 115°C for 10 min.

(Shaw and Clarke, 1955)

Phosphate Buffer, 0.1 M

Solution A

Dissolve 13.6 g of potassium dihydrogen phosphate (KH_2PO_4) in distilled water, and make up to 1 litre of solution.

Solution B

Dissolve 26.8 g of disodium hydrogen phosphate heptahydrate ($Na_2HPO_4 \cdot 7H_2O$) in distilled water, and make up to 1 litre of solution.

	Volume of solution (ml)	
pH	A	B
6.70	52	48
6.81	48	52
7.00	34	66
7.10	28	72
7.30	20	80
7.42	16	84

Distribute and sterilize as required.

Physiological Saline

Sodium chloride	8.5 g
Distilled water	1 litre

Dissolve the sodium chloride in the water, distribute as required, and sterilize by autoclaving at 121°C for 15 min.

Plate Count Agar (Tryptone Glucose Yeast Extract Agar)

(A non-selective medium for general viable counts of bacteria in foods)

Tryptone	5.0 g
Yeast extract	2.5 g
D-Glucose	1.0 g
Agar	15.0 g
Distilled water	1 litre

pH 7.0 ± 0.2

Dissolve the ingredients in the water by steaming. Adjust to pH 7.0, dispense in 10-ml or 100-ml amounts in screw-capped bottles, and sterilize by autoclaving at 121°C for 15 min.

Polymyxin Acriflavin Lithium Chloride Ceftazidime Aesculin Mannitol (PALCAM) Agar

(For the isolation of *Listeria monocytogenes*)

Basal medium

Proteose peptone	23.0 g
Yeast extract	3.0 g
D-Glucose	0.5 g
Starch	1.0 g
Sodium chloride	5.0 g
Aesculin	0.8 g
Ammonium iron (III) citrate*	0.5 g
Lithium chloride	15.0 g
Phenol red	0.08 g
Agar	15.0 g
Distilled water	960 ml

pH 7.2 ± 0.1

* The ammonium iron (III) citrate used should be the more pure green form (around 18% iron content).

Dissolve the ingredients in the water, with steaming or boiling. Adjust the pH so that after sterilization it will be 7.2 ± 0.2 at 25°C. Dispense in 96-ml amounts in screw-capped bottles and sterilize at 121°C for 15 min.

Polymyxin B solution

Polymyxin B sulphate	0.1 g
Distilled water	100 ml

Dissolve the polymyxin B sulphate in the water and sterilize by membrane filtration.

Acriflavine solution

Acriflavine hydrochloride	50 mg
Ethanol (95%)	50 ml
Distilled water	50 ml

Dissolve the acriflavine hydrochloride in the ethanol/water mixture, and sterilize by membrane filtration.

Ceftazidime solution

Sodium ceftazidime pentahydrate (Glaxo)	0.116 g
Distilled water	100 ml

Dissolve the sodium ceftazidime pentahydrate in the water and sterilize by membrane filtration.

Preparation of complete medium

To a bottle of 96 ml of basal medium, melted and cooled to 47–50°C, add with aseptic precautions 1.0 ml of the sterile polymyxin B solution, 1.0 ml of the sterile acriflavine solution and 2.0 ml of sterile ceftazidime solution. Mix well and pour plates. The prepared plates may be kept in plastic bags (to minimize dehydration), refrigerated and in the dark for up to 4 weeks.

(van Netten *et al.*, 1989)

Polymyxin Pyruvate Egg-yolk Mannitol Bromthymol Blue Agar (PPEMBA)

(For the enumeration of *Bacillus cereus* in foods)

Basal medium

Peptone	1.0 g
Mannitol	10.0 g
Sodium chloride	2.0 g
Magnesium sulphate, hydrated ($MgSO_4 \cdot 7H_2O$)	0.1 g
Disodium hydrogen phosphate	2.5 g
Potassium dihydrogen phosphate	0.25 g
Bromthymol blue	0.12 g
Agar	15.0 g
Distilled water	1 litre

pH 7.2 ± 0.2

Add ingredients to the water, allow to soak for 10 min and steam or boil to dissolve. Adjust pH to be 7.2 ± 0.2 at 25°C. Dispense in 90-ml amounts and autoclave at 121°C for 15 min.

Polymyxin B solution

Polymyxin B sulphate	1 million units
Distilled water	10 ml

Dissolve the polymyxin B sulphate in the water and sterilize by membrane filtration.

Sodium pyruvate solution

Sodium pyruvate	20.0 g
Distilled water	to 100 ml

Dissolve the sodium pyruvate in the water and sterilize by membrane filtration.

Cycloheximide solution

Cycloheximide[H]	0.4 g
Distilled water	100 ml

Dissolve the cycloheximide in the water and sterilize by membrane filtration.

Preparation of complete medium

To a bottle of molten basal medium, cooled to 50°C, add with aseptic precautions:

(a) 1.0 ml of sterile polymyxin B solution
(b) 1.0 ml of sterile cycloheximide solution
(c) 5.0 ml of sodium pyruvate solution
(d) 5.0 ml of sterile egg-yolk emulsion (see p. 412)

Mix well and pour into Petri dishes. The prepared plates of complete medium may be stored in plastic bags in the refrigerator for up to 5 days.

(Holbrook and Anderson, 1980; Corry *et al.*, 1995a)

Polypectate Gel Medium

(For the detection of pectinolytic microorganisms)

Peptone	5.0 g
Dipotassium hydrogen phosphate	5.0 g
Potassium dihydrogen phosphate	1.0 g
Calcium chloride, hydrated ($CaCl_2 \cdot 2H_2O$)	0.6 g
Sodium polypectate (polygalacturonic acid, sodium salt)	70.0 g
Distilled water	to 1 litre

Heat 500 ml of water using a magnetic stirrer–hotplate, and dissolve the first four ingredients. Next gradually add the sodium polypectate and add water to make 1 litre of medium. Heat the medium in a steamer for 15 min, then distribute as required and sterilize by autoclaving at 121°C for 15 min.

The medium may be made partially selective for Gram-negative bacteria by incorporating crystal violet (see p. 84), and yeasts and moulds may be inhibited by the addition of cycloheximide (see p. 285).

Note. After pouring plates, they should be stored overnight before use.

(American Public Health Association, 1966)

Potato Dextrose Agar

(For the growth of microfungi)

Potatoes, peeled and diced	200 g
D-Glucose	20 g
Agar	15 g
Distilled water	to 1 litre

Boil 200 g of peeled, diced potatoes for 1 h in 1 litre of distilled water. Filter, and make up the filtrate to 1 litre. Add the glucose and agar and dissolve by steaming. Distribute in 10- or 100-ml amounts in screw-capped bottles and sterilize by autoclaving at 121°C for 20 min.

Note. Potato carrot dextrose agar can be prepared by substituting 50 g carrot for 50 g potato in the above recipe.

Rappaport–Vassiliadis Enrichment Broth

(For the enrichment of *Samonella*)

Solution A

Soya peptone	4.5 g
Sodium chloride (AR grade)	7.2 g
Potassium dihydrogen phosphate	1.44 g
Distilled water	1 litre

This must be made on the day of preparation of the complete medium. Mix the ingredients into the water, and heat at 70°C until dissolved.

Solution B

Magnesium chloride hexahydrate ($MgCl_2 \cdot 6H_2O$)	400 g
Distilled water	1 litre

The total volume of this stock solution will be 1260 ml, and will contain 31.7 g of $MgCl_2 \cdot 6H_2O$ per 100 ml (Peterz *et al.*, 1989). Magnesium chloride is very hygroscopic, so Vassiliadis (1983) recommends preparing this from a newly opened container of the chemical. The solution may be kept in a dark bottle at ambient temperature for at least 1 year.

Solution C

Malachite green oxalate (analytically pure)	0.4 g
Distilled water	100 ml

Dissolve the dye in the water. The solution may be stored in a dark bottle for at least 6 months at ambient temperature.

Preparation of complete medium

To 1 litre of Solution A add 100 ml of solution B and 10 ml of solution C. (The final concentration of $MgCl_2 \cdot 6H_2O$ in the complete medium will be 28.6 g per litre (Peterz *et al.*, 1989). This concentration is critical.)

Distribute as required (e.g. 10-ml amounts in test-tubes, or 500-ml amounts in screw-capped bottles). Sterilize at 115°C for 15 min. The medium should be stored refrigerated in the dark and used within 1 month.

This medium is available as a dehydrated medium. However, Peterz *et al.* (1989) found that a number of dehydrated media performed poorly compared with the 'home-made' version. This has been discussed by Busse (1995).

(Vassiliadis, 1983; Peterz *et al.*, 1989)

Reinforced Clostridial Medium

(For the growth of anaerobes, and for use as a diluent in the enumeration of anaerobes)

Yeast extract	3.0 g
Peptone	10.0 g
Lab-Lemco meat extract (Oxoid)	10.0 g
D-Glucose	5.0 g
Sodium acetate, hydrated	5.0 g
L-Cysteine hydrochloride	0.5 g
Soluble starch	1.0 g
Distilled water	1 litre

pH 7.1 ± 0.1

If required as an agar medium, agar to 1.5% should be included; if required as a semi-solid medium, include agar to 0.5%. The medium without agar is suitable as a diluent.

Add all the ingredients to the water, and dissolve by steaming. Filter through hot paper pulp (see p. 445). Adjust to pH 7.4, dispense in screw-capped bottles as required, and sterilize by autoclaving at 121°C for 15 min. The pH after autoclaving should be 7.1–7.2.

(Hirsch and Grinsted, 1954; Gibbs and Hirsch, 1956)

Ringer's Solution, Quarter-strength

(Diluent and suspending liquid)

Sodium chloride	2.25 g
Potassium chloride	0.105 g
Calcium chloride, hydrated	0.12 g
Sodium hydrogen carbonate	0.05 g
Distilled water	1 litre

Dissolve the salts in the water, distribute as required, and sterilize by autoclaving at 121°C for 20 min.

(Wilson, 1935)

Robertson's Cooked Meat Medium

(For growth of anaerobes, and maintenance of stock cultures of bacteria, especially microaerophilic and anaerobic bacteria)

Mince 500 g of fresh fat-free bullock's heart and simmer for 20 min in 500 ml of boiling 0.05 N sodium hydroxide. After cooking, adjust to pH 7.4. Strain off the liquid and dry the meat by spreading on filter paper. Distribute the meat in 30-ml (1-oz McCartney) screw-capped bottles to a depth of 3 cm. Add 10 ml of peptone water or nutrient broth. A 2-cm depth of liquid paraffin may be added if required. Sterilize, with the screw-caps slightly loosened, by autoclaving at 121°C for 20 min. After autoclaving, tighten the caps.

(Lepper and Martin, 1929)

Note. This medium will often be found more satisfactory than equivalent dehydrated formulations, particularly for growth and maintenance of *Clostridium*.

Rogosa Agar

(For the isolation and enumeration of lactobacilli)

Tryptone or trypticase	10.0 g
Yeast extract	5.0 g
D-Glucose	20.0 g
Tween 80	1.0 g
Potassium dihydrogen phosphate	6.0 g
Ammonium citrate	2.0 g
Sodium acetate, hydrated	25.0 g
Glacial acetic acid[H]	1.32 ml

Magnesium sulphate, hydrated ($MgSO_4 \cdot 7H_2O$)	0.575 g
Manganese sulphate, hydrated ($MnSO_4 \cdot 4H_2O$)	0.14 g
Ferrous sulphate, hydrated ($FeSO_4 \cdot 7H_2O$)	0.034 g
Agar	15.0 g
Distilled water	to 1 litre

Final pH 5.4

Suspend all the ingredients in the water and dissolve by heating in a steamer. Distribute into sterile containers as required and heat in a steamer for a further 50 min. Store the medium under refrigeration until required. Use the minimum amount of heating necessary to melt the medium and avoid overheating (e.g. avoid melting more medium than is required, and attempting to re-store the surplus).

(Rogosa *et al.*, 1951)

Rogosa Agar Modified (Mabbitt and Zielinska)

(For the isolation and enumeration of lactobacilli in fermented milk products)

Milk digest

Mix 1 litre of separated milk or reconstituted milk, adjusted to pH 8.5, 5 g of trypsin[H], and 10 ml of chloroform[H]. Incubate at 37°C for 24 h, then steam for 20 min and filter while hot. Adjust the pH to 6.65 ± 0.02 with glacial acetic acid[H] (about 0.5 ml per litre) using a pH meter.

Nutrient solution

Yeast extract	12.0 g
Di-ammonium hydrogen citrate	4.8 g
Potassium dihydrogen phosphate	14.4 g
D-Glucose	48.0 g
Tween 80	2.4 g
Salts solution§	12.0 ml
Distilled water	200 ml

§ Salts solution

Magnesium sulphate, hydrated ($MgSO_4 \cdot 7H_2O$)	11.5 g
Manganese sulphate, hydrated ($MnSO_4 \cdot 4H_2O$)	2.8 g
Ferrous sulphate, hydrated ($FeSO_4 \cdot 7H_2O$)	0.08 g
Distilled water	100 ml

Dissolve ingredients in water by gentle heating. Add 60 ml of 4 M sodium acetate–acetic acid buffer at pH 5.37 ± 0.02. Make up the volume to 200 ml with distilled water.

The salts solution has a final pH of 5.0 and is not sterilized; it should be stored in the refrigerator until required.

Preparation of complete medium

Dissolve 19 g agar in 700 ml of milk digest at 121°C for 20 min, and while hot mix with 185 ml of nutrient solution previously warmed to 50°C. Make up the volume to 1 litre with hot digest.

Distribute the medium aseptically in 10-ml amounts into sterile screw-capped bottles, and store in the refrigerator (without sterilizing) until required.

The medium should be prepared for use with as little heating as possible to avoid darkening and the formation of a precipitate.

(Mabbitt and Zielinska, 1956)

Rogosa Broth Modified (Mabbitt and Zielinska)

This is prepared in a similar manner to Rogosa agar (modified), described above, but the agar is omitted.

(Mabbitt and Zielinska, 1956)

Salmonella–Shigella Agar

(For the isolation of salmonellae and shigellae)

Proteose peptone or polypeptone	5.0 g
Meat extract	5.0 g
Lactose	10.0 g
Bile salts§	8.5 g
tri-Sodium citrate	8.5 g
Sodium thiosulphate	8.5 g
Ammonium iron (III) citrate	1.0 g
Neutral red	0.025 g
Brilliant green (0.1% solution)	0.3 ml
Agar	13.0 g

pH 7.0 ± 0.2

§ Different proprietary preparations may not have equivalent activity at the same concentration.

Mix the ingredients with the water, allow to soak for 15 min. Dissolve by heating to boiling. Do not autoclave. Cool to 50°C and pour into Petri dishes.

Salt Meat Broth

(For the liquid enrichment of staphylococci from low-salt foods)

Peptone	10.0 g
Beef extract	10.0 g
Sodium chloride	100.0 g
Distilled water	1 litre

pH 7.3

Dissolve the ingredients in the water by steaming. Adjust to pH 7.3, distribute as required, and sterilize by autoclaving at 121°C for 20 min.

Salt Polymyxin B Broth

(For the enrichment of *Vibrio*)

Basal medium

Peptone	10.0 g
Yeast extract	3.0 g
Sodium chloride	20.0 g
Water	1 litre

pH 7.4 ± 0.2

Dissolve the ingredients in the water. Dispense as required (e.g. in 90- or 225-ml amounts) and sterilize by autoclaving at 121°C for 15 min.

Polymxin B solution

Polymyxin B sulphate	100 000 units
Distilled water	100 ml

Dissolve the polymyxin B in the water and sterilize by membrane filtration.

Preparation of complete medium

Add, with aseptic precautions, 10 ml of sterile polymyxin B solution to 90 ml of sterile basal medium (or 25 ml to 225 ml). Use the complete medium on the day of preparation.

(ISO 8914:1990)

Sea Water Yeast Peptone Agar

(For the growth of marine organisms in general, and the detection of luminous marine bacteria in particular)

Yeast extract	3.0 g
Peptone	5.0 g
Agar	15.0 g
Aged sea water	750 ml
Distilled water	250 ml

pH 7.4

Dissolve the ingredients in the water by steaming. Adjust to pH 7.4. Distribute as required and sterilize by autoclaving at 121°C for 20 min.

Note. The sea water should be aged for at least 3 weeks in the dark, and then filtered.

(Hendrie *et al*., 1970)

Selenite Cystine Broth

(An enrichment medium for *Salmonella*)

Peptone	4.0 g
Lactose	4.0 g
Sodium hydrogen selenite[H]	5.0 g
Disodium hydrogen phosphate, anhydrous	5.0 g
Potassium dihydrogen phosphate	5.0 g
Cystine	10 mg
Distilled water	1 litre

pH 6.9

Dissolve the ingredients in the water. The pH of the medium should be 6.9 without the need for adjustment. Dispense in suitable amounts (depending on the amount of inoculum to be used) and sterilize in a boiling water-bath for 10 min only.

(North and Bartram, 1953)

Semi-solid Citrate Milk Agar

(For the detection of citrate fermentation by lactic streptococci)

1. Dispense 500 ml of reconstituted skim milk into 150 × 16-mm test-tubes in 10.5-ml quantities. Sterilize in the autoclave by heating at 100°C for 30 min and then at 115°C for 10 min.

2. Prepare 100 ml of a 10% solution of sodium citrate dihydrate in distilled water, distribute in 10-ml amounts and sterilize by autoclaving at 121°C for 20 min.
3. Dissolve 2.0 g of agar in 100 ml of distilled water by steaming, distribute in 4-ml amounts in 150 × 16-mm test-tubes. Close loosely with Astell rubber stoppers and sterilize at 121°C for 20 min.

Preparation and use of complete medium

Add 0.5 ml of sterile 10% sodium citrate solution to each tube of skim milk. Invert to mix and allow to stand for 30 min. When required for use, inoculate 0.1 ml of the test culture into the citrated milk and mix by rotation of the tube. Pour the citrated milk culture on to 4 ml of molten 2% agar (cooled to 45–50°C), replace the Astell seal, invert to mix and incubate at the desired temperature.

(Crawford, 1962)

Serum Agar

(For the growth of fastidious animal parasites)

Sterile serum	7.0 ml
Nutrient agar, sterile	100 ml

Liquefy the basal medium and cool to 45–50°C. Add the serum with aseptic precautions, mix well and pour plates as required.

Shahidi and Ferguson's Polymyxin Kanamycin Sulphite Agar (SFA)

(For the isolation of *Clostridium perfringens*)

Basal medium

Tryptose	15.0 g
Soya peptone	5.0 g
Yeast extract	5.0 g
Ammonium iron (III) citrate*	1.0 g
Sodium metabisulphite[H]	1.0 g
Agar	15.0 g
Distilled water	890 ml

pH 7.6 ± 0.2

* The ammonium iron (III) citrate used should be the more pure green form (around 18% iron content).

Dissolve the ingredients in the water by steaming or boiling, adjust pH to 7.6 (at 25°C). Distribute in 89-ml amounts in screw-capped bottles, and autoclave at 121°C for 10 min.

Polymyxin B solution

Polymyxin B sulphate	0.1 g
Distilled water	100 ml

Dissolve, and sterilize by membrane filtration.

Kanamycin solution

Kanamycin sulphate	0.12 g
Distilled water	100 ml

Dissolve, and sterilize by membrane filtration.

Preparation of complete medium

To a bottle of 89 ml of molten basal medium, cooled to 50°C, add with aseptic precautions 0.3 ml of sterile polymyxin solution, 1.0 ml of sterile kanamycin solution, and 10 ml of sterile egg-yolk emulsion (see p. 412). Mix and pour plates.

Preparation of overlay agar

If the medium is being used as layer plates, the overlay medium should omit the egg-yolk emulsion so that the upper layer of medium is transparent. In this case to a bottle of 89 ml of molten basal medium, cooled to 50°C, add with aseptic precautions 0.3 ml of sterile polymyxin solution, 1.0 ml of sterile kanamycin solution, and 10 ml of sterile distilled water prewarmed to 50°C. (10 ml of this medium is poured over each inoculated plate, and allowed to set before incubation.)

(ICMSF, 1978)

Simmon's Citrate Agar

(For the determination of citrate utilization, particularly for differentiation of enterobacteria)

Sodium chloride	5.0 g
Magnesium sulphate, hydrated ($MgSO_4 \cdot 7H_2O$)	0.2 g
Ammonium dihydrogen phosphate	1.0 g

Dipotassium hydrogen phosphate	1.0 g
Sodium citrate	5.0 g
Bromthymol blue (1% aqueous solution)	8 ml
Agar	15.0 g
Distilled water	990 ml

pH 7.0

Add all the ingredients except the indicator solution to the water and dissolve by steaming. Adjust to pH 7.0 and add the bromthymol blue. Mix and dispense in test-tubes or screw-capped bottles with sufficient medium in each tube or bottle to form a slope with a 3-cm butt. Sterilize by autoclaving at 121°C for 15 min, and allow to set in a sloped position.

(Report, 1958)

Sodium Chloride (15%) Diluent

(A diluent for detection or enumeration of halophiles)

Sodium chloride	150.0 g
Distilled water	to 1 litre

Dissolve the sodium chloride in distilled water, and make up to 1 litre. Distribute in 9- or 90-ml amounts in screw-capped bottles and sterilize at 121°C for 20 min.

Sorbitol MacConkey's Agar (SMAC)

(For the isolation and differentiation of *Escherichia coli* O157)

Peptone	20.0 g
Sorbitol	10.0 g
Sodium desoxycholate	1.0 g
Sodium chloride	5.0 g
Crystal violet (0.1% aqueous solution)	1.0 ml
Neutral red (1.0% aqueous solution)	3.0 ml
Agar	15.0 g
Distilled water	1 litre

pH 7.1 ± 0.2

Dissolve the ingredients in the water in a steamer or by boiling. Distribute in 100-ml amounts in screw-capped bottles, and sterilize at 121°C for 15 min.

This medium is used as the base for Cefixime–Tellurite Sorbitol MacConkey's Agar (CT-SMAC) (see p. 403).

Starch Agar

(For the detection of starch-hydrolysing ability)

Soluble starch	0.2 g
Yeast extract agar or nutrient agar	100 ml

Dissolve the starch in the molten basal medium, and sterilize by autoclaving at 121°C for 15 min.

Alternatively add aseptically 0.2 ml of a sterile 10% solution of soluble starch to 10 ml of sterile molten basal medium, mix well and pour into a Petri dish.

Starch Milk Agar

(For the detection of spores in heated milk and milk products)

Nutrient agar	100 ml
Reconstituted skim milk	1 ml
Soluble starch (10% solution)	1 ml

Add the skim milk and the soluble starch solution to molten nutrient agar. Mix well and sterilize by autoclaving at 121°C for 20 min.

(Grinsted and Clegg, 1955)

Stuart's Transport Medium

(For transport of swabs, etc., particularly for examination for fastidious animal parasites and for anaerobes)

Sodium thioglycollate	1.0 g
Agar	3.0 g
Sodium glycerophosphate (20% solution)	50 ml
Calcium chloride (1% solution)	10 ml
Methylene blue (0.1% solution)	2 ml
Distilled water	950 ml

Add the agar and sodium thioglycollate to the water and dissolve by steaming. Adjust to pH 7.2, then add the sodium glycerophosphate and calcium chloride solutions. Mix and adjust to pH 7.4. Add the methylene blue solution, mix and place in the steamer for 10 min. Distribute in 30-ml (1-oz) bijou or screw-capped bottles, filling them completely. Screw on the caps tightly and sterilize by autoclaving at 115°C for 20 min.

This medium should keep well, but occasionally an insufficiently tightened cap will allow oxygen to be absorbed, indicated by the methylene blue regaining its blue colour. When in this condition the medium should not be used until the oxygen has been expelled by steaming with the cap slightly loosened, after which the cap is tightened fully once more until use.

(Moffett *et al.*, 1948; Stuart *et al.*, 1954)

Sucrose Agar (1)

(For detecting polysaccharide production from sucrose)

Sucrose	10 g
Yeast extract agar or nutrient agar	100 ml

Add the sucrose to the molten basal medium, mix well and sterilize at 115°C for 20 min.

(Evans *et al.*, 1956)

Sucrose Agar (2)

(For detecting dextran production by *Leuconostoc*)

Tryptone	10.0 g
Yeast extract	5.0 g
Dipotassium hydrogen phosphate	5.0 g
Triammonium citrate	5.0 g
Sucrose	50.0 g
Agar	15.0 g
Distilled water	1 litre

Dissolve the ingredients in distilled water by heating in a steamer, with frequent mixing. Distribute as required and sterilize at 121°C for 15 min.

(Garvie, 1960)

Sucrose Diluent (20%)

(For studies involving osmophilic or osmotolerant organisms)

Sucrose	200.0 g
Distilled water	to 1 litre

Dissolve the sucrose in distilled water and make up to 1 litre. Distribute in 90- or 9-ml amounts in screw-capped bottles, and sterilize by autoclaving at 115°C for 20 min.

Taylor's Xylose Lysine Desoxycholate (XLD) Agar

(For the selective isolation of *Salmonella* and *Shigella*)

Basal medium

Yeast extract	3.0 g
Xylose	3.75 g
L-Lysine monohydrochloride	5.0 g
Lactose	7.5 g
Sucrose	7.5 g
Sodium chloride	5.0 g
Phenol red (0.2% solution)	40 ml
Agar	15.0 g
Distilled water	960 ml

Dissolve the ingredients in the water by steaming, cool to 50°C, add the indicator solution, and adjust to pH 6.9. Distribute in 100-ml amounts in screw-capped bottles and sterilize by autoclaving at 121°C for 15 min.

Thiosulphate–citrate solution

Sodium thiosulphate, hydrated	34.0 g
Ammonium iron (III) citrate*	4.0 g
Distilled water	to 100 ml

* The ammonium iron (III) citrate used should be the more pure green form (around 18% iron content).

Dissolve the salts in the water, make up to 100 ml and sterilize by filtration.

Sodium desoxycholate solution

Dissolve 10 g of sodium desoxycholate in sufficient distilled water to make 100 ml of solution and sterilize at 121°C for 15 min.

Preparation of complete medium

To 100 ml of basal medium, melted and cooled to 50°C, add aseptically 2.0 ml of thiosulphate–citrate solution and 2.5 ml of sodium desoxycholate solution. If necessary, the pH should be readjusted to pH 6.9 using sterile reagents, the medium mixed, and plates poured as required. Since it may be necessary to readjust pH just before pouring plates, the procedure is made easier if the basal medium and supplements are made in reasonably large batches and kept under refrigeration until required, so that any adjustment necessary may be determined with the first bottle of medium, and all other bottles in the batch marked accordingly.

(Taylor, 1965; Taylor and Schelhart, 1967)

Thiosulphate Citrate Bile-salts Sucrose Agar (TCBS)

(For the selective isolation of *Vibrio*)

Peptone	10.0 g
Yeast extract	5.0 g
Sodium thiosulphate pentahydrate	10.0 g
Trisodium citrate dihydrate	10.0 g
Bile salts	8.0 g
Sucrose	20.0 g
Sodium chloride	10.0 g
Iron (III) citrate	1.0 g
Bromthymol blue	0.04 g
Thymol blue	0.04 g
Agar	15.0 g
Distilled water	1 litre

pH 8.6 ± 0.2

Dissolve the ingredients in the water by boiling with constant mixing or stirring. Do not autoclave. Pour into Petri dishes. The poured plates may be stored in plastic bags in the refrigerator for 4 weeks.

(MacFaddin, 1985)

Thornley's Semi-solid Arginine Medium

(For the production of ammonia from arginine by pseudomonads)

Peptone	1.0 g
Dipotassium hydrogen phosphate	0.3 g
Sodium chloride	5.0 g
L-Arginine monohydrochloride	10.0 g
Phenol red (0.2% solution)	5 ml
Agar	3.0 g
Distilled water	1 litre

pH 7.2 ± 0.2

Dissolve the ingredients in the water by steaming. Mix well, and dispense in 7-ml (¼-oz bijou) screw-capped bottles to a depth of 2 cm. Sterilize at 121°C for 15 min.

(Thornley, 1960)

Tomato Juice Lactate Agar

(For the detection and enumeration of citrate-fermenting lactic acid bacteria)

Basal medium

Tomato juice agar, dehydrated (Oxoid, Difco or BBL)	15.0 g
Calcium lactate, hydrated	5.0 g
Agar	9.0 g
Distilled water	1 litre

Dissolve the ingredients in the distilled water by heating at 121°C for 5 min. Distribute the medium in 15-ml amounts in 30-ml screw-capped bottles (Universal containers). Sterilize at 121°C for 15 min.

Calcium citrate suspension

Add 10 g of calcium citrate to 100 ml of distilled water. Shake well and, keeping the particles in suspension, distribute in 10-ml amounts in 30-ml screw-capped bottles (Universal containers). Sterilize at 121°C for 15 min.

Preparation of complete medium

The complete medium is prepared by adding aseptically 1 ml of well-shaken calcium citrate suspension to 15 ml of molten tomato juice lactate agar, inverting to mix and pouring into Petri dishes.

(Skean and Overcast, 1962)

Note. It is not necessary to suspend the calcium citrate in a 1.5% solution of carboxymethylcellulose as used in the original version of this medium by Galesloot *et al.* (1961).

Tributyrin Agar

(For the detection of lipolytic activity)

Tributyrin	10.0 g
Yeast extract agar or nutrient agar, adjusted to pH 7.5	100 ml

Add the tributyrin to the molten basal medium in an electric blender and mix until completely emulsified. Dispense into screw-capped bottles in 10-ml amounts and sterilize by steaming for 30 min on each of three successive days.

Triple Sugar Iron Agar

(For differentiation of members of the Enterobacteriaceae)

Peptone	20.0 g
Lactose	10.0 g
Sucrose	10.0 g
D-Glucose	1.0 g
Sodium chloride	5.0 g
Sodium thiosulphate pentahydrate	0.2 g
Diammonium iron (II) sulphate hexahydrate	0.2 g
Phenol red (1% aqueous solution)	2.5 ml
Agar	13.0 g
Distilled water	litre

pH 7.4 ± 0.2

Dissolve the ingredients in the water by steaming or boiling. Dispense in 7-ml amounts in test-tubes. Sterilize by autoclaving at 121°C for 15 min. Allow to set as slopes with a 3-cm butt.

(Hajna, 1945)

Tryptone Glucose Yeast Extract Broth

(Non-selective medium for general viable counts by the multiple tube technique)

Prepare to the same recipe as Plate Count Agar, except that the agar should be omitted. Distribute in bottles or test-tubes as required (double-strength medium can be prepared for large volumes of inocula) and sterilize by autoclaving at 121°C for 20 min.

Tryptone Soya Agar

Tryptone soya broth	1 litre
Agar	15.0 g

Dissolve the agar in the broth by steaming or boiling. Dispense as required and sterilize at 121°C for 15 min.

Tryptone Soya Broth

(General purpose non-selective liquid medium)

Tryptone or trypticase	17.0 g
Soya peptone or phytone	3.0 g
Dipotassium hydrogen phosphate	2.5 g
Sodium chloride	5.0 g
D-Glucose	2.5 g
Distilled water	1 litre

pH 7.2 ± 0.2

Dissolve ingredients in the water by steaming. Cool and adjust to pH 7.3. Distribute as required and sterilize at 121°C for 15 min.

The medium may also be prepared without the glucose, or as a double-strength medium for membrane filtration.

Tryptone Soya Yeast Extract Agar

Tryptone soya yeast extract broth	1 litre
Agar	15 g

Dissolve the agar in the broth by steaming or boiling. Distribute as required and sterilize at 121°C for 15 min.

Tryptone Soya Yeast Extract Broth

Tryptone (pancreatic digest of casein)	17.0 g
Soya peptone (papaic digest of soybean meal)	3.0 g
Yeast extract	6.0 g
Sodium chloride	5.0 g
Dipotassium hydrogen phosphate	2.5 g
D-Glucose	2.5 g
Water	1 litre

pH 7.3 ± 0.2

Dissolve the ingredients in the water. Distribute and sterilize at 121°C for 15 min.

Tween Agar

(Lipolysis test medium)

Peptone	10.0 g
Calcium chloride, hydrated	0.1 g
Sodium chloride	5.0 g
Tween	10.0 g
Agar	15.0 g
Distilled water	1 litre

pH 7.0–7.4

Dissolve all the ingredients in the water by steaming. Check that the pH is within the indicated range, and adjust if necessary. Dispense as required and sterilize by autoclaving at 115°C for 20 min.

Tween 80 (polyoxyethylene(20) sorbitan monooleate) is the Tween most often used. Although liquid at ambient temperatures, it is fairly viscous, and it will prove easier to dispense by warming to 45–50°C.

(Sierra, 1957)

V-8 Agar

(A medium for the inducement of sporulation by yeasts)

Adjust the pH of canned V-8 vegetable juice (manufactured by Campbell's Soup Company) to pH 6.8. To 100 ml of the juice add 40 g of baker's yeast, and heat for 10 min in a steamer. Readjust the pH to 6.8 and add an equal amount of a melted 4% solution of agar in distilled water. Mix, and dispense in 7-ml amounts in 150 × 16-mm test-tubes and sterilize at 121°C for 15 min. Not more than 8 h before use, prepare as slopes in the test-tubes by melting the medium in a steamer, and allowing to set in a slanted position.

(Wickerham *et al.*, 1946)

Vaspar

(For sealing media for anaerobic growth studies)

Melt and mix together equal amounts of petroleum jelly ('Vaseline') and paraffin wax. Distribute into wide-mouthed screw-capped bottles, and sterilize by autoclaving at 121°C for 15 min.

Victoria Blue Butter-fat (or Margarine) Agar

(For the detection of lipolytic activity by microorganisms)

Preparation of Victoria blue base

Boil 2 g of powdered Victoria blue in 200 ml of distilled water until thoroughly dispersed. Slowly add a 10% solution of sodium hydroxide[H] with constant mixing until the colour disappears from the solution. Allow the water-insoluble precipitate (the basic dye) to settle out. Filter off the precipitated basic dye and wash with distilled water made slightly alkaline with ammonium hydroxide. Dry the dye at 30°C.

Preparation of fat

Obtain separated butter-fat or margarine fat as already described (see Butter-fat Agar).

Preparation of dye–fat mixture

Heat 100 g of fat in a conical flask with 100 ml of distilled water and some glass beads. When boiling, slowly add Victoria blue base with constant mixing until the fat is saturated with dye (the fat will be deep red, with some particles of undissolved dye at the bottom of the flask). Boil for 30 min. Separate the fat from the bulk of the water, and filter overnight at 37°C. Separate the filtered fat from any residual water, using a separating funnel, dispense in 30-ml screw-capped bottles in 10-ml amounts, and sterilize by autoclaving at 121°C for 15 min.

Preparation of basal medium

The basal medium may consist of yeast extract agar, nutrient agar or Tryptose Blood Agar Base (Oxoid). Whichever basal medium is used, the agar content should be increased to 2% (w/v), and the pH adjusted to 7.8. Distribute the medium in 20 ml amounts in 30-ml screw-capped bottles, and sterilize by autoclaving at 121°C for 20 min.

Preparation of the complete medium

Add aseptically 1 ml of the molten dye–fat mixture to 20 ml of sterile molten basal medium which has been cooled to 45°C. Emulsify by shaking vigorously for 1 min and pour into a Petri dish. Surface air bubbles can be eliminated by immediately flaming the surface rapidly with a Bunsen flame.

If the plates must be stored before use, they should be refrigerated. The colour of the medium may change on storage, or during incubation (an uninoculated plate should always be incubated as a control). The medium should be pinkish-mauve; if it is bright blue it is unsuitable for use.

(Jones and Richards, 1952; Paton and Gibson, 1953; Harrigan, 1967)

Violet Red Bile Glucose Agar

(For the enumeration of the Enterobacteriaceae in foods)

This medium should be made to the same recipe as Violet Red Bile Lactose Agar (see below), but 10 g of glucose per litre of medium incorporated in place of the lactose.
This medium should be melted once only for use.

(Mossel *et al.*, 1962)

Violet Red Bile Lactose Agar

(For the enumeration of lactose-fermenting enterobacteria in foods)

Peptone	7.0 g
Yeast extract	3.0 g
Bile salts[§] (No. 3, Oxoid L56 or equivalent)	1.5 g
Sodium chloride	5.0 g
Lactose	10.0 g
Neutral red (1% aqueous solution)	3 ml
Crystal violet (0.05% solution)	4 ml
Agar	15.0 g
Distilled water	1 litre

pH 7.4

[§] Different proprietary preparations may not have equivalent activity at the same concentration.

Dissolve the peptone, yeast extract, bile salts, agar and sodium chloride in the water by steaming. Cool to 50°C and adjust to pH 7.4. Add the lactose, neutral red and crystal violet, and mix until dissolved. Distribute in 15-ml amounts in screw-capped bottles and sterilize by autoclaving at 115°C for 15 min. This medium should not be reheated to melt more than once.

Willis and Hobbs' Lactose Egg-yolk Milk Agar

(For the differentiation of clostridia)

Basal medium

Nutrient broth	800 ml
Agar	12.0 g
Lactose	9.6 g
Neutral red (1% solution)	2.6 ml

Dissolve the agar in the nutrient broth by heating, then add the lactose and neutral red solution. Mix well and distribute in 80-ml amounts in screw-capped bottles. Sterilize by autoclaving at 121°C for 20 min.

Egg-yolk emulsion

Separate aseptically an egg-yolk from the egg-white and drain the yolk into a sterile measuring cylinder. Mix the yolk with an equal volume (approx. 20 ml) of 0.85% sterile saline solution, and transfer aseptically to a sterile screw-capped bottle. This emulsion has been found to be satisfactory for routine purposes; it can be tested for sterility by plating out 1-ml quantities. Alternatively, egg-yolk emulsion is available from manufacturers of media and media ingredients.

Preparation of complete medium

Melt the basal medium and cool to 50–55°C. Add 3 ml of egg-yolk emulsion and 12 ml of sterile reconstituted skim milk to 80 ml of molten basal medium. Mix well and pour into Petri dishes.

(Willis, 1962, 1965)

Wilson and Blair's Bismuth Sulphite Agar (for *Salmonella* and *Shigella*)

Nutrient agar (containing 3% agar)	100.0 ml
Sulphite–bismuth–phosphate solution	20.0 ml
Iron (III) citrate–brilliant green solution	4.5 ml

The basal medium consists of a nutrient agar containing 3% agar to ensure solidification after the addition of the stock solutions.

To the molten nutrient agar at 50°C add the two stock solutions previously warmed to 50°C, mix well and pour plates containing 20 ml of medium per plate. Refrigerate the plates for 4–5 days at 4–7°C before use. This storage before use is advisable as it reduces the inhibitory action shown by the freshly prepared medium towards certain strains of *Salmonella*.

Preparation of the sulphite–bismuth–phosphate solution

Ammonium bismuth citrate scales	6.0 g
Sodium sulphite, anhydrous	20.0 g
Disodium hydrogen phosphate, anhydrous	10.0 g

D-Glucose	10.0 g
Distilled water	200.0 ml

Dissolve the ammonium bismuth citrate in 50 ml of boiling distilled water, and dissolve the sodium sulphite in 100 ml of boiling distilled water. Mix and, while still boiling, add the disodium hydrogen phosphate, and stir until dissolved. Cool and then add the glucose previously dissolved by warming in the remaining 50 ml of distilled water.

Preparation of iron (III) citrate–brilliant green solution

Iron (III) citrate	2.0 g
Brilliant green (1% solution)	25.0 ml
Distilled water	200.0 ml

These stock solutions may be stored refrigerated for months without significant deterioration.

(Wilson, 1938)

Wilson and Blair's Sulphite Medium (for *Clostridium*)

Nutrient agar (containing 3% agar)	100 ml
Sodium sulphite–glucose solution	15 ml
Iron (II) sulphate, hydrated (8% solution)	1 ml

The nutrient agar base is of the usual composition except that the concentration of agar is increased to 3%. To 100 ml of the molten nutrient agar cooled to 55°C, add, immediately before use, the solutions of sodium sulphite–glucose and iron (II) sulphate, and mix. In use the predetermined amount of water sample or other inoculum to be examined should be mixed with an equal quantity of the molten medium (just prepared) in sterile Miller–Prickett tubes or other suitable containers.

Preparation of sodium sulphite-glucose solution

Sodium sulphite, anhydrous	20.0 g
D-Glucose	10.0 g
Distilled water	150.0 ml

Dissolve the sodium sulphite in 100 ml of boiling distilled water, and dissolve the glucose in the remaining 50 ml of boiling distilled water. When cool, mix the two solutions together.

(Wilson, 1938)

Wort Agar

(For growth of yeasts and moulds)

Malt extract	15.0 g
Peptone	0.78 g
Maltose	12.75 g
Dextrin	2.75 g
Glycerol	2.35 g
Dipotassium hydrogen phosphate	1.0 g
Ammonium chloride	1.0 g
Agar	20.0 g
Distilled water	1 litre

pH 4.8

Dissolve all the ingredients in the water by steaming. Adjust to pH 4.8, and distribute in 15-ml amounts in test-tubes or screw-capped containers. Sterilize by autoclaving at 115°C for 15 min.

This medium will normally prove too soft for inoculation by streaking, and the low pH makes it unsuitable for repeated reheating.

For Osmophilic Agar (see p. 443) the ingredients of the medium should be dissolved in a 45°C Brix syrup consisting of 350 g of sucrose and 100 g of glucose in 1 litre of solution; sterilize at 108°C for 20 min.

Yeast Extract Agar

Peptone	5.0 g
Yeast extract	3.0 g
Agar	15.0 g
Distilled water	1 litre

pH 7.2 ± 0.2

Dissolve the ingredients in the water by steaming or boiling. Distribute in 100-ml amounts in screw-capped bottles and sterilize by autoclaving at 121°C for 15 min.

Yeast Extract Lactate Broth

(For detection and isolation of propionibacteria)

Yeast extract	5.0 g
Sodium lactate	20.0 g
Distilled water	1 litre

pH 7.0

Dissolve the ingredients and adjust to pH 7.0. Distribute as required and sterilize by autoclaving at 121°C for 20 min.

If sodium lactate is not available, it may be prepared by adding 6.2 g of sodium hydroxide[H] pellets to 14 g of lactic acid[H] or, alternatively, neutralizing the lactic acid with 155 ml of N sodium hydroxide and making up the volume to 1 litre with distilled water.

This medium may be converted into the *semi-solid medium* by the inclusion of 3.0 g of agar per litre, or into the *solid agar medium* by the inclusion of 15.0 g of agar per litre.

(van Neil, 1928)

Yeast Extract Milk Agar

(General non-selective medium for detecting bacteria in milk and milk products)

Yeast extract	3.0 g
Peptone	5.0 g
Agar	15.0 g
Fresh whole or skim milk	10.0 g
Distilled water	1 litre

Dissolve the yeast extract and peptone in distilled water by steaming, then cool and adjust to pH 7.4. Add the agar and milk to the broth and autoclave at 121°C for 20 min. Filter the medium while hot through hot paper pulp (see p. 445). Determine the pH of the filtrate at 50°C and adjust if necessary to pH 7.0. Distribute the medium as required and sterilize by autoclaving at 121°C for 15 min. The final reaction of the medium at room temperature should be pH 7.2.

This medium is available dehydrated as Milk Agar (Oxoid). Its composition differs from that of Wilson (1935) and the similar formulation of the British Standard (1968) in the substitution of Yeastrel by the generic yeast extract.

(Wilson, 1935)

Yeast Glucose Chalk Litmus Milk

(For maintenance of cultures of lactic acid bacteria)

Prepare 1 litre of yeast glucose litmus milk. Distribute calcium carbonate (precipitated) in 0.5-g quantities in 30-ml (Universal) screw-capped containers. Sterilize in the autoclave at 121°C for 20 min. Allow to cool and add 10 ml of yeast glucose litmus milk to each container. Sterilize in the steamer at 100°C for 30 min on each of three successive days. Incubate at 37°C for 2 days and at 30°C for 2 days (or at 55°C if to be used for thermophilic organisms) before use to check sterility of the medium.

Yeast Glucose Lemco Agar

Peptone	10.0 g
Lab-Lemco meat extract (Oxoid)	10.0 g
Sodium chloride	5.0 g
D-Glucose	5.0 g
Yeast extract	3.0 g
Agar	15.0 g
Distilled water	1 litre

pH 7.0

Dissolve the ingredients, except the glucose and agar, by steaming for ½–1 h. Cool to room temperature and adjust to pH 7.0. Add the agar and dissolve at 121°C for 15 min. Filter through hot paper pulp (see p. 445) and dissolve the glucose in the filtrate. Distribute as required and sterilize by autoclaving at 121°C for 15 min.

(Naylor and Sharpe, 1958a)

Yeast Glucose Lemco Broth

Prepare in a similar manner to yeast glucose lemco agar, except that the agar is omitted from the recipe.

(Naylor and Sharpe, 1958a)

This medium is also used as the basal medium for determining the growth characteristics of streptococci in the identification of isolates to primary physiological groups (see p. 344).

Yeast Glucose Lemco Broth plus 4% (or 6.5%) sodium chloride

Dissolve 4 g (or 6.5 g) of sodium chloride in about 70 ml of yeast glucose lemco broth and make up volume to 100 ml with more broth. Distribute in 7-ml amounts in 150 × 16-mm test-tubes and sterilize by autoclaving at 115°C for 15 min.

Yeast Glucose Litmus Milk

Litmus milk (see p. 429)	1 litre
Yeast extract	3.0 g
D-Glucose	10.0 g

Dissolve the yeast extract and glucose in previously bulk-sterilized litmus milk by steaming for ½–1 h. Distribute in test-tubes or as required and sterilize by steaming for

30 min on each of three successive days. Incubate at 37°C for 2 days and at 30°C for 2 days (and at 55°C if required for work with thermophiles) before use to check the sterility of the medium.

(Wheater, 1955)

Yeast Glucose Lemco Broth, pH 9.6

(For the differentiation of the primary groups of streptococci)

Glucose lemco broth

D-Glucose	10.0 g
Lab-Lemco meat extract (Oxoid)	10.0 g
Peptone	10.0 g
Sodium chloride	5.0 g
Distilled water	1 litre

Dissolve ingredients and sterilize by autoclaving at 121°C for 20 min.

Buffer solution

Glycine	7.505 g
Sodium chloride	5.85 g
Distilled water, freshly boiled	1 litre

Dissolve ingredients and sterilize by autoclaving at 121°C for 20 min. Mix six parts with four parts of 0.1 N sodium hydroxide by volume.

Preparation of complete medium

To 900 ml of glucose lemco broth add 100 ml of buffer solution and adjust to pH 9.8 with N sodium hydroxide. Store overnight in a stoppered flask in the cold to complete precipitation. Sterilize by filtration and dispense immediately and aseptically into sterile screw-capped bottles to leave a minimum of air space. The medium should be used within 48 h of preparation, and the pH of uninoculated control tubes should be checked electrometrically immediately before and after incubation. The initial pH should be 9.6 ± 0.02 and should not drop during incubation by more than 0.04 units.

(Shattock and Hirsch, 1947)

APPENDIX 2

Probability Tables for the Estimation of Microbial Numbers by the Multiple Tube Technique

Introduction

In the case of samples for which it is impossible to make an informed guess about the approximate concentration of organisms likely to be found, it is sometimes helpful to set up a two-tube series (Table A.1) over a very wide range of dilutions. Such tests may enable the correct range of dilutions to be chosen for future samples which can be used in the preparation of three- or five-tube series in order to obtain better estimates of the microbial load (Tables A.2 and A.3).

It should be noted that these tables assume that adequate mixing of samples and dilutions has taken place, that the microorganisms are randomly distributed, that one or more organisms in a tube always results in growth, and that there are no disturbing influences. These assumptions may sometimes be incorrect. If, for example, results are obtained in which no growth occurs in the lower dilutions and growth occurs in the higher dilutions, this *could* be indicative of an antimicrobial inhibitory substance present in the food which becomes diluted below a threshold minimum inhibitory concentration before the microorganisms have been diluted out to undetectably low numbers. It is extremely unlikely that results such as 'no tubes positive at the first dilution, no tubes positive at the second dilution and all tubes positive at the third dilution' would be obtained in an experiment in statistical control (i.e. one in which the stated assumptions are correct). Consequently not all combinations of positive and negative tubes are listed. Even so, the tables list a number of combinations that are very unlikely, so in the principal tables (Tables A.2 and A.3) the listed combinations are categorized in terms of likelihood, indicating those that should not be used as the basis for quality assurance decisions, and for which retesting should be undertaken.

Determination of MPNs from Series of More than Three Dilutions

If more than three dilutions have been used, choose the three dilutions for determining the MPN from the Tables according to the following rules:

1. If more than one dilution has all its tubes positive, choose the set of three to include only the most dilute of these.
2. If more than one dilution contains no positive tubes, choose the set of three dilutions so that only the most concentrated of these is included.
3. If only three dilutions have been left by steps 1 and 2, the MPN may be determined from the Tables as indicated. If four dilutions remain, take both sets of three consecutive dilutions, determine the $\log_{10}$ (MPN) of each and take the average of the logarithms to determine a final $\log_{10}$ (MPN) and hence the MPN. In this case the result should, however, be taken as only a rough indication of the microbial load, and should not be used as the basis for quality assurance decisions.
4. If more than four sets of tubes remain, the Tables cannot be used. Stevens (1958) described a procedure for determining the probability that any particular range of

results may occur (the range being the number of dilution levels from the lowest dilution at which at least one tube shows no growth to the highest dilution at which at least one tube shows growth). Where the range of results left by steps 1–3 exceeds the number that can be dealt with by these Tables, an estimate may be obtained using the procedure described in Table 8,2 of Fisher and Yates (1963), together with an indication of the significance of the estimate so obtained. Usually, however, it is preferable to review the experimental procedure in order to identify possible deficiencies or faults in technique, and to repeat the test on a further sample.

TABLE A.1 Values of the MPN for two tubes inoculated from each of three successive tenfold dilutions

No. of positive tubes observed at each dilution			MPN of microorganisms per inoculum of the first dilution
1st dilution	2nd dilution	3rd dilution	
0	0	0	0
0	0	1	0.45
0	1	0	0.46
1	0	0	0.6
1	0	1	1.2
1	1	0	1.3
1	1	1	2.0
1	2	0	2.1
2	0	0	2.3
2	0	1	5.0
2	1	0	6.2
2	1	1	13
2	1	2	21
2	2	0	24
2	2	1	70
2	2	2	100+

Approximate 95% confidence limits may be calculated (Cochran, 1950) as:

$$\frac{\text{MPN}}{6.61} \text{ to MPN} \times 6.61$$

TABLE A.2 Values of the MPN for three tubes inoculated from each of three successive tenfold dilutions (after Demeter *et al.*, 1933)

No. of positive tubes observed at each dilution			MPN of microorganisms per inoculum of 1st dilution	Category*	95% C.I.[†]
1st dilution	2nd dilution	3rd dilution			
0	0	0	0	–	
0	0	1	0.3	3	
0	1	0	0.3	2	<0.1–1.7
0	1	1	0.6	4	
0	2	0	0.6	4	
1	0	0	0.4	1	<0.1–2.1
1	0	1	0.7	3	
1	0	2	1.1	4	
1	1	0	0.7	2	0.2–2.8
1	1	1	1.1	4	
1	2	0	1.1	3	
1	2	1	1.5	4	
1	3	0	1.6	4	
2	0	0	0.9	1	0.2–3.8
2	0	1	1.4	3	
2	0	2	2.0	4	
2	1	0	1.5	2	0.5–5.0
2	1	1	2.0	4	
2	1	2	3.0	4	
2	2	0	2.0	3	
2	2	1	3.0	4	
2	2	2	3.5	4	
2	2	3	4.0	4	
2	3	0	3.0	4	
2	3	1	3.5	4	
2	3	2	4.0	4	
3	0	0	2.5	1	<1–13
3	0	1	4.0	2	1–18
3	0	2	6.5	4	
3	1	0	4.5	1	1–21
3	1	1	7.5	2	2–28
3	1	2	11.5	3	
3	1	3	16.0	4	
3	2	0	9.5	1	3–38
3	2	1	15.0	2	5–50
3	2	2	20.0	3	
3	2	3	30.0	4	
3	3	0	25.0	1	<10–140
3	3	1	45.0	1	10–240
3	3	2	110.0	1	30–480
3	3	3	140.0 +	–	

* In the long run, category 1 combinations may be expected to constitute 67.5% of test results containing both postive and negative tubes; categories (1 + 2) 91%; and categories (1 + 2 + 3) 99% of such test results. Category 4 and unlisted combinations are highly unlikely (Woodward, 1957). Combinations from categories 3 and 4 should not be used as the basis for quality assurance decisions involving the rejection and/or reprocessing of batches of food: samples should be retested. In the event of large numbers of improbable combinations being obtained, experimental procedures should be examined closely, as this is indicative of the presence of disturbing influences (e.g. improper mixing of samples and/or dilutions, presence of antagonistic substances in the foodstuff, etc.).

[†] The 95% confidence intervals (as calculated by de Man (1975)) are given for category 1 and 2 results. Since it is likely that the experimental conditions for a microbiological analysis resulting in a category 3 or 4 combination will not have been in statistical control (i.e. the assumptions described in the introduction to this Appendix are incorrect), 95% confidence intervals are not given for these combinations.

TABLE A.3 Values of the MPN for five tubes inoculated from each of three successive tenfold dilutions (after Woodward 1957; Report, 1969)

No. of positive tubes observed at each dilution			MPN of microorganisms per inoculum of 1st dilution	Category*	95% C.I.†
1st dilution	2nd dilution	3rd dilution			
0	0	0	0	–	
0	0	1	0.2	3	
0	0	2	0.4	4	
0	1	0	0.2	3	
0	1	1	0.4	4	
0	1	2	0.6	4	
0	2	0	0.4	3	
0	2	1	0.6	4	
0	3	0	0.6	4	
1	0	0	0.2	1	<0.1–1.1
1	0	1	0.4	3	
1	0	2	0.6	4	
1	1	0	0.4	2	0.1–1.5
1	1	1	0.6	4	
1	1	2	0.8	4	
1	2	0	0.6	3	
1	2	1	0.8	4	
1	3	0	0.8	4	
1	3	1	1.0	4	
2	0	0	0.4	1	0.1–1.7
2	0	1	0.7	3	
2	0	2	0.9	4	
2	0	3	1.2	4	
2	1	0	0.7	2	0.2–2.1
2	1	1	0.9	3	
2	1	2	1.2	4	
2	2	0	0.9	3	
2	2	1	1.2	4	
2	2	2	1.4	4	
2	3	0	1.2	4	
2	3	1	1.4	4	
3	0	0	0.8	1	0.3–2.4
3	0	1	1.1	3	
3	0	2	1.3	4	
3	1	0	1.1	2	0.4–2.9
3	1	1	1.4	3	
3	1	2	1.7	4	
3	1	3	2.0	4	
3	2	0	1.4	3	
3	2	1	1.7	4	
3	2	2	2.0	4	
3	3	0	1.7	3	
3	3	1	2.1	4	
3	4	0	2.1	4	
3	4	1	2.4	4	
4	0	0	1.3	1	0.5–3.8
4	0	1	1.7	3	
4	0	2	2.1	4	

No. of positive tubes observed at each dilution			MPN of microorganisms per inoculum of 1st dilution	Category*	95% C.I.[†]
1st dilution	2nd dilution	3rd dilution			
4	0	3	2.5	4	
4	1	0	1.7	2	0.7–4.6
4	1	1	2.1	3	
4	1	2	2.6	4	
4	2	0	2.2	2	0.9–5.6
4	2	1	2.6	3	
4	2	2	3.2	4	
4	3	0	2.7	3	
4	3	1	3.3	3	
4	3	2	3.9	4	
4	4	0	3.4	3	
4	4	1	4.0	4	
4	4	2	4.7	4	
4	5	0	4.1	4	
4	5	1	4.8	4	
4	5	2	5.6	4	
5	0	0	2.3	1	0.9–8.6
5	0	1	3.1	3	
5	0	2	4.3	3	
5	0	3	5.8	4	
5	1	0	3.3	1	1–12
5	1	1	4.6	2	2–15
5	1	2	6.4	4	
5	1	3	8.4	4	
5	1	4	11.5	4	
5	2	0	4.9	1	2–17
5	2	1	7.0	2	3–21
5	2	2	9.5	3	
5	2	3	12.0	4	
5	2	4	15.0	4	
5	3	0	7.9	1	3–25
5	3	1	11.0	2	4–30
5	3	2	14.0	3	
5	3	3	17.5	3	
5	3	4	20.0	4	
5	3	5	25.0	4	
5	4	0	13.0	1	5–39
5	4	1	17.0	2	7–48
5	4	2	22.0	2	10–58
5	4	3	28.0	3	
5	4	4	35.0	3	
5	4	5	42.5	4	
5	5	0	24.0	1	10–94
5	5	1	35.0	1	10–130
5	5	2	54.0	1	20–200
5	5	3	92.0	1	30–290
5	5	4	160.0	1	60–530
5	5	5	180.0 +	–	

* See footnote to Table A.2.
[†] See footnote to Table A.2.

ISO Standards for Microbiological Methods

1. General Methods (i.e. Horizontal Methods)

ISO 4831:1991 Microbiology – General guidance for the enumeration of coliforms – Most probable number technique

ISO 4832:1991 Microbiology – General guidance for the enumeration of coliforms – Colony count technique

ISO 4833:1991 Microbiology – General guidance for the enumeration of micro-organisms – Colony count technique at 30°C

ISO 6579:1993 Microbiology – General guidance on methods for the detection of *Salmonella*

ISO 6887:1983 Microbiology – General guidance for the preparation of dilutions for microbiological examinations

ISO/DIS 6887-1 Microbiology of food and animal feedingstuffs – Preparation of test samples, initial suspension and decimal dilutions for microbiological examination – Part 1: General rules for the preparation of the initial suspension and decimal dilutions

(ISO 6888:1983 Microbiology – General guidance for enumeration of *Staphylococcus aureus* – Colony count technique)

ISO/DIS 6888-1 Microbiology of food and animal feedingstuffs – Horizontal method for the enumeration of coagulase-positive staphylococci (*Staphylococcus aureus* and other species) – Part 1: Technique including confirmation of colonies

ISO/DIS 6888-2 Microbiology of food and animal feedingstuffs – Horizontal method for the enumeration of coagulase-positive staphylococci (*Staphylococcus aureus* and other species) – Part 2: Technique without confirmation of colonies

ISO 7002:1986 Agricultural food products – Layout for a standard method of sampling from a lot

ISO 7218:1996 Microbiology of food and animal feedingstuffs – General rules for microbiological examination

ISO 7251:1993 Microbiology – General guidance for enumeration of presumptive *Escherichia coli* – Most probable number technique

ISO 7402:1993	Microbiology – General guidance for the enumeration of Enterobacteriaceae without resuscitation – MPN technique and colony count technique
ISO 7667:1983	Microbiology – Standard layout for methods of microbiological examination
ISO 7932:1993	Microbiology – General guidance for the enumeration of *Bacillus cereus* – Colony count technique at 30°C
ISO 7937:1997	Microbiology of food and animal feedingstuffs – Horizontal method for enumeration of *Clostridium perfringens* – Colony count technique
ISO 7954:1987	Microbiology – General guidance for enumeration of yeasts and moulds – Colony count technique at 25°C
ISO 8523:1991	Microbiology – General guidance for the detection of Enterobacteriaceae with pre-enrichment
ISO 8914:1990	Microbiology – General guidance for the detection of *Vibrio parahaemolyticus*
ISO 10272:1995	Microbiology of food and animal feedingstuffs – Horizontal method for detection of thermotolerant *Campylobacter*
ISO 10273:1994	Microbiology – General guidance for the detection of presumptive pathogenic *Yersinia enterocolitica*
ISO 11290-1:1996	Microbiology of food and animal feedingstuffs – Horizontal method for the detection and enumeration of *Listeria monocytogenes* – Part 1: Detection method
ISO/DIS 11290-2	Microbiology of food and animal feedingstuffs – Horizontal method for the detection and enumeration of *Listeria monocytogenes* – Part 2: Enumeration method
ISO/DIS 15214	Microbiology of food and animal feedingstuffs – Horizontal method for the enumeration of mesophilic lactic acid bacteria – Colony count technique at 30°C

2. Vertical Methods for Milk and Milk Products

ISO 707:1997	Milk and milk products – Methods of sampling
ISO 5538:1987	Milk and milk products – Sampling – Inspection by attributes
ISO 5541-1:1986	Milk and milk products – Enumeration of coliforms – Part 1: Colony count technique at 30°C
ISO 5541-2:1986	Milk and milk products – Enumeration of coliforms – Part 2: Most probable number technique at 30°C
ISO 6610:1992	Milk and milk products – Enumeration of colony-forming units of microorganisms – Colony count technique at 30°C
ISO 6611:1992	Milk and milk products – Enumeration of colony-forming units of yeasts and/or moulds – Colony count technique at 25°C
ISO 6730:1992	Milk – Enumeration of colony-forming units of psychrotrophic microorganisms – Colony count technique at 6.5°C

ISO 6785:1985 Milk and milk products – Detection of *Salmonella*

ISO 8086:1986 Dairy plant – Hygiene conditions – General guidance on inspection and sampling procedures

ISO 8197:1988 Milk and milk products – Sampling – Inspection by variables

ISO 8261:1989 Milk and milk products – Preparation of test samples and dilutions for microbiological examination

ISO 10560:1993 Milk and milk products – Detection of *Listeria monocytogenes*

ISO 11866-1:1997 Milk and milk products – Enumeration of presumptive *Escherichia coli* – Part 1: Most probable number technique

ISO 11866-2:1997 Milk and milk products – Enumeration of presumptive *Escherichia coli* – Part 2: Most probable number technique using 4-methylumbelliferyl-β-D-glucuronide (MUG)

ISO 11866-3:1997 Milk and milk products – Enumeration of presumptive *Escherichia coli* – Part 3: Colony count technique at 44°C using membranes

ISO 13366-1:1997 Milk – Enumeration of somatic cells – Part 1: Microscopic method (reference method)

ISO 13366-2:1997 Milk – Enumeration of somatic cells – Part 2: Method using an electronic particle counter

ISO 13366-3:1997 Milk – Enumeration of somatic cells – Part 3: Fluoro-opto-electronic method

ISO/DIS 14501 Milk and milk powder – Determination of aflatoxin M_1 content – Clean-up by immunoaffinity chromatography and determination by HPLC

ISO/DIS 14674 Milk and milk powder – Determination of aflatoxin M_1 content – Thin-layer chromatographic method

ISO/DIS 14675 Milk and milk powder – Determination of aflatoxin M_1 content – ELISA screening method

3. Vertical Methods for Meat and Meat Products

ISO 3100-1:1991 Meat and meat products – Sampling and preparation of test samples – Part 1: Sampling

ISO 3100-2:1988 Meat and meat products – Sampling and preparation of test samples – Part 2: Preparation of test samples for microbiological examination

ISO 5552:1997 Meat and meat products – Detection and enumeration of Enterobacteriaceae without resuscitation – MPN technique and colony count technique

ISO 6391:1997 Meat and meat products – Enumeration of *Escherichia coli* – Colony count technique at 44°C using membranes

ISO 13681:1995 Meat and meat products – Enumeration of yeasts and moulds – Colony count technique

ISO 13720:1995 Meat and meat products – Enumeration of *Pseudomonas* spp.

ISO 13721:1995 Meat and meat products – Enumeration of lactic acid bacteria – Colony count technique at 30°C

ISO 13722:1996 Meat and meat products – Enumeration of *Brochothrix thermosphacta* – Colony count technique

4. Methods for Water

ISO 5667-2:1991 Water quality – Sampling – Part 2: Guidance on sampling techniques

ISO 5667-3:1994 Water quality – Sampling – Part 3: Guidance on the preservation and handling of samples

ISO 5667-5:1991 Water quality – Sampling – Part 5: Guidance on sampling of drinking water and water used for food and beverage processing

ISO 5667-6:1990 Water quality – Sampling – Part 6: Guidance on sampling of rivers and streams

ISO 5667-9:1992 Water quality – Sampling – Part 9: Guidance on sampling from marine waters

ISO 6060:1989 Water quality – Determination of the chemical oxygen demand

ISO/DIS 6222 Water quality – Enumeration of culturable microorganisms – Colony count by inoculation in a nutrient agar culture medium (Revision of ISO 6222:1988)

ISO 6340:1995 Water quality – Detection of Salmonella species

ISO 7704:1985 Water quality – Evaluation of membrane filters used for microbiological analyses

ISO 7899-1:1984 Water quality – Detection and enumeration of faecal streptococci – Part 1: Method by enrichment in a liquid medium

ISO/DIS 7899-1 Water quality – Enumeration of intestinal enterococci in surface and waste water – Part 1: Miniaturized method (most probable number) by inoculation in liquid medium

ISO 7899-2:1984 Water quality – Detection and enumeration of faecal streptococci – Part 2: Method by membrane filtration

ISO 9308-1:1990 Water quality – Detection and enumeration of coliform organisms, thermotolerant coliform organisms and presumptive *Escherichia coli* – Part 1: Membrane filtration method

ISO 9308-2:1990 Water quality – Detection and enumeration of coliform organisms, thermotolerant coliform organisms and presumptive *Escherichia coli* – Part 2: Multiple tube (most probable number) method

ISO/DIS 9308-3 Water quality – Detection and enumeration of *Escherichia coli* and coliform bacteria – Part 3: Miniaturized method (most probable number (MPN)) for surface and waste water

ISO/DIS 11348-1 Water quality – Determination of the inhibitory effect of water samples on the light emission of *Vibrio fischeri* (luminescent bacteria test) – Part 1: Method using freshly prepared bacteria

ISO/DIS 11348-2	Water quality – Determination of the inhibitory effect of water samples on the light emission of *Vibrio fischeri* (luminescent bacteria test) – Part 2: Method using liquid-dried bacteria
ISO/DIS 11348-3	Water quality – Determination of the inhibitory effect of water samples on the light emission of *Vibrio fischeri* (luminescent bacteria test) – Part 3: Method using freeze-dried bacteria
ISO/FDIS 11731-1	Water quality – Detection and enumeration of *Legionella*

5. Standards Relating to Quality Management

ISO 8402:1994	Quality management and quality assurance – vocabulary
ISO 9000-1:1994	Quality management and quality assurance standards. Part 1: Guidelines for selection and use
ISO 9000-2:1997	Quality management and quality assurance standards. Part 2: Generic guidelines for the application of ISO 9001, ISO 9002 and ISO 9003
ISO 9001:1994	Quality systems – Model for quality assurance in design, development, production, installation and servicing
ISO 9002:1994	Quality systems – Model for quality assurance in production, installation and servicing
ISO 9003:1994	Quality systems – Model for quality assurance in final inspection and test
ISO 9004-1:1994	Quality management and quality system elements – Part 1: Guidelines
ISO 9004-2:1991	Quality management and quality system elements. Part 2: Guidelines for services
ISO 9004-3:1993	Quality management and quality system elements. Part 3: Guidelines for processed materials
ISO 9004-4:1993	Quality management and quality system elements. Part 4: Guidelines for quality improvement
ISO 10011-1:1990	Guidelines for auditing quality systems. Part 1: Auditing
ISO 10011-2:1991	Guidelines for auditing quality systems. Part 2: Qualification criteria for quality systems auditors
ISO 10011-3:1991	Guidelines for auditing systems. Part 3: Management of audit programmes
ISO 10013:1995	Guidelines for developing quality manuals
ISO/IEC Guide 2: 1996	Standardization and related activities – general vocabulary
ISO/IEC Guide 25: 1990	General requirements for the competence of calibration and testing laboratories

References

ACDP (1990). *Categorisation of Pathogens According to Hazard and Categories of Containment*, 2nd edn. Advisory Committee on Dangerous Pathogens. London: HMSO.
ACMSF (1992). *Report on Vacuum Packaging and Associated Processes*. Report by the Advisory Committee on the Microbiological Safety of Food. London: HMSO.
ACMSF (1993). *Report on* Salmonella *in Eggs*. Report by the Advisory Committee on the Microbiological Safety of Food. London: HMSO.
ACMSF (1995). *Report on Verocytotoxin-Producing* Escherichia coli. Report by the Advisory Committee on the Microbiological Safety of Food. London: HMSO.
ACMSF (1996). *Report on Poultry Meat*. Report by the Advisory Committee on the Microbiological Safety of Food. London: HMSO.
AOAC (1995). *Official Methods of Analysis of AOAC International*, 16th edn. Arlington, Virginia: AOAC International.
Abd-el-Malek, Y. and Gibson, T. (1948). Studies in the bacteriology of milk. 1. The streptococci of milk. *Journal of Dairy Research* **15**, 233–248.
Akman, M. and Park, R. W. A. (1974). The growth of salmonellas on cooked cured pork. *Journal of Hygiene, Cambridge* **72**, 369–377.
Albert, J. O., Long, H. F. and Hammer, B. W. (1944). Classification of the organisms important in dairy products. *Bulletin of the Iowa Agricultural Experimental Station* 328.
Aldred, J. B., Evans, A. F. and Husbands, V. (1971). Aspects of the Howard mould count. *Journal of the Association of Public Analysts* **9**, 47–52.
Alton, G. G. and Jones, L. (1963). *Laboratory Techniques in Brucellosis (FAO Animal Health Branch, Monograph No. 7)*. Rome: Food and Agriculture Organisation of the United Nations.
Altwegg, M. and Geiss, H. K. (1989). *Aeromonas* as a human pathogen. *CRC Critical Reviews in Microbiology* **16**, 253–286.
American Public Health Association (1966). *Recommended Methods for the Microbiological Examination of Foods*, 2nd edn. New York: APHA.
American Public Health Association (1995). *Standard Methods for the Examination of Water and Wastewater*, 19th edn. Washington, DC: APHA.
Amos, A. J. (1968). Flour and bread. In *Quality Control in the Food Industry, Vol. 2* (ed. S. M. Herschdoerfer), pp. 195–217. London: Academic Press.
Anderson, E. C. and Hobbs, B. (1973). Studies of the strain of *Salmonella typhi* responsible for the Aberdeen typhoid outbreak. *Israel Journal of Medical Sciences* **9**, 162.
Anderson, W. A., McClure, P. J., Baird-Parker, A. C. *et al.*, (1996). The application of a log-logistic model to describe the thermal inactivation of *Clostridium botulinum* 213B at temperatures below 121.1°C. *Journal of Applied Bacteriology* **80**, 283–290.
Andrew, M. H. E. and Russell, A. D. (1984). *The Revival of Injured Microbes (Society for Applied Bacteriology, Symposium No. 12)*. London: Academic Press.
Angelotti, R. and Foter, M. J. (1958). A direct surface agar plate laboratory method for quantitatively detecting bacterial contamination on nonporous surfaces. *Food Research* **23**, 170–174.

Anonymous (1989). *Chilled and Frozen: Guidelines on Cook–Chill and Cook–Freeze Catering Systems*. Department of Health. London: HMSO.

Anonymous (1995). *The Dairy Products (Hygiene) Regulations 1995. (S.I. No. 1086)*. London: HMSO.

Anonymous (1997). COSHH Regulation 1994: Risk assessment and VTEC producing organisms. *PHLS Microbiology Digest* **14**, 62–63.

Anonymous (1998). *The Fishery Products and Live Shellfish (Hygiene) Regulations 1998*. London: HMSO.

D'Aoûst, J. Y., Sewell, A. M., Greco, P. *et al.* (1995). Performance assessment of the Gene-Trak® colorimetric probe assay for the detection of food borne *Salmonella* spp. *Journal of Food Protection* **58**, 1069–1076.

Asai, T. (1968). *Acetic Acid Bacteria: Classification and Biochemical Activity.* Tokyo: University of Tokyo Press.

Atlas, R. M. and Bej, A. K. (1994). Polymerase chain reaction. In *Methods for General and Molecular Bacteriology* (ed. P. Gerhardt *et al.*), pp. 418–435. Washington, DC: American Society for Microbiology.

Atmar, R. L., Neill, F. H., Woodley, C. M. *et al.* (1996). Collaborative evaluation of a method for the detection of Norwalk virus in shellfish tissues by PCR. *Applied and Environmental Microbiology* **62**, 254–258.

Aulisio, C. C. G., Mehlman, I. J. and Saunders, A. C. (1980). Alkali method for rapid recovery of *Yersinia enterocolitica* and *Yersinia pseudotuberculosis* from foods. *Applied and Environmental Microbiology* **39**, 135–140.

Austin, B. and Lee, J. V. (1992). Aeromonadaceae and Vibrionaceae. In *Identification Methods in Applied and Environmental Microbiology (Society for Applied Bacteriology, Technical Series No. 29)* (ed. R. G. Board, D. Jones and F. A. Skinner), pp. 163–182. London: Academic Press.

Baird, R. M., Corry, J. E. L. and Curtis, G. D. W. (1987). Pharmacopœia of culture media for food microbiology. Appendix 1. Testing methods for use in quality assurance of culture media. *International Journal of Food Microbiology* **5**, 291–296.

Baird-Parker, A. C. (1962). An improved diagnostic and selective medium for isolating coagulase positive staphylococci. *Journal of Applied Bacteriology* **25**, 12–19.

Baird-Parker, A. C. (1969). The use of Baird-Parker's medium for the isolation and enumeration of *Staphylococcus aureus*. In *Isolation Methods for Microbiologists (Society for Applied Bacteriology, Technical Series, No. 3)* (ed. D. A. Shapton and G. W. Gould), pp. 1–8. London: Academic Press.

Baird-Parker, A. C. and Kooiman, W. J. (1980). Natural Mineral Waters. In *Microbial Ecology of Foods, Vol. 2: Food Commodities* (ICMSF), pp. 834–837. New York: Academic Press.

Baker, D. A. and Park, R. W. A. (1975). Changes in morphology and cell wall structure that occur during growth of *Vibrio* sp. NCTC 4716 in batch culture. *Journal of General Microbiology* **86**, 12–28.

Baker, J. H. and Orr, D. R. (1979). The construction and application of temperature gradient incubators. In *Cold Tolerant Microbes in Spoilage and the Environment (Society for Applied Bacteriology, Technical Series No. 13)* (ed. A. D. Russell and R. Fuller), pp. 25–38. London: Academic Press.

Ballongue, J. (1993). Bifidobacteria and probiotic action. In *Lactic Acid Bacteroa* (ed. S. Salminen and A. von Wright), pp. 357–428. New York: Marcel Dekker.

Balows, A., Hauser, W. J., Herrmann, K. L. *et al.* (1991). *Manual of Clinical Microbiology*, 5th edn. Washington, DC: American Society for Microbiology.

Banks, J. G., Board, R. G., Carter, J. *et al.* (1985). The cytotoxic and photodynamic inactivation of micro-organisms by rose bengal. *Journal of Applied Bacteriology* **58**, 391–400.

Barber, M. and Kuper, S. W. A. (1951). Identification of *Staphylococcus pyogenes* by the phosphatase reaction. *Journal of Pathology and Bacteriology* **63**, 65–68.

Barnes, E. M. (1956). Methods for the isolation of faecal streptococci (Lancefield Group D) from bacon factories. *Journal of Applied Bacteriology* **19**, 193–203.

Barnes, E. M. and Impey, C. S. (1968). Psychrophilic spoilage bacteria of poultry. *Journal of Applied Bacteriology* **31**, 97–107.

Barnes, E. M., Impey, C. S. and Parry, R. T. (1973). The sampling of chickens, turkeys, ducks and game birds. In *Sampling – Microbiological Monitoring of Environments (Society for Applied Bacteriology, Technical Series No. 7)* (ed. R. G. Board and D. W. Lovelock), pp. 63–75. London: Academic Press.

Barnett, J. A., Payne, R. W. and Yarrow, D. (1990). *Yeasts: Characteristics and Identification*, 2nd edn. Cambridge: Cambridge University Press.

Barritt, M. M. (1936). The intensification of the Voges–Proskauer reaction by the addition of α-naphthol. *Journal of Pathology and Bacteriology* **42**, 441–454.

Bartholomew, J. W. and Mittwer, T. (1950). A simplified bacterial spore stain. *Stain Technology* **25**, 153–156.

Bean, P. G. and Everton, J. R. (1969). Observations on the taxonomy of chromogenic bacteria isolated from cannery environments. *Journal of Applied Bacteriology* **32**, 51–59.

Beech, F. W. and Davenport, R. R. (1969). The isolation of non-pathogenic yeasts. In *Isolation Methods for Microbiologists (Society for Applied Bacteriology, Technical Series No. 3)* (ed. D. A. Shapton and G. W. Gould), pp. 71–88. London: Academic Press.

Beerens, H., Sugama, S. and Tahon-Castel, M. (1965). Psychrotrophic clostridia. *Journal of Applied Bacteriology* **28**, 36–48.

Bell, C., Rhoades, J. R., Neaves, P. *et al.* (1995). An evaluation of the IDEXX SNAP test for the detection of beta-lactam antibiotics in ex-farm raw milks. *Netherlands Milk and Dairy Journal* **49**, 15–25.

Bender, J. B., Hedberg, C. W., Besser, J. M. *et al.* (1997). Surveillance for *Escherichia coli* O157:H7 infections in Minnesota by molecular subtyping. *New England Journal of Medicine* **337**, 388–394.

Benedict, S. R. (1908–1909). A reagent for the detection of reducing sugars. *Journal of Biological Chemistry* **5**, 485.

Bernaerts, M. J. and De Ley, J. (1963). A biochemical test for crown gall bacteria. *Nature (London)* **197**, 406–407.

Berry, J. A. (1933). Detection of microbial lipase by copper soap formation. *Journal of Bacteriology* **25**, 433–434.

Besser, R. E., Lett, S. M., Weber, J. T. *et al.* (1993). An outbreak of diarrhea and hemolytic uremic syndrome from *Escherichia coli* O157:H7 in fresh-pressed apple cider. *Journal of the American Medical Association* **269**, 2217–2220.

Bettes, D. C. (1965). Canning in the dairy industry. *Journal of the Society of Dairy Technology* **18**, 224–229.

Bielaszewska, M., Janda, J., Bláhová, K. *et al.* (1997). Human *Escherichia coli* O157:H7 infection associated with the consumption of unpasteurized goat's milk. *Epidemiology and Infection* **119**, 299–305.

Billing, E. and Cuthbert, W. A. (1958). Bitty cream: the occurrence and significance of *Bacillus cereus* spores in raw milk supplies. *Journal of Applied Bacteriology* **21**, 65–78.

Billing, E. and Luckhurst, E. R. (1957). A simplified method for the preparation of egg yolk media. *Journal of Applied Bacteriology* **20**, 90.

Blackburn, C de W. and Patel, P. D. (1991). *Brief Evaluation of the* Salmonella *Rapid Test (Oxoid) and Hydrophobic Grid Membrane Filter Technique for the Detection of Salmonellae in Foods (Technical Notes No. 91)*. Leatherhead, UK: Leatherhead Food Research Association.

Blackburn, P. S. and Macadam, I. (1954). The cells in bovine milk. *Journal of Dairy Research* **21**, 31–36.

Blackburn, P. S., Laing, C. M. and Malcolm, J. F. (1955). A comparison of the diagnostic value of the total and differential cell counts of bovine milk. *Journal of Dairy Research* **22**, 37–42.

Blake, P. A., Weaver, R. E. and Hollis, D. G. (1980). Diseases of humans (other than cholera) caused by vibrios. *Annual Review of Microbiology* **34**, 341–367.

Blake, P. A., Rosenberg, M. L., Florencia, J. *et al.* (1977). Cholera in Portugal 1974. II. Transmission by bottled mineral water. *American Journal of Epidemiology* **105**, 344–348.

Blood, R. M., Abbiss, J. S. and Jarvis, B. (1981). Assessment of two methods for testing disinfectants and sanitizers for use in the meat processing industry. In *Disinfectants: Their Use and Evaluation of Effectiveness (Society for Applied Bacteriology, Technical Series No. 16)* (ed. C. H. Collins *et al.*), pp. 17–31. London: Academic Press.

Board, R. G. (1966). Review article: the course of microbial infection of the hen's egg. *Journal of Applied Bacteriology* **29**, 319–341.
Board, R. G. and Fuller, R. (eds) (1994). *Microbiology of the Avian Egg*. London: Chapman and Hall.
Board, R. G. and Holding, A. J. (1960). The utilization of glucose by aerobic Gram-negative bacteria. *Journal of Applied Bacteriology* **23**, xi.
Bolton, F. J. and Robertson, L. (1982). A selective medium for isolating *Campylobacter jejuni/coli*. *Journal of Clinical Pathology* **35**, 462–467.
Bolton, F. J., Crozier, L. and Williamson, J. K. (1995). New technical approaches to *Escherichia coli* O157. *PHLS Microbiology Digest* **12**, 67–73.
Bolton, F. J., Crozier, L. and Williamson, J. K. (1996). Isolation of *Escherichia coli* O157 from raw meat products. *Letters in Applied Microbiology* **23**, 317–321.
Bonde, G. J. (1977). Bacterial indication of water pollution. In *Advances in Aquatic Microbiology, Vol. 1*, (ed. by M. R. Droop and H. W. Jannasch), pp. 273–364. London: Academic Press.
Booth, C. (1971). Fungal culture media. In *Methods in Microbiology, Vol. 4* (ed. C. Booth), pp. 49–94. London: Academic Press.
Botana, L. M., Rodriguez-Vieytes, M., Alfonso, A. *et al.* (1996). Phycotoxins: paralytic shellfish poisoning and diarrhetic shellfish poisoning. In *Handbook of Food Analysis, Vol. 2*, (ed. L. M. L. Nollet), pp. 1147–1169. New York: Marcel Dekker.
Bousfield, I. J. and Callely, A. G. (Eds.) (1978). *Coryneform Bacteria*. London: Academic Press.
Boyce, T. G., Koo, D., Swerdlow, D. L. *et al.* (1996). Recurrent outbreaks of *Salmonella enteritidis* infections in a Texas restaurant: phage type 4 arrives in the United States. *Epidemiology and Infection* **117**, 29–34.
Brailsford, M. A. and Gatley, S. (1993). Industrial application of flow cytometry for the rapid detection of microorganisms. In *New Techniques in Food and Beverage Microbiology (Society for Applied Bacteriology, Technical Series No. 31)* (ed. R. G. Kroll, A. Gilmour and M. Sussman), pp. 87–100. Oxford: Blackwell Scientific.
Breckinridge, J. C. and Bergdoll, M. S. (1971). Outbreak of food-borne gastroenteritis due to a coagulase-negative enterotoxin producing staphylococcus. *New England Journal of Medicine* **284**, 541–543.
Brenner, D. J., Hickman-Brenner, F. W., Lee, J. V. *et al.* (1983). *Vibrio furnissii* (formerly aerogenic biogroup of *Vibrio fluvialis*), a new species isolated from human feces and the environment. *Journal of Clinical Microbiology* **18**, 816–824.
Brewer, J. H. and Allgeier, D. L. (1966). Safe self-contained carbon dioxide–hydrogen anaerobic system. *Applied Microbiology* **14**, 985–988.
Breznak, J. A. and Costilow, R. N. (1994). Physicochemical factors in growth. In *Methods for General and Molecular Bacteriology* (ed. P. Gerhardt *et al.*), pp. 137–154. Washington, DC: American Society for Microbiology.
Bridson, E. Y. and Brecker, A. (1970). Design and formulation of microbial culture media. In *Methods in Microbiology, Vol. 3A* (ed. J. R. Norris and D. W. Ribbons), pp. 229–295. London: Academic Press.
Briggs, M. (1953). The classification of lactobacilli by means of physiological tests. *Journal of General Microbiology* **9**, 234–248.
Brinley Morgan, W. J. and Corbel, M. J. (1990). *Brucella* infections in man and animals. In *Topley and Wilson's Principles of Bacteriology, Virology and Immunity, Vol. 3, Bacterial Diseases*, 8th edn (ed. G. R. Smith and C. S. F. Easman), pp. 547–570. London: Edward Arnold.
Brock, T. D. (1971). Microbial life at 90°C: the sulfur bacteria of Boulder Spring. *Journal of Bacteriology* **107**, 303–314.
Brock, T. D. and Brock, M. L. (1971). Temperature optimum of non-sulphur bacteria from a spring at 90°C. *Nature (London)* **233**, 494–495.
Brown, K. L. (1994) Spore resistance and ultra heat treatment processes. *Journal of Applied Bacteriology Symposium Supplement* **76s**, 67S–80S.
Brown, K. L., Ayres, C. A., Gaze, J. E. *et al.* (1984). Thermal destruction of bacterial spores immobilized in food/alginate particles. *Food Microbiology* **1**, 187–198.
Bryan-Jones, G. (1975). Lactic acid bacteria in distillery fermentations. In *Lactic Acid Bacteria in*

Beverages and Food (ed. J. G. Carr, C. V. Cutting and G. C. Whiting), pp. 165–175. London: Academic Press.

Buchanan, R. E., Gibbons, N. E., Cowan, S. T. *et al.* (1974). *Bergey's Manual of Determinative Bacteriology,* 8th edn. Baltimore: Williams and Wilkins.

Buckle, K. A. and Kartadarma, E. (1989). *Pseudomonas cocovenenans*. In *Foodborne Microorganisms of Public Health Significance,* 4th edn. (ed. K. A. Buckle *et al.*), pp. 328–332. North Ryde, NSW, Australia: AIFST (NSW Branch) Food Microbiology Group.

Buckley, H. R., Campbell, C. K. and Thompson, J. C. (1969). Techniques for the isolation of pathogenic fungi. In *Isolation Methods for Microbiologists (Society for Applied Bacteriology, Technical Series, No. 3)*, (ed. D. A. Shapton and G. W. Gould), pp. 113–126. London: Academic Press.

Busse, M. (1995). Media for salmonella. In *Culture Media for Food Microbiology (Progress in Industrial Microbiology, Vol. 34)* (ed. J. E. L. Corry, G. D. W. Curtis and R. M. Baird), pp. 187–201. Amsterdam: Elsevier.

CDC (1991). Multistate outbreak of *Salmonella poona* infections – United States and Canada, 1991. *Morbidity and Mortality Weekly Report* **40**(32), 549–552.

CDC (1993). *Salmonella* serotype Tennessee in powdered milk products and infant formula – Canada and United States 1993. *Morbidity and Mortality Weekly Report* **42**, 516–517.

CDC (1994). Foodborne outbreaks of enterotoxigenic *Escherichia coli* – Rhode Island and New Hampshire, 1993. *Morbidity and Mortality Weekly Report* **43**(5), 81, 87–89.

CDC (1995). Assessing the public health threat associated with waterborne cryptosporidiosis. *MMWR Recommendations and Reports* **44**, No. RR–6.

CDC (1996). Foodborne outbreak of diarrheal illness associated with *Cryptosporidium parvum. Morbidity and Mortality Weekly Report* **45**(36), 783–784.

CDC (1997a). Outbreaks of *Escherichia coli* O157:H7 infection and cryptosporidiosis associated with drinking unpasteurized apple cider. *Morbidity and Mortality Weekly Report* **46**(1), 4–8.

CDC (1997b). Update: outbreaks of cyclosporiasis – United States, 1997. *Morbidity and Mortality Weekly Report* **46**(21), 461–462.

CDC (1997c). Update: Outbreaks of cyclosporiasis – United States and Canada, 1997. *Morbidity and Mortality Weekly Report* **46** (23), 521–523.

CDSC (1992). An outbreak of foodborne *Shigella sonnei* infection. *CDSC Communicable Disease Report Weekly* **2**(8), 33.

CDSC (1993). Shellfish-associated illness. *CDSC Communicable Disease Report Weekly* **3**(7), 29.

CDSC (1995). Brucellosis associated with unpasteurised milk products abroad. *CDSC Communicable Disease Report Weekly* **5**(32), 151.

CDSC (1996a). Salmonella gold-coast [food poisoning in which Cheddar cheese was implicated] *CDSC Communicable Disease Report Weekly* **6**(51), 443.

CDSC (1996b). General outbreaks of foodborne illness, England and Wales: weeks 37–40/96. *CDSC Communicable Disease Report Weekly* **6**(41), 358.

CDSC (1997a). *Salmonella anatum* infection in infants linked to dried milk. *CDSC Communicable Disease Report Weekly* **7**(5), 33.

CDSC (1997b). *Salmonella gold-coast* and cheddar cheese: update. *CDSC Communicable Disease Report Weekly* **7**, 93.

Carr, J. G. (1968). Methods for identifying acetic acid bacteria. In *Identification Methods for Microbiologists, Part B (Society for Applied Bacteriology, Technical Series No. 2)*, (ed. B. M. Gibbs and D. A. Shapton), pp. 1–8. London: Academic Press.

Cerf, O. (1977). Tailing of survival curves of bacterial spores. *Journal of Applied Bacteriology* **42**, 1–19.

Chalmers, C. H. (1962). *Bacteria in Relation to the Milk Supply,* 4th edn. London: Edward Arnold.

Champsaur, H., Andremont, A., Mathieu, D. *et al.* (1982). Cholera-like illness due to *Aeromonas sobria*. Journal of Infectious Diseases **145**, 248–254.

Charlett, S. M. (1954). An improved staining method for the direct microscopical counting of bacteria in milk. *Dairy Industries* **19**, 652–653.

Charney, J., Fisher, W. P. and Hegarty, C. P. (1951). Manganese as an essential element for sporulation in the genus *Bacillus. Journal of Bacteriology* **62**, 145–148.

Christensen, S. G. (1981). The *Yersinia enterocoliticia* prevalence in slaughter animals, water and raw milk in Denmark. In *Psychrotrophic Micro-organisms in Spoilage and Pathogenicity* (ed. T. A. Roberts *et al.*), pp. 439–445. London: Academic Press.

Christensen, W. B. (1946). Urea decomposition as a means of differentiating *Proteus* and paracolon cultures from each other and from *Salmonella* and *Shigella* types. *Journal of Bacteriology* **52**, 461–466.

Christie, R., Atkins, N. E. and Munch-Petersen, E. (1944). A note on a lytic phenomenon shown by group B streptococci. *Australian Journal of Experimental Biology and Medical Science* **22**, 197–200.

Chu, F. S. and Li, G. Y. (1994). Simultaneous occurrence of fumonisin B_1 and other mycotoxins in moldy corn collected from the People's Republic of China in regions with high incidences of esophageal cancer. *Applied and Environmental Microbiology* **60**, 847–852.

Clarke, P. H. and Steel, K. J. (1966). Rapid and simple biochemical tests for bacterial identification. In *Identification Methods for Microbiologists, Part A (Society for Applied Bacteriology, Technical Series No. 1)* (ed. B. M. Gibbs and F. A. Skinner) pp. 111–115. London: Academic Press.

Cochran, W. G. (1950). Estimation of bacterial densities by means of the 'most probable number'. *Biometrics* **6**, 105–116.

Codd, G. A. and Roberts, C. (eds) (1991). Public health aspects of cyanobacteria (blue-green algae). *PHLS Microbiology Digest* **8**, 78–100.

Collins, C. H. and Grange, J. M. (1983). A review: The bovine tubercle bacillus. *Journal of Applied Bacteriology* **55**, 13–29.

Collins, C. H., Allwood, M. C., Bloomfields, S. F. *et al.* (1981). *Disinfectants: Their Use and Evaluation of Effectiveness (Society for Applied Bacteriology, Technical Series No. 16)*. London: Academic Press.

Collins, M. D., Shah, H. N. and Minnikin, D. E. (1980). A note on the separation of natural mixtures of bacterial menaquinones using reverse phase thin-layer chromatography. *Journal of Applied Bacteriology* **48**, 277–282.

Collins, V. G. (1964). The fresh water environment and its significance in industry. *Journal of Applied Bacteriology* **27**, 143–150.

Collins, V. G. (1970). Recent studies of bacterial pathogens of freshwater fish. *Water Treatment and Examination* **19**, 3.

Collins, V. G., Jones, J. G., Hendrie, M. S. *et al.* (1973). Sampling and estimation of bacterial populations in the aquatic environment. In *Sampling–Microbiological Monitoring of Environments (Society for Applied Bacteriology Technical Series, No. 7)* (ed. R. G. Board and D. W. Lovelock). pp. 77–110. London: Academic Press.

Committee on Salmonella (1969). *An Evaluation of the* Salmonella *problem. A report of the US Department of Agriculture and the US Food and Drug Administration (Publication No. 1683)*. Washington, DC: National Academy of Sciences.

Cook, P. E. (1995). Fungal ripened meats and meat products. In *Fermented Meats* (ed. G. Campbell-Platt and P. E. Cook), pp. 110–129. Glasgow: Blackie.

Cooper, K. E. (1955). Theory of antibiotic inhibition zones in agar media. *Nature (London)* **176**, 510–511.

Corry, J. E. L., Roberts, D. and Skinner, F. A. (1982). *Isolation and Identification Methods for Food Poisoning Organisms (Society for Applied Bacteriology, Technical Series No. 17)*. London: Academic Press.

Corry, J. E. L., Curtis, G. D. W. and Baird, R. M. (Eds.) (1995a). *Culture Media for Food Microbiology (Progress in Industrial Microbiology, Vol. 34)*. Amsterdam: Elsevier.

Corry, J. E. L., Post, D. E., Colin, P. *et al.* (1995b). Culture media for the isolation of campylobacters. In *Culture Media for Food Microbiology (Progress in Industrial Microbiology, Vol. 34)* (ed. J. E. L. Corry, G. D. W. Curtis and R. M. Baird), pp. 129–162. Amsterdam: Elsevier.

Cowan, S. T. (1974). *Cowan and Steel's Manual for the Identification of Medical Bacteria*, 2nd edn. Cambridge: Cambridge University Press.

Cowan, S. T. and Steel, K. J. (1965). *Manual for the Identification of Medical Bacteria*. Cambridge: Cambridge University Press.

Cox, L. J., Keller, N. and van Schothorst, M. (1988). The use and misuse of quantitative determinations of Enterobacteriaceae in food microbiology. *Journal of Applied Bacteriology Symposium Supplement* **65S**, 237S–249S.

Craven, P. C., Mackel, D. C., Baine, W. B. *et al.* (1975). International outbreak of *Salmonella* eastbourne infection traced to contaminated chocolate. *Lancet* **i**, 788–793.

Crawford, R. J. M. (1962). Citrate-utilizing activity of certain starter bacteria. *Proceedings of the 16th Inernational Dairy Congress, Vol. B*, 322–330.

Crawford, R. J. M. and Galloway, J. H. (1964). Testing milk for antibiotic residues. *Dairy Industries* **29**, 256–262.

Croft, C. C. and Miller, M. J. (1956). Isolation of *Shigella* from rectal swabs with Hajna 'GN' broth. *American Journal of Clinical Pathology* **26**, 411.

Crombach, W. H. J. (1974). Morphology and physiology of coryneform bacteria. *Antonie van Leeuwenhoek* **40**, 361–376.

Croshaw, B. (1981). Disinfectant testing – with particular reference to the Rideal–Walker and Kelsey–Sykes tests. In *Disinfectants: Their Use and Evaluation of Effectiveness (Society for Applied Bacteriology, Technical Series No. 16)* (ed. C. H. Collins *et al*). pp. 1–15. London: Academic Press.

Cross, T. (1968). Thermophilic actinomycetes. *Journal of Applied Bacteriology* **31**, 36–53.

Crossley, E. L. (1941). The routine detection of certain spore-forming anaerobic bacteria in canned foods. *Journal of the Society of the Chemical Industry, London, Transactions* **60**, 131–136.

Cruickshank, R., Duguid, J. P., Marmion, B. P. *et al.* (1975). *Medical Microbiology, Vol. 2: The Practice of Medical Microbiology*, 12th edn. Edinburgh: Churchill Livingstone.

Cure, G. L. and Keddie, R. M. (1973). Methods for the morphological examination of aerobic coryneform bacteria. In *Sampling – Microbiological Monitoring of Environments Society for Applied Bacteriology, Technical Series No. 7)* (ed. R. G. Board and D. W. Lovelock), pp. 123–135. London: Academic Press.

Curtis, G. D. W., Mitchell, R. G., King, A. F. *et al.* (1989). A selective differential medium for the isolation of *Listeria monocytogenes*. *Letters in Applied Microbiology* **8**, 95–98.

DoE/SCA (1989). *Isolation and Identification* of Giardia *Cysts,* Cryptosporidium *Oocysts and Free Living Pathogenic Amoebae in Water etc., 1989 (Methods for the Examination of Waters and Associated Materials)*. Department of the Environment Standing Committee of Analysts. London: HMSO.

DoE/SCA (1995). *Methods for the Isolation and Identification of Human Enteric Viruses from Waters and Associated Materials 1995 (Methods for the Examination of Waters and Associated Materials)*. Department of the Environment Standing Committee of Analysts. London: HMSO.

DoE/SCA (1996). *Methods for the Isolation and Identification of* Escherichia coli *O157:H7 from Waters 1996 (Methods for the Examination of Waters and Associated Materials)*. Department of the Environment Standing Committee of Analysts. London: HMSO.

Dainty, R. H., Shaw, B. G. and Roberts, T. A. (1983). Microbial and chemical changes in chill-stored red meats. In *Food Microbiology: Advances and Prospects (Society for Applied Bacteriology, Symposium Series No. 11)* (ed. T. A. Roberts and F. A. Skinner), pp. 151–178. London: Academic Press.

Damgaard, P. H., Larsen, H. D., Hansen B. M. *et al.* (1996). Enterotoxin-producing strains of *Bacillus thuringiensis* isolated from food. *Letters in Applied Microbiology* **23**, 146–150.

Darlow, H. M. (1969). Safety in the microbiological laboratory. In *Methods in Microbiology, Vol. 1* (ed. J. R. Norris and D. W. Ribbons), pp. 169–204. London: Academic Press.

Darlow, H. M. (1972). Safety in the microbiological laboratory: an introduction. In *Safety in Microbiology (Society for Applied Bacteriology, Technical Series No. 6)* (ed. D. A. Shapton and R. G. Board), pp. 1–20. London: Academic Press.

Davies, F. L. and Wilkinson, G. (1973). *Bacillus cereus* in milk and dairy products. In *The Microbiological Safety of Food* (ed. by B. C. Hobbs and J. H. B. Christian), pp. 57–67. London: Academic Press.

Davis, B. R., Fanning, G. R., Madden, J. M. *et al.* (1981). Characterization of biochemically atypical *Vibrio cholerae* strains and designation of new pathogenic species, *Vibrio mimicus*. *Journal of Clinical Microbiology* **14**, 631–639.

Davis, J. G. (1958). A convenient semi-synthetic medium for yeast and mould counts. *Laboratory Practice* **7**, 30.

Davis, J. G. (1963). Microbiological standards for dairy products. *Journal of the Society of Dairy Technology* **16**, 150–155, 224–231.

Davis, J. G. (1968). Dairy products. In *Quality Control in the Food Industry, Vol. 2* (ed. S. M. Herschdoerfer), pp. 29–194. London: Academic Press.

Davis, J. G. (1970). Laboratory control of yoghurt. *Dairy Industries* **35**, 139–144.

Davis, J. G. and Wilbey, R. A. (1990). Microbiology of cream and dairy desserts. In *Dairy Microbiology, Vol. 2: The Microbiology of Milk Products* (ed. R. K. Robinson), pp. 41–108. London: Elsevier Applied Science.

Deak, T. and Beuchat, L. R. (1993). Comparison of the SIM, API20C and ID32C systems for identification of yeasts isolated from fruit juice concentrates and beverages. *Journal of Food Protection* **56**, 585–592.

Demeter, K. J. Sauer, F. and Miller, M. (1933). Vergleichende Untersuchungen über verschiedene Methoden zur coli-aerogenes-Titerbestimmung in Milch. *Milchwirtschaftliche Forschungen* **15**, 265–280.

DeSmedt, J., Bolderdijk, R. and Milas, R. (1994). *Salmonella* detection in cocoa and chocolate by motility enrichment on modified semi-solid Rappaport–Vassiliadis medium: collaborative study. *Journal of AOAC International* **77**, 365–373.

Devriese, L. A., Pot, B. and Collins, M. D. (1993). Phenotypic identification of the genus *Enterococcus* and differentiation of phylogenetically distinct enterococcal species and species groups. *Journal of Applied Bacteriology* **75**, 399–408.

Diaper, J. P. and Edwards, C. (1994). The use of fluorogenic esters to detect viable bacteria by flow cytometry. *Journal of Applied Bacteriology* **77**, 221–228.

Dixon, J. M. S. and Wilson, F. N. (1960). Salmonellae in fertilizers containing superphosphate. *Monthly Bulletin of the Ministry of Health and the Public Health Laboratory Service* **19**, 79–82.

Dowdell, M. J. and Board, R. G. (1968). A microbiological survey of British fresh sausage. *Journal of Applied Bacteriology* **31**, 378–396.

Dowdell, M. J. and Board, R. G. (1971). The microbial associations in British fresh sausages. *Journal of Applied Bacteriology* **34**, 317–337.

Druce, R. G. and Thomas, S. B. (1959). The microbiological examination of butter. *Journal of Applied Bacteriology* **22**, 52–56.

Druggan, P., Forsythe, S. J. and Silley, P. (1993). Indirect impedance for microbial screening in the food and beverage industries. In *New Techniques in Food and Beverage Microbiology (Society for Applied Bacteriology, Technical Series No. 31)* (ed. R. G. Kroll, A. Gilmour and M. Sussman), pp. 115–130. Oxford: Blackwell Scientific.

Duguid, J. P. and Wilkinson, J. F. (1961). Environmentally induced changes in bacterial morphology. In *Microbial Reaction to Environment (11th Symposium of the Society for General Microbiology)* (ed. G. G. Meynell and H. Gooder), pp. 69–99. Cambridge: Cambridge University Press.

Duncan, C. L. (1973). Time of enterotoxin formation and release during sporulation of *Clostridium perfringens* type A. *Journal of Bacteriology* **113**, 932–936.

EC (1980a). Council Directive of 15 July 1980 on the approximation of the laws of the Member States relating to the exploitation and marketing of natural mineral waters (80/777/EEC). *Official Journal of the European Communities* **23**, L229, 1–10.

EC (1980b). Council Directive of 15 July 1980 relating to the quality of water intended for human consumption. (Council Directive 80/778/EEC). *Official Journal of the European Communities* **23**, L229, 11–29.

EC (1991). Council Directive of 15 July 1991 laying down the health conditions for the production and the placing on the market of live bivalve molluscs (Council Directive 91/492/EEC). *Official Journal of the European Communities* **34**, L268, 1–14.

EC (1993). Council of the European Communities Commission Decision of 15 December 1992 on the microbiological criteria applicable to the production of cooked crustaceans and molluscan shellfish (93/51/EEC). *Official Journal of the European Communities* **36**, L13, 11–13.

EC (1996). Directive 96/70/EC of the European Parliament and of the Council of 28 October 1996 amending Council Directive 80/777/EEC on the approximation of the laws of the Member States relating to the exploitation and marketing of natural mineral waters. *Official Journal of the European Communities* **39**, L299, 26–28.

Eddy, B. P. (1960). The use and meaning of the term 'psychrophilic'. *Journal of Applied Bacteriology* **23**, 189–190.

Eddy, B. P. and Carpenter, K. P. (1964). Further studies on *Aeromonas*. II. Taxonomy of *Aeromonas* and C.27 strains. *Journal of Applied Bacteriology* **27**, 96–109.

Edwards, S. J. (1933). Studies on bovine mastitis. IX. A selective medium for the diagnosis of streptococcus mastitis. *Journal of Comparative Pathology and Therapeutics* **46**, 211–217.

Egdell, J. W., Thomas, S. B., Clegg, L. F. L. and Cuthbert, W. A. (1950). Thermoduric organisms in milk. Part III. Provisional standard test etc. *Proceedings of the Society for Applied Bacteriology* **13**, 132–134.

Ellner, P. D. (1956). A medium promoting rapid quantitative sporulation in *Clostridium perfringens. Journal of Bacteriology* **71**, 495–496.

Eloy, C. and Lacrosse, R. (1976). Composition d'un milieu de culture destine a effectuer le denombrement des micro-organismes thermophiles du yoghurt. *Bulletin de la Recherche Agronomique Gemblou* **11**, 83–86.

Euzeby, J. P. (1997). List of bacterial names with standing in nomenclature: a folder available on the Internet. *International Journal of Systematic Bacteriology* **47**, 590–592 (and on the Internet at URL/:ftp://ftp.cict.fr/pub/bacterio/).

Evans, H. J., Kwantes, W., Jenkins, D. C. *et al.* (1956). Sucrose loss from ice-cream on storage. *Analyst (London)* **81**, 204.

FAO (1991). *Manual of Food Quality Control. 12. Quality Assurance in the Food Control Microbiological Laboratory.* Rome: Food and Agriculture Organization of the United Nations.

FDA (1992). *Bacteriological Analytical Manual*, 7th edn. Arlington, Virginia: AOAC International.

Farrell, J. and Rose, A. H. (1967). Temperature effects in micro-organisms. In *Thermobiology* (ed. A. H. Rose), pp.147–218. London: Academic Press.

Favero, M. S. *et al.* (1968). Microbiological sampling of surfaces. *Journal of Applied Bacteriology* **31**, 336–343.

Feng, P. (1995). *Escherichia coli* serotype O157:H7: novel vehicles of infection and emergence of phenotypic variants. *Emerging Infectious Diseases* **1**, 47–52.

Fidler, J. C., Wilkinson, B. G., Edney, K. L. *et al.* (1973). *The Biology of Apple and Pear Storage (Commonwealth Bureau of Horticulture and Plantation Crops, Research Review No. 3).* Farnham Royal, Slough: Commonwealth Agricultural Bureaux.

Fifield, C. W. and Hoff, J. E. (1957). Dilute malachite green: a background stain for the Millipore filter. *Stain Technology* **32**, 95–96.

Findlay, W. P. K. (ed.) (1971). *Modern Brewing Technology.* London: Macmillan.

Finley, N. and Fields, M. L. (1962). Heat activation and heat-induced dormancy of *Bacillus stearothermophilus* spores. *Applied Microbiology* **10**, 231–236.

Firstenberg-Eden, R. and Eden, G. (1984). *Impedance Microbiology.* Letchworth, UK: Research Studies Press.

Fisher, P. J. (1963). The effect of freeze drying on the viability of *Chromobacterium lividum. Journal of Applied Bacteriology* **26**, 502–503.

Fisher, R. A. and Yates, F. (1963). *Statistical Tables for Biological, Agricultural and Medical Research*, 6th edn. Edinburgh: Oliver and Boyd.

Flanagan, P. A. (1992). *Giardia* – diagnosis, clinical course and epidemiology. A review. *Epidemiology and Infection* **109**, 1–22.

Franklin, J. G., Williams, D. J. and Clegg, L. F. L. (1956). A survey of the number and types of aerobic mesophilic spores in milk before and after commercial sterilization. *Journal of Applied Bacteriology* **19**, 46–53.

Fratamico, P. M., Buchanan, R. L. and Cooke, P. H. (1993). Virulence of an *Escherichia coli* O157:H7 sorbitol-positive mutant. *Applied and Environmental Microbiology* **59**, 4245–4252.

Freame, B. and Fitzpatrick, B. W. F. (1971). The use of differential reinforced clostridial medium for the isolation and enumeration of clostridia from food. In *Isolation of Anaerobes (Society for Applied Bacteriology, Technical Series No. 5)* (ed. D. A. Shapton and R. G. Board), pp. 49–55. London: Academic Press.

Fricker, C. R. (1987). The isolation of salmonellas and campylobacters. *Journal of Applied Bacteriology* **63**, 99–116.

Futter, B. V. (1967). *The Detection and Viability of Anaerobic Spores Surviving Bactericidal Influences*. PhD thesis (CNAA), Portsmouth College of Technology.

Futter, B. V. and Richardson, G. (1970). Viability of clostridial spores and the requirements of damaged organisms. II. Gaseous environment and redox potentials. *Journal of Applied Bacteriology* **33**, 331–341.

Futter, B. V. and Richardson, G. (1971). Anaerobic jars in the quantitative recovery of clostridia. In *Isolation of Anaerobes (Society for Applied Bacteriology, Technical Series No. 5)* (ed. D. A. Shapton and R. G. Board), pp. 81–91. London: Academic Press.

Gabis, D. A. and Silliker, J. H. (1974). ICMSF Methods Studies. II. Comparison of analytical schemes for detection of *Salmonella* in high-moisture foods. *Canadian Journal of Microbiology* **20**, 663–669.

Galesloot, T. E. and Hassing, F. (1962). Een snelle en gevoelige methode om met papierschijfjes penicilline in melk aante tonen. *Nederlands Melk-en Zuiveltijdschrift* **16**, 89–95.

Galesloot, T. E. and Stadhouders, J. (1968). The microbiology of spray-dried milk products with special reference to *Staphylococcus aureus* and salmonellae. *Nederlands Melk-en Zuiveltijdschrift* **22**, 158–172.

Galesloot, T. E., Hassing, F. and Stadhouders, J. (1961). Agar media voor net isoleren en tellen van aromabacteriën in zuursels. *Nederlands Melk-en Zuiveltijdschrift* **15**, 127–150.

Gardner, G. A. (1966). A selective medium for the enumeration of *Microbacterium thermosphactum* in meat and meat products. *Journal of Applied Bacteriology* **29**, 455–460.

Gardner, G. A. (1969). Physiological and morphological characteristics of *Kurthia zopfii* isolated from meat products. *Journal of Applied Bacteriology* **32**, 371–380.

Gardner, G. A. (1981). *Brochothrix thermosphacta* (*Microbacterium thermosphactum*) in the spoilage of meats. A review. In *Psychrotrophic Microorganisms in Spoilage and Pathogenicity* (ed. T. A. Roberts *et al.*), pp. 139–173. London: Academic Press.

Garin-Bastuji, B. and Verger, J. M. (1994). *Brucella abortus* and *Brucella melitensis*. In *The Significance of Pathogenic Microorganisms in Raw Milk (International Dairy Federation Monograph)*, pp. 167–185. Brussels: IDF.

Garvie, E. I. (1960). The genus *Leuconostoc* and its nomenclature. *Journal of Dairy Research* **27**, 283–292.

Geldreich, E. E. (1972). Water-borne pathogens. In *Water Pollution Microbiology* (ed. R. Mitchell), pp. 207–241. New York: Wiley–Interscience.

Georgala, D. L. and Boothroyd, M. (1969). Methods for the detection of salmonellae in meat and poultry. In *Isolation Methods for Microbiologists* (*Society for Applied Bacteriology, Technical Series, No. 3)* (ed. D. A. Shapton and G. W. Gould), pp. 29–39. London: Academic Press.

Gerhardt, P., Murray, R. G. E., Wood, W. A. *et al.* (1994). *Methods for General and Molecular Bacteriology.* Washington, DC: American Society for Microbiology.

Gibbs, B. M. and Freame, B. (1965). Methods for the recovery of clostridia in foods. *Journal of Applied Bacteriology* **28**, 95–111.

Gibbs, B. M. and Hirsch, A. (1956). Spore formation by *Clostridium* species in an artificial medium. *Journal of Applied Bacteriology* **19**, 129–141.

Gibson, T. and Abd-el-Malek, Y. (1945). The formation of carbon dioxide by lactic acid bacteria and *Bacillus licheniformis* and a cultural method of detecting the process. *Journal of Dairy Research* **14**, 35–44.

Gilbert, R. J., Kendall, M. and Hobbs, B. C. (1969). Media for the isolation and enumeration of coagulase-positive staphylococci from foods. In *Isolation Methods for Microbiologists (Society for Applied Bacteriology, Technical Series No. 3)* (eds. D. A. Shapton and G. W. Gould), pp. 9–15. London: Academic Press.

Gilbert, R. J., Stringer, M. F. and Peace, T. C. (1974). The survival and growth of *Bacillus cereus* in boiled and fried rice in relation to outbreaks of food poisoning. *Journal of Hygiene (Cambridge)* **73**, 433–444.

Giolitti, G. and Cantoni, C. (1966). A medium for the isolation of staphylococci from foodstuffs. *Journal of Applied Bacteriology* **29**, 395–398.

Goepfert, J. M., Spira, W. M. Glatz, B. A. *et al.*, (1973). Pathogenicity of *Bacillus cereus*. In *The Microbiological Safety of Food* (eds. B. C. Hobbs and J. H. B. Christian), pp. 69–75. London: Academic Press.

Golden, D. A., Rhodehamel, E. J. and Kautter, D. A. (1993). Growth of *Salmonella* spp. in canteloupe, watermelon, and honeydew melon. *Journal of Food Protection* **56**, 194–196.

Gomez, R. F., Sinskey, A. J., Davies, R. *et al.* (1973). Minimal medium recovery of heated *Salmonella typhimurium* LT 2. *Journal of General Microbiology* **74**, 267–274.

Goodfellow, M. and Board, R. G. (eds) (1980). *Microbiological Classification and Identification (Society for Applied Bacteriology, Symposium No. 8)*. London: Academic Press.

Gopal Rao, G., Saunders, B. P. and Masterton, R. G. (1996). Laboratory acquired verotoxin producing *Escherichia coli* (VTEC) infection. *Journal of Hospital Infection* **33**, 228–230.

Gordon, R. E. and Mihm, J. M. (1957). A comparative study of some strains received as Nocardiae. *Journal of Bacteriology* **73**, 15–27.

Gordon, R. E. and Mihm, J. M. (1959). A comparison of four species of mycobacteria. *Journal of General Microbiology* **21**, 736–748.

Gordon, R. E. and Mihm, J. M. (1962). The type species of the genus *Nocardia. Journal of General Microbiology* **27**, 1–10.

Gordon, R. E., Haynes, W. C. and Pang, C. H.-N. (1973). *The Genus* Bacillus *(Agriculture Handbook No. 427)*. Washington, DC: US Department of Agriculture.

Gorrill, R. H. and McNeil, E. M. (1960). The effect of cold diluent on the viable count of *Pseudomonas pyocyanea. Journal of General Microbiology* **22**, 437–442.

Goudkov, A. V. and Sharpe, M. E. (1965). Clostridia in dairying. *Journal of Applied Bacteriology* **28**, 63–73.

Gould, G. W. (1971). Methods for studying bacterial spores. In *Methods in Microbiology, Vol. 6A* (ed. J. R. Norris and D. W. Ribbons), pp. 327–381. London: Academic Press.

Grange, J. M. (1990). Tuberculosis. In *Topley and Wilson's Principles of Bacteriology, Virology and Immunity, Volume 3, Bacterial Diseases*, 8th ed., edited by G. R. Smith and C. S. F. Easman, pp. 93–121 London: Edward Arnold.

Grange, J. M., Fox, A. and Morgan, N. L. (eds.) (1987). *Immunological Techniques in Microbiology (Society for Applied Bacteriology, Technical Series No. 24)*. Oxford: Blackwell Scientific.

Grant, K. A., Dickinson, J. H. and Kroll, R. G. (1993). Specific rapid detection of foodborne bacteria with rRNA sequences and the polymerase chain reaction. In *New Techniques in Food and Beverage Microbiology (Society for Applied Bacteriology, Technical Series No. 31)* (eds. Kroll, R. G., Gilmour, A. and Sussman, M.), pp. 147–162. Oxford: Blackwell Scientific.

Greenwood, M. H. and Hooper, W. L. (1983). Chocolate bars contaminated with Salmonella napoli: an infectivity study. *British Medical Journal* **286**, 1394.

Gregory, P. H., Lacey, M. E., Festenstein, G. N. *et al.* (1963). Microbial and biochemical changes during the moulding of hay. *Journal of General Microbiology* **33**, 147–174.

Griffith, C. J., Blucher, A., Fleri, J. *et al.* (1994). An evaluation of luminometry as a technique in food microbiology and a comparison of six commercially available luminometers. *Food Science and Technology Today* **8**(4), 209–216.

Griffiths, M. W., Phillips, J. D. and Muir, D. D. (1981). Thermostability of proteases and lipases from a number of species of psychrotrophic bacteria of dairy origin. *Journal of Applied Bacteriology* **50**, 289–303.

Grinsted, E. and Clegg, L. F. L. (1955). Spore-forming organisms in commercial sterilized milk. *Journal of Dairy Research* **22**, 178–190.

Guerrant, R. L. (1997). Cryptosporidiosis: an emerging, highly infectious threat. *Emerging Infectious Diseases* **3**, 51–57.

Günther, H. L. (1959). Mode of division of pediococci. *Nature (London)* **183**, 903.

Günther, H. L. and White, H. R. (1961). The cultural and physiological characters of the pediococci. *Journal of General Microbiology* **26**, 185–197.

Günzler, H. (ed.) (1996). *Accreditation and Quality Assurance in Analytical Chemistry* (trans. by G. Lapitajs). Berlin: Springer.

Hadfield, S. G., Jouy, N. F. and McIllmurray, M. B. (1987). The application of a novel coloured latex test to the detection of *Salmonella*. In *Immunological Techniques in Microbiology (Society for Applied Bacteriology, Technical Series No. 24)* (ed. J. M. Grange, A. Fox and N. L. Morgan), pp. 145–151. Oxford: Blackwell Scientific.

Hajna, A. A. (1945). Triple-Sugar Iron Agar medium for the identification of the intestinal group of bacteria. *Journal of Bacteriology* **49**, 516–517.

Hajna, A. A. (1955). A new enrichment broth medium for Gram-negative organisms of the intestinal group. *Public Health Laboratory* **13**, 83.

Hajna, A. A. and Perry, C. A. (1943). Comparative study of presumptive and confirmative media for bacteria of the coliform group and for fecal streptococci. *American Journal of Public Health* **33**, 550.

Hanks, J. H. and James, D. F. (1940). The enumeration of bacteria by the microscopic method. *Journal of Bacteriology* **39**, 297–305.

Hannay, C. L. and Norton, I. L. (1947). Enumeration, isolation and study of faecal streptococci from river water. *Proceedings of the Society for Applied Bacteriology* No. 1, 39–45.

Hargrove, R. E. (1959). A simple method for eliminating and controlling bacteriophage in lactic starters. *Journal of Dairy Science* **42**, 906.

Harrigan, W. F. (1966). The nutritional requirements and biochemical reactions of *Corynebacterium bovis*. *Journal of Applied Bacteriology* **29**, 380–394.

Harrigan, W. F. (1967). *A Study of Coryneform Bacteria from Milk and Other Sources with Particular Reference to* Corynebacterium bovis. PhD thesis, University of Glasgow.

Harrigan, W. F. and Park, R. W. A. (1991). *Making Safe Food: A Management Guide for Microbiological Quality*. London: Academic Press.

Harris, R. F. and Somers, L. E. (1968). Plate-dilution frequency technique for assay of microbial ecology. *Applied Microbiology* **16**, 330–334.

Harrison, J. (1938). Numbers and types of bacteria in cheese. *Proceedings of the Society of Agricultural Bacteriologists*, 12–14.

Harvey, R. W. S. (1965). *A Study of the Factors Governing the Isolation of Salmonellae from Infected Materials and the Application of Improved Techniques to Epidemiological Problems*. MD thesis, University of Edinburgh. *Cited by* Harvey and Price (1974).

Harvey, R. W. S. and Phillips, W. P. (1961). An environmental survey of bakehouses and abattoirs for salmonellae. *Journal of Hygiene (Cambridge)* **59**, 93–103.

Harvey, R. W. S. and Price, T. H. (1961). An economical and rapid method for H antigen phase change in the *Salmonella* group. *Monthly Bulletin of the Ministry of Health and the Public Health Laboratory Service* **20**, 11–13.

Harvey, R. W. S. and Price, T. H. (1967). The isolation of salmonellas from animal feeding stuffs. *Journal of Hygiene (Cambridge)* **65**, 237–244.

Harvey, R. W. S. and Price, T. H. (1968). Elevated temperature incubation of enrichment media for the isolation of salmonellas from heavily contaminated materials. *Journal of Hygiene (Cambridge)* **66**, 377–381.

Harvey, R. W. S. and Price, T. H. (1970). Sewer and drain swabbing as a means of investigating salmonellosis. *Journal of Hygiene (Cambridge)* **68**, 611–624.

Harvey, R. W. S. and Price, T. H. (1974). *Isolation of Salmonellas (Public Health Laboratory Service, Monograph Series, No. 8)*. London: HMSO.

Hauge, S. (1955). Food poisoning caused by aerobic spore-forming bacilli. *Journal of Applied Bacteriology* **18**, 591–595.

Hayes, G. L. and Reister, D. W. (1952). The control of 'off-odor' spoilage in frozen concentrated orange juice. *Food Technology* **6**, 386.

Hayes, P. R. (1963). Studies on marine flavobacteria. *Journal of General Microbiology* **30**, 1–19.

Hendrie, M. S., Hodgkiss, W. and Shewan, J. M. (1970). The identification, taxonomy and classification of luminous bacteria. *Journal of General Microbiology* **64**, 151–169.

Hersom, A. C. and Hulland, E. D. (1969). *Canned Foods (Baumgartner)*, 6th edn. London: JA Churchill.

Hess, E. (1934). Effects of low temperatures on the growth of marine bacteria. In *Contributions to Canadian Biology and Fisheries, No. 34 (Series C Industrial No. 22)*, Vol. 8, pp. 489–505.

Hickman, F. W., Farmer, J. J., Hollis, D. G. *et al.* (1982). Identification of *Vibrio hollisae* sp. nov. from patients with diarrhea. *Journal of Clinical Microbiology* **15**, 395–401.

Higgins, M. (1950). A comparison of the recovery rate of organisms from cotton-wool and calcium alginate wool swabs. *Monthly Bulletin of the Ministry of Health and the Public Health Laboratory Service* **9**, 50–51.

Hill, E. C. and Wenzel, F. W. (1957). The diacetyl test as an aid for quality control of citrus products. I. Detection of bacterial growth in orange juice during concentration. *Food Technology* **11**, 240.

Hirsch, A. and Grinsted, E. (1954). Methods for the growth and enumeration of anaerobic spore-formers from cheese, with observations on the effect of nisin. *Journal of Dairy Research* **21**, 101–110.

Hobbs, G. (1981). The ecology and taxonomy of psychrotrophic strains of *Clostridium botulinum*. In *Psychrotrophic Microorganisms in Spoilage and Pathogenicity* (ed. T. A. Roberts *et al.*), pp. 449–462. London: Academic press.

Hobbs, G. (1983). Microbial spoilage of fish. In *Food Microbiology: Advances and Prospects (Society for Applied Bacteriology, Symposium No. 11)* (ed. T. A. Roberts and F. A. Skinner), pp. 217–229. London: Academic Press.

Hobbs, G., Williams, K. and Willis, A. T. (1971). Basic methods for the isolation of clostridia. In *Isolation of Anaerobes (Society for Applied Bacteriology, Technical Series No. 5)* (ed. D. A. Shapton and R. G. Board), pp. 1–23. London: Academic Press.

Hoben, D. A., Ashton, D. H. A. and Peterson, A. C. (1973). A rapid, presumptive procedure for the detection of *Salmonella* in foods and food ingredients. *Applied Microbiology* **25**, 123–129.

Hocking, A. D. and Pitt, J. I. (1980). Dichloran–glycerol medium for enumeration of xerophilic fungi from low-moisture foods. *Applied and Environmental Microbiology* **39**, 488–492.

Holbrook, R. and Anderson, J. M. (1980). An improved selective and diagnostic medium for the isolation and enumeration of *Bacillus cereus* in foods. *Canadian Journal of Microbiology* **26**, 753–759.

Holbrook, R., Anderson, J. M. and Baird-Parker, A. C. (1969). The performance of a stable version of Baird-Parker's medium for isolating *Staphylococcus aureus. Journal of Applied Bacteriology* **32**, 187–192.

Holbrook, R., Anderson, J. M., Baird-Parker, A. C. *et al.* (1989a). Rapid detection of salmonella in foods – a convenient two-day procedure. *Letters in Applied Microbiology* **8**, 139–142.

Holbrook, R., Anderson, J. M., Baird-Parker, A. C. *et al.* (1989b). Comparative evaluation of the Oxoid Salmonella Rapid Test with three other rapid salmonella methods. *Letters in Applied Microbiology* **9**, 161–164.

Holding, A. J. (1960). The properties and classification of the predominant Gram-negative bacteria occurring in soil. *Journal of Applied Bacteriology* **23**, 515–525.

Holmes, B. and Costas, M. (1992). Identification and typing of Enterobacteriaceae by computerized methods. In *Identification Methods in Applied and Environmental Microbiology (Society for Applied Bacteriology, Technical Series No. 29)* (ed. R. G. Board, D. Jones and F. A. Skinner), pp. 127–149. Oxford: Blackwell Scientific.

Holt, J. G., Kreig, N. R., Sneath, P. H. A. *et al.* (1994). *Bergey's Manual of Determinative Bacteriology*, 9th edn. Baltimore, Maryland: Williams and Wilkins.

Hoyle, D. (1994). *ISO 9000 Quality Systems Handbook*, 2nd edn. Oxford: Butterworth–Heinemann.

Howie, J. W. (1968). Typhoid in Aberdeen, 1964. *Journal of Applied Bacteriology* **31**, 171–178.

Hugh, R. and Leifson, E. (1953). The taxonomic significance of fermentative versus oxidative metabolism of carbohydrate by various Gram negative bacteria. *Journal of Bacteriology* **66**, 24–26.

Huhtanen, C. N. (1975). Use of pH gradient plates for increasing the acid tolerance of salmonellae. *Applied Microbiology* **29**, 309–312.

Humphrey, T. J. (1986). Techniques for the optimum recovery of cold injured *Campylobacter jejuni* from milk or water. *Journal of Applied Bacteriology* **61**, 125–132.

Hungate, R. E. (1969). A roll-tube method for cultivation of strict anaerobes. In *Methods in Microbiology, Vol. 3B* (ed. J. R. Norris and D. W. Ribbons), pp. 117–132. London: Academic Press.

Huq, M. I., Alam, A. K. M. J., Brenner, D. J. *et al.* (1980). Isolation of *Vibrio*-like group, EF6, from patients with diarrhea. *Journal of Clinical Microbiology* **11**, 621–624.

Hutchinson, D. N. and Bolton, F. J. (1983). Improved blood-free selective medium for the isolation of *Campylobacter jejuni* from faecal specimens. *Journal of Clinical Pathology* **37**, 956–957.

ICMSF (1974). *Microorganisms in Foods. 2. Sampling for Microbiological Analysis: Principles and Specific Applications.* International Commission on Microbiological Specifications for Foods of the International Association of Microbiological Societies. Toronto: University of Toronto Press.

ICMSF (1978). *Microorganisms in Foods. 1. Their Significance and Methods of Enumeration*, 2nd edn. Toronto: University of Toronto Press.

ICMSF (1980a). *Microbial Ecology of Foods. Vol. 1: Factors Affecting Life and Death of Microorganisms*. New York: Academic Press.

ICMSF (1980b). *Microbial Ecology of Foods. Vol. 2: Food Commodities*. New York: Academic Press.

ICMSF (1986). *Microorganisms in Foods. 2. Sampling for Microbiological Analysis: Principles and Specific Applications*, 2nd edn. Oxford: Blackwell Scientific.

ICMSF (1996). *Microorganisms in Foods. 5. Microbiological Specifications of Food Pathogens.* London: Blackie.

IFST (1992). *Guidelines to Good Catering Practice*. London: Institute of Food Science and Technology.

IFST (1993). *Shelf Life of Foods – Guidelines for its Determination and Prediction*. London: Institute of Food Science and Technology.

IPCS (1990). *Selected Mycotoxins: Ochratoxins, Trichothecenes, Ergot (International Programme on Chemical Safety: Environmental Health Criteria No. 105)*. Geneva: World Health Organization.

Ingraham, J. L. and Stokes, J. L. (1959). Psychrophilic bacteria. *Bacteriological Reviews* **23**, 97–108.

Ingram, M. and Kitchell, A. G. (1970). Introductory paper to a symposium on microbiological standards for foods. *Chemistry and Industry* **2**, 186–188.

International Dairy Federation (1962). *Standard Suspension Test for the Evaluation of the Disinfectant Activity of Dairy Disinfectants*. FIL-IDF19:1962. Brussels: IDF.

International Dairy Federation (1993). *Analytical Quality Assurance and Good Laboratory Practice in Dairy Laboratories (Proceedings of an International Seminar, Sonthofen, Germany, May 1992)*. Brussels: IDF.

Jackson, G. J., Bier, J. W., Payne, W. L. *et al.* (1981). Recovery of parasitic nematodes from fish by digestion or elution. *Applied and Environmental Microbiology* **41**, 912–914.

Jackson, S. G., Goodbrand, R. B., Ahmed, R. *et al.* (1995). *Bacillus cereus* and *Bacillus thuringiensis* isolated in a gastroenteritis outbreak investigation. *Letters in Applied Microbiology* **21**, 103–105.

Jarvis, B., Lach, V. H. and Wood, J. M. (1977). Evaluation of the spiral plate maker for the enumeration of micro-organisms in foods. *Journal of Applied Bacteriology* **43**, 149–157.

Jayne-Williams, D. J. and Skerman, T. M. (1966). Comparative studies on coryneform bacteria from milk and dairy sources. *Journal of Applied Bacteriology* **29**, 72–92.

Jeppesen, C. (1995). Media for *Aeromonas* spp., *Plesiomonas shigelloides* and *Pseudomonas* spp. from food and environment. In *Culture Media for Food Microbiology (Progress in Industrial Microbiology, Vol. 34)* (ed. J. E. L. Corry, G. D. W. Curtis and R. M. Baird), pp. 111–127. Amsterdam: Elsevier.

Jinneman, K. C. and Hill, W. E. (1997). Applications of gene probes for the detection of foodborne pathogens. In *Food Microbiological Anslysis: New Technologies* (ed. M. L. Tortorello and S. M. Gendel), pp. 115–181. New York: Dekker.

Johnson, E. A. (1946). An improved slide culture technique for the study and identification of pathogenic fungi. *Journal of Bacteriology* **51**, 689–694.

Jones, D. (1975). A numerical taxonomic study of coryneform and related bacteria. *Journal of General Microbiology* **87**, 52–96.

Jones, D. (1978). An evaluation of the contribution of numerical taxonomic studies to the classification of coryneform bacteria. In *Coryneform Bacteria* (ed. I. J. Bousfield and A. G. Callely), pp. 13–46. London: Academic Press.

Jones, A. and Richards, T. (1952). Night blue and Victoria blue as indicators in lipolysis media. *Proceedings of the Society of Applied Bacteriology* **15**, 82–93.

Kapikian, A. Z. (1994). *Viral Infections of the Gastrointestinal Tract*, 2nd edn. New York: Marcel Dekker.

Kapperud, G. (1991). *Yersinia enterocolitica* in food hygiene. *International Journal of Food Microbiology* **12**, 53–65.

Kapperud, G. and Bergan, T. (1984). Biochemical and serological characterization of *Yersinia enterocolitica*. In *Methods in Microbiology, Vol. 15* (ed. T. Bergan), pp. 295–344. London: Academic Press.

Karmali, M. A., Steele, B. J., Petric, M. *et al.* (1983). Sporadic cases of haemolytic–uraemic syndrome associated with faecal cytotoxin and cytotoxin-producing *Escherichia coli* in stools. *Lancet* **i**, (8325), 619–620.

Kasuga, F., Hara-Kudo, Y. and Machii, K. (1996) Evaluation of enzyme-linked immunosorbent assay (ELISA) kit for paralytic shellfish poisoning toxins. *Journal of the Food Hygiene Society of Japan* **37**, 407–410.

Keddie, R. M. (1951). The enumeration of lactobacilli on grass and in silage. *Proceedings of the Society for Applied Bacteriology* **14**, 157–160.

Keddie, R. M. and Bousfield, I. J. (1980). Cell wall composition in the classification and identification of coryneform bacteria. In *Microbiological Classification and Identification (Society for Applied Bacteriology, Symposium No. 8)* (ed. M. Goodfellow and R. G. Board), pp. 167–189. London: Academic Press.

Keddie, R. M. and Cure, G. L. (1978). Cell wall composition of coryneform bacteria. In *Coryneform Bacteria* (ed. I. J. Bousfield and A. G. Callely), pp. 47–83. London: Academic Press.

Kenner, B. A., Clark, H. F. and Kabler, P. W. (1961). Fecal streptococci. I. Cultivation and enumeration of streptococci in surface waters. *Applied Microbiology* **9**, 15–20.

King, E. O., Ward, M. K. and Raney, D. E. (1954). Two simple media for the demonstration of pyocyanin and fluorescin. *Journal of Laboratory and Clinical Medicine* **44**, 301.

King, S. and Metzger, W. (1968). A new plating medium for the isolation of enteric pathogens. I. Hektoen enteric Agar. *Applied Microbiology* **16**, 577–578.

Kirov, S. M. (1993). The public health significance of *Aeromonas* spp. in foods. *International Journal of Food Microbiology* **20**, 179–198.

Kirsop, B. H. and Dolezil, L. (1975). Detection of lactobacilli in brewing. In *Lactic Acid Bacteria in Beverages and Food* (ed. J. G. Carr, C. V. Cutting and G. C. Whiting), pp. 159–164. London: Academic Press.

Kirsop, B. E. and Snell, J. J. S. (1984). *Maintenance of Microorganisms: A Manual of Laboratory Methods*. London: Academic Press.

Klaenhammer, T. R. and Fitzgerald, G. F. (1994). Bacteriophages and bacteriophage resistance. In *Genetics and Biotechnology of Lactic Acid Bacteria*, (ed. M. J. Gasson and W. M. de Vos), pp. 106–168. Glasgow: Blackie.

Kliks, M. M. (1983). Anisakiasis in the western United States: four new case reports from California. *American Journal of Tropical Medicine and Hygiene* **32**, 526–532.

Knisely, R. F. (1965). Differential media for the identification of *Bacillus anthracis. Journal of Bacteriology* **90**, 1778–1783.

Koch, A. L. (1984). Turbidity measurements in microbiology. *ASM News* **50**, 473–477.

Kogure, K., Simidu, U. and Taga, N. (1979). A tentative direct microscopic method for counting living marine bacteria. *Canadian Journal of Microbiology* **25**, 415–420.

Kogure, K., Simidu, U. and Taga, N. (1984). An improved direct viable count method for aquatic bacteria. *Archiv für Hydrobiologie* **102**, 117–122.

Konowalchuk, J., Speirs, J. I. and Stavric, S. (1977). Vero response to a cytotoxin of *Escherichia coli. Infection and Immunity* **18**, 775–9.

Koper, J. W., Hagenaars, A. M. and Notermans, S. (1980). Prevention of cross-reactions in the

enzyme-linked immunosorbent assay (ELISA) for the detection of *Staphylococcus aureus* enterotoxin Type B in culture filtrates and food. *Journal of Food Safety* **2**, 35–45.
Kovacs, N. (1956). Identification of *Pseudomonas pyocyanea* by the oxidase reaction. *Nature (London)* **178**, 703.
Kramer, J. M., Turnbull, P. C. B., Munshi, G. *et al.* (1982). Identification and characterization of *Bacillus cereus* and other *Bacillus* species associated with foods and food poisoning. In *Isolation and Identification Methods for Food Poisoning* (*Society for Applied Bacteriology, Technical Series No. 17*) (ed. J. E. L. Corry, D. Roberts and F. A. Skinner), pp. 261–286. London: Academic Press.
Kreger-van Rij, N. J. W. (ed.) (1984). *The Yeasts: A Taxonomic Study*, 3rd edn. Amsterdam: Elsevier.
Krieg, N. R. *et al.* (1984). *Bergey's Manual of Systematic Bacteriology, Vol. 1.* Baltimore, Maryland: Williams and Wilkins.
Kroll, R. G., Gilmour, A. and Sussman, M. (eds) (1993). *New Techniques in Food and Beverage Microbiology (Society for Applied Bacteriology, Technical Series No. 31).* Oxford: Blackwell Scientific.
Kudva, I. T., Hatfield, P. G. and Hovde, C. J. (1996). *Escherichia coli* O157:H7 in microbial flora of sheep. *Journal of Clinical Microbiology* **34**, 431–433.
Kuiper-Goodman, T. (1994). Prevention of human mycotoxicoses through risk assessment and risk management. In *Mycotoxins in Grain: Compounds Other Than Aflatoxin* (ed. J. D. Miller and H. L. Trenholm), pp. 439–469. St Paul, Minnesota: Eagan Press.
Laing, C. M. and Malcolm, J. F. (1956). The incidence of bovine mastitis with special reference to the non-specific condition. *Veterinary Record* **68**, 447–455.
Lapage, S. P., Shelton, J. E. and Mitchell, T. G. (1970a). Media for the maintenance and preservation of bacteria. In *Methods in Microbiology, Vol. 3A* (ed. J. R. Norris and D. W. Ribbons), pp. 1–133. London: Academic Press.
Lapage, S. P., Shelton, J. E., Mitchell, T. G. *et al.* (1970b). Culture collections and the preservation of bacteria. In *Methods in Microbiology, Vol. 3A* (ed. J. R. Norris and D. W. Ribbons), pp. 135–228. London: Academic Press.
Lauwers, S. De Boeck, M. and Butzler, J. P. (1978). *Campylobacter* enteritis in Brussles. *Lancet* **i**, 604–605.
Lederberg, J. and Lederberg, E. M. (1952). Replica plating and indirect selection of bacterial mutants. *Journal of Bacteriology* **63**, 399–406.
Lee, J. V., Donovan, T. J. and Furniss, A. L. (1978). Characterization, taxonomy, and emended description of *Vibrio metschnikovii*. *International Journal of Systematic Bacteriology* **28**, 99–111.
Lee, J. V., Shread, P., Furniss, A. L. *et al.* (1981). Taxonomy and description of *Vibrio fluvialis* sp. nov. (synonym group F vibrios, group EF6). *Journal of Applied Bacteriology* **50**, 73–94.
Lee, R., Peppe, J. and George, H. (1998). Pulsed-field gel electrophoresis of genomic digests demonstrates linkages among food, food handlers, and patrons in a food-borne *Salmonella javiana* outbreak in Massachusetts. *Journal of Clinical Microbiology* **36**, 284–285.
Leifson, E. (1951). Staining, shape, and arrangement of bacterial flagella. *Journal of Bacteriology* **62**, 377–389.
Leifson, E. (1958). Timing of the Leifson flagella stain. *Stain Technology* **33**, 249.
Lepper, E. and Martin, C. J. (1929). The chemical mechanisms exploited in the use of meat media for the cultivation of anaerobes. *British Journal of Experimental Pathology* **10**, 327–334.
Levine, M. M. (1987). *Escherichia coli* that cause diarrhea: enterotoxigenic, enteropathogenic, enteroinvasive, enterohemorrhagic and enteroadherent. *Journal of Infectious Diseases* **155**, 377–389.
Lewis, D. A., Paramathasan, R., White, D. G. *et al.* (1996). Marshmallows cause an outbreak of infection with *Salmonella enteritidis* phage type 4. *CDSC Communicable Disease Report Review* **6**, R183–R186.
Linton, A. H. (1961). Interpreting antibiotic sensitivity tests. *Journal of Medical Laboratory Technology* **18**, 1–20.

Lodder, J. (1970). *The Yeasts: A Taxonomic Study*. Amsterdam: North-Holland.

Lodder, J. and Kreger-van Rij, N. J. W. (1952). *The Yeasts – A Taxonomic Study*. Amsterdam: North-Holland.

Logan, N. A. and Berkeley, R. C. W. (1981). Classifcation and identification of members of the genus *Bacillus* using API tests. In *The Aerobic Endospore-Forming Bacteria: Classification and Identification* (ed. R. C. W. Berkeley and M. Goodfellow), pp. 105–140. London: Academic Press.

Logan, N. A. and Moss, M. O. (1992). Identification of *Chromobacterium*, *Janthinobacterium* and *Iodobacter* species. In *Identification Methods in Applied and Environmental Microbiology (Society for Applied Bacteriology, Technical Series No. 29)* (ed. R. G. Board, D. Jones and F. A. Skinner), pp. 182–192. Oxford: Blackwell Scientific.

Lowe, G. H. (1962). The rapid detection of lactose fermentation in paracolon organisms by the demonstration of β-D-galactosidase. *Journal of Medical Laboratory Technology* **19**, 21–25.

Luxon, S. G. (1992). *Hazards in the Chemical Laboratory*, 5th edn. Cambridge: Royal Society of Chemistry.

Lück, H. (1972). Bacteriological quality tests for bulk-cooled milk. *Dairy Science Abstracts* **34**, 101–122.

Mabbitt, L. A. and Zielinska, M. (1956). The use of a selective medium for the enumeration of lactobacilli in Cheddar cheese. *Journal of Applied Bacteriology* **19**, 95–101.

MacFaddin, J. F. (1980). *Biochemical Tests for Identification of Medical Bacteria*, 2nd edn. Baltimore, Maryland: Williams and Wilkins.

MacFaddin, J. F. (1985). *Media for Isolation – Cultivation – Identification – Maintenance of Medical Bacteria*. Baltimore, Maryland: Williams and Wilkins.

MacKelvie, R. M., Gronlund, A. F. and Campbell, J. J. R. (1968). Influence of cold shock on the endogenous metabolism of *Pseudomonas aeruginosa. Canadian Journal of Microbiology* **14**, 633–638.

Mackenzie, E. F. W., Taylor, E. W. and Gilbert, W. E. (1948). Recent experiences in the rapid identification of *Bacterium coli* type 1. *Journal of General Microbiology* **2**, 197–204.

Mackey, B. M. (1984). Lethal and sublethal effects of refrigeration, freezing and freeze-drying on micro-organisms. In *The Revival of Injured Microbes (Society for Applied Bacteriology Symposium Series No. 12)* (ed. M. H. E. Andrew and A. D. Russell), pp. 45–75. London: Academic Press.

McLean, J. and Henderson, A. (1966). Test for the presence of nitrite not involving carcinogenic reagents. *Journal of Clinical Pathology* **19**, 632.

MacLeod, R. A. (1965). The question of the existence of specific marine bacteria. *Bacteriological Reviews* **29**, 9–23.

McLauchlin, J. (1997), The identification of *Listeria* species. *International Journal of Food Microbiology* **38**, 77–81.

McMeekin, T. A. and Shewan, J. M. (1978). Taxonomic strategies for *Flavobacterium* and related genera. *Journal of Applied Bacteriology* **45**, 321–332.

McMeekin, T. A., Patterson, J. T. and Murray, J. G. (1971). An initial approach to the taxonomy of some Gram negative yellow pigmented rods. *Journal of Applied Bacteriology* **34**, 699–716.

McMeekin, T. A., Stewart, D. B. and Murray, J. G. (1972). The Adansonian taxonomy and the deoxyribonucleic acid base composition of some Gram negative, yellow pigmented rods. *Journal of Applied Bacteriology* **35**, 129–137.

Madden, J. M. (1992). Microbial pathogens in fresh produce – the regulatory perspective. *Journal of Food Protection* **55**, 821–823.

Mallidis, C. G. and Scholefield, J. (1985). Determination of the heat resistance of spores using a solid heating block system. *Journal of Applied Bacteriology* **59**, 407–411.

Mallman, W. L. and Darby, C. W. (1941). Uses of lauryl sulphate tryptose broth for the detection of coliform organisms. *American Journal of Public Health* **31**, 127.

Man, C. M. D. and Jones, A. A. (1994). *Shelf Life Evaluation of Foods*. Glasgow: Blackie.

de Man, J. C. (1975). The probability of most probable numbers. *European Journal of Applied Microbiology* **1**, 67–78.

de Man, J. C. (1977). MPN tables for more than one test. *European Journal of Applied Microbiology* **4**, 307–316.

de Man, J. C. Rogosa, M. and Sharpe, M. E. (1960). A medium for the cultivation of lactobacilli. *Journal of Applied Bacteriology* **23**, 130–135.

Manson-Bahr, P. E. C. and Bell, D. R. (1995). *Manson's Tropical Diseases*, 20th edn. London: WB Saunders.

Massad, G. and Oliver, J. (1987). New selective and differential medium for *Vibrio cholerae* and *Vibrio vulnificus*. *Applied and Environmental Microbiology* **53**, 2262–2264.

Mäyrä-Mäkinen, A. and Bigret, M. (1993). Industrial use and production of lactic acid bacteria. In *Lactic Acid Bacteria* (ed. S. Salminen and A. von Wright), pp. 65–95. New York: Marcel Dekker.

Mazigh, D. (1994). Microbiology of chocolate. In *Industrial Chocolate Manufacture and Use*, 2nd edn (ed. S. T. Beckett), pp. 312–320. Glasgow: Blackie.

Mead, G. C. (1985). Enumeration of pseudomonads using cephaloridine–fucidine–cetrimide agar (CFC). *International Journal of Food Microbiology* **2**, 21–26.

Meers, P. D. and Goode, D. (1965). The influence of preservative salts on the anaerobic growth of *Salmonella typhi*. *Monthly Bulletin of the Ministry of Health and Public Health Laboratory Service, London* **24**, 334–336.

Merino, S. Rubires, X., Knøchel, S. *et al.* (1995). Emerging pathogens: *Aeromonas* spp. *International Journal of Food Microbiology* **28**, 157–168.

Meynell, G. G. (1958). The effect of sudden chilling on *Escherichia coli*. *Journal of General Microbiology* **19**, 380–389.

Meynell, G. G. and Meynell, E. (1970). *Theory and Practice in Experimental Bacteriology*, 2nd edn. Cambridge: Cambridge University Press.

Miles, A. A. and Misra, S. S. (1938). The estimation of the bactericidal power of the blood. *Journal of Hygiene (Cambridge)* **38**, 732–749.

Miller, J. D. and Trenholm, H. L. (1994). *Mycotoxins in Grain: Compounds Other Than Aflatoxin.* St Paul, Minnesota: Eagan Press.

Minnikin, D. E. and Goodfellow, M. (1980). Lipid composition in the classification and identification of acid-fast bacteria. In *Microbiological Classification and Identification (Society for Applied Bacteriology, Symposium No. 8)* (ed. M. Goodfellow and R. G. Board), pp. 189–256. London: Academic Press.

Minnikin, D. E., Goodfellow, M. and Collins, M. D. (1978). Lipid composition in the classification and identification of coryneform and related taxa. In *Coryneform Bacteria* (ed. I. J. Bousfield and A. G. Calleley), pp. 85–160. London: Academic Press.

Moake, J. L. (1994). Haemolytic–uraemic syndrome: basic science. *Lancet* **343**, 393–397.

Moats, W. A. (1971). Kinetics of thermal death of bacteria. *Journal of Bacteriology* **105**, 165–171.

Moats, W. A., Dabbah, R. and Edwards, V. M. (1971). Interpretation of nonlogarithmic survivor curves of heated bacteria. *Journal of Food Science* **36**, 523–526.

Moffett, M., Young, J. L. and Stuart, R. D. (1948). Centralized gonococcus culture for dispersed clinics. *British Medical Journal* **ii**, 421–424.

Møller, V. (1955). Simplified tests for some amino acid decarboxylases and for the arginine dihydrolase system. *Acta Pathologica Microbiologica Scandinavica* **36**, 158.

Moreau, C. (trans. and ed. M. O. Moss 1979). *Moulds, Toxins and Food*. Chichester, UK: John Wiley.

Morgan, D. R. and Wood, L. V. (1988). Is *Aeromonas* sp. a foodborne pathogen? *Journal of Food Safety* **9**, 59–72.

Morita, R. Y. (1975). Psychrophilic bacteria. *Bacteriological Reviews* **39**, 144–167.

Morris, E. O. and Eddy, A. A. (1957). Method for the measurement of wild yeast infection in pitching yeast. *Journal of the Institute of Brewing* **63**, 34–35.

Mortimore, S. and Wallace, C. (1994). *HACCP: A Practical Approach*. London: Chapman and Hall.

Moss, M. O. (1980). Mycotoxins in foods In *Food and Health: Science and Technology* (ed. G. G. Birch and K. J. Parker), pp. 115–127. London: Applied Science.

Mossel, D. A. A. (1957). The presumptive enumeration of lactose negative as well as lactose positive Enterobacteriaceae in foods. *Applied Microbiology* **5**, 379–381.

Mossel, D. A. A. (1964). Essentials of the assessment of the hygienic condition of food factories and their products. *Journal of the Science of Food and Agriculture* **15**, 349–362.

Mossel, D. A. A. (1967). Ecological principles and methodological aspects of the examination of foods and feeds for indicator micro-organisms. *Journal of the Association of Official Analytical Chemists* **50**, 91–104.

Mossel, D. A. A. (1978). Streptococci in food. In *Streptococci (Society for Applied Bacteriology, Symposium Series No. 7)* (ed. F. A. Skinner and L. B. Quesnel), p. 392. London: Academic Press.

Mossel, D. A. A. (1981). Coliform test for cheese and other foods. *Lancet* **ii**, 1425.

Mossel, D. A. A. (1982). Marker (index and indicator) organisms in food and drinking water. Semantics, ecology, taxonomy and enumeration. *Antonie van Leeuwenhoek* **48**, 609–611.

Mossel, D. A. A., Bijker, P. G. H. and Eelderink, I. (1978). Streptococci of Lancefield groups A, B and D and those of buccal origin in foods: their public health significance, monitoring and control. In *Streptococci (Society for Applied Bacteriology, Symposium Series No. 7)* (ed. F. A. Skinner and L. B. Quesnel), pp. 315–334. London: Academic Press.

Mossel, D. A. A., van Diepen, H. M. J. and de Bruin, A. S. (1957). The enumeration of faecal streptococci in foods, using Packer's crystal violet sodium azide blood agar. *Journal of Applied Bacteriology* **20**, 265–272.

Mossel, D. A., Mengerink, W. H. J. and Scholts, H. H. (1962). Use of a modified MacConkey agar medium for the selective growth and enumeration of Enterobacteriaceae. *Journal of Bacteriology* **84**, 381.

Mossel, D. A. A., Visser, M. and Cornelissen, A. M. R. (1963). The examination of foods for Enterobacteriaceae using a test of the type generally adopted for the detection of salmonellae. *Journal of Applied Bacteriology* **26**, 444–452.

Mossel, D. A. A., Koopman, M. J. and Jongerius, E. (1967). Enumeration of *Bacillus cereus* in foods. *Applied Microbiology* **15**, 650–653.

Mossel, D. A. A., Shennan, J. L., Meursing, F. H. *et al.* (1973). The enumeration of 'all' spore-bearing cells of Bacillaceae in heat-processed foods. *Antonie van Leeuwenhoek* **39**, 656.

Mossel, D. A. A., Bonants-Van Laarhoven, T., Ligtenberg-Markus, A. M. Th. *et al.* (1983). Quality assurance of selective culture media for bacteria, moulds and yeasts: an attempt at standardization at the international level. *Journal of Applied Bacteriology* **54**, 313–327.

Mossel, D. A. A., Correy, J. E. L., Struijk, C. B. *et al.* (1995). *Essentials of the Microbiology of Foods*. Chichester, UK: John Wiley.

Mozola, M. A. (1997). Detection of microorganisms in foods using DNA probes targeted to ribosomal RNA sequences. In *Food Microbiological Analysis: New Technologies* (ed. M. L. Tortorello and S. M. Gendel), pp. 207–228. New York: Marcel Dekker.

Mulder, E. G. (1964). Iron bacteria, particularly those of the *Sphaerotilus–Leptothrix* group, and industrial problems. *Journal of Applied Bacteriology* **27**, 151–173.

Muller, E. G. (1972). The sugar industry. In *Quality Control in the Food Industry, Vol. 3* (ed. S. M. Herschdoerfer), pp. 229–258. London: Academic Press.

Murdock, D. I. (1968). Diacetyl test as a quality control tool in processing frozen concentrated orange juice. *Food Technology* **22**, 90–94.

Murdock, D. I., Folinazzo, J. F. and Troy, V. S. (1952). Evaluation of plating media for citrus concentrates. *Food Technology* **6**, 181.

Murray, R. G. E. and Robinow, C. F. (1994). Light microscopy. In *Methods for General and Molecular Bacteriology* (ed. P. Gerhardt *et al.*), pp. 7–20. Washington, DC: American Society for Microbiology.

Naylor, J. and Sharpe, M. E. (1958a). Lactobacilli in Cheddar cheese. I. The use of selective media for isolation and of serological typing for identification. *Journal of Dairy Research* **25**, 92–103.

Naylor, J. and Sharpe, M. E. (1958b). Lactobacilli in Cheddar cheese. III. The source of lactobacilli in cheese. *Journal of Dairy Research* **25**, 431–438.

Neal, C. E. and Calbert, H. E. (1955). The use of 2,3,5-triphenyltetrazolium chloride as a test for antibiotic substances in milk. *Journal of Dairy Science* **38**, 629–633.

Neild, G. H. (1994). Haemolytic–uraemic syndrome in practice. *Lancet* **343**, 398–401.

Netten, P. van, Perales, I., Moosdijk, A. van de *et al.* (1989). Liquid and solid selective differential media for the detection and enumeration of *L. monocytogenes* and other *Listeria* species. *International Journal of Food Microbiology* **8**, 299–316.

Newell, K. W. (1955). Outbreaks of paratyphoid B fever associated with imported frozen egg. I. Epidemiology. *Journal of Applied Bacteriology* **18**, 462–470.

Niimura, Y., Koh, E., Yanagida, F. *et al.* (1990). *Amphibacillus xylanus* gen. nov., sp. nov., a facultatively anaerobic sporeforming xylan-digesting bacterium which lacks cytochrome, quinone and catalase. *International Journal of Systematic Bacteriology* **40**, 297–301.

Nir, R., Yisraeli, Y., Lamed, R. *et al.* (1990). Flow cytometry sorting of viable bacteria and yeasts according to β-galactosidase activity. *Applied and Environmental Microbiology* **56**, 3861–3866.

North, W. R. and Bartram, M. T. (1953). The efficiency of selenite broth of different compositions in the isolation of *Salmonella. Applied Microbiology* **1**, 130–134.

Oliver, J. D., Guthrie, K., Preyer, J. *et al.* (1992). Use of colistin–polymyxin B–cellobiose agar for isolation of *Vibrio vulnificus* from the environment. *Applied and Environmental Microbiology* **58**, 737–739.

Onions, A. H. S., Allsopp, D. and Eggins, H. O. W . (1981). *Smith's Introduction to Industrial Mycology*, 7th edn. London: Edward Arnold.

Onishi, H., McCance, M. E. and Gibbons, N. E. (1965). A synthetic medium for extremely halophilic bacteria. *Canadian Journal of Microbiology* **11**, 365–373.

Orla-Jensen, S. (1943). *The Lactic Acid Bacteria* (Erganzungsband). Copenhagen: Ejnar Munksgaard.

Ormay, L. and Novotny, T. (1969). The significance of *Bacillus cereus* food poisoning in Hungary. In *The Microbiology of Dried Foods (Proceedings of 6th International Symposium on Food Microbiology, 1968)*, pp. 279–285. Haarlem, The Netherlands: International Association of Microbiological Societies.

Owens, J. D., Thomas, D. S., Thompson, P. S. *et al.* (1989). Indirect conductimetry: a novel approach to the conductimetric enumeration of microbial populations. *Letters in Applied Microbiology* **9**, 245–9.

Packer, R. A. (1943). The use of sodium azide (NaN_3) and crystal violet in a selective medium for streptococci and *Erysipelothrix rhusiopathiae. Journal of Bacteriology* **46**, 343–349.

Palmer, S. R. and McGuirk, S. M. (1995). Bird attacks on milk bottles and campylobacter infection. *The Lancet* **345**, 326–327.

Park, C. E., Landgraf, M. and Stankiewicz, Z. K. (1981). A new sensitive and rapid procedure for the isolation of *Yersinia enterocolitica* from food, particularly from low calory (*sic*) food such as vegetables. In *Psychrotrophic Micro-organisms in Spoilage and Pathogenicity* (ed. T. A. Roberts *et al.*), pp. 425–429. London: Academic Press.

Park, C. E. and Sanders, G. W. (1991). A sensitive enrichment procedure for the isolation of *Campylobacter jejuni* from frozen foods. In *Campylobacter V (Proceedings of the 5th International Workshop on Campylobacter infections, Mexico)* (ed. G. M. Ruiz-Palacios, F. Calva and B. R. Ruiz-Palacios), p. 102. Mexico City: Department of Infectious Diseases, Institute Nacional de la Nutricion.

Park, R. W. A. (1967). A comparison of two methods for detecting attack on glucose by pseudomonads and achromobacters. *Journal of General Microbiology* **46**, 355–360.

Paton, A. M. and Gibson, T. (1953). The use of hydrogenated fats in tests for the detection of microbiological lipases. *Proceedings of the Society for Applied. Bacteriology* **16**, iii–iv.

Patterson, J. T. (1969). Salmonellae in meat and poultry, poultry plant cooling waters and effluents and animal feeding stuffs. *Journal of Applied Bacteriology* **32**, 329–337.

Patton, J. (1950). Bacteriological testing of ice-cream in Northern Ireland. *Proceedings of the Society for Applied Bacteriology* **13**, 100–107.

Payne, W. J. (1973). Reduction of nitrogenous oxides by microorganisms. *Bacteriological Reviews* **37**, 409–452.

Perkin, A. G., Davies, F. L., Neaves, P. *et al.* (1980). Determination of bacterial spore inactivation at high temperatures. In *Microbial Growth and Survival in Extremes of Environment (Society for Applied Bacteriology, Technical Series 15)* (ed. G. W. Gould and J. E. L. Corry), pp. 173–188. London: Academic Press.

Pérombelon, M. C. M. and Burnett, E. M. (1991). Two modified crystal violet pectat (CVP) media for the detection, isolation and enumeration of soft rot erwinias. *Potato Research* **34**, 79–85.

Perry, L. B. (1973). Gliding motility in some non-spreading flexibacteria. *Journal of Applied Bacteriology* **36**, 227–232.

Peterz, M., Wiberg, C. and Norberg, P. (1989). The effects of incubation temperature and magnesium chloride concentration on growth of salmonella in home-made and in commercially available dehydrated Rappaport–Vassiliadis broths. *Journal of Applied Bacteriology* **66**, 523–528.

Pettipher, G. L. (1983). *The Direct Epifluorescent Filter Technique for the Rapid Enumeration of Microorganisms*. Letchworth, UK: Research Studies Press.

Pettipher, G. L. (1986). The direct epifluorescent filter technique. *Journal of Food Technology* **21**, 535–546.

Pettipher, G. L. (1991). Preliminary evaluation of flow cytometry for the detection of yeasts in soft drinks. *Letters in Applied Microbiology* **12**, 109–112.

Pettipher, G. L. and Rodrigues, U. M. (1981). Rapid enumeration of bacteria in heat-treated milk and milk products using a membrane filtration–epifluorescent microscopy technique. *Journal of Applied Bacteriology* **50**, 157–166.

Pettipher, G. L., Mansell, R., McKinnon, C. H. *et al.* (1980). Rapid membrane filtration–epifluorescent microscopy technique for direct enumeration of bacteria in raw milk. *Applied and Environmental Microbiology* **39**, 423–429.

Pettipher, G. L., Kroll, R. G., Farr, L. J. *et al.* (1989). DEFT: recent developments for food and beverages. In *Rapid Microbiological Methods for Foods, Beverages and Pharmaceuticals (Society for Applied Bacteriology, Technical Series No. 25)* (ed. C. J. Stannard, S. B. Petitt and F. A. Skinner), pp. 33–45. Oxford: Blackwell Scientific.

Petts, D. N. (1984). Colistin–oxolinic acid–blood agar: a new selective medium for streptococci. *Journal of Clinical Microbiology* **19**, 4–7.

Pike, E. B. (1962). The classification of staphylococci and micrococci from the human mouth. *Journal of Applied Bacteriology* **25**, 448–455.

Pin, C., Marín, M. L., Garcia, M. L. *et al.* (1994). Comparison of different media for the isolation and enumeration of *Aeromonas* spp. in foods. *Letters in Applied Microbiology* **18**, 190–192.

Pinder, A. C., Edwards, C., Clarke, R. G. *et al.* (1993). Detection and enumeration of viable bacteria by flow cytometry. In *New Techniques in Food and Beverage Microbiology (Society for Applied Bacteriology, Technical Series No. 31)* (ed. R. G. Kroll, A. Gilmour and M. Sussman), pp. 67–86. Oxford: Blackwell Scientific.

Pitt, J. I. and Hocking, A. D. (1985). *Fungi and Food Spoilage*. Sydney: Academic Press.

Pitt, J. I. and Hocking, A. D. (1997). *Fungi and Food Spoilage*, 2nd edn. London: Chapman and Hall.

Pitt, J. I., Hocking, A. D., Samson, R. A. *et al.* (1992). Recommended methods for mycological examination of foods. In *Modern Methods in Food Mycology)Development in Food Science, Vol. 31)* (ed. R. A. Samson, *et al.*), pp. 365–368. Amsterdam: Elsevier.

Pivnick, H. (1980). Spices. In *Microbial Ecology of Foods, Vol. 2: Food Commodities*, pp. 731–751. New York: Elsevier.

Poelma, P. L. (1968). Recommended changes in the method for detection and identification of *Salmonella* in egg products. *Journal of the Association of Official Analytical Chemists* **51**, 870–872.

Post, F. J. and Krishnamurty, G. B. (1964). Suggested modifications of the calcium alginate swab technique. *Journal of Milk and Food Technology* **27**, 62–65.

Post, F. J., Krishnamurty, G. B. and Flanagan, M. D. (1963). Influence of sodium hexametaphosphate on selected bacteria. *Applied Microbiology* **11**, 430–435.

Powell, E. O. (1963). Photometric methods in bacteriology. *Journal of the Science of Food and Agriculture.* **14**, 1–8.

Quesnel, L. B. (1971). Microscopy and micrometry. In *Methods in Microbiology, Vol. 5A* (ed. J. R. Norris and D. W. Ribbons), pp. 1–103. London: Academic Press.

Rabinowitz, R. P. and Donnenberg, M. S. (1996). *Escherichia coli*. In *Enteric Infections and*

Immunity (ed. L. J. Paradise, M. Bendinelli and H. Friedman), pp. 101–131. New York: Plenum Press.

Ranken, M. D. and Kill, R. C. (eds.) (1993). Meat and meat products. In *Food Industries Manual*, 23rd edn, pp. 1–41. Glasgow: Blackie.

Recommendations (1965). Recommendations of the Subcommittee on taxonomy of staphylococci and micrococci. *International Bulletin of Bacteriological Nomenclature and Taxonomy* **15**, 109–110.

Reed, G. and Peppler, H. J. (1973). *Yeast Technology*. Westport, Connecticut: Avi.

Reid, T. M. S. and Robinson, H. G. (1987). Frozen raspberries and hepatitis A. *Epidemiology and Infection* **98**, 109–112.

Report (1950). The bacteriological examination and grading of ice-cream. *Monthly Bulletin of the Ministry of Health and the Public Health Laboratory Service* **9**, 231–239.

Report (1958). Report of the Enterobacteriaceae Subcommittee of the Nomenclature Committee of the International Association of Microbiological Societies. *International Bulletin of Bacteriological Nomenclature and Taxonomy* **8**, 25–70.

Report (1959). Sterilisation by steam under increased pressure. *Lancet* **i**, 425–435.

Report (1964). *The Aberdeen Typhoid Outbreak, 1964*. Report of the Committee of Enquiry. Command 2542. Edinburgh: HMSO.

Report (1969). *The Bacteriological Examination of Water Supplies*, 4th edn. Report of the Public Health Laboratory Service Standing Committee on the Bacteriological Examination of Water Supplies. *Reports on Public Health and Medical Subjects, No. 71*. London: HMSO.

Report (1994). *The Microbiology of Water 1994. Part 1 – Drinking Water. Reports on Public Health and Medical Subjects*, No. 71. London: HMSO.

Rhodes, M. E. (1958). The cytology of *Pseudomonas* spp. as revealed by a silver-plating method. *Journal of General Microbiology* **18**, 639–648.

Riley, L. W., Remis, R. S., Helgerson, S. D. *et al.* (1983). Hemorrhagic colitis associated with a rare *Escherichia coli* serotype. *New England Journal of Medicine* **308**, 681–685.

Riordan. T., Humphrey, T. J. and Fowles, A (1993). A point source outbreak of campylobacter infection related to bird-pecked milk. *Epidemiology and Infection* **110**, 261–265.

Roberts, C. and Gross, R. J. (eds) (1990). Verocytotoxin-producing *Escherichia coli* O157 (Proceedings of a Seminar, London, 1 June 1990). *PHLS Microbiology Digest* **7** (4), Supplement, 116–170.

Roberts, D., Hooper, W. and Greenwood, M. (1995). *Practical Food Microbiology*. London: Public Health Laboratory Service.

Roberts, T. A., Hobbs, G., Christian, J. N. B. *et al.* (eds) (1981). *Psychrotrophic Microorganisms in Spoilage and Pathogenicity*. London: Academic Press.

Robertson, D. S. F. (1970). Selenium – a possible teratogen. *Lancet* **i**, 518.

Rogosa, M., Mitchell, J. A. and Wiseman, R. F. (1951). A selective medium for the isolation and enumeration of oral and fecal streptococci. *Journal of Bacteriology* **62**, 132–133.

Rosenblum, L. S. *et al.* (1990). A multifocal outbreak of hepatitis A traced to commercially distributed lettuce. *American Journal of Public Health* **80**, 1075–1079.

Rosenthal, S. L. (1974). A simplified method for single carbon source tests with *Pseudomonas* species. *Journal of Applied Bacteriology* **37**, 437–441.

Roszak, D. B., Grimes, D. J. and Colwell, R. R. (1984). Viable but nonrecoverable stage of *Salmonella enteritidis* in aquatic systems. *Canadian Journal of Microbiology* **30**, 334–338.

Rubbo, S. D. and Gardner, J. F. (1965). *A Review of Sterilization and Disinfection, as Applied to Medical, Industrial and Laboratory Practice*. London: Lloyd-Luke (Medical Books).

Rusul, G. and Yaacob, N. H. (1995). Prevalence of *Bacillus cereus* in selected foods and detection of enterotoxin using TECRA-VIA and BCET-RPLA. *International Journal of Food Microbiology* **25**, 131–139.

Ruttenber, A. J., Weniger, B. G., Sorvillo, F. *et al.* (1984). Diphyllobothriasis associated with salmon consumption in Pacific coast states. *American Journal of Tropical Medicine and Hygiene* **33**, 455–459.

Sacks, L. E. (1956). A pH gradient agar plate. *Nature (London)* **178**, 269–270.

Salamina, G. *et al.* (1996). A foodborne outbreak of gastroenteritis involving *Listeria monocytogenes*. *Epidemiology and Infection* **117**, 429–436.

Salminen, S. and von Wright, A. (1993). *Lactic Acid Bacteria*. New York: Marcel Dekker.

Salminen, S., Deighton, M. and Gorbach, S. (1993). Lactic acid bacteria in health and disease. In *Lactic Acid Bacteria* (ed. S. Salminen and A. von Wright), pp. 199–225. New York: Marcel Dekker.

Samson, R. A. and van Reenen-Hoekstra, E. S. (1988). *Introduction to Food-borne Fungi*, 3rd edn. Baarn, Delft: Centraalbureau voor Schimmelcultures.

Samson, R. A., Hocking, A. D., Pitt. J. I. *et al.* (eds) (1992). *Modern Methods in Food Mycology (Developments in Food Science. Vol. 31)*. Amsterdam: Elsevier.

Saunders, G. C. and Bartlett, M. L. (1977). Double-antibody solid-phase enzyme immunoassay for the detection of staphylococcal enterotoxin A. *Applied and Environmental Microbiology* **34**, 518–522

Scarr, M. P. (1959). Selective media used in the microbiological examination of sugar products. *Journal of the Science of Food and Agriculture* **10**, 678–681.

Schiemann, D. A. (1979). Synthesis of a selective agar medium for *Yersinia enterocolitica*. *Canadian Journal of Microbiology* **25**, 1298–1304.

Schiemann, D. A. (1982). Development of a two-step enrichment procedure for recovery of *Yersinia enterocolitica* from food. *Applied and Environmental Microbiology* **43**, 14–27.

Schleifer, K. H. and Kilpper-Bälz, R. (1984). Transfer of *Streptococcus faecalis* and *Streptococcus faecium* to the genus *Enterococcus* nom. rev. as *Enterococcus faecalis* comb. nov. and *Enterococcus faecium* comb. nov. *International Journal of Systematic Bacteriology* **34**, 31–34.

Schleifer, K. H. and Kilpper-Bälz, R. (1987). Molecular and chemotaxonomic approaches to the classification of streptococci, enterococci and lactococci: a review. *Systematic and Applied Microbiology* **10**, 1–19.

Scholefield, J. (1964). *A Comparison of the Biochemical Activities of Psychrophilic Bacteria*. MSc thesis, University of Leeds.

Scotter, S. L. (1996). Proficiency testing in food microbiology: MAFF Central Science Laboratory 'Quality Assessment Scheme'. *Food Science and Technology Today* **10**(4), 227–230.

Shandera, W. X., Johnston, J. M., Davis, B. R. *et al.* (1983). Disease from infection with *Vibrio mimicus*, a newly recognized *Vibrio* species. *Annals of Internal Medicine* **99**, 169–171.

Shapton, D. A. and Shapton, N. F. (1991). *Principles and Practices for the Safe Processing of Foods*, pp. 38–197. Oxford: Butterworth–Heinemann.

Sharman, M., Patey, A. L., Bloomfield, D. A. *et al.* (1991). Surveillance and control of aflatoxin contamination of dried figs and fig paste imported into the United Kingdom. *Food Additives and Contaminants* **8**, 299–304.

Sharpe, M. E. (1962). Taxonomy of the lactobacilli. *Dairy Science Abstracts* **24**, 109–118.

Sharpe, A. N. (1973). Automation and instrumentation developments for the bacteriology laboratory. In *Sampling – Microbiological Monitoring of Environments (Society for Applied Bacteriology Technical Series, No. 7* (ed. R. G. Board and D. W. Lovelock), pp. 197–232. London: Academic Press.

Sharpe, A. N. (1980). *Food Microbiology: A Framework for the Future*. Springfield, Illinois: Charles C. Thomas.

Sharpe, A. N. (1994). Preparing samples for rapid detection of microbes. In *Rapid Methods and Automation in Microbiology and Immunology (RAMI-93)* (ed. R. C. Spencer, E. P. Wright and S. W. B. Newsom), pp. 97–105. Andover, UK: Intercept.

Sharpe, A. N. (1997). Separation and concentration of pathogens from foods. In *Food Microbiological Analysis: New Technologies* (ed. M. L. Tortorello and S. M. Gendel), pp. 27–44. New York: Marcel Dekker.

Sharpe, A. N. and Jackson, A. K. (1972). Stomaching: a new concept in bacteriological sample preparation. *Applied Microbiology* **24**, 175–178.

Sharpe, A. N. and Kilsby, D. C. (1971). A rapid, inexpensive bacterial count technique using agar droplets. *Journal of Applied Bacteriology* **34**, 435–440.

Sharpe, A. N. and Peterkin, P. I. (1988). *Membrane Filter Food Microbiology.* Letchworth, UK: Research Studies Press.

Sharpe, A. N., Woodrow, M. N. and Jackson, A. K. (1970). Adenosinetriphosphate (ATP) levels in foods contaminated by bacteria. *Journal of Applied Bacteriology* **33**, 758–767.

Shattock, P. M. F. and Hirsch, A. (1947). A liquid medium buffered at pH 9.6 for the differentiation of *Streptococcus faecalis* from *Streptococcus lactis. Journal of Pathology and Bacteriology* **59**, 495–497.

Shaw, B. G. and Farr, L. J. (1989). The rapid estimation of bacterial counts on meat and poultry by the direct epifluorescent filter technique. In *Rapid Microbiological Methods for Foods, Beverages and Pharmaceuticals (Society for Applied Bacteriology, Technical Series No. 25)* (ed. C. J. Stannard, S. B. Petitt and F. A. Skinner), pp. 47–57. Oxford: Blackwell Scientific.

Shaw, C. and Clarke, P. H. (1955). Biochemical classification of Proteus and Providence cultures. *Journal of General Microbiology* **13**, 155–161.

Shewan, J. M. (1970). Bacteriological standards for fish and fishery products. *Chemistry and Industry* **2**, 193–199.

Shewan, J. M. (1971). The microbiology of fish and fishery products – a progress report. *Journal of Applied Bacteriology* **34**, 299–315.

Shewan, J. M. and Hobbs, G. (1967). The bacteriology of fish spoilage and preservation. *Progress in Industrial Microbiology* **6**, 169–208.

Sierra, G. (1957). A simple method for the detection of lipolytic activity of micro-organisms and some observations on the influence of the contact between cells and fatty substrates. *Antonie van Leeuwenhoek* **23**, 15–22.

Sierra, G. (1964). Hydrolysis of triglycerides by a bacterial proteolytic enzyme. *Canadian Journal of Microbiology* **10**, 926–928.

Silva, G. A. N. da and Holt, J. G. (1965). Numerical taxonomy of certain coryneform bacteria. *Journal of Bacteriology* **90**, 921–927.

Silverstolpe, L. *et al.* (1961). An epidemic among infants caused by *Salmonella muenchen. Journal of Applied Bacteriology* **24**, 134–142.

Silverton, R. E. and Anderson, M. J. (1961). *Handbook of Medical Laboratory Formulae*. London: Butterworths.

Sinclair, N. A. and Stokes, J. L. (1963). Role of oxygen in the high cell yields of psychrophiles and mesophiles at low temperatures. *Journal of Bacteriology* **85**, 164–167.

Sizmur, K. and Walker, C. W. (1988). Listeria in prepacked slads. *Lancet* **i**, 1167.

Skean, J. D. and Overcast, W. W. (1962). Another medium for enumerating citrate-fermenting bacteria in lactic cultures. *Journal of Dairy Science* **45**, 1530–1531.

Skerman, V. B. D. (1959). *A Guide to the Identification of the Genera of Bacteria*. Baltimore, Maryland: Williams and Wilkins.

Skirrow, M. B. (1977). Campylobacter enteritis: a 'new' disease. *British Medical Journal* **ii**, 9–11.

Skirrow, M. B. (1982). Isolation, cultivation, and identification of *Campylobacter jejuni* and *C. coli.* In *Isolation and Identification Methods for Food Poisoning Organisms (Society for Applied Bacteriology, Technical Services No. 17)* (ed. J. E. L. Corry, D. Roberts and F. A. Skinner), pp. 313–328. London: Academic Press.

Slanetz, L. W., Chichester, C. O., Gaufin, A. R. *et al.* (eds) (1963). *Microbiological Quality of Foods*. New York: Academic Press.

Smith, B. A. and Baird-Parker, A. C. (1964). The use of sulphamezathine for inhibiting *Proteus* spp. on Baird-Parker's isolation medium for *Staphylococcus aureus. Journal of Applied Bacteriology* **27**, 78–82.

Smith, D. G. (1975). Inhibition of swarming in *Proteus* spp. by tannic acid. *Journal of Applied Bacteriology* **38**, 29–32.

Smith, H. R. and Scotland, S. M. (1994). Recent developments in laboratory techniques for the detection of diarrhoeagenic *Escherichia coli. PHLS Microbiology Digest* **11**, 7–12.

Smith, N. R., Gordon, R. E. and Clark, F. E. (1952). *Aerobic Spore-Forming Bacteria (US Department of Agriculture, Monograph No. 16)*. Washington, DC: US Department of Agriculture.

Sneath, P. H. A. (1966). Identification methods applied to *Chromobacterium*. In *Identification*

Methods for Microbiologists, Part A (Society for Applied Bacteriology, Technical Series No. 1) (ed. B. M. Gibbs and F. A. Skinner), pp. 15–20. London: Academic Press.

Sneath, P. H. A. (1974). Test reproducibility in relation to identification. *International Journal of Systematic Bacteriology* **24**, 508–523.

Sneath, P. H. A. and Collins, V. G. (eds) (1974). A study in test reproducibility between laboratories: report of a Pseudomonas Working Party. *Antonie van Leeuwenhoek* **40**, 481–527.

Sneath, P. H. A., Mair, N. S., Sharpe, M. E. *et al.* (eds) (1986). *Bergey's Manual of Systematic Bacteriology, Vol. 2*. Baltimore, Maryland: Williams and Wilkins.

Splittstoesser, D. F. (1996). Microbiology of fruit products. In *Processing Fruits: Science and Technology. Vol. 1: Biology, Principles and Applications* (ed. L. P. Somogyi, H. S. Ramaswamy and Y. H. Hui), pp. 261–292. Basel: Technomic.

Staley, J. T., Bryant, M. P., Pfennig, N. *et al.* (1989). *Bergey's Manual of Systematic Bacteriology, Vol. 3*. Baltimore, Maryland: Williams and Wilkins.

Stamer, J. R., Albury, M. N. and Pederson, C. S. (1964). Substitution of manganese for tomato juice in the cultivation of lactic acid bacteria. *Applied Microbiology* **12**, 165–168.

Stanley, P. E., McCarthy, B. J. and Smither, R. (eds) (1989). *ATP Luminescence: Rapid Methods in Microbiology (Society for Applied Bacteriology, Technical Series No. 26)*. Oxford: Blackwell Scientific.

Stanley, S. O. and Rose, A. H. (1967). Bacteria and yeasts from lakes on Deception Island. *Philosophical Transactions of the Royal Society of London, Series B* **252**, 199–207.

Stannard, C. J., Petitt, S. B. and Skinner, F. A. (1989). *Rapid Microbiological Methods for Foods, Beverages and Pharmaceuticals (Society for Applied Bacteriology, Technical Series No. 25)*. Oxford: Blackwell Scientific.

Steel, K. J. (1961). The oxidase reaction as a taxonomic tool. *Journal of General Microbiology* **25**, 297–306.

Steel, K. J. (1962). The oxidase activity of staphylococci. *Journal of Applied Bacteriology* **25**, 445–447.

Steiner, E. H. (1984). Statistical methods in quality control. In *Quality Control in the Food Industry, Vol. 1*, 2nd edn. (ed. S. M. Herschdoerfer), pp. 169–298, London: Academic Press.

Steinkraus, K. H., Cullen, R. E., Pederson, C. S. *et al.* (1983). *Handbook of Indigenous Fermented Foods*. New York: Marcel Dekker.

Stephens, P. J., Cole, M. B. and Jones, M. V. (1994). Effect of heating rate on the thermal inactivation of *Listeria monocytogenes*. *Journal of Applied Bacteriology* **77**, 702–708.

Stevens, W. L. (1958). Dilution series: a statistical test of technique. *Journal of the Royal Statistical Society* **B20**, 205–214.

Stevenson, I. L. (1963). Some observations on the so-called cystites' of the genus *Arthrobacter. Canadian Journal of Microbiology* **9**, 467–472.

Stickler, D. J. (1989). The microbiology of bottled natural mineral waters. *Journal of the Royal Society of Health* **109**, 118–124.

Stiffler-Rosenberg, G. and Fey, H. (1978). Simple assays for Staphylococcal enterotoxins A, B and C: modifications of enzyme-linked immunosorbent assay. *Journal of Clinical Microbiology* **8**, 473–479.

Straka, R. P. and Stokes, J. L. (1957). Rapid destruction of bacteria in commonly used diluents and its elimination. *Applied Microbiology* **5**, 21–25.

Straka, R. P. and Stokes, J. L. (1959). Metabolic injury to bacteria at low temperature. *Journal of Bacteriology* **78**, 181–185.

Strange, R. E. and Dark, F. A. (1962). Effect of chilling on *Aerobacter aerogenes* in aqueous suspension. *Journal of General Microbiology* **29**, 719–730.

Stuart, R. D., Toshach, S. R. and Patsula, T. M. (1954). The problem of transport of specimens for culture of gonococci. *Canadian Journal of Public Health* **45**, 73–83.

Stumbo, C. R. (1973). *Thermobacteriology in Food Processing*, 2nd edn. New York: Academic Press.

Swaminathan, B., Harmon, M. C. and Mehlman, I. J. (1982). A review: *Yersinia enterocolitica*. *Journal of Applied Bacteriology* **52**, 151–183.

Sykes, G. (1965). *Disinfection and Sterilization*, 2nd edn. London: Spon.

Talon, R., Grimont, P. A. D., Grimont, F. *et al.* (1988). *Brochothrix campestris* sp. nov. *International Journal of Systematic Bacteriology* **38**, 99–102.

Taylor, E. W., Burman, N. P. and Oliver, C. W. (1955). Membrane filtration technique applied to the routine bacteriological examination of water. *Journal of the Institution of Water Engineers* **9**, 248.

Taylor, J. L., Tuttle, J., Pramukel, T. *et al.* (1993). An outbreak of cholera in Maryland associated with imported commercial frozen fresh coconut milk. *Journal of Infectious Diseases* **167**, 1330–1335.

Taylor, W. I. (1965). Isolation of shigellae. I. Xylose lysine agars; new media for isolation of enteric pathogens. *American Journal of Clinical Pathology* **44**, 471–475.

Taylor, W. I. and Achanzar, D. (1972). Catalase test as an aid to the identification of Enterobacteriaceae. *Applied Microbiology* **24**, 58.

Taylor, W. I. and Harris, B. (1965). Isolation of shigellae. II. Comparison of plating media and enrichment broths. *American Journal of Clinical Pathology* **44**, 476–479.

Taylor, W. I. and Schelhart, D. (1967). Isolation of shigellae. IV. Comparison of plating media with stools. *American Journal of Clinical Pathology* **48**, 356–362.

Taylor, W. I. and Schelhart, D. (1971). Isolation of shigellae. VIII. Comparison of xylose lysine deoxycholate agar, Hektoen enteric agar, *Salmonella–Shigella* agar and eosin methylene blue agar with stool specimens. *Applied Microbiology* **21**, 32–37.

Ten Cate, L. (1965). A note on a simple and rapid method of bacteriological sampling by means of agar sausages. *Journal of Applied Bacteriology* **28**, 221.

Thatcher, F. S. and Clark, D. S. (eds) (1968). *Micro-organisms in Foods: Their Significance and Methods of Enumeration*. Recommendations of the International Committee on Microbiological Specifications for Foods, a Standing Committee of the International Association of Microbiological Societies. Toronto: University of Toronto Press.

Thomas, M. (1961). The sticky film method of detecting skin staphylococci. *Monthly Bulletin of the Ministry of Health and the Public Health Laboratory Service* **20**, 37–40.

Thomas, M. (1966). Bacterial penetration in raw meats: comparisons using a new technique. *Monthly Bulletin of the Ministry of Health and the Public Health Laboratory Service* **25**, 42–51.

Thomson-Carter, F. M., Carter, P. E. and Pennington. T. H. (1993). Pulsed-field gel electrophoresis for the analysis of bacterial populations. In (eds. Kroll, Gilmour and Sussman), pp. 251–264.

Thornley, M. J. (1960). The differentiation of *Pseudomonas* from other Gram-negative bacteria on the basis of arginine metabolism. *Journal of Applied Bacteriology* **23**, 37–52.

Tomlins, R. I. and Ordal, Z. J. (1976). Thermal injury and inactivation in vegetative bacteria. In *Inhibition and Inactivation of Vegetative Microbes (Society for Applied Bacteriology, Symposium Series No. 5)* (ed. F. A. Skinner and W. B. Hugo), pp. 153–190. London: Academic Press.

Tortorello, M. L. and Gendel, S. M. (1997). *Food Microbiological Analysis: New Technologies*. New York: Marcel Dekker.

Trust, T. J. (1975). Bacteria associated with the gills of salmonid fishes in freshwater. *Journal of Applied Bacteriology* **38**, 225–233.

Tschäpe. H., Prager, R., Streckel, W. *et al.*, (1995). Verotoxinogenic *Citrobacter freundii* associated with severe gastroenteritis and cases of haemolytic uraemic syndrome in a nursery school: green butter as the infection source. *Epidemiology and Infection* **114**, 441–450.

Turner, N., Sandine, W. E., Elliker, P. R. *et al.* (1963). Use of tetrazolium dyes in an agar medium for the differentiation of *Streptococcus lactis* and *Streptococcus cremoris*. *Journal of Dairy Science* **46**, 380–385.

Turpin, P. E., Maycroft, K. A., Rowlands, C. L. *et al.* (1993). Viable but non-culturable salmonellas in soil. *Journal of Applied Bacteriology* **74**, 421–427.

Tuttlebee, J. W. (1975). The Stomacher – its use for homogenization in food microbiology. *Journal of Food Technology* **10**, 113–122.

van Niel, C. B. (1928). *The Propionic Bacteria*. Haarlem: N. V. Uitgeverszaak J. W. Boissevain.

van Niel, C. B. (1955). Natural selection in the microbial world. *Journal of General Microbiology* **13**, 201–217.

Varnam, A. H. and Grainger, J. M. (1972). Enumeration of certain lactic acid bacteria from Wiltshire bacon-curing brines. *Journal of the Science of Food and Agriculture* **23**, 546–547.

Varnam, A. H. and Grainger, J. M. (1973). Methods for the general microbiological examination of Wiltshire bacon curing brines. In *Sampling – Microbiological Monitoring of Environments (Society of Applied Bacteriology Technical Series No. 7)* (ed. R. G. Board and D. W. Lovelock), pp. 29–41. London: Academic Press.

Varnam, A. H. and Grainger, J. M. (1975). The nature of the stimulatory effect of pork extract on the growth of bacteria of Wiltshire bacon curing brines. *Journal of Applied Bacteriology* **39**, vii.

Vassiliadis, P. (1983). The Rappaport–Vassiliadis (RV) enrichment medium for the isolation of salmonellas: an overview. *Journal of Applied Bacteriology* **54**, 69–76.

Vernon, E. and Tillett, H. E. (1972). Food poisoning and *Salmonella* infections in England and Wales, 1969–1972. *Public Health (London)* **88**, 225.

Vidon, D. J.-M. and Delmas, C. (1981). The incidence of *Yersinia enterocolitica* in raw milk of Eastern France. In *Psychrotrophic Microorganisms in Spoilage and Pathogenicity* (ed. T. A. Roberts *et al.*), pp. 431–438. London: Academic Press.

WHO (1990). *Selected Mycotoxins: Ochratoxins, Trichothecenes, Ergot (Environmental Health Criteria No. 105)* Geneva: World Health Organization.

WHO (1993). *Guidelines for drinking-water quality, Vol. 1 Recommendations*, 2nd edn. Geneva: World Health Organization.

WHO (1996). *Guidelines for drinking-water quality, Vol. 2, Health Criteria and Other Supporting Information*, 2nd edn. Geneva: World Health Organization.

Wadström, T. and Ljungh, Å. (1991). *Aeromonas* and *Plesiomonas* as food- and waterborne pathogen. *International Journal of Food Microbiology* **12**, 303–311.

Waes, G. (1968). The enumeration of aromabacteria in B D starters. *Nederlands Melk-en Zuiveltijdschrift* **22**, 29–39.

Wang, R.-F., Cao, W.-W and Cerniglia, C. E. (1997). A universal protocol for PCR detection of 13 species of foodborne pathogens in foods. *Journal of Applied Microbiology* **83**, 727–736.

Wauters, G. *et al.* (1988). New enrichment method for isolation of pathogenic *Yersinia enterocolitica* serogroup O:3 from pork. *Applied and Environmental Microbiology* **54**, 851–854.

Weibull, C. (1960). Movement. In *The Bacteria: A Treatise on Structure and Function, Vol. 1. Structure* (ed. I. C. Gunsalus and R. Y. Stanier), pp. 153–205. New York: Academic Press.

Weltman, A. C., Bennett, N. M., Ackman, D. A. *et al.* (1996). An outbreak of hepatitis A associated with a bakery, New York, 1994: the 1968 'West Branch, Michigan' outbreak repeated. *Epidemiology and Infection* **117**, 333–341.

Wetherill, G. B. (1977). *Sampling Inspection and Quality Control*, 2nd edn. London: Chapman and Hall.

Wheater, D. M. (1955). The characteristics of *Lactobacillus plantarum, L. helveticus* and *L. casei*. *Journal of General Microbiology* **12**, 133–139.

Whitmore, T. N. and Sidorowicz, S. (1995). Novel methods for the concentration and selective purification of microorganisms and their potential application for water analysis. *Microbiology Europe* **3**, 16–22.

Whittenbury, R. (1963). The use of soft agar in the study of conditions affecting the utilization of fermentable substrates by lactic acid bacteria. *Journal of General Microbiology* **32**, 375–384.

Whittenbury, R. (1964). Hydrogen peroxide formation and catalase activity in the lactic acid bacteria. *Journal of General Microbiology* **35**, 13–26.

Wickerham, L. J., Flickinger, M. H. and Burton, K. A. (1946). A modification of Henrici's vegetable-juice sporulation medium for yeasts. *Journal of Bacteriology* **52**, 611–612.

Wilkinson, J. F. (1958). The extracellular polysaccharides of bacteria. *Bacteriological Reviews* **22**, 46–73.

Williams, H. A. (1968). The detection of rot in tomato products. *Journal of the Association of Public Analysts* **6**, 69–84.

Williams, S. T. and Wellington, E. M. H. (1980). Micromorphology and fine structure of Actinomycetes. In *Microbiological Classification and Identification (Society for Applied Bacteriology, Symposium No. 8)* (ed. M. Goodfellow and R. G. Board), pp. 139–165. London: Academic Press.

Williams, S. T., Davies, F. L. and Cross, T. (1968). Identification of genera of the Actinomycetales. In *Identification Methods for Microbiologists, Part B (Society for Applied Bacteriology Technical Series No. 2)* (ed. B. M. Gibbs and D. A. Shapton), pp. 111–124. London: Academic Press.

Williams, S. T., Sharpe, M. E., Holt, J. G. *et al.* (1989). *Bergey's Manual of Systematic Bacteriology, Vol. 4*. Baltimore: Williams and Wilkins.

Willis, A. T. (1962). Some diagnostic reactions of clostridia. *Laboratory Practice* **11**, 526–530.

Willis, A. T. (1965). Media for clostridia. *Laboratory Practice* **14**, 690–696.

Willis, A. T. (1969). Techniques for the study of anaerobic spore-forming bacteria. In *Methods in Microbiology, Vol. 3B* (ed. J. R. Norris and D. W. Ribbons), pp. 79–115. London: Academic Press.

Willis, A. T. (1977). *Anaerobic Bacteriology: Clinical and Laboratory Practice*, 3rd edn. London: Butterworths.

Wilson, G. S. (1922). The proportion of viable bacteria in young cultures with especial reference to the technique employed in counting. *Journal of Bacteriology* **7**, 405–446.

Wilson, G. S. (1935). *The Bacteriological Grading of Milk. Medical Research Council Special Report Series, No. 206*. London: HMSO.

Wilson, G. S. (1955). Symposium on Food Microbiology and Public Health: general conclusions. *Journal of Applied Bacteriology* **18**, 629–630.

Wilson, G. S., Miles, A. A. and Parker, M. T. (1983). *Topley and Wilson's Principles of Bacteriology, Virology and Immunity*, 7th edn. London: Edward Arnold.

Wilson, J. M. and Davies, R. (1976). Minimal medium recovery of thermally injured *Salmonella seftenberg* 4969. *Journal of Applied Bacteriology* **40**, 365–374.

Wilson, M. M. and MacKenzie, E. F. (1955). Typhoid fever and salmonellosis due to the consumption of infected desiccated coconut. *Journal of Applied Bacteriology* **18**, 510–521.

Wilson, S. and Weir, G. (1995). *Food and Drink Laboratory Accreditation: A Practical Approach*. London: Chapman and Hall.

Wilson, W. J. (1938). Isolation of *Bact. typhosum* by means of bismuth sulphite medium in water- and milk-borne epidemics. *Journal of Hygiene (Cambridge)* **38**, 507–519.

Wood, D. (1969). Isolation of *Listeria monocytogenes*. In *Isolation Methods for Microbiologists (Society for Applied Bacteriology, Technical Series No. 3)* (ed. D. A. Shapton and G. W. Gould), pp. 63–69. London: Academic Press.

Wood, G. M., Mann, P. J., Lewis, D. F. *et al.* (1990). Studies on a toxic metabolite from the mould *Wallemia*. *Food Additives and Contaminants* **7**, 69–77.

Woodward, R. L. (1957). How probable is the most probable number? *Journal of the American Water Works Association* **49**, 1060–1068.

Woolaway, M. C., Bartlett, C. L. R., Wilneke, A. *et al.* (1986). International outbreak of staphylococcal food poisoning caused by contaminated lasagne. *Journal of Hygiene (Cambridge)* **96**, 67–73.

Wreghitt, T. G. and Morgan-Capner, P. (eds) (1990). *ELISA in the Clinical Microbiology Laboratory*. London: Public Health Laboratory Service.

Wright, R. C. and Tramer, J. (1961). The estimation of penicillin in milk. *Journal of the Society of Dairy Technology* **14**, 85.

Xu, H.-S., Roberts, N., Singleton, F. L. *et al.* (1982). Survival and viability of nonculturable *Escherichia coli* and *Vibrio cholerae* in the estuarine environment. *Microbial Ecology* **8**, 313–323.

Zadik, P. M., Chapman, P. A. and Siddons, C. A. (1993). Use of tellurite for the selection of verocytotoxigenic *Escherichia coli* O157. *Journal of Medical Microbiology* **39**, 155–158.

Index